Methods in Cell Biology

VOLUME 70

Cell Biological Applications of Confocal Microscopy
Second Edition

Series Editors

Leslie Wilson

Department of Biological Sciences
University of California
Santa Barbara, California

Paul Matsudaira

Whitehead Institute for Biomedical Research and
Department of Biology
Massachusetts Institute of Technology
Cambridge, Massachusetts

Methods in Cell Biology

Prepared under the auspices of the American Society for Cell Biology

VOLUME 70

Cell Biological Applications of Confocal Microscopy

Second Edition

Edited by

Brian Matsumoto

Microscopy Facility
Neuroscience Research Institute
University of California
Santa Barbara, California

ACADEMIC PRESS

An imprint of Elsevier Science

Amsterdam   Boston   London   New York   Oxford   Paris
San Diego   San Francisco   Singapore   Sydney   Tokyo

Cover image (paperback only): Imaging *Drosophila* embryos. (a) though (f), (g) and (g'): Two magnifications of a 22C10-labeled embryo (h) and (h'): *ftz* mRna in red and wheat germs. (For more details see figure 7 in chapter 9).

This book is printed on acid-free paper. ∞

Academic Press
An imprint of Elsevier Science.
525 B Street, Suite 1900, San Diego, California 92101-4495, USA.
http://www.academicpress.com

Academic Press
84 Theobalds Road, London WC1X 8RR, UK
http://www.academicpress.com

International Standard Book Number: 0-12-480277-X (case)
International Standard Book Number: 0-12-580445-8 (pb)

PRINTED IN THE UNITED STATES OF AMERICA
02 03 04 05 06 07   9 8 7 6 5 4 3 2 1

CONTENTS

CONTRIBUTORS

Numbers in parentheses indicate the pages on which the authors' contributions begin.

Serge Arnaudeau (453), Department of Physiology, University of Geneva Medical Center, CH–1211 Geneva 4, Switzerland

K. Aufderheide (337), Department of Biology, Texas A & M University, College Station, Texas 77843

T. Clark Brelje (165), Department of Cell Biology and Neuroanatomy, University of Minnesota Medical School, Minneapolis, Minnesota 55455

Paul Campagnola (429), Department of Physiology, Center for Biomedical Imaging Technology, University of Connecticut Health Center, Farmington, Connecticut 06030

Victoria E. Centonze (129), Department of Cellular and Structural Biology, University of Texas Health Science Center at San Antonio, San Antonio, Texas 78229-3900

Eric S. Cole (337), Biology Department, St. Olaf College, Northfield, Minnesota 55057

Susan DeMaggio (475), President, Flocyte Associates, Irvine, California 92612

Nicolas Demaurex (453), Department of Physiology, University of Geneva Medical Center, CH–1211 Geneva 4, Switzerland

Per-Ola Forsgren (149), Molecular Dynamics, Sunnyvale, California 94086

David L. Gard (379), Department of Biology, University of Utah, Salt Lake City, Utah 84112

Irene L. Hale (301), Department of Biological Sciences and Neuroscience Research Institute, University of California, Santa Barbara, Santa Barbara, California 93106

Shinya Inoué (87), Marine Biological Laboratory, Woods Hole, Massachusetts 02543

Ted Inoué (87), Universal Imaging Corporation, West Chester, Pennsylvania 19380

Ira Kurtz (417), Division of Nephrology, Department of Medicine, UCLA School of Medicine, Los Angeles, California 90024

Aaron Lewis (429), Division of Applied Physics, Hebrew University, Jerusalem, Israel

Leslie M. Loew (429), Department of Physiology, Center for Biomedical Imaging Technology, University of Connecticut Health Center, Farmington, Connecticut 06030

Lars Majlof (149), Molecular Dynamics, Sunnyvale, California 94086

T. C. Marsh (337), Department of Pharmacology, University of Minnesota, Minneapolis, Minnesota 55455

Brian Matsumoto (301), Department of Biological Sciences and Neuroscience Research Institute, University of California, Santa Barbara, Santa Barbara, California 93106

Michal Opas (453), Department of Laboratory Medicine and Pathobiology, University of Toronto, Toronto, Ontario, M5S 1A8 Canada

Stephen W. Paddock (361), Department of Molecular Biology, University of Wisconsin, Madison, Wisconsin 53706

W. Ringlien (337), Carleton College, Northfield, Minnesota 55057

Robert L. Sorenson (165), Department of Cell Biology and Neuroanatomy, University of Minnesota Medical School, Minneapolis, Minnesota 55455

K. R. Stuart (337), Biology Department, St. Olaf College, Northfield, Minnesota 55057

Martin W. Wessendorf (165), Department of Cell Biology and Neuroanatomy, University of Minnesota Medical School, Minneapolis, Minnesota 55455

David J. Wright (1), Department of Biology, University of Dayton, Dayton, Ohio 45469-2320

Shirley J. Wright (1), Department of Biology, University of Dayton, Dayton, Ohio 45469-2320

Joseph P. Wuskell (429), Department of Physiology, Center for Biomedical Imaging Technology, University of Connecticut Health Center, Farmington, Connecticut 06030

Kay-Pong Yip (417), Department of Molecular Pharmacology, Physiology and Biotechnology, Brown University, Providence, Rhode Island

PREFACE FOR SECOND EDITION

The light microscope has always been a major tool of the cell biologist. In preserved cells, it can be used to localize specific proteins with fluorescent antibodies and in live cells it can be used to follow dynamic processes with vital dyes. With every advance in optical design, the microscopist has had an opportunity to see more and to obtain a greater understanding of cellular structure. Recently, another new advance in optical design as become available to the researcher, the confocal microscope. Its advantages include increased resolution, higher contrast, and greater depth of field. Together these features promise to increase our understanding of the cell's three-dimensional structure as well as its physiology and motility.

These advantages can be realized only if the samples are properly prepared for observation. It has been my experience that a major limitation in the effective application of confocal technology does not have to do with the microscope, but with the manner in which the sample has been prepared for imaging. An investigator must first decide whether the sample, if living, can withstand the intense illumination from the incident beam. Or, in fixed preparations, the investigator must ensure that the specimen is properly preserved to yield accurate images of its architecture. Failure to properly prepare the sample may cause the confocal microscopist to obtain images and data that are inferior to that obtained by more conventional techniques. The intent of this volume is to show the proper techniques to yield superior samples.

Additionally, this volume was prepared for the relatively inexperienced microscopist who has limited time to develop his expertise. It is unfortunate that the expense of these microscopes, coupled with the poor funding environment, has restricted the instruments to shared instrumentation facilities. Thus, the majority of new users have less time than they would like to develop their expertise with the microscope and to get feedback about their sample preparation techniques. For such investigators, this volume should prove especially advantageous. To simplify specimen preparation, specific chapters have detailed protocols so that they can be used at the laboratory bench as "cook book" recipes.

It is hoped that the more experienced investigator may also benefit from this volume. By combining the information from different contributors, it is hoped that the knowledgeable researcher can develop "hybrid" protocols for preparing samples not specifically described within this volume. Each chapter of this book is an independent unit; however, its utility can be enhanced by reading and collating the information from other parts of the volume. For example, the chapter on multicolor imaging has useful information that is applicable to single-label studies and thus, to nearly all the protocols within this volume.

In closing, I would like to thank the contributors for devoting their time and energy to the completion of this volume. This book would not have been possible without the kind understanding and sympathy of the contributing authors and they deserve recognition for their professionalism and enthusiasm. In addition, I wish to acknowledge the staff of Elsevier Science/Academic Press, especially Ms. Mica Haley, for showing extraordinary patience and encouragement in this work. Finally, I would like to thank Dr. Leslie Wilson and Dr. Paul Matsudaira, the series editors for their support in this project.

Acknowledgments

We would like to offer special thanks to the following sponsors for their generous contributions toward publication of this volume:

Nikon Instrument Group
Optronics Engineering

Brian Matsumoto

CHAPTER 1

Introduction to Confocal Microscopy

Shirley J. Wright[1] **and David J. Wright**

Department of Biology
University of Dayton
Dayton, Ohio 45469-2320

[1]To whom correspondence should be addressed.

METHODS IN CELL BIOLOGY, VOL. 70

I. Introduction

Confocal microscopes have become a powerful tool of choice for researchers interested in serious imaging of cell structure and function. These microscopes have revolutionized our view of cells and have been a major instrument in unraveling the complexities of the morphology and dynamics of cells and tissues. Confocal microscopy has a rich history (see review by Inoué, 1995). The concept of a confocal microscope was patented by Marvin Minsky in 1957. Approximately 10 years later, Egger and Petráň (Egger and Petráň, 1967; Petráň *et al.*, 1968) developed a spinning disk, multiple-beam confocal microscope (see Section IV.C). As a direct continuation of research on this confocal microscope, the first mechanically scanned confocal laser microscope was built in the early 1970s by Davidovits and Egger (1971, 1972) at Yale University. The first confocal laser photomicrographs of histologically recognizable cells were published in 1973 (Davidovits and Egger, 1973; Egger, 1989). Advances in computers, lasers, and digital imaging software in the late seventies allowed development of confocal microscopes in several cutting-edge laboratories (cf. Shotton, 1989). Confocal laser microscopes were first shown to have biological applications in the mid-1980s and became commercially available in the late 1980s (Amos *et al.*, 1987; Carlsson *et al.*, 1985; White *et al.*, 1987).

During the 1990s, there were several additional advances in confocal microscopy and in the high-technology industry. Advances in optics and electronics used in confocal microscope design, more stable and powerful lasers, more efficient mirrors, and lower-noise photodetectors have led to an improved generation of confocal microscopes. Newly developed fluorochromes were better matched to the laser lines and exhibited brighter fluorescence. Higher resolution monitors and printers have also given rise to even more stunning and informative confocal images.

Ten years ago confocal microscopes were greatly limited by the existing computer technology. Although hundreds of digital confocal images could be generated in one afternoon, storage of these images became a real problem. Back then, a large hard disk was considered to be 350 megabytes (MB; see Section X.D. for abbreviations used in this chapter). The fastest computer CPU speeds were only 66 MHz. There were few useful mass storage devices. Images were rarely if ever compressed so that depending on the image format, only 4–6 images (optical slices) would fit on a floppy disk. Since then, computer CPU speeds have greatly improved to 1 GHz or more. Hard disks readily contain 20 gigabytes of storage space. New removable mass storage media such as rewritable CD-ROMs, Zip disks, and Jaz disks, which did not exist 10 years ago, have replaced the cumbersome streaming tapes. These new storage media, together with more efficient networking capabilities, have greatly improved long-term image storage. In addition to better and faster computing power, there is better software to control the microscope and perform image processing. These developments in computer technology have resulted in improved confocal microscope capabilities and more in-depth applications.

In the past decade, interests and applications of confocal microscopists have also changed emphasis, as microscopists continue to push confocal technology to its limits. Initially, a major thrust of the research was to observe the general spatial distribution of usually one or two fluorescently labeled structures or cell populations in living or fixed samples. Now researchers can distinguish between two closely spaced structures within a specific organelle. For example, Wei *et al.* (1999) have used confocal microscopy to show that RNA polymerase II-mediated transcription sites, but not replication sites, are associated with speckles in the nucleus. High-resolution confocal microscopy has been used during *in situ* hybridization to colocalize specific genes or sequences to a specific chromosome. For example, telomeric and centromeric sequences have been localized to chromosomes of the human sperm to reveal a well-defined higher order structure of chromatin in the sperm nucleus (Zalensky *et al.*, 1995). Green fluorescent protein (GFP) constructs can be observed as they move through an entire living cell (Chapter 3, this volume). Propagation of ion fluxes can be followed through entire living cells and tissues in three dimensions over time at video rates (Chapters 12 and 13, this volume). With the recent technological advances, confocal microscopes have opened a new world to biologists interested in the dynamic complexities of the cell.

This chapter introduces the principle of confocal microscopy and describes the latest generation of confocal microscopes. Considerations for specimen preparation and some of the various applications of confocal microscopy are briefly described. Alternatives to confocal microscopy are also discussed. The chapter is aimed at novice users and those looking to set up confocal systems at their own institutions. The reader is encouraged to consult the references and subsequent chapters in this volume for additional technical information.

A. Advantages of Confocal Microscopy

When deciding which type of microscopy to use, one may ask, "Why confocal microscopy?" Transmission electron microscopy offers superb resolution; however, it is damaging to living specimens and suffers from fixation and sectioning artifacts. Conventional light microscopy allows examination of living and fixed cells with a variety of imaging modes including fluorescence. However, ultrastructural details cannot be obtained because of the relatively low resolution of the light microscope (0.2 μm). Another difficulty with conventional light microscopy is that it suffers from out-of-focus

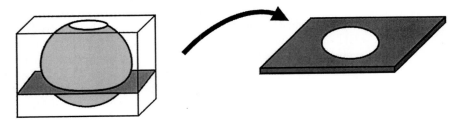

Fig. 1 Diagram showing generation of optical sections in the confocal microscope. A fluorescent structure of interest is embedded in a tissue. Instead of physically sectioning the specimen, the confocal microscope optically sections the specimen in a thin focal plane (dark plane) while leaving the sample intact. The optical section shows only the structures in the focal plane of interest and excludes most of the out-of-focus information above and below the focal plane.

information which often blurs the image. Video image processing can help "clean up" the images by improving contrast and detection, but does not completely eliminate the problem.

Confocal microscopy offers several advantages over conventional light and electron microscopy. First, the confocal microscope optically sections the specimen (Fig. 1). This eliminates any physical sectioning artifacts observed with conventional light and electron microscopic techniques. Since optical sectioning is relatively noninvasive, living as well as fixed samples can be observed with greater clarity. The optical sections have a relatively shallow depth of field (0.5–1.5 μm) which allows information to be collected from a well-defined plane, rather than from the entire specimen thickness. Therefore, instead of collecting out-of-focus light when the focal plane is above or below the specimen, the image is black (Fig. 2). This elimination of out-of-focus light results in an increase in contrast, clarity, and detection sensitivity. Another advantage of confocal microscopy is that optical sections can be taken in different planes, that is, the xy plane (perpendicular to the optical axis of the microscope) and the xz and yz planes (parallel to the optical axis of the microscope) as shown in Fig. 3. With vertical sectioning (xz and yz planes), cells are scanned in depth (z axis) as well as laterally (x or y axis). Figure 4 shows an example of an xy section. Another advantage of confocal microscopy is that the optical sections are obtained in a digital format. This greatly facilitates image processing and makes it very easy for the user to obtain a digital print of the image, eliminating the need for darkroom chemicals and lengthy waiting periods to obtain photographs. An additional advantage of using the confocal microscope is that all the fluorochromes of multiple-labeled specimens are in register, unlike in conventional fluorescence microscopy.

B. Disadvantages of Confocal Microscopy

Although confocal microscopes have many advantages, they do have some disadvantages. Confocal laser scanning microscopes (CLSMs) are limited by the available wavelengths of light produced by lasers (*laser lines*). This is unlike the conventional

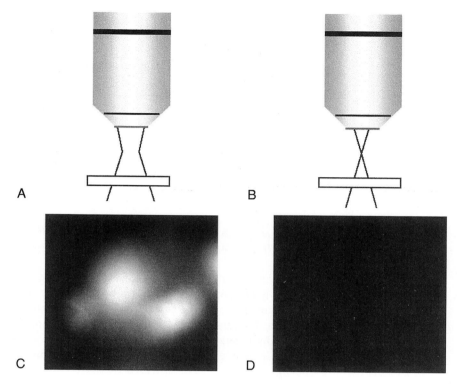

Fig. 2 Comparison of conventional (A, C) and confocal (B, D) light microscopes when the objective lens is focused well above the fluorescent specimen, depicted by the small rectangular box (A, B). (A) and (B) were taken at one instant in time with the same high-numerical aperture oil immersion objective. When the objective lens is focused above the specimen (A, B) in a conventional light microscope, the sample appears blurry (C), whereas it appears totally black in a confocal fluorescence microscope (D) because of optical sectioning and rejection of out-of-focus information. Therefore, it is very helpful to first focus on the sample using conventional fluorescence microscopy, and then switch to the confocal imaging mode once the focal plane has been identified.

fluorescence light microscope which uses a mercury or xenon arc lamp as the illumination source and therefore offers a full range of excitation wavelengths including UV. The intensity of the laser beam can be harmful to living cells, if the laser light has not been suitably attenuated. However, the new multiphoton microscope systems have effectively solved this problem (Section VIII.B and Chapter 3). For example, mitochondria in living preimplantation hamster embryos could not be imaged well with conventional microscopes because of out-of-focus blur and photodamage, and the laser intensity of older CLSMs would inhibit development. With two-photon imaging, the distribution of mitochondria in living hamster embryos has been observed at frequent intervals over 24 h, while the developmental competence of the embryos has been maintained (Squirrell *et al.,* 1999).

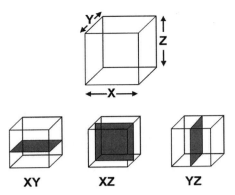

Fig. 3 Optical sectioning in the confocal microscope. Optical sections can be generated in three different planes in the confocal microscope: in the xy or horizontal direction which is parallel to the microscope stage, and in the xz and yz axes which are in the vertical direction. Sectioning in the xy plane is most common because the lateral spatial resolution (xy plane) is better than the axial spatial resolution (xz and yz planes).

An unfortunate disadvantage of confocal microscopes is their price. Depending on the setup, a confocal microscope can be over 10 times the price of a typical conventional fluorescence light microscope. The cost of owning and operating a confocal microscope can be minimized if several researchers share the microscope in a centralized "core" facility. Another disadvantage of some confocal microscope systems is that they often take up more space than a conventional fluorescence light microscope since they also

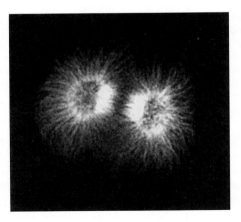

Fig. 4 Confocal fluorescence micrograph of the mitotic spindle in a sea urchin embryo undergoing first mitosis. One optical section is shown which was taken in the xy plane with a Bio-Rad MRC-1024 confocal laser scanning microscope using a 60×1.4 NA objective on a Nikon Diaphot. Microtubules were labeled with a mouse monoclonal E7 anti-tubulin antibody and FITC-conjugated goat anti-mouse IgG. The E7 anti-tubulin monoclonal antibody developed by Dr. Michael Klymkowsky was obtained from the Developmental Studies Hybridoma Bank developed under the auspices of the NICHD and maintained by The University of Iowa, Iowa City, IA 52242.

require space for a laser, scan head, and computer hardware. Recent confocal microscope systems are more compact than earlier versions. Although most confocal systems come with easy-to-use software, users require training and an understanding of the confocal principle to obtain the best confocal images. Personal confocal systems are usually easier to use than the more sophisticated multiuser confocal systems because fewer features are available on the personal systems.

II. Principle of Confocal Microscopy

One way to understand the principle of confocal microscopy is to compare confocal microscopes to conventional (wide field) light microscopes. In conventional microscopy, when the objective lens is focused at a specific focal plane in the specimen, the entire depth or volume of the specimen is uniformly and simultaneously illuminated, and most emitted light is collected (Fig. 5). This causes out-of-focus blur or "flare" from areas above and below the desired plane of focus. The background fluorescence from either labeled structures or naturally occurring autofluorescence results in a blurry image, since the out-of-focus light reduces contrast and decreases resolution, making it difficult to discern details in the specimen (Fig. 6). In confocal microscopy, most of this out-of-focus light has been eliminated. Several features of the confocal microscope make this possible. First, the illumination in a confocal microscope is focused as a very small spot (or series of spots) on the specimen at any one time (Fig. 5). The size of the illuminating spot may be as small as 0.25 μm in diameter and 0.5 μm deep at its brightest intensity (Wright *et al.*, 1993). This depends on the specific confocal microscope design, wavelength of incident (illuminating) light, objective lens, confocal microscope settings, and characteristics of

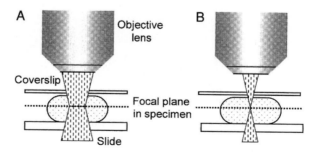

Fig. 5 Comparison of specimen illumination in conventional (A) and confocal (B) microscopes. (A) and (B) were taken at one instant in time with the same high-numerical aperture oil immersion objective. The whole depth of the sample is continuously illuminated with the conventional light microscope. Out-of-focus signals are detected as well as the focal plane of interest. In contrast, with a confocal microscope, the sample is scanned with one or more finely focused spots of light that illuminate only a portion of the sample at one time. Eventually a complete image of the sample is produced. Thus conventional fluorescence images of cells and tissues often suffer from out-of-focus information, whereas this problem is greatly reduced in confocal fluorescence images.

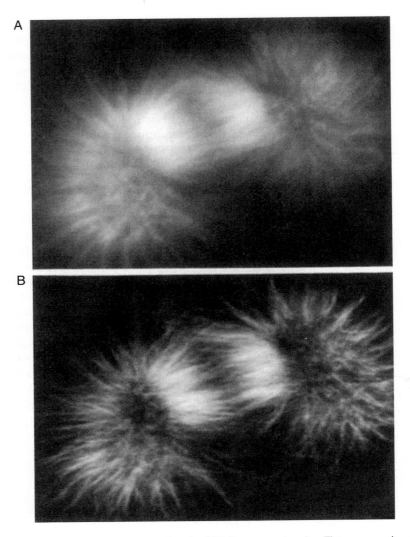

Fig. 6 Comparison of conventional (A) and confocal (B) fluorescence imaging. Fluorescence micrographs of the same mitotic sea urchin embryo stained for microtubules and taken with a NORAN Odyssey video-rate confocal laser scanning microscope. A conventional fluorescence image was obtained in (A) by removing the slit aperture, whereas in (B) the slit aperture was present, showing a confocal fluorescence image of a single xy optical section with improved contrast and resolution. Microtubules were labeled with a mouse monoclonal E7 anti-tubulin antibody and an FITC-conjugated goat anti-mouse secondary antibody. Reprinted from Wright, S. J., *et al.* (1993). *Meth. Cell Biol.* **38,** 1–45, with permission of Academic Press, copyright © 1993.

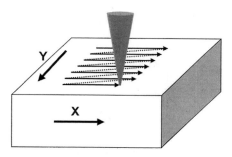

Fig. 7 The raster scan pattern that confocal laser scanning microscopes commonly use to generate an image. The emission patterns of a fluorochrome are detected as the laser beam (cone) moves across the sample (rectangle) from the left to right ($\rightarrow$) and from top to bottom. "Flyback" occurs when the laser beam rapidly moves to the left (dotted lines) to begin a new scan on the next line. Most confocal microscopes do not collect information from the sample during flyback. A complete image of the fluorescence emitted by the sample is displayed on the image monitor when the scan is finished.

the sample. Second, the illumination in the confocal microscope is sequential rather than simultaneous. Therefore, the specimen is scanned with one (or more) finely focused spot(s) of light that illuminates only a relatively small portion of the specimen at any one time. Eventually the spot(s) of incident light is rapidly scanned over the entire plane of focus in a raster pattern so that a complete image of the focal plane is generated (Fig. 7). As the illumination beam diverges above and below the plane of focus in the specimen, areas away from the focal plane receive less illumination, thus reducing some of the out-of-focus information.

A third feature of the confocal microscope is the use of spatial filters (apertures) to eliminate out-of-focus light (Figs. 6 and 8). Confocal imaging systems are based on Marvin Minsky's (1957, 1988) design in which both the illumination and detection systems are focused on the same single area of the specimen. The specimen is imaged in such a way that only the light coming from the focal plane of interest is collected, and signals coming from above and below the plane of focus are blocked by the spatial filter (Figs. 8 and 9). This further reduces the out-of-focus information. Therefore the illumination, specimen, and detector all have the same focus, that is, they are confocal. The result of these special features of confocal microscopes is a greatly improved, crisp image with an increase in contrast and resolution (Fig. 6).

When light passes through the objective lens, diffraction causes the light focused to a point in the focal plane to appear as a bright spot surrounded by rings of decreasing intensity (Fig. 10). The brightest, central spot is called an Airy disk and is named after the 19th-century British astronomer G. B. Airy. Thus, a typical biological specimen is represented in the image by many overlapping Airy disks and their surrounding rings. Theoretically, the best confocal imaging is obtained if the size of the confocal aperture matches the size of the Airy disk so that the dimmer rings of light are blocked from detection (Snyman *et al.,* 1999). Reducing the size of the confocal aperture to the size of the Airy disk projected on the confocal aperture improves axial (z axis) resolution, with

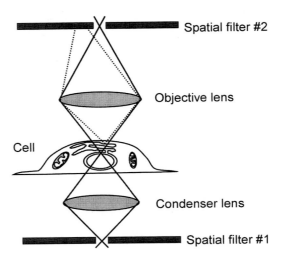

Fig. 8 Apertures reduce out-of-focus light in confocal microscopes. Cell membranes are labeled with a membrane-permeant fluorescent dye. Although the illumination beam emanating from spatial filter 1 is focused on the nuclear envelope, structures from above and below this focal plane also intercept the incident light and fluoresce. The objective lens collects fluorescence emitted by the nuclear envelope (solid lines) and other focal planes (dotted lines). In-focus light from the nuclear envelope passes through spatial filter 2 and is detected. Out-of-focus light from the other focal planes such as the endoplasmic reticulum is essentially blocked by spatial filter 2, so that mainly in-focus fluorescence is detected. Both the illumination and detection (imaging) systems are focused on the same single volume element of the cell, that is, they are confocal. Adapted from Shuman, H., *et al.* (1989). *BioTechniques* **7**, 154–163, with permission of Eaton Publishing Co., copyright © 1989.

a decrease in overall image brightness. In very dim specimens, increasing the confocal aperture increases the image brightness with a reduction in axial resolution.

The confocal microscope provides improved lateral and axial resolution. Resolution is the ability to resolve or discern two point objects from each other. It refers to the ability to distinguish fine details in a sample. In the light microscope, there are three types of resolution to consider: (1) lateral spatial resolution which occurs in the xy axis, (2) axial spatial resolution in the z axis, and (3) temporal resolution which occurs over time (Fig. 11). The lateral spatial resolution is usually 2–3 times better than the axial spatial resolution under the same imaging conditions (Brelje *et al.,* 1993). For example, the theoretical lateral resolution is 0.17 μm using 514 nm excitation with a 60 × 1.4 numerical aperture (NA) objective on a confocal fluorescence microscope, while the axial resolution under the same conditions is 0.58 μm. Spatial resolution (both lateral and axial) depends on the imaging conditions including the NA, which is a measure of the objective lens aperture taking into account the wavelength of the illuminating light and refractive properties of the medium. The smaller the objective lens aperture, the lower the NA and the lower the resolution (Inoué and Spring, 1997). Conversely, a large objective lens aperture results in a high NA and high resolution (Fig. 12). The higher the NA, the greater the lateral spatial resolution and the thinner the optical section.

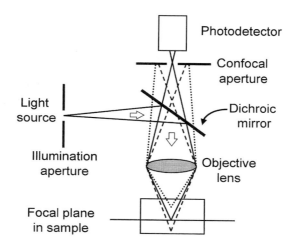

Fig. 9 Schematic diagram of the confocal principle in an epifluorescence microscope. Excitation light enters through an illumination aperture, reflects off a dichroic mirror, and is focused on the specimen by the objective lens. The longer wavelength fluorescence emissions generated by the sample return via the objective and pass through the dichroic mirror. Light primarily from the focal plane of the objective passes through the confocal aperture and is detected by a photodetector such as a photomultiplier tube. In contrast, light from above (dotted lines) and below (dashed lines) the plane of focus is not focused on the confocal aperture, but is attenuated by it. As a result, most of the out-of-focus light is eliminated from the final confocal image. Adapted from Shotton, D. M. (1989). *J. Cell Sci.* **94,** 175–206, with permission of The Company of Biologists, Ltd., copyright © 1989.

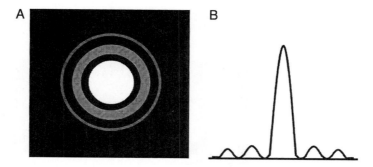

Fig. 10 Airy disk (A) and its intensity distribution (B). Diffraction induced by the objective's internal aperture causes an intensely fluorescent 0.17 μm bead to appear as a bright spot surrounded by dimmer rings. The bright central spot is called an Airy disk. A mathematical representation of (A) is seen as a line profile or a point spread function in (B). Theoretically, the best images are obtained in the confocal microscope when the confocal aperture is set to include light from the Airy disk, while excluding light from the rings. Adapted from Keller, H. E. (1995). *In* "Handbook of Biological Confocal Microscopy" (J. B. Pawley, ed.), 2nd Ed., pp. 111–126, with permission of Plenum Press, copyright © 1995.

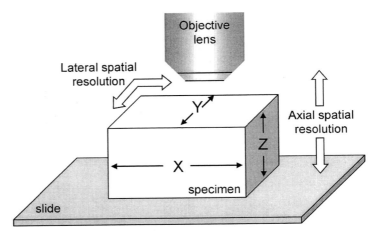

Fig. 11 Resolution in the light microscope. Specimen on a microscope slide with the coverslip removed for clarity. Each specimen that is imaged by the objective of the light microscope has three dimensions, one each in the x axis, y axis and z axis. Lateral spatial resolution refers to the resolution in the horizontal plane (xy axis), whereas axial spatial resolution refers to resolution in the vertical plane (z axis).

Therefore, optical section thickness relates inversely to the NA of the objective lens and relates directly to the wavelength of the incident light.

To achieve confocal fluorescence imaging, the incident light passes through an illumination aperture and becomes a point source of excitation light (Fig. 9). The light is then reflected by a dichroic mirror and directed to the objective lens, which focuses the incident beam as a small spot of light on the desired focal plane of the specimen. As the fluorochrome in the specimen is illuminated, it gives off fluorescence emissions which scatter in all directions. Fluorescence from the sample returns via the objective lens, and

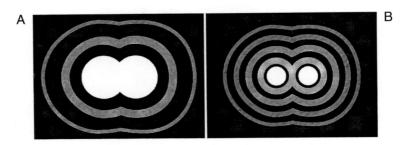

Fig. 12 Effect of numerical aperture (NA) on lateral spatial resolution in the light microscope. Two closely spaced, fluorescent microbeads are imaged in (A) and (B) with the same magnification. In (A), an objective lens with a low NA is unable to resolve the two fluorescent beads and they appear as one, whereas in (B), a higher NA objective lens shows the beads as two separate spots. Therefore, a high NA objective lens produces higher resolution images.

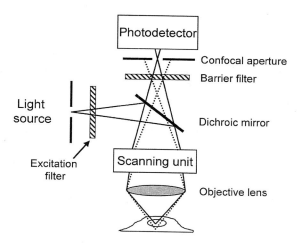

Fig. 13 Basic design of a confocal laser scanning microscope for use in fluorescence microscopy. An excitation filter is used to select a specific wavelength of excitation light from the laser. The laser beam is then directed onto the specimen by reflection off of a highly reflective dichroic mirror onto a scanning unit which scans the beam across the specimen in a raster pattern. The laser light is focused on the nucleus of the cell by a high-numerical-aperture objective lens. Induced fluorescent or reflected light from the nucleus is scattered in all directions. Some of the light is collected by the objective lens and is directed through the scanning system toward the dichroic mirror. The longer wavelength induced fluorescence passes through the dichroic mirror toward the photodetector. Specific fluorescence emission wavelengths are selected when the light next passes through a barrier filter which functions to block out unwanted wavelengths of light. In-focus light from the nucleus then passes through the confocal aperture to the photodetector. In contrast, out-of-focus light from the cytoplasm (dotted lines) has a different primary image plane of focus and is attenuated by the confocal aperture. Thus most out-of-focus information is not detected in the final confocal image.

the longer wavelength light from the sample passes through the dichroic mirror toward the detector. In front of the detector is a spatial filter containing a confocal aperture which can be a pinhole diaphragm or slit type aperture. The in-focus light from the sample passes through the aperture of the spatial filter, while the out-of-focus light from above and below the plane of focus is blocked by the spatial filter. As a result, very little out-of-focus light reaches the detector. Thus, the spatial filter not only provides continuous access to the detector for in-focus light, but also effectively suppresses light from nonfocal planes. A scanning system is placed between the light source and objective lens so that the incident spot of light is scanned across the specimen (Fig. 13). To form a two-dimensional (2-D) image in a CLSM, the laser beam scans the specimen in a raster pattern (Fig. 7), that is, in the x and y axes, and the induced fluorescent light is detected and converted to a video signal for display on a computer screen. The intensity of the bright spots on the computer screen corresponds to the intensity of the fluorescent light generated during the raster scan. The result of confocal imaging provides optical sections with exceptional contrast and often with details that were previously hidden by out-of-focus blur (Fig. 6). Therefore, confocal microscopy is a technique where out-of-focus background information is rejected by the confocal aperture to produce thin optical sections.

III. Major Components of a Confocal Microscope

The basic components of a confocal microscope highlighted in Fig. 14 are briefly described below. The differences between commercially available microscopes are reflected in their different components used to achieve confocal imaging, and in their performance. In a complete confocal system, all the components are designed to work well together to provide superior images. Although a confocal attachment can be placed onto a preexisting microscope in the researcher's laboratory, one has to consider the best attachment design that will work with the particular model of microscope.

A. Microscope

The microscope used for confocal imaging can be an upright or an inverted model. It can use the 160-mm fixed-tube-length optics or the newer infinity optics. Most confocal attachments are configured to work with either type of microscope. An upright model is useful for fixed specimens, and living specimens when used with a water immersion objective. An inverted model is better for living samples that need to be kept sterile, or if access to the samples is required when using micromanipulators and microinjectors.

An important feature is that the confocal attachment must be able to fit onto the video port of the microscope. In designing a confocal microscope, many companies have worked specifically with a particular type or make of microscope, and thus the best images are observed with that model of microscope. For example, the Olympus Fluoview confocal attachment works best with an Olympus microscope and is not interchangeable

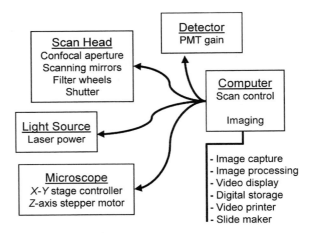

Fig. 14 Basic components of a confocal laser scanning microscope. A laser and scan head are connected to a conventional light microscope base. A photodetector such as a photomultiplier tube (PMT) is used to detect the in-focus light. All components are designed to work well together as a confocal system. This task is achieved by the computer which is used to control not only the scan and detection of the image, but also the imaging functions such as image capture, processing, and storage. Stored images can then be printed for dissemination.

with other microscope models, whereas NORAN's Oz confocal system can be fitted to more than 26 different microscopes. Another important attachment to the confocal microscope is the motorized z axis stage controller or stepper motor. This unit is attached to the fine focusing knob of the microscope so that optical sections can be taken at precise intervals in a *Z-series* (stack of optical sections). The stepper motor moves the sample in a vertical direction by adjusting the focus precisely in increments as small as 0.025 μm. Some confocal microscopes also contain a motorized *xy* stage controller which will move the sample in a horizontal plane. Placing the microscope on a vibration-free table is very useful, since image acquisition is sensitive to vibrations, and full-format Z-series can take several minutes.

B. Objective Lens

The choice of objective lens is critical in obtaining the best confocal images. For example, a high-NA water immersion lens is essential for studying living specimens with an upright microscope. Factors that come in to play when choosing an objective include optical aberrations and flatness of field. Each objective lens must be matched to the other optical components of the microscope. For example, a 160 mm tube length objective lens from an older microscope will not function optimally on a newer, infinity optics microscope. The objective lens should be able to transmit fluorescent light efficiently without any aberrations and little loss of light (Hale and Matsumoto, 1993; Keller, 1995). A fluor type, Plan-Apo lens works well for this. Since the NA relates to the ability of the objective lens to collect light, an objective with a high NA produces a bright image with the best resolution (Fig. 12). Therefore, high NA objective lenses provide the thinnest optical sections and are most desirable.

If the objective lens is not corrected for aberrations, the confocal images can become distorted (Brelje *et al.*, 1993; Dunn and Wang, 2000; Hale and Matsumoto, 1993; Keller, 1995). Spherical aberrations result when the ray paths of the illuminating light do not converge to a single focus (Fig. 15). This is especially a problem in thick specimens. These aberrations can introduce error in volume measurements of images generated from multilabeled samples. During chromatic aberration, different wavelengths (colors) of light do not come to the same focal point in the sample. The shorter wavelengths of light are refracted more and may focus in a higher plane, while the longer wavelengths are refracted less and may focus in a lower focal plane (Fig. 15). The end result is a loss of intensity from the sample. New lenses designed for confocal microscopy are usually corrected for these aberrations. See also Chapters 4, 6 and 7 of this volume for additional discussions of aberrations in thick specimens.

Another consideration is the *working distance* of the objective, which is the distance between the sample focal plane and the lowest structural component of the objective. This distance sets the limit of the deepest focal plane that can be properly imaged in the specimen. Working distance can greatly vary between objective lenses. The working distance is shorter for high NA lenses. For example, a 10 × 0.3 NA objective lens can have a working distance of 2–3 mm, whereas that of a 40 × 0.7 NA objective is 0.57 mm. A larger working distance permits imaging of thick specimens. A coverslip alone can have a thickness of 160–170 μm. Thus, the working distance must be greater than

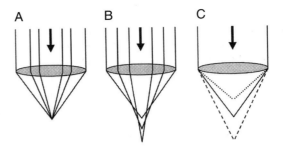

Fig. 15 Aberrations in an objective lens focused deep in the sample. The same magnification of objective lens is used, and the same white light source is used as the incident light (arrow). For clarity, effects of aberrations are exaggerated. The objective lens in (A) produces high-quality images since it has been corrected for aberrations, and thus all incident light passing through it comes to focus at the same point. In (B), spherical aberrations in the objective lens cause the incident light to come to focus at slightly different focal planes, thus reducing image quality. Chromatic aberrations in the objective lens (C) prevent incident light from converging to a single focal point. The longer wavelength, red light (dashed lines) converges below the plane of focus since it scatters less; the intermediate-wavelength green light (solid lines) converges at the plane of focus; and the shorter wavelength blue light (dotted lines) converges above the plane of focus since it is more susceptible to scattering induced by the sample. Therefore to obtain the best images, it is important to use objective lenses that have been corrected for aberrations.

this. However, with the larger working distance objectives comes a reduction in image quality.

Choice of magnification is also critical in obtaining excellent confocal images. Since image brightness is inversely proportional to objective magnification, lower magnification objectives appear to produce better images. For example, rather than using a 100×1.4 NA oil immersion objective, one may obtain better quality confocal images with a 63×1.4 NA oil immersion objective. But this cannot be carried to an extreme. For example, a 10×0.3 NA objective lens with an electronic zoom carried out by the confocal microscope will produce an image of poorer quality than an unzoomed image taken directly with the 63×1.4 NA oil immersion objective. In CLSMs that produce very bright images at high magnification, the 100×1.4 NA objective lens is a good choice.

Thus, the specific choice of objective lens that is used for a particular application depends on a series of choices, often compromises, that have a specific combination which yields the best confocal image. Usually during an imaging session, a combination of objective lenses is used. One may first use a 10×0.45 NA objective to locate an area of interest, and then perform the final confocal imaging using a 60×1.4 NA oil immersion objective lens. It is not beneficial to purchase an expensive confocal microscope, and then buy inexpensive, uncorrected objective lenses. A more detailed discussion on the selection of objective lenses for confocal microscopy can be found in Chapters 4 and 6.

C. Light Source

Conventional fluorescence microscopes use white light produced by either a mercury or xenon arc lamp as an excitation light source, which produces a broad emission

Table I

Lasers Commonly Used in Confocal Laser Scanning Microscopes

Type of laser	Power (mW)[a]	Emission wavelength (nm)				
		UV	Blue	Green	Yellow	Red
Argon Ion	10		488			
Argon Ion	25		488	514		
Argon Ion	100		457 488	514		
Argon Ion UV/Visible	250	351 363	488	514		
Krypton Argon	15		488		568	647
Krypton	10				568	
Helium Cadmium	10		442			
Green Helium Neon	1			543		
Red Helium Neon	10					633
Red Laser Diode	5					635

[a]Power levels may vary depending on the laser manufacturer.

Abbreviations: mW, milliwatt; nm, nanometer; UV, ultraviolet.

Modified from Sheppard, C. J. R., and Shotton, D. M. (1997). "Confocal Laser Scanning Microscopy." Copyright © 1997, with permission from BIOS Scientific Publishers Ltd. and the Royal Microscopical Society.

spectrum. Confocal microscopes use either a laser (CLSMs) or a mercury (or xenon) arc lamp (spinning disk confocal microscopes) as a light source. Lasers are ideal as a light source for the confocal microscope because they have high brightness, low noise, and low beam divergence and can be focused into a very small spot. There are several different laser light sources to choose from which fall in the range of ultraviolet (UV), visible, and infrared wavelengths of light (Brelje *et al.,* 1993; Gratton and vandeVen, 1995). The lasers vary in their gas content, available laser lines, cooling system, output power, and operational lifetime (Table I). The most common lasers include air-cooled or water-cooled argon ion lasers, krypton–argon ion lasers, helium–neon lasers, helium–cadmium lasers, UV lasers, and infrared lasers. Some lasers emit multiple wavelengths that can be operated in either the multiline mode or the single-wavelength mode. Many lasers have a limited number of laser lines so that not all fluorescent dyes can be excited by the laser. For example, an argon ion laser may emit at 488 nm and 514 nm, and thus is unable to excite UV fluorochromes which require excitation wavelengths less than 400 nm. New generation CLSMs use a combination of lasers such as an argon ion laser for fluorescein, a helium–neon laser for rhodamine, and a red diode laser for Cy-5. Fluorochromes that can be excited by these laser lines are listed in Table II. Tunable dye lasers have also been used with confocal microscopes to give a wider range of wavelengths (Brelje *et al.,* 1993; Gratton and vandeVen, 1995). This feature depends on the quality of the filter sets and strength of the laser lines. Thus the type of fluorochromes that will be used dictates the type of laser needed for the confocal system.

Table II
Fluorescent Probes Commonly Used in Confocal Fluorescence Microscopy

Fluorochrome[a]	Absorption maximum	Emission maximum	Application
Blue probes requiring UV excitation			
AMCA[b]	349	448	Covalent labeling agent
DAPI	358	461	DNA stain
Fura-2	363	512	Calcium ions
Hoechst 33258	352	461	DNA stain
Hoechst 33342	350	461	DNA stain
Indo-1	346	475	Calcium ions
Green/yellow probes requiring blue excitation			
Acridine orange	500, 460[c]	526, 650[c]	DNA, RNA stain
BCECF	482	520	pH-sensitive dye
BODIPY	503	511	Covalent labeling agent
Calcein	494	517	Cell viability
Calcium green	506	532	Calcium ions
FITC	494	518	Covalent labeling agent
Fluo-3	506	526	Calcium ions
GFP	395[d]	508[d]	Gene expression, tracing
Lucifer yellow	428	536	Neuronal tracer
TOTO-1	514	533	Nuclear stain
YOYO-1	491	509	Nuclear stain
Red probes requiring green/yellow excitation			
DiI	549	565	Neuronal tracer, lipophilic tracer
Cy-3	556[d]	574[d]	Covalent labeling agent
Propidium iodide	535	617	Nuclear stain
Texas Red	595	615	Covalent labeling agent
TRITC	555	580	Covalent labeling agent
Far red probes requiring red excitation			
Cy-5	649[d]	666[d]	Covalent labeling agent

[a] Although the maximum excitation and emission wavelengths are listed here, the actual wavelengths are a abroad peak which depends on the chemical environment and/or linkage to proteins such as antibodies. All values except those indicated by *d* are taken from Haugland (1999).

[b] Abbreviations: AMCA, 7-Amino-4-methylcoumarin-3-acetic acid; BODIPY, 4,4-difluora-5,7-dimethyl-4-bora-3*a*,4*a*-diazaindacene 3-propionic acid; DiI, 1,1′-dioctadecyl-3,3,3′,3′-tetramethylindocarbocyanine perchlorate; FITC, fluorescein isothiocyanate; GFP, green fluorescent protein (wild type); TRITC, tetramethyl rhodamine isothiocyanate; UV, ultraviolet.

[c] First value, when bound to DNA, second value when bound to RNA.

[d] Values obtained from Sheppard, C. J. R., and Shotton, D. M. (1997). "Confocal Laser Scanning Microscopy," p. 65.

Modified from Sheppard, C. J. R., and Shotton, D. M. (1997). "Confocal Laser Scanning Microscopy." Copyright © 1997, with permission from BIOS Scientific Publishers Ltd. and the Royal Microscopical Society.

Older CLSMs used excitation and emission filter sets common to conventional epifluorescence microscopes for selecting the appropriate laser wavelength. Neutral density filters were used to attenuate the laser beam. Often these filters were manually changed by the user and required alignment each time. Many newer CLSMs use a software-driven acoustooptical tunable filter (AOTF) which provides appropriate laser attenuation and selection of excitation wavelength for the various fluorochromes one may wish to use (Hoyt, 1996; Inoué and Spring, 1997). The maximum excitation wavelength of each fluorochrome can be chosen through the AOTF. This helps reduce background noise from autofluorescence and bleed-through or cross-talk from other channels in multiple-labeled samples. Thus, AOTFs remove the need for line selection filters for excitation wavelength. An added feature of AOTFs is that they provide the ability to individually attenuate the lines of a laser that emits multiple wavelengths. This feature is essential in samples labeled with multiple fluorochromes, since each dye has a different excitation wavelength and may photobleach at a different rate. Individual laser line attenuation in multiple-labeled specimens is not possible to perform with a neutral density filter. AOTFs also provide fast (microsecond) switching between different imaging modes. For example, some confocal systems such as the Bio-Rad Radiance2000 and the Zeiss LSM 510 use the AOTF to blank (turn off) the laser beam during flyback of the raster scan to eliminate unnecessary exposure of the sample. In addition, the AOTF in the Zeiss LSM 510, permits the user to choose a region of interest (ROI) of any desired size and shape by quickly changing the laser intensity on a pixel-by-pixel basis. Thus photobleaching can be restricted to the ROI so that photodamage is minimized in the specimen. AOTFs are especially important for ratio measurements since they are fast enough to switch between different excitation wavelengths. In addition, AOTFs can prevent cross-talk or bleed-through from other channels in multiple-labeled samples. For example, when a sample double-labeled with fluorescein and propidium iodide is scanned simultaneously for both channels at a high laser intensity, the resulting optical section may show considerable penetration of the fluorescein signal into the other channel (*cross-talk, bleed-through*). To avoid this problem, some CLSMs containing AOTFs (Zeiss LSM 510) have dual direction scan which allows one signal to be collected as the scanner moves forward along the scan line, while the other channel is recorded as the scanner returns back horizontally along the scanning line. Therefore, CLSMs containing AOTFs are very useful to have in a multiuser core facility since various fluorescent samples can be easily accommodated.

Lasers can last from approximately 2000 to 15,000 h depending on the laser type and tube current used during its operation (Gratton and vandeVen, 1995). For example, a krypton argon ion laser (2000 h) has a much shorter lifetime than an argon ion laser (5000–7000 h) or a helium neon laser (8000–15,000 h). To avoid unnecessary load on the laser, some CLSMs, such as the Zeiss LSM 510, use software to control the laser load so that it is at the lowest tube current when in standby mode. Although the lifetime of a laser can be extended by running it at less power (e.g., 1%, 10%, 25%, 50%) when imaging bright samples, users should make plans to replace the laser in the future so that funds are available for its replacement.

Unlike older generation CLSMs, the laser has been separated from the scan head in some newer CLSMs. This feature reduces vibrations coming from the laser and its cooling fan. The laser light is delivered to the scan head by a fiber optic cable. This makes the laser more accessible, easier to align, and more mobile so that it can be quickly changed or moved between confocal systems. See also Chapter 5 for a detailed discussion of the dyes and lasers used in confocal microscopy.

D. Scan Head

This is one of the most variable parts of the confocal system. It houses several important components including the scanning system which scans one or more fine spots of light onto the specimen in a raster pattern (Fig. 14). In most CLSMs, a single spot of finely focused light is used for raster scanning (Fig. 7). When the specimen is scanned horizontally with the laser beam, the resulting horizontal scan is called a *line scan* and is very rapid as compared to the slower vertical scan, which is also called a *frame scan*. In some CLSMs (Bio-Rad, Zeiss) the scanning system is able to change the orientation of the entire raster scan so that a specimen can be scanned with greater temporal resolution. This feature is called scan rotation or raster rotation and is especially useful in capturing rapid, cellular processes in elongated specimens. It is also useful to optimize the position of structures and cells in the image frame.

Another important component of the scan head in a CLSM is the scanning mirrors, which function to scan light onto the specimen in a raster pattern (Fig. 7) and bring light from the focal plane of interest to the photodetector. These scanning mirrors are not required in the spinning disk type of confocal microscope since scanning is achieved with a different method (Section IV.C). Different mechanisms have been explored to achieve confocal imaging, including: (1) a rotating polygon mirror such as the kind used in a bar code reader, (2) a single galvanometer mirror mounted on a gimbal, (3) two closely coupled galvanometer mirrors in which one mirror oscillates rapidly to produce the line scan and the other mirror produces the slower frame scan, and (4) a system of four mirrors in which two concave mirrors image one galvanometer mirror onto a second galvanometer mirror (Amos, 2000). Some of these designs produce higher quality images than others.

The scan head also contains the confocal aperture which can be either a pinhole or slit type. Depending on the microscope model, the confocal aperture can be placed between the dichroic mirror and the detection system, or closer to the detector. The confocal aperture comes in several sizes, so that the aperture can be matched to the application of the user. During confocal imaging, the user chooses between the different apertures for optimal imaging. Confocal aperture choice may be software driven, or the user may need to adjust this manually. The number of available confocal apertures varies with the model of confocal microscope. Some models have a continuously variable confocal aperture, whereas others have fixed position apertures. Several types of filters can be located in or near the scan head depending on the confocal microscope system. These include AOTFs, excitation, emission, and neutral density filters, and dichroic mirrors. The scan head usually contains one or more barrier filters that are used to block out unwanted wavelengths of light before they reach the photodetector (Figs. 13 and 14).

E. Photodetector

The photodetector can be located within the scan head or outside of it depending on the confocal system. Some confocal systems rely on the human eye for detection (Section IV.C). However, because the laser light can be damaging to the eye, the detector is usually a low-noise, highly sensitive photomultiplier tube (PMT) that efficiently converts light energy into electrical energy (Majlof and Forsgren, 1993). Light striking a phosphor causes electrons to be released, which are then collected and multiplied into a signal for display on a video monitor. PMTs are excellent detectors because of their speed, sensitivity, high signal-to-noise ratio, and wide dynamic range. Some confocal systems have several PMTs which allow for simultaneous imaging of different fluorochromes in multiple-labeled specimens, as well as reflected light. For example, the Zeiss LSM 510 and Bio-Rad Radiance2000 can be configured with up to four detectors for fluorescent or reflected light imaging and all four channels can be recorded and displayed simultaneously. Care must be taken so that bleed-through is prevented. Usually the AOTF will minimize cross-talk between channels. If three PMTs [one each for red, green, and blue (RGB) channels] are used to collect information simultaneously from the sample, the images can be merged into a single RGB image that represents the real colors of the sample, similar to a digital video camera (Inoué and Spring, 1997). Transmitted light is usually detected with a photodiode.

An important recent development in photodetection is the spectrophotometric detection system developed by Leica Microsystems for use in its spectral confocal microscopes (Calloway, 1999). Light emitted from the sample contains both intensity and spectral information. This light is passed through a prism which splits it into a spectrum from 400 nm to 750 nm (Fig. 16). This spectrum is directed to a photodetector, PMT1 (Fig. 16). In front of PMT1 is a slit (S1) formed by two mirrored plates (Fig. 16). The slit can be narrowed or widened to detect any portion of the spectrum. This is analogous to changing the barrier filter to one with a different *bandwidth,* which is the filter's range of specific accepted

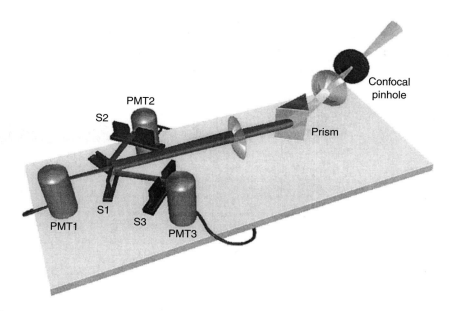

Fig. 16 The spectrophotometric photodetection system. PMT1–3, photomultiplier tubes 1–3; S1–3, slit systems 1–3. See text for details. (Courtesy of Leica Microsystems, Heidelberg.)

wavelengths of light. The slit in front of the PMT can also be moved along the spectrum so that the center of the bandwidth can be positioned anywhere on the spectrum. Thus, the system allows the creation of filters of any bandwidth anywhere across the spectrum. This is an important feature as new fluorochromes are discovered with nontraditional emission wavelengths. The system is designed to work simultaneously with multiple detectors (Fig. 16, PMT1–3) as follows: Wavelengths shorter than the bandwidth selected by PMT1 are excluded by the plate on one side of slit S1, while wavelengths longer than the selected bandwidth are excluded by the plate on the opposite side of slit S1. The surface of each plate is mirrored and angled so that the light excluded from PMT1 is directed to the other PMTs (Fig. 16, PMT2, PMT3). These PMTs (PMT2, PMT3) each have a variable slit system (S2, S3) that can be set by the user to select the desired emission wavelengths to be detected. For example, when imaging a sample triple-labeled with fluorescein, rhodamine, and Cy-5, light emitted from the fluorochromes that passes through the confocal pinhole is directed to the prism where it is split into a spectrum (Fig. 16). The spectrum is directed to PMT1 which is set to accept only the rhodamine signal (centered at 580 nm). The shorter wavelength fluorescein signal (centered at 520 nm) is rejected by the mirrored plate (S1) on one side of PMT1 and is reflected to PMT2. The variable slit (S2) in front of PMT2 is set to accept only the fluorescein signal. The longer wavelength Cy-5 signal (centered at 667 nm) is rejected by the mirrored plate on the opposite side of S1 and is reflected to PMT3. The variable slit (S3) in front of PMT3 is set to accept only the Cy-5 signal. The system can also be configured with a fourth PMT that functions simultaneously with the other three PMTs. The prism in combination with the slits in front of the detectors both split the wavelengths of emitted light between the detectors and select the wavelengths of light to be collected by the detectors. Thus the spectrophotometric detection system replaces the dichroic and barrier filters used in confocal microscopes with filtered detection systems. This results in an increase in the efficiency of detection, and consequently, in brighter confocal images.

F. Computer Hardware

Most confocal microscopes use a computer as an integral part of the system. An exception is some of the spinning disk confocal microscopes (Section IV.C). The computer must be powerful enough to control several devices and analyze the complex data acquired (Fig. 14). The most common type of computer used with confocal microscopes is a PC or workstation. Some computers use reduced instruction set computing (RISC) microprocessors to handle the computation intensive three-dimensional (3-D) imaging applications. Many confocal systems are designed so that the computer is used to change the laser intensity (power), choice of confocal aperture, stepper motor to control focusing through the sample at defined intervals, shutter to block the laser light, gain (sensitivity) of the photomultiplier tube, selection of filter wheels, and movement of the scanning mirrors (Fig. 14). It also is a host for the software which allows for image acquisition, storage, display, and further processing. The computer monitor is usually a high-resolution color type. It may be used to display not only the software user interface, but also the recently collected images. Some models of confocal microscope systems employ two monitors, one to display the images, and the other to display menu functions. This is especially prevalent in older types of confocal systems. The computer should be networked to free up the core system for more users. With a networked system, the digital images can be exported to several other systems for further image processing, or to be printed on a remote digital printer, or even submitted to the World Wide Web.

G. Computer Software

Each computer-controlled confocal microscope comes with a custom software package that is used to control the microscope (Fig. 14). For example, Bio-Rad, Leica, and NORAN confocal microscopes come with software programs called LaserSharp, PowerScan, and Intervision, respectively. In addition, the computer comes with an operating system that varies between different confocal companies. For example, the Windows NT operating system is used by Leica, Nikon, Olympus, and Carl Zeiss. Bio-Rad Laboratories uses OS/2 for their MRC-1024, and Windows NT for their Radiance2000 and RTS2000 confocal microscopes. Images may be digitally saved as full-resolution images in several formats including Bitmap or TIFF file formats. Bio-Rad uses a proprietary file format that is unable to be read by most other programs. These files can be converted to a more popular file format using Confocal Assistant or NIH Image software. In addition to file conversion, this software can also colorize images and create 3-D reconstructions.

Most confocal companies offer a variety of image resolutions to choose from. Image resolution relates to the number of pixels that are used to create the image. A *pixel* is a picture element; pixels are the basic units that make up an image. Two closely spaced objects can be more easily distinguished when the image's pixels are very small. Each image on the computer monitor is made of many pixels in both horizontal and vertical directions. Each pixel is assigned a different gray level or color. The higher the total number of pixels, the greater the image resolution, and the larger the scanned area that can be covered at a given pixel size. High image resolution is considered to be 1024×1024 or 2048×2048 pixels as observed in many CLSMs. The newest models of CLSMs are even offering images with a 4096×4096 pixel array. The time it takes to generate an image is called the *scan speed* or *scan rate*. Scan speeds can vary from 2.08 ms to 158 s. This depends on the image resolution and confocal microscope system (Section IV.B). The larger the pixel array of the image, the longer it takes to scan the sample to generate the image. It is important that the computer monitor (and digital printer) have a high enough resolution to display the images properly. For example, resolution is lost if a 2048×2048 image is displayed on a monitor that can handle only a 800×600 pixel array.

IV. Designs of Confocal Microscopes

Several different versions of confocal microscopes are currently available on the market (Table III). There are three major types of confocal microscope designs based on the methods of scanning that are used to achieve confocal imaging. Confocal microscopes employ a laterally moving stage and stationary beam of light (stage scanning), a scanning

Table III

Major Types of Biological Confocal Microscopes and Their Manufacturers

Manufacturer	Confocal microscope[a]	Scan mechanism	Scan type
Advanced Scanning	ASL-100	Nipkow disk	Tandem scanning spinning disk
Atto Bioscience	CARV	Nipkow disk	Monoscanning spinning disk
Bio-Rad	Radiance2000	2 galvanometers	Beam scanning
Laboratories	MicroRadiance	2 galvanometers	Beam scanning
	MRC-1024	2 galvanometers	Beam scanning
	RTS2000	2 galvanometers[b]	Beam scanning
Carl Zeiss, Inc.	LSM 5 Pascal	2 galvanometers	Beam scanning
	LSM 510	2 galvanometers	Beam scanning
Leica Microsystems	TCS NT2	Galvanometer-driven K scanner	Beam scanning
	TCS SP2	Galvanometer-driven K scanner	Beam scanning
	TCS SL	Galvanometer-driven K scanner	Beam scanning
Nikon Inc.	PCM 2000	2 galvanometers	Beam scanning
NORAN Instruments	Oz	1 AOD (horizontal), 1 galvanometer (vertical)	Beam scanning
Olympus	FluoView 300	2 galvanometers	Beam scanning
	FluoView 500	2 galvanometers	Beam scanning
PerkinElmer Life Sciences	UltraVIEW[c]	Nipkow disk	Monoscanning spinning disk
Solamere Technology Group	QLC100[c]	Nipkow disk	Monoscanning spinning disk
Technical Instrument San Francisco	K2-S Bio	Nipkow disk	Monoscanning Spinning disk

[a] Based on confocal microscopes available in the United States in 2000.

[b] One resonant galvanometer (horizontal), one galvanometer (vertical).

[c] Modified version of the Yokogawa Electric Company's CSU10 monoscanning spinning disk confocal microscope.

beam of light and stationary stage (beam scanning), or both a stationary stage and light source (spinning disk). Each instrument design has characteristics that are desirable for specific confocal applications.

A. Stage Scanning Confocal Microscope

Specimens on the stage are slowly moved during the scan to produce a confocal image with a stage scanning confocal microscope. Therefore, this confocal microscope has the advantage of scanning a field of view much larger than the actual field of view of the objective with high resolution (Brakenhoff, 1979; Cogswell and Sheppard, 1990; Sheppard and Choudhury, 1977; Wijnaendts van Resandt et al., 1985; Wilson and Sheppard, 1984;

see also Shotton, 1989). Because the optical arrangement is stationary (only the sample is moved) and the sample is scanned with a single beam, this confocal system has constant axial illumination. This minimizes optical aberrations from the objective and condenser and provides an even optical response across the entire scanned field. The scanning speed is similar to the slowest rates of slow-scan CLSMs (Section IV.B.1). This can be improved by scanning a smaller pixel field. Since the stage is moved, rather than the light beam, cells such as those living in aqueous media must be firmly attached to the slide or they may shift during image acquisition, resulting in a distorted image. The stage scanning confocal microscope works well with slowly changing living samples and fixed samples. Because the image area is not limited by the field of view of the microscope, but rather by the range of movement of the stage, this type of confocal microscope may be useful to analyze gene expression patterns on "gene chips" or microarrays which are glass or nylon wafer supports coated with fluorescently labeled DNA oligonucleotides in precise, sequence-specific arrays (Bowtell, 1999; DeRisi *et al.*, 1996; Golub *et al.*, 1999; Kawasaki *et al.*, 1999). Because of inherent surface variations of the microarrays, care must be taken so that the signals from the microarray are not attenuated as a result of incorrect focus or too thin of an optical section (Cheung *et al.*, 1999).

B. Beam Scanning Confocal Microscope

The beam scanning type of confocal microscope uses rapidly scanning mirrors to scan a laser beam across a stationary specimen. This allows detection of induced fluorescence or reflected light (Carlsson *et al.*, 1985; White *et al.*, 1987; Wilke, 1985; Wright *et al.*, 1993; see also Shotton, 1989). This type of confocal microscope is most commonly employed to examine biological samples that are fluorescently labeled. The beam scanning confocal system has a higher resolution and a faster scan rate than the stage scanning type of confocal microscope. There are two major types of CLSM: a slow-scan and a video-rate confocal microscope. They differ in the way specimens are scanned, and images are acquired. Each has characteristics that are advantageous for different biological applications.

1. Slow–Scan Confocal Laser Scanning Microscope

The slow-scan CLSM works well with slowly changing, living specimens such as dividing cells, and fixed samples. The slow-scan CLSM uses highly reflective, independent galvanometer mirrors to produce the horizontal and vertical patterns of the raster scan (Fig. 7). Mechanical limitations of the galvanometers restrict the speed at which a single frame (image) can be scanned. A full, high resolution 2048×2048 image may take as long as 37 s to acquire. As a result, rapid cellular processes such as ionic fluctuations cannot be imaged at high resolution without blur or are entirely missed. Fast cellular events may be captured with a slow-scan CLSM by lowering the pixel resolution, by changing the scan rate (pixel dwell time), and by using a smaller image size (256×256 at 0.5 s). Slow-scan CLSMs use a digital frame buffer to collect the image line-by-line and store it in computer memory. The time it takes to store an image may also reduce

Fig. 17 The Bio-Rad MRC-1024 confocal laser scanning microscope which is a single point-scanning system. (Courtesy of Bio-Rad Microscopy Division, Hemel Hempstead, UK.)

acquisition speed of rapid events. Once assembled, a 2-D image is displayed on the computer monitor until a new optical section is collected. Highlights of the most popular models of slow-scan confocal microscopes are described next (see also Table III and Section X.A).

a. Bio-Rad Laboratories

Bio-Rad Laboratories offers several CLSMs which are single-point scanning systems that use paired galvanometer mirrors to scan the sample with a laser beam (Figs. 17–20). They differ in their cost, number of PMT detectors, and number of lasers that can function at any one time. The MRC-1024 has up to three confocal PMTs each with a continuously variable, pinhole type, confocal aperture (Figs. 17 and 18). Up to three photodiodes can be used to detect transmitted light for real time color transmission imaging. Several laser options are available including visible and UV lasers (Table I). Several filter sets are used to control the excitation and emission wavelengths, and neutral density filters are used to attenuate the laser beam. The scan head can be quickly mounted on the video port of upright and inverted microscopes from Leica, Nikon, Olympus, and Zeiss. Several image sizes are available with the highest resolution at 1280×1024 pixels. LaserSharp software running under OS/2 Warp allows a wide range of image processing functions including 2-D, 3-D, and four-dimensional (4-D) imaging. The MicroRadiance CLSM has similar features as the MRC-1024 except that it is less complex and costs less. It uses up to two lasers and two PMTs at any one time and relies on computer-controlled laser alignment, choice of confocal aperture size, and filter selection. It also uses LaserSharp software running under OS/2 Warp to perform image processing.

The newer Radiance2000 system differs from the MRC-1024 and MicroRadiance CLSMs in several ways (Figs. 19 and 20). All laser beams (of which there are several choices) pass through an AOTF and then a single fiber so that the laser beams are aligned. The AOTF is also used to turn off the beam during flyback to eliminate any unnecessary exposure of the sample so that photobleaching and photodamage are reduced. The Radiance2000 is more compact and the scan head can be mounted to the video port of a variety of upright and inverted microscopes from different manufacturers. Up to three PMTs each with a continuously variable confocal aperture are available on the Radiance2000. LaserSharp2000 software running on Windows NT is used to control the choice of filters (excitation, emission, neutral density, and polarization), PMT channels and their sensitivity, laser intensity, and image acquisition (speed, collection filter such as Kalman averaging, zoom, scan mode, and image size). The scan axis can be freely rotated 360° without moving the sample. This allows for the best positioning of the sample on the screen without physically moving the microscope stage. The LaserSharp2000 software also has a depth correction algorithm to compensate for mismatches in refractive index among the oil, water, and glass interfaces.

b. Carl Zeiss

The LSM 510 is a compact, modular CLSM that uses two galvanometer scanning mirrors to scan a single point across the specimen (Fig. 21). The scan head can be mounted to the video port of upright and inverted Zeiss microscopes. A variety of lasers including visible and UV lasers can be coupled to the scan head using fiber optic cables. A photodiode monitors the laser intensity and the software controls fluctuations in laser intensity. AOTFs handle wavelength selection and attenuation of the visible and UV laser beams. A variety of scans are possible including unidirectional scan, dual-directional scanning in which the laser beam scans the sample during flyback, and scan rotation in which the scan can be rotated up to 360° one degree at a time so that the sample does not need to be repositioned for the best image. Dual-directional scanning prevents bleed-through between different channels since fluorescence from one channel, e.g., fluorescein, can be detected during the scan in one direction, whereas another channel, e.g., rhodamine, can be detected as the sample is scanned in the reverse direction. The AOTF allows changes in intensity on a pixel-by-pixel basis so that an ROI scan can be generated. An ROI scan can be an irregular shape, since the AOTF can limit the scan to a user-defined area. This is an important feature for experiments using bleaching, uncaging, or fluorescence recovery after photobleaching (FRAP). Four confocal PMTs (three for fluorescence and one for reflected light), each with its own variable confocal pinhole aperture, can be recorded and displayed simultaneously. A sensitive transmitted light PMT is also available to detect brightfield, phase contrast, and Nomarski (differential interference contrast) images. Software which runs on Windows NT controls the emission filters, beam splitters, confocal pinhole apertures, and excitation wavelength selection. Several image sizes are available including a high-resolution, 2048 × 2048 maximum scan and a faster 512 × 512 scan that can be generated at 2.6 frames per second. Both 2-D and 3-D image processing software is available for image analysis.

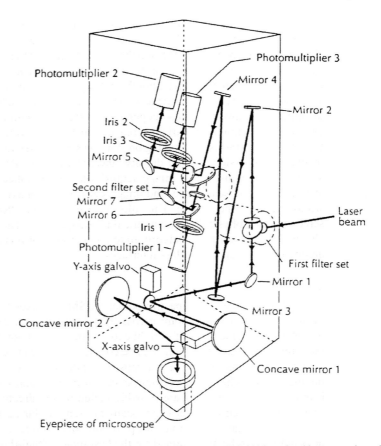

Fig. 18 Optical light path in the scan head of the Bio-Rad MRC-1024 confocal laser scanning microscope. The scan head can be mounted on the video port of many models of microscope (not shown). Light from a variety of laser combinations including a multiline krypton–argon ion laser passes through an excitation filter in the laser box that allows specific selection of excitation wavelengths such as 488 nm light (blue) for fluorescein fluorescence, 568 nm light (green) for rhodamine fluorescence, and 647 nm light (far red) for Cy-5 fluorescence. In the MRC-1024ES model (not shown), the excitation filters and neutral density filters are replaced by an acoustooptical tunable filter (AOTF). In both models, incident laser light (laser beam) is reflected off a dichroic mirror and directed onto the scanning unit by a highly reflective mirror (Mirror 1). Two oscillating galvanometer mirrors in the scanning unit generate the x and y scanning movements of the laser beam. Two curved mirrors (Concave mirrors 1 and 2) direct the scanning beam between the x- and y-scanning galvanometer mirrors. The x axis-scanning galvanometer mirror (x-axis galvo) directs the beam to the microscope which focuses the light as a scanning spot moving across the focal plane of the specimen. Induced fluorescence generated by the specimen retraces its path back through the scanning system. The emitted fluorescence then passes through the dichroic mirror, which cuts off wavelengths close to the excitation light source. The emitted fluorescence is then directed toward a six-position filter wheel (First filter set) which contains barrier filters that select the desired wavelengths. The longer wavelength fluorescence that passes through the first barrier filter is directed by a series of highly reflective mirrors (Mirrors 2–4) to another dichroic mirror that distinguishes between various wavelengths (e.g., fluorescein and rhodamine fluorescence). Green light (fluorescein) is reflected by the dichroic mirror and passes through a barrier filter (Second filter set) toward Mirror 5 which directs the

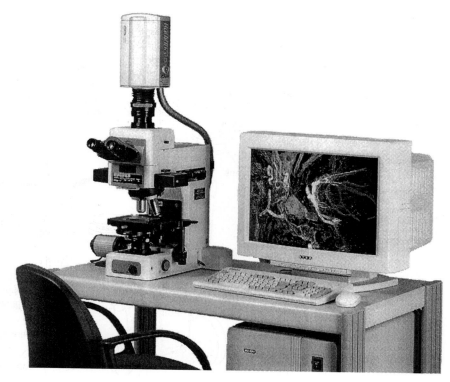

Fig. 19 The Bio-Rad Radiance2000 confocal laser scanning microscope which is a single point-scanning system. (Courtesy of Bio-Rad Microscopy Division, Hemel Hempstead, UK.)

For the researcher who does not need all of these features, a personal CLSM called the LSM 5 PASCAL is available at a lower price (Fig. 22). A major difference is that the LSM 5 PASCAL has only one or two confocal channels for fluorescence or reflected light rather than the four channels in the LSM 510. Thus the PASCAL has only one variable confocal pinhole aperture. The modular design, flexibility for upright and inverted Zeiss microscopes, and software control of the filters and mode of scanning are many of the features shared between the two instruments.

---

light through a continuously variable confocal aperture (Iris 2). On the other side of the confocal aperture is a photomultiplier tube (Photomultiplier 2) which detects the in-focus green light. Red emission light (rhodamine) passes through the dichroic mirror, barrier filter, and Mirror 6 before entering its own confocal aperture (Iris 1). The in-focus red light is detected by Photomultiplier 1. Far red emission light (Cy-5) which also passes through the dichroic mirror and barrier filter is directed by Mirrors 6 and 7 onto its own confocal aperture (Iris 3). In-focus far red light is detected by Photomultiplier 3. The confocal microscope also detects confocal reflected light and nonconfocal (conventional) transmitted light. The photomultiplier tubes detect and convert the fluorescence emissions into a video signal that is displayed on the computer monitor as a confocal image. (Courtesy of Bio-Rad Microscopy Division, Hemel Hempstead, UK.)

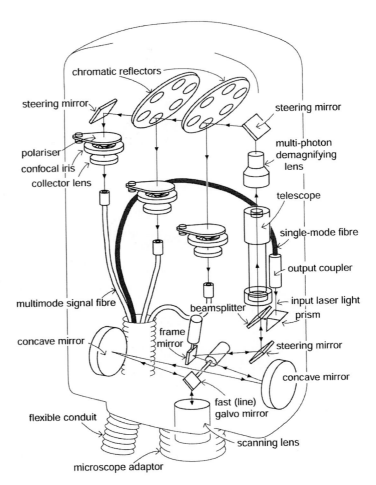

Fig. 20 Optical light path in the scan head of the Bio-Rad Radiance2000 confocal laser scanning microscope. This confocal microscope system consists of a scan head, microscope, lasers, instrument control unit (ICU), and computer system. The microscope adaptor allows the scan head to be mounted on any video port (top, side, or base port) of the microscope. The scan head houses the confocal apertures, scanning mirrors, filters, and laser alignment jigs which are computer-controlled, whereas the ICU houses the computer-controlled acoustooptical tunable filter (AOTF), lasers, laser control optics, and remote detector modules including the confocal photomultiplier tube (PMT) detectors and transmitted light photodiode detector. All laser excitation light from an argon ion laser, a green helium–neon laser, or a red laser diode first passes through an AOTF in the ICU and is coaligned in a single illumination fiber which is connected to the scan head with an output coupler. The incident laser light (input laser light) enters the scan head through the single-mode fiber and is directed toward a beam splitter by a prism. The incident light then reflects off of the beam splitter and a steering mirror before being directed toward the vertical scanning galvanometer mirror (frame mirror). The laser light then reflects off of two curved mirrors (concave mirror) before reaching the horizontal scanning galvanometer mirror (fast galvo mirror) which directs the laser light through the microscope toward the sample. Induced fluorescent emissions from the sample retrace their path back through the scanning system before passing through the beam splitter toward the emission telescope unit. If the scan head is configured as a multiphoton unit, a multiphoton

c. Leica Microsystems

As shown in Figs. 23 and 24, a recent design in slow-scan CLSMs is the spectral confocal microscope (TCS SP) which uses proprietary filterless detector optics (Fig. 16, Section III.E). The confocal scanner head can be mounted on upright or inverted Leica microscopes and can be quickly switched between the two by the user. The TCS SP spectral confocal microscope system consists of the confocal scanner head, an upright and/or inverted Leica microscope, a high-precision galvanometer focusing stage, system electronics, computer system, and up to four lasers including UV, visible, and infrared lasers. TCS systems have been designed to: (1) provide the highest optical resolution including lateral spatial, axial spatial, temporal, and spectral resolution, (2) maximize light-gathering efficiency using the spectrophotometric detection system, (3) provide maximum stability over time (a single detector pinhole ensures perfect registration channel to channel with no alignment requirements), and (4) optimize ease of use both in the user interface and in the hardware (no alignment by the user is needed). The TCS SP is designed to work simultaneously with up to four spectrophotometric detectors as described in Section III.E.

The spectrophotometric detection system has several advantages. It can be used in both confocal and multiphoton microscopes. It permits a wide variety of wavelengths to be specifically chosen for detection without the need to add new filter sets as new dyes are developed. It also minimizes crosstalk between fluorochromes by changing the bandwidth of the photodetector to exclude the emission of one fluorochrome from another. Another advantage of the system is that it can determine the spectral characteristics of emitted light such as for autofluorescence or new dyes, since the detection system is a spectrophotometer. The images are brighter and clearer and contain less noise because the detection dichroic and barrier filters which reduce light intensity have been eliminated; also, because of the high transmittance of the detection system (primarily a prism and mirrors) with an increase in the efficiency of detection (i.e., more signal is transmitted to the detector), the intensity of laser illumination can be reduced, decreasing photobleaching and photodamage.

---

demagnifying lens is used to focus the rays which would otherwise be blocked by the confocal aperture. The fluorescence is then reflected off another steering mirror and directed toward two emission filter wheels (chromatic reflectors) which are barrier filters that select the desired wavelengths emitted from the sample. Green fluorescent emissions from the sample are reflected off the first filter wheel and are directed through a confocal collector module consisting of a polarizer, continuously variable confocal aperture (confocal iris), and collector lens which function to block out-of-focus fluorescence from the sample. The in-focus green rays then enter a multimode signal fiber that delivers light from the scan head to the ICU where it is detected by a PMT. Red and far-red emissions from the sample pass through the first filter wheel. Red emissions from the sample are reflected by the second filter wheel and directed toward the middle confocal collector module and through the multimode signal fiber to a PMT in the ICU. Far-red emissions pass through the second filter wheel and are reflected off a steering mirror before entering a third confocal collector module and multimode signal fiber. The far-red emissions are then detected in the ICU by either a third PMT or a photodiode which is up to five times more sensitive to far-red light than a PMT. The confocal microscope also detects confocal reflected light and nonconfocal (conventional) transmitted light. Signals from the PMTs are then displayed on the computer monitor as a confocal image. (Courtesy of Bio-Rad Microscopy Division, Hemel Hempstead, UK.)

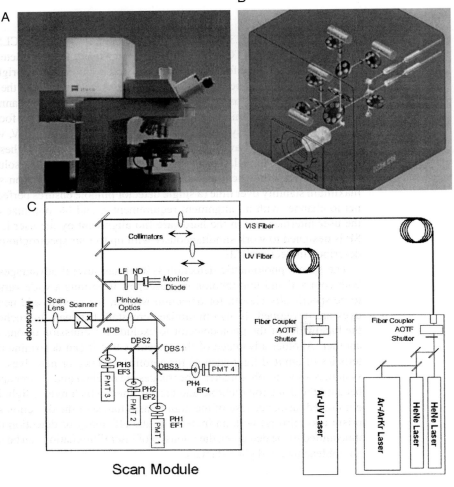

Fig. 21 The LSM 510 confocal laser scanning microscope from Carl Zeiss. (A) The scan module (white) can be mounted on upright (A) or inverted Zeiss microscopes. (B, C) Schematic drawing of the optical light path in the LSM 510 single point-scanning confocal microscope. A choice of laser lines from three visible and one ultraviolet laser is shown. An acoustooptical tunable filter (AOTF) is used to control the laser intensity and excitation wavelengths. The laser light is directed to the scan module via separate fiber optic cables and passes adjustable collimators which ensure that the focal planes of the various wavelengths exactly coincide. The laser light is combined and reflected by the main dichroic beam splitter (MDB) on to the scanning galvanometer mirrors (Scanner) which deflect the laser beam across the specimen in the *x* and *y* axis after the laser light passes through a scan lens and the microscope optics. Induced fluorescent light emitted from the specimen retraces its path back to the scanning module where it passes through the main dichroic beam splitter (MDB). After this point, fluorescence emissions from the sample are spectrally separated by three secondary dichroic beam splitters (DBS1–3). The scan module contains up to four confocal detectors each of which has its own adjustable pinhole confocal aperture (PH1–4) and emission filters (EF1–4) which ensure each photomultiplier tube (PMT1–4) detects only desired wavelengths. As an option, some of the collimated incident laser light can be reflected on to a monitor diode (not shown) for attenuation with neutral density filters if necessary. The LSM 510 also detects confocal reflected light and nonconfocal (conventional) transmitted light. Signals from the PMTs are then displayed on the computer monitor as a confocal image. (Courtesy of Carl Zeiss, Inc., Thornwood, NY.)

The Leica TCS NT confocal microscope is a point scanning system that has similar features to the TCS SP except that instead of using the spectrophotometric detection system, it uses dichroic mirrors on a filter wheel to separate the wavelengths of fluorescent light emitted from the sample (or reflected light from the sample), and barrier filters to select the wavelength directed to the detector. As with the Leica TCS SP, this systems uses both a source pinhole and a single detector (confocal) pinhole which ensure alignment-free long-term stability of the system. The confocal pinhole is variable in size for each objective lens and is under software control. Scanning in the TCS SP and TCS NT is achieved with a single scan mirror which provides both unidirectional and bidirectional scanning. The lateral spatial resolution of the TCS NT is 0.18 μm, and the axial spatial resolution is 0.35 μm using 488-nm excitation light and a 1.32 NA objective lens on both the upright and inverted models. Up to four visible and one UV laser can be connected to work simultaneously with the TCS NT to provide multiwavelength detection of up to four fluorescence/reflected light channels and one nonconfocal transmitted light channel. The maximum image size is 1024 × 1024 pixels. The software includes 3-D reconstruction capabilities.

The latest models in the Leica line, the TCS NT2 and TCS SP2, incorporate new hardware and software. Like the TCS NT and TCS SP models that they replace, the new TCS NT2 and TCS SP2 models feature filtered and spectrophotometric detection, respectively. New hardware features include the "K" scanner, a new scanner design that enables very high frame rates (line frequency: up to 2000 lines/second; frame rate: 3 frames/s at 512 × 512 pixels), high-speed strip scanning (25 strips/s at 512 × 32 pixels), and scan rotation without optical/mechanical compromise. A new adjustable pupil illumination system automatically adjusts the illumination to fill the pupil of the objective to optimize the illumination efficiency of the system. The scan field of the instrument has been significantly increased to enable large-field scanning at low magnifications. Other new features include ROI scanning, 12 bit A/D conversion per channel, an expanded spectral range of detection from UV to IR in the SP2 detection system, and wavelength switching line scan for crosstalk reduction. Multiphoton, UV, and visible confocal microscopy can be combined in these new systems. New software features include a fully operator-configurable user interface that enables individual user logon, a context-sensitive online help system, multidimensional image series recording ($x, y, z, t, \alpha, \lambda, i$), a user-configurable VBA macro programming function, and new 3-D, physiology, surface reconstruction, and multicolor analysis software packages. The largest available image size is 4096 × 4096 pixels.

d. Nikon

The Nikon PCM 2000 confocal laser scanning microscope uses a modular design in which the PMT detectors and lasers are located outside the scan head to reduce noise in the images (Figs. 25 and 26). The PCM 2000 consists of a single or multiple laser light source, the xy scanner module, an upright or inverted Nikon microscope, detection and control electronics, and a computer system. Its compact, modular design allows the system to be quickly moved from upright to inverted Nikon microscopes. The PCM 2000 CLSM can be fitted with different lasers including argon ion, green helium–neon,

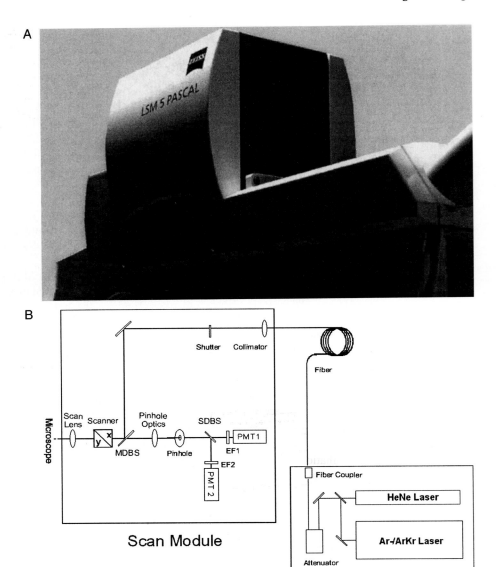

Fig. 22 Scan module of the LSM 5 Pascal confocal laser scanning microscope from Carl Zeiss. The LSM 5 Pascal is a personal laser scanning confocal microscope which is a single point-scanning system that can be mounted on upright (A) and inverted Zeiss microscopes. (B) Schematic diagram of the optical beam path of the scan module. Laser light from a choice of two different visible lasers is directed to the scan module via a fiber optic cable. After passing through the collimator and shutter in the scan module, the incident laser light is directed to the main dichroic beam splitter (MDBS) which reflects the rays toward the scanner. The scanner contains the x and y scanning galvanometer mirrors that direct the laser beam across the specimen, after the

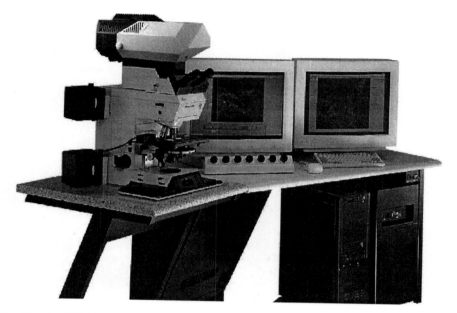

Fig. 23 The TCS SP spectral confocal laser scanning microscope from Leica. The TCS SP, which is a single point-scanning system, combines spectrophotometric detection with confocal microscopy through the use of a prism and variable slit mechanisms. (Courtesy of Leica Microsystems, Heidelberg.)

and red helium–neon lasers which last approximately 10,000 h. The laser beam can be attenuated with a choice of neutral density filters. Specific laser lines can be manually selected with a choice of excitation, dichroic, and emission filters. The scan head contains paired-galvanometer scanning mirrors that raster scan the specimen with a focused point of laser light. Since a single pinhole aperture is used for both excitation and emission light, the PCM 2000 does not require alignment. Three fixed confocal pinhole sizes are available to maximize the signal-to-noise ratio. Several scan rates and image sizes are available. The standard 640×480 pixels takes 1 s, the higher resolution 1024×1024 pixel image takes 2 s, and the line scans take only 2 ms to produce.

---

laser beam passes through the scanning lens. Induced fluorescent light emitted from the specimen retraces its path back to the scanning module where it passes through the MDBS and pinhole optics toward the adjustable confocal pinhole aperture which ensures that mainly in-focus light is detected. On the other side of the confocal aperture is a second dichroic beam splitter (SDBS) which spectrally separates the fluorescence emissions from the sample and directs the in-focus light to the eight-position emission filter wheels (EF1–2) which ensure that each photomultiplier tube (PMT1–2) detects only desired wavelengths. Signals from the PMTs are then displayed on the computer monitor as a confocal image. (Courtesy of Carl Zeiss, Inc., Thornwood, NY.)

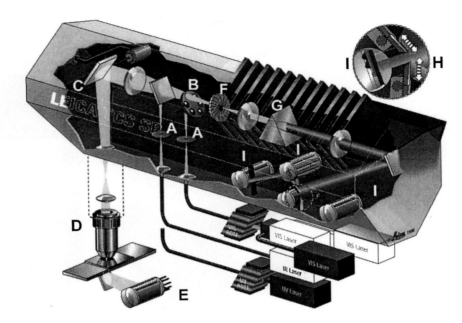

Fig. 24 Optical light path diagram of the Leica TCS SP spectral confocal laser scanning microscope. A choice of laser lines from three visible (VIS) and one ultraviolet (UV) or infrared (IR) laser is shown. Acoustooptical tunable filters (AOTF) are used to control the laser intensity and excitation wavelengths. Light from the visible light lasers is introduced into the scan head via one optical fiber, passed through an illumination pinhole (A), and reflected to the single scan mirror (C) by the main beamsplitter (B). Several main beamsplitters for different applications are present on a motorized computer-controlled filter wheel. Light from the UV or IR laser is introduced to the scan head via a second optical fiber, passed through a separate illumination pinhole (A), and reflected to the single scan mirror (C) by a fixed beamsplitter. The single scanning mirror (C), moved by two galvanometers, deflects the excitation light across the specimen in the x and y axes after the laser light passes through a scanning lens and the microscope optics (D). Induced fluorescent light emitted from the specimen retraces its path back to the scanner head, passes through the main dichroic beam splitter (B), and is directed to the confocal aperture (F) which blocks the out-of-focus light from the sample. The in-focus fluorescent light is separated into a spectrum by a prism (G), and the spectrum is directed to a series of four photomultiplier tubes or PMTs (I). A variable slit system (H) in front of each PMT specifically selects the desired emission wavelengths to be detected, and mirrors which form the edges of each variable slit system direct the other emission wavelengths to other variable slit systems in front of other PMTs (up to a total of four PMTs) until all desired wavelengths have been detected (see text sections III.E and IV.B.1.c for details). The confocal microscope also detects nonconfocal laser-scanned transmitted light via a transmitted light PMT (E). Signals from the PMTs are then displayed on the computer monitors as either merged or separate confocal images. (Courtesy of Leica Microsystems, Heidelberg.)

Although two PMTs for confocal fluorescence imaging come with the system, an optional third PMT for nonconfocal transmitted light allows for simultaneous three-channel imaging and three-color acquisition. The software which runs off the Windows NT operating system allows a variety of functions including image acquisition, processing, and morphometry.

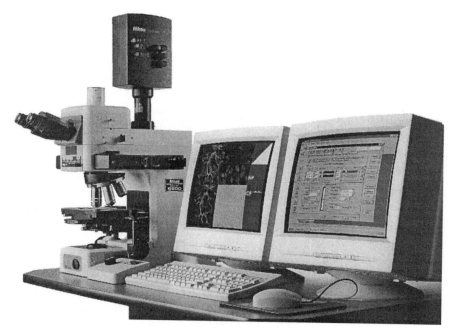

Fig. 25 The Nikon PCM 2000 is a personal confocal laser scanning microscope that uses a single point-scanning system. (Courtesy of Nikon Inc., Melville, NY.)

e. Olympus

The FluoView 300 and FluoView 500 CLSMs are single point-scanning systems that use a variable pinhole confocal aperture, and two scanning galvanometer mirrors to scan and descan the specimen (Figs. 27 and 28). These confocal systems consist of the scanning unit, microscope, choice of visible and ultraviolet lasers, computer system, and electronics. The scan unit is configured for both upright and inverted Olympus microscopes so that it can be quickly interchanged between microscopes. Each CLSM has several options of visible wavelength lasers, and the FluoView 500 has optional fiber optic coupling ports for up to four UV and infrared lasers. The FluoView 500 has software-driven control of filter settings, laser lines, laser intensity, and confocal aperture settings which are continuously variable. The FluoView 300 relies on manual control of the filter settings and the five fixed-position confocal apertures which range in size from 60 to 300 μm. Images may be scanned at three scan speeds up to 2048 × 2048 pixels, which takes 37 s at the slowest scan rate, and smaller image sizes such as 256 × 256 pixels are available which take only 0.45 s to collect at the fastest scan speed. Two channels of confocal fluorescence can be detected in addition to nonconfocal transmitted light simultaneously with the FluoView 300, whereas the FluoView 500 allows four-channel simultaneous detection (three confocal fluorescence and one nonconfocal transmitted channel). The information from these channels can be

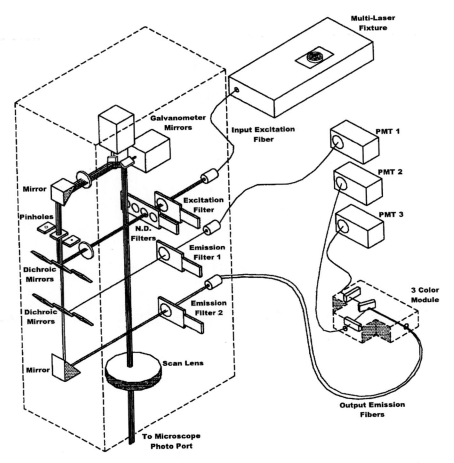

Fig. 26 Optical light path of the scanner module of the Nikon PCM 2000 confocal laser scanning microscope. Laser light from an argon ion, green helium–neon, or red helium–neon visible laser is directed to the scanner module with a single-mode fiber optic cable via a laser coupler. The incident laser light passes through an excitation filter and neutral density filter which is used to attenuate the laser beam as necessary. The laser light is reflected by a dichroic mirror through a confocal pinhole aperture and is directed by a highly reflective mirror to the galvanometer mirrors which, after directing the excitation laser light through the scan lens and microscope optics, scan the laser beam in the x and y axes across the specimen. Induced fluorescent light emitted from the specimen returns to the scanning module where it reflects off the mirror and passes through the same confocal pinhole aperture before passing through the dichroic mirror. The confocal aperture blocks unwanted, out-of-focus emissions from the sample. A second dichroic mirror separates the in-focus green fluorescence emissions from the longer wavelength orange and red light rays. The green light is directed toward emission filter 1 which ensures that the photomultiplier tube (PMT 1) detects only desired green wavelengths. The in-focus orange and red rays are reflected off a mirror and pass through emission filter 2, and then are directed outside of the scanner module to a three-color module via output emission fibers where the orange and red rays are directed to separate photomultiplier tubes (PMT 2 and 3, respectively). Signals from the PMTs are then displayed on the computer monitor as a confocal image. (Courtesy of Nikon, Inc., Melville, NY.)

displayed on a 19- or 21-inch monitor singly, side-by-side, or overlaid (merged) into one image.

2. Video–Rate Confocal Laser Scanning Microscope

This type of CLSM is designed to capture images of fast, dynamic cellular signals such as fluctuations in calcium ions, pH, and membrane potential which can occur in milliseconds. This application requires a microscope that can capture and store images as close to real time as possible—at video rates (30 Hz) or even faster. The video-rate CLSM differs from the slow-scan CLSM in its mode of scanning the specimen and in image formation. See also Chapter 2 for an in-depth discussion of video-rate confocal microscopy.

One method to achieve video-rate scanning with the CLSM is through the use of an acoustooptical deflector (AOD). The AOD is an electrically controllable diffraction grating that generates the high frequency (15.75 kHz) horizontal scan of the laser beam (Art and Goodman, 1993; Draaijer and Houpt, 1988, 1993; Tsien and Bacskai, 1995; Wright *et al.,* 1993). As piezoelectric transducers attached to the glass AOD cause acoustic waves to propagate throughout the glass; the waves, acting as gratings, diffract the incident laser light. To produce the vertical scan, the diffracted beam then strikes a galvanometer driven mirror that oscillates at the slower 60 Hz. Both the amplitude and frequency of the sound waves can be precisely controlled. The intensity of the incident laser light can be varied by adjusting the amplitude of the sound waves. The speed of the scan can be changed by varying the frequency of the sound waves.

In the reflectance mode, reflected light returns to the detector through the AOD. In the fluorescence mode, the induced fluorescent light has a longer emission wavelength than the excitation wavelength (Stokes shift) of the laser beam. As a result, the entire spectral width of the fluorescence cannot be descanned back through the wavelength-specific AOD without a significant loss of signal intensity. To compensate for this, the partially descanned fluorescence originating from the sample is directed by a dichroic mirror between the galvanometer mirror and the AOD. The fluorescence passes through a barrier filter and is imaged through a variable-slit aperture before detection with a PMT and conversion into a video signal. Thus, video output from the microscope can be directly connected to a television monitor to produce live images and does not require computer storage, as do slow-scan CLSMs.

a. NORAN Instruments

Several advances have been made in the video-rate CLSMs that use AODs. For example, in the past decade, NORAN Instruments has greatly improved its former Odyssey video-rate CLSM into a newer video-rate CLSM called the Oz confocal laser scanning imaging system (Fig. 29). Although the Oz is based on the same optical principles of using an AOD-driven horizontal scan with a galvanometer-driven vertical scan, the Oz has numerous improvements in overall design. For example, the Oz uses a new tellurium dioxide glass AOD which is 10% more efficient at 488 nm and provides for a 50% increase in resolution. Other improvements include better antireflectance coatings on

A

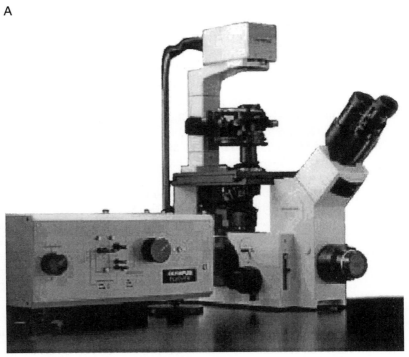

B

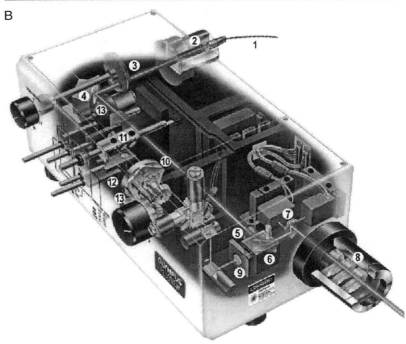

mirrors and lenses, more efficient light path, and PMTs with up to 45% higher quantum efficiency at 488 nm. The use of a motorized primary dichroic filter wheel that matches each dichroic to the corresponding laser excitation line allows users to maximize both excitation and emission efficiency in multiple-labeled samples. Additional improvements in the electronics and computer power of the Oz have led to more scan speeds and image sizes including the larger 1024×960 pixels. The Oz now uses a powerful Silicon Graphics workstation rather than a PC which limited the Odyssey to a single image format (512×512) and single pixel dwell time (100 ns/pixel). Image acquisition speeds of the Oz can vary from very slow (6400 ns/pixel; 8 s for a 1024×960 image) to extremely fast (100 ns/pixel; 480 images per second of images size 220×30 pixels). This allows more versatility to meet the needs of a multiuser core facility. The Oz also allows digital capture of full-size images (512×480) at a rate of 30 frames per second (fps), directly to the hard disk for over 120 s. Built-in video compression can sustain rates of image acquisition to 30 fps. The data can be stored in a variety of file formats and video formats. Unlike the Odyssey, the Oz is designed to be modular. This offers several advantages including a reduction in thermal noise and vibration, more accessibility for maintenance, and easier upgrades. The scan head is highly mobile and can fit more than 26 different microscopes (including upright and inverted). Three fiber-optic input ports also allow for multiple lasers. Since the lasers are separated from the scan head in the new modular design, this also greatly reduces thermal noise at the PMTs and results in better image quality.

The use of the AOD in the video-rate CLSMs meets the needs of physiological imaging studies that require high temporal resolution such as signaling pathways or ion dynamics, and uncaging caged compounds. Combining the video-rate CLSM with a piezo-driven lens z axis positioning device and fast sequential volume acquisition makes this system a very powerful tool to obtain real-time 4-D imaging of ion, membrane, and organelle dynamics. For example, the Oz is capable of acquiring up to 15 slices per second over a 100-μm range at 0.01 μm accuracy. The new Silicon Graphics workstation can handle the tremendous volumes of data generated in a short time and facilitates 4-D reconstructions. See the NORAN Web site for sample 4-D movie clips (Section X).

Fig. 27 The Olympus Fluoview FV300 confocal laser scanning microscope. (A) FV300 single point-scanning system. (B) Optical light path diagram of the FV300 scan unit. Incident laser light is delivered to the scan unit via an optical fiber (1) connected to the laser port. After collimation by the beam collimator (2) and attenuation by the laser adjustment neutral density filter turret (3), the incident laser light is reflected off a dichroic mirror (4), polarized (5), and reflected off an excitation dichroic mirror (6) toward the xy galvanometer scanning mirrors (7) which direct the scanning beam through the pupil lens (8) and microscope optics to the specimen. Fluorescence emitted from the sample returns back through the scan unit and is directed by the scanning mirrors to the confocal lens (9) after which it is directed to the confocal pinhole turret (10) which blocks unwanted out-of-focus light. The in-focus fluorescence from the sample is then directed to the emission filter in the emission beam splitter slider (11) which contains a choice of emission filters. The fluorescence is then directed to the barrier filter slider (12) which selects the desired emission wavelength that is directed to the photodetector (13). Signals from the photodetector are then displayed on the computer monitor as a confocal image. (Courtesy of Olympus Optical Co., Melville, NY.)

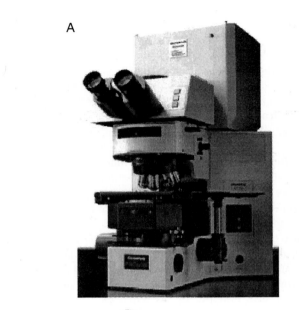

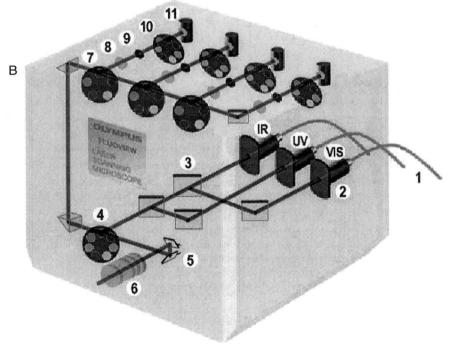

b. Bio–Rad Laboratories

The Real Time Scanner RTS2000 CLSM manufactured by Bio-Rad uses a resonant galvanometer with bidirectional scanning and descanning to scan at video rate, 30 frames/s at 512 × 480 pixels. One 512 × 480 frame can be generated in 33 ms. The scan head can be mounted on an upright (Nikon E600 FN) or inverted (Nikon TE300) microscope. A choice of visible laser lines (488, 568, 637 nm) is generated by a krypton–argon laser. The software running under Windows NT allows simultaneous image acquisition of three channels with the ability to average and store the images arriving on two independent channels and display their ratio as pseudocolor in a continuous real-time display. This feature is important for fast physiological experiments using ratio-indicator dyes. Images generated by the RTS2000 can be saved in TIFF file format. The RTS2000 MP can be fitted with a variety of femtosecond, pulsed, infrared lasers for real time, multiphoton imaging.

C. Spinning Disk Confocal Microscope

The third type of confocal microscope employs a spinning Nipkow disk and stationary light source and stage (Kino, 1989; Petráň *et al.,* 1968, 1985). A Nipkow disk is a black, metallic-coated, glass disk that has several thousand apertures 20–60 μm in diameter arranged in Archimedian spirals or in an interlacing pattern. The Nipkow disk is used to raster scan the specimen. A special feature of the spinning disk confocal microscope is that samples can be seen directly by eye in their true colors as a confocal image in both the reflected and fluorescence modes (Boyde, 1990; Boyde *et al.,* 1990; Paddock, 1989; Wright *et al.,* 1989, 1993). Because the disk spins rapidly (up to 3600 rpm in the UltraView), images can be observed in real time. These features allow direct confocal focusing through the specimen. Thus, all of the spinning disk confocal microscopes can capture images at video rates or faster.

Unlike CLSMs which use a single spot of light to scan the specimen, spinning disk confocal microscopes scan the specimen with thousands of spots of light simultaneously. Light from each spot traces a single scan over successive parts of the specimen. Spinning disk confocal microscopes also differ from CLSMs in that instead of using coherent

---

Fig. 28 The Olympus Fluoview FV500 confocal laser scanning microscope. (A) FV500 single point-scanning system. (B) Optical light path diagram of the FV500 scan unit which is equipped with three fiber optic ports (1) for a choice of infrared (IR), ultraviolet (UV), and visible (VIS) lasers. After passing through the beam collimator (2), the incident laser light is directed by dichroic mirrors (3) to the excitation dichroic mirror turret (4) which houses excitation filters that select the excitation wavelength. The laser light is directed toward the xy galvanometer scanning mirrors (5) which direct the scanning beam through the pupil lens (6) and microscope optics to the specimen. Fluorescence emitted from the sample returns through the scan unit and is directed by dichroic mirrors to the emission beam splitter turrets (7) which select for the desired fluorescence emissions from the sample, and direct them through a confocal lens (8) and adjustable confocal pinhole (9) before further selection by the appropriate choice of barrier filter in the barrier filter turret (10) and detection by the photomultiplier tube (PMT) type of photodetector (11). The four PMTs for confocal fluorescence and a separate external PMT (not shown) for transmitted light detection allow for recording of up to five channels simultaneously. (Courtesy of Olympus Optical Co., Melville, NY.)

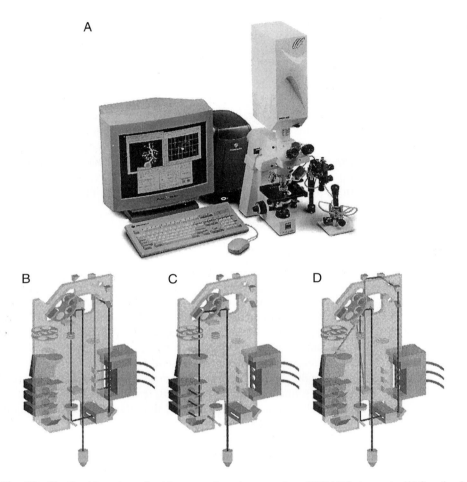

Fig. 29 The Oz video-rate confocal laser scanning microscope from NORAN Instruments. (A) Scan head of this slit-scanning system mounted on an upright microscope equipped with a microinjection system. (B–D) Schematic diagrams of the excitation (B), emission (C), and reflected (D) optical light paths in the scan head unit of the Oz. Incident laser light from a variety of visible and UV lasers is transmitted to the scan head via fiber optic cables (B). The light is then reflected and combined by highly reflective mirrors toward an excitation filter wheel which selects the illumination wavelength. The incident laser light is then reflected by a mirror and a polarizing beam splitter that separates excitation light from returning reflected light. The incident laser light is then directed toward an acoustooptic deflector (AOD) which rapidly scans the excitation light along the horizontal x axis. The laser beam is then reflected off a mirror, passes through an astigmator lens which corrects astigmatism for each illumination wavelength, and passes through a collimating lens which corrects axial chromatic aberrations so that excitation wavelengths are parfocal. The laser light is then directed toward the galvanometer-driven scanning mirror which scans the specimen along the vertical y axis after the laser light passes through the microscope optics represented by the objective lens. Fluorescent light emitted from the specimen (C) is reflected off the scanning galvanometer mirror through a primary dichroic mirror which separates incident and reflected light, and toward a barrier filter which blocks unwanted illumination. The fluorescent light is then focused by a focusing lens onto a slit-type confocal aperture which blocks

laser light to scan the specimen, a broad-spectrum, noncoherent light (white light) is the illumination source as in conventional light and fluorescence microscopy. The white light source can be a mercury arc lamp or a xenon lamp, permitting the selection of any filter combination employed for conventional epifluorescence. Dyes excited by UV can be imaged confocally in addition to fluorescein, rhodamine, and infrared dyes without the added expense of UV and infrared lasers. Thus cells labeled with multiple fluorochromes can be imaged confocally in their true colors (Wright *et al.,* 1989, 1990, 1993). Some spinning disk systems use laser light for fluorochrome excitation (Ohta *et al.,* 1999). The design of the spinning disk confocal microscopes also differs from the CLSMs in that only xy sectioning, not orthogonal xz sectioning, is possible. Another difference from CLSMs is that a computer is not required for most spinning disk confocal microscopes to function.

In older spinning disk systems, the Nipkow disk typically transmitted only 1% of the illumination, resulting in dim fluorescence images. Unless fluorescent samples were bright, an intensifying video camera or cooled CCD camera was needed to generate sufficient signals and improve the signal-to-noise ratio (Wright *et al.,* 1989, 1993). The relatively large size (60 μm) of the confocal apertures in the Nipkow disk as compared to CLSMs compromised optical sectioning so that optical section thickness was approximately 1 μm for spinning disk confocal microscopes as compared to 0.5 μm for CLSMs with the same objective (Wright *et al.,* 1993). Newer models of confocal microscopes have addressed the lack of brightness using microlenses (see later discussion). There are two types of spinning disk confocal microscopes: the tandem scanning and monoscanning microscopes.

1. Tandem Scanning Confocal Microscope

Because the illumination and detection occur in tandem, this confocal microscope has been referred to as the tandem scanning confocal microscope (TSM). The Nipkow disk of the TSM contains thousands of apertures (20–60 μm in diameter) that are in tandem or diametrically opposed pairs arranged in spirals (Fig. 30) (Egger and Petráň, 1967; Kino, 1989; Kino and Xiao, 1990; Petráň *et al.,* 1968, 1985, 1990; Wright *et al.,* 1993). Each aperture on one side of the Nipkow disk serves as an illumination aperture and acts as a small diffraction-limited spot in the focal plane of the objective, which focuses the

---

out-of-focus light rays. The in-focus fluorescent light is then separated by three secondary dichroic mirrors to as many as three different photomultiplier tube (PMT) detector channels. Signals from the PMTs are then displayed on the computer monitor as a confocal image. Reflected light coming from the specimen takes a different route in the scan head (D). A quarter-wave plate changes the polarization of the reflected light after it passes through the objective lens so that unwanted surface reflections are eliminated from the light path. The reflected light retraces its path through the scan head and is descanned by the AOD onto a reflective mirror. The returning reflected light then passes through the polarizing beam splitter and is then directed by a series of highly reflective mirrors to a pinhole-type confocal aperture that blocks out-of-focus light reflected from planes above and below the desired focal plane within the specimen. The in-focus reflected light is detected by a PMT located on the other side of the confocal aperture. (Courtesy of NORAN Instruments, Middleton, WI.)

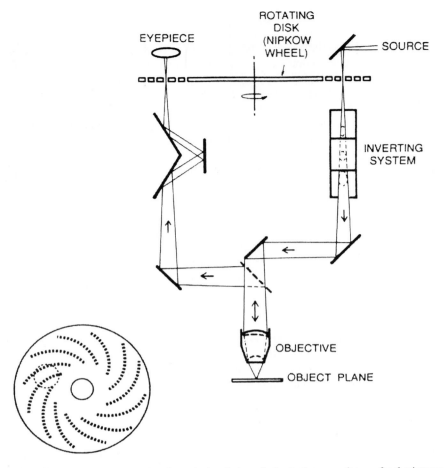

Fig. 30 Schematic diagram showing the optical path through the tandem scanning confocal microscope which is a multibeam point-scanning system. White light from a conventional light source (typically a mercury or xenon arc lamp) passes through an excitation filter (not shown) before being directed onto a spinning Nipkow disk that contains thousands of 20- to 60-μm apertures arranged in a precise spiral pattern. As the light falls on one side of the disk, it passes through the apertures and then through an inverting system. The spots of light are then directed to the objective, where they are focused on the specimen. High-speed rotation (up to 2400 rpm) of the Nipkow disk causes the spots of light to scan the specimen. The induced fluorescent or reflected light produced by the specimen is collected by the objective lens and reflected off a beam splitter and highly reflective mirrors to the conjugate apertures on the other side of the spinning disk. The in-focus fluorescence from the specimen passes through the conjugate apertures and a barrier filter (not shown) before being detected. In contrast, out-of-focus light from the specimen does not have a conjugate aperture to pass through and is stopped by the disk, contributing little to the confocal image. The time taken to traverse the field of view is so short that an apparent real-time image is produced at the eyepiece. Thus the specimen can be observed confocally by eye if the sample is of sufficient brightness, or with cameras (e.g., 35 mm, video, digital, or cooled charge-coupled device). Because the apertures in the disk are arranged in spirals, an inverting system is placed under the illumination side of the Nipkow disk in order for the conjugate apertures to line up, and a series of mirrors directs the spots of light to the beam splitter. On the detection side of the Nipkow disk, a system of mirrors maintains an equal distance for the optical path on both sides of the disk to create the confocal effect. Reprinted with permission from Kino, G. S. (1989). In "The Handbook of Biological Confocal Microscopy" (J. Pawley, ed.), IMR Press, copyright © 1989.

multiple illuminating spots on the specimen (Fig. 30). In-focus light from the specimen is detected after passing thorough the conjugate aperture on the opposite side of the disk. High-speed rotation (up to 2400 rpm) of the Nipkow disk causes the spots of light to scan the specimen many times a second and produce an apparent real-time image. This yields a stable, continuous confocal image in fluorescence and reflectance modes that can be directly viewed by eye in real time. A video camera mounted on the microscope also allows direct confocal recording and digital image processing of the specimen. Therefore the TSM offers the advantage of real-time imaging and direct confocal viewing of true image colors by eye in both reflected and fluorescence modes. This confocal microscope has been used to measure tissue thickness in the cornea of the eye (Jester *et al.,* 1999).

2. Monoscanning Confocal Microscope

The monoscanning spinning disk confocal microscope also uses a spinning Nipkow disk with spirally arranged apertures (Kino, 1989, 1995; Kino and Xiao, 1990; Lichtman *et al.,* 1989; Xiao and Kino, 1987; Xiao *et al.,* 1988). It produces images in real time and uses noncoherent white light for illumination. Unlike the TSM, the monoscanning confocal microscope detects light through the same aperture from which a given area of the specimen was illuminated (Fig. 31). Thus illumination and imaging are performed simultaneously through the same aperture. This design permits easier alignment of the Nipkow disk. To eliminate unwanted light reflected by the Nipkow disk, one version of this confocal microscope tilts the Nipkow disk and uses polarized light (Kino, 1995). This system, known as the K2-S Bio, is available from Technical Instrument San Francisco (Table III).

Although older models suffered from light loss, newer models of the monoscanning confocal microscope (Fig. 32) use a second disk fitted with 20,000 microlenses that are aligned and spin together with the apertures in the Nipkow disk (Figs. 1–3, Chapter 2). This amplifies the light transmitted to and from the Nipkow disk (Ichihara *et al.,* 1996; Inoué and Spring, 1997; Yamaguchi *et al.,* 1997). Moreover, in the monoscanning microlens confocal microscope (CSU10) developed by the Yokogawa Electric Company, background light is reduced since the dichroic mirror is placed between the Nipkow disk and the disk containing the microlens array. Confocal microscopes using the microlens array have been used to examine microcirculatory events and calcium ion dynamics (Genka *et al.,* 1999; Hama *et al.,* 1998; Yamaguchi, 1999). For a detailed discussion of the Yokogawa monoscanning confocal microscope and additional applications, see Chapter 2.

The PerkinElmer UltraView confocal microscope system is a modified version of the Yokogawa unit that uses dual spinning disks with microlens technology coupled to a choice of argon ion or krypton–argon ion laser, and digital cooled CCD camera detection to produce images (Fig. 32). The basic optical ray path diagram of this system can be seen in Fig. 2 of Chapter 2. Both fast and slow scan rates are available. Fast temporal data acquisition with frame transfer or interline cameras at 12 or 14 bits per pixel allows images to be captured at video rates or faster (50 frames/s and as fast as

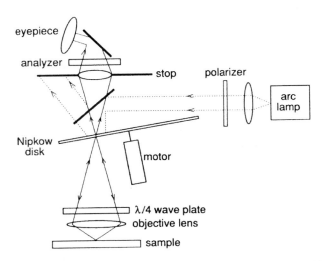

Fig. 31 Schematic diagram of the optical path in a monoscanning confocal microscope which is a multibeam point-scanning system. White light from a conventional light source (e.g., mercury or xenon arc lamp) is passed through a polarizer and directed on to the Nipkow disk that contains 200,000 pinholes that are 20–25 μm in diameter. The Nipkow disk is rotated at ~2000 rpm to produce a 5000-line image at 700 frames/s. Thus thousands of spots of light are focused by the objective lens and scan the specimen at any one time. Reflected light or fluorescence emitted from the sample passes through a quarter-wave plate and retraces its path through the same illumination aperture. Light from the focal plane of the specimen passes through an analyzer before being detected at the eyepiece. Out-of-focus light does not enter the apertures and is not imaged. The specimen can be viewed confocally by eye if the sample is sufficiently bright. A camera (35 mm, video, digital, cooled charge-coupled device) may also be used to observe specimens confocally. To reduce unwanted reflected light, the Nipkow disk is tilted and is made of a low-reflective material, and a stop is placed where light reflected from the disk is focused. To eliminate the remaining reflected light, the input light is polarized by the polarizer, rotated 90° by the quarter-wave plate, and observed through an analyzer with its plane of polarization at right angles to that of the input light. Reprinted with permission from Kino, G. S. (1989). In "The Handbook of Biological Confocal Microscopy" (J. Pawley, ed.), IMR Press, copyright © 1989.

Fig. 32 The UltraView multibeam point scanning confocal microscope from PerkinElmer Life Sciences. From left to right, the system consists of a visible multiwavelength laser as an illumination source, a cooled CCD camera as the detector, a confocal scanner module, an inverted or upright microscope, and a computer system. This confocal microscope uses spinning disk and microlens technology which allows confocal viewing directly by eye as well as high-speed, real-time image acquisition. (Courtesy of PerkinElmer Life Sciences, Gaithersburg, MD.)

120 frames/s). This is especially useful for imaging ionic fluctuations in living cells and tissues. Exciting images of calcium waves and sparks can be viewed at the PerkinElmer Web site (http://lifesciences.perkinelmer.com/products/ultraview2.asp).

V. Choosing and Setting Up a Confocal Microscope System

Several features need to be considered in choosing and setting up a confocal microscope system. These considerations revolve around researchers' needs and available resources. Time spent in carefully determining the potential applications of the confocal microscope before a system is chosen will maximize the possible number of users and successful applications for the microscope.

A. Considerations in Choosing a Confocal Microscope

Since the currently available confocal microscopes have a variety of features, how does one choose the best confocal system to meet the needs of the users? A number of features should be considered when choosing a confocal microscope. These choices relate to the anticipated applications for which the microscope will be used. For example, confocal microscopes are designed to work with an upright and/or an inverted microscope system. Choice in objective lens is also critical (Section III.B). If there will be multiple applications and/or users, an interchangeable, modular system may be required. If fast-changing living samples will be imaged such as during intracellular ion imaging, a fast scan rate is required. A video-rate confocal imaging system may be the best choice for this application. The type of fluorochromes that will be used to stain samples will dictate the type of lasers and number of photodetectors that will be needed. For example, if samples stained with UV-excitable fluorochromes are to be imaged, a CLSM with a UV laser or a spinning disk confocal microscope will be needed. If simultaneous imaging of several fluorescent signals is required, the confocal system will need multiple photodetectors that can collect signals at the same time, rather than sequentially. If real colors are important to determine by eye from the sample, then a direct-view confocal system is essential.

If the samples are living, the temperature and CO_2 levels of the tissue need to be maintained. This becomes especially important not only in preventing artifacts introduced by poor culture conditions in the sample, but also during 3-D reconstruction. For example, if the temperature shifts by 1°C during collection at 2.5-s intervals of a Z-series in a sample which is viewed under immersion oil at 37°C, the focal plane can shift as much as 0.25 μm (Kong et al., 1999). Since the temperature fluctuations induced by the stage temperature controller affect the refractive index of the immersion oil, the resulting 3-D reconstruction would be affected.

Other features to be considered when selecting a confocal microscope system deal with the software and computer hardware of the microscope. Digital image capture may be critical to the applications of the various users of the system. Additional image storage space may be needed, depending on the number of users of the system and the types of images (Z-series or individual optical sections) that will be collected. Compatibility of the image storage system with personal computer systems of the users outside of the confocal microscope will also be important. Supplemental software may be needed if the confocal system software does not perform all the required tasks. If morphometry will be performed on a Z-series, additional 3-D software may be required. If deconvolution will be performed on the confocal optical sections, the confocal microscope may not come with a deconvolution software package. The computer hardware must be robust enough to handle all the computational requirements including 3-D reconstructions and deconvolutions. Software such as Adobe Photoshop (Adobe Systems, Inc., Mountain View, CA) and Corel PhotoPaint (Corel Corporation, Ottawa, Ontario, Canada) can be used to crop and label images for publication.

Often overlooked, but very important, is the technical support and training available from the confocal microscope manufacturer. Will the users be adequately trained on the microscope? A well-designed system has on-line help, manuals, and tutorials for self-training (in addition to the initial training at setup) so that a multiuser core facility manager's time is not continually taken up with these duties. Service is also key— technical support and repair turnaround time is very important. If the laser or another component of the system goes down, turnaround time for repair becomes critical. Another possible expense to consider is a service contract, if available. It may be especially useful in systems with multiple lasers.

Even after all of the foregoing considerations have been addressed, the best way to choose a confocal microscope system is to examine the user's own samples with the microscope. For example, permanently mounted slides can be taken to a trade show or large professional conference (such as The Microscopy Society of America, Society for Neuroscience, and American Society for Cell Biology) for an evaluation of confocal systems from many competing manufactures. If multiple applications greatly differ, more than one confocal system may be required. Anticipating as many uses as possible for the microscope will ensure that the possible number of different applications have been maximized for the system.

B. Choosing a Location for the Confocal Microscope

The environment in which the confocal microscope system is placed is very important in obtaining the best images. The images can suffer if vibrations become a problem. If the imaging room is located near the loading dock of a building, or the building is located near railroad tracks or near a construction site, image quality may be reduced and 3-D reconstructions may be inaccurate. The best performance can be obtained on a vibration-free isolation table. Rooms being considered to house the confocal microscope system can be checked for vibrations ahead of time with vibration detectors (such as an IRD 401 microanalyzer), monitoring possible horizontal and vertical vibrations for several days. Confocal imaging is even susceptible to air currents from an air conditioner or heating system. Therefore, care must be taken not to place the microscope near these currents.

Room size must be adequate to house all of the confocal microscope equipment such as the microscope, scan head, laser, fluorescence light source, computer, computer monitor, printers, table/desktop space, and chairs for users. The equipment can generate heat. Thus, temperature of the room may also be critical especially if live samples will be imaged. Some rooms housing confocal systems used for imaging living samples are maintained at 37°C. Room lighting is also important. For many applications, the confocal room should be as light-tight as a darkroom. Therefore, a room with a window is inappropriate to house the confocal system unless the window is blocked off from light.

Power requirements are critical. This includes an adequate number of electrical outlets with a stable line of the right type. Stability of the power line can be monitored with a Power Line Disturbance Monitor (Model 3600A; Liebert Corporation, Fremont, CA). In addition to considering all of the microscope components that plug into electrical outlets (incandescent microscope light sources, laser or fluorescence light source, computer, computer monitors, printers, etc.), one must not forget that a small-wattage tabletop lamp is very helpful to allow the user to perform specimen manipulations and to take notes. Having the computer networked to the users' individual laboratory computers and to the Internet is very beneficial and eases image transfer. It is also helpful to have a phone directly in the room for safety and service questions.

C. Personal Confocal Systems versus Multiuser Confocal Systems

A personal confocal system is used by an individual or small number of researchers with similar applications. It may not need as many features and functions as a multiuser confocal microscope in a core facility, and thus has a relatively low price. Confocal microscope system requirements can greatly differ between users. For example, a researcher who only uses low magnification (20×) and fixed fluorescein- and/or rhodamine-labeled samples can get superb images with a lower-priced confocal system which has only an argon ion laser. On the other hand, a researcher who performs ion imaging of living cells has very different system requirements. This user may need a more sophisticated confocal system with more laser lines and an exceptionally fast scan rate.

A personal system may be easier to attach to a preexisting microscope than a full system. In addition, space and electrical requirements may be easier to meet with a personal system. Before choosing a system, one must keep in mind not only cost, but also all the potential applications that will be used with the confocal system. One may find that an elaborate full system is not necessary to answer biological questions about cellular microarchitecture, and the user may achieve spectacular confocal images with a personal system.

When one uses a confocal system in a multiuser or core facility, there are usually specific rules governing the use of the equipment. First, the researcher will be charged for the microscope imaging time, training, assistance of an experienced microscopist to "drive" the microscope, and/or for image processing time. Although the charges may seem high, they are for upkeep of the service contract and for future replacement of the laser(s). The user may also find very strict rules about image storage in a multiple user facility. Users should come with their own Zip disks (or equivalent) to copy their new images before they complete their imaging session. Most facilities will not allow the hard disk to be used for permanent image storage. If anything is left on the hard disk after an imaging session, it may be deleted. A facility rarely has enough room for all of the imaging needs of each researcher. For a detailed look at managing and setting up a confocal microscope core facility see Chapter 14.

VI. Specimen Preparation

Like any type of microscopy, confocal microscopy requires adequate sample preparation (Wright *et al.,* 1993). Care must be taken to choose proper fixatives and specific antibody probes with minimal cross-reactivity. Choice of fluorescent dyes that will be used is critical. The best fluorescent signals come from dyes in which the excitation wavelength is closely matched to the laser line. Filter sets in the microscope must be precise enough to prevent bleed-through from different channels in multilabeled specimens. For example, care must be taken to prevent green fluorescein signals from being detected in the red channel and vice versa (Section III.C).

Photobleaching occurs when the fluorochromes fade upon exposure to light. Specimens are usually placed in mounting media to prevent dehydration and fading during storage and confocal viewing. To reduce photobleaching, mounting medium often contains antioxidants such as 1,4-diazabicyclo[2.2.2]octane (DABCO; Johnson *et al.,* 1982; Langanger *et al.,* 1983), *p*-phenylenediamine (Johnson and de C Nogueira Arajo, 1981; Johnson *et al.,* 1982), or *n*-propyl gallate (Giloh and Sedat, 1982). These antioxidants are dissolved in a buffer such as 10% phosphate-buffered saline (PBS), pH 7.2, and 90% glycerol, at concentrations of 100 mg/ml, 2–7 mM, and 3–9 mM, respectively. The antioxidants, also called antifade reagents, are commercially available individually, or premixed for immediate use (Berrios *et al.,* 1999; Longin *et al.,* 1993). See Chapters 4 and 5 for information on specimen mounting media and antifade reagents.

Care must be taken to not introduce artifacts such as tissue compression into the specimens during their preparation. The microscopist must prevent the artificial flattening of a sample during mounting with spacers such as broken coverslips, fishing line, or double-stick tape to preserve the 3-D shape of the sample. The space between the top of the specimen and coverslip should be minimized, or a reduction in resolution may result. Coverslips must be sealed to the slide to prevent the sample from drying out. Nail polish works well for this purpose; care must be taken to give the nail polish sufficient time to dry completely before viewing the samples to prevent it from adhering to the objectives. See Chapters 7–13 for a discussions of sample preparation for the confocal microscope. Detailed procedures for immunolabeling and mounting invertebrate eggs and embryos can be found in Wright *et al.* (1993).

VII. Applications of Confocal Microscopy

Confocal microscopy offers a powerful means to address biological problems and gives new understanding of cellular structure and function. There is a wide variety of fluorescent dyes and fluorochromes to choose from that work well with living and fixed specimens imaged with confocal microscopes (Table II). Some applications of confocal microscopy include: (1) determining the cellular localization of ions, macromolecules (such as proteins, RNA, and DNA), cytoskeletal elements, and organelles; (2) tracing specific cells or structures through a tissue; (3) producing optical sections for stereo image production, 3-D reconstruction, and 4-D imaging; and (4) uncaging photoactivatible caged compounds (Table IV). Fluorescence recovery after photobleaching and ion imaging are also possible with confocal microscopy. When performed with conventional fluorescence microscopy, these techniques yield limited information, but when used with confocal microscopy, they show whether a structure has been photobleached completely, and whether ionic changes occur throughout a whole cell or tissue or only in a specific region. Confocal microscopes can also be used in the reflectance mode. This allows reduction of out-of-focus blur from nonfluorescent labels such as diaminobenzidine reaction products formed during enzyme cytochemical reactions, and from silver grains present in autoradiograms produced during *in situ* hybridization (Wright *et al.,* 1993). In addition to confocal fluorescence images, CLSMs can also collect nonconfocal transmitted light images including brightfield, differential interference contrast (Nomarski), phase contrast and dark field concurrent with, or separately from the fluorescence images. See Chapters 8–13 for details about specific applications of confocal microscopy.

Table IV
Common Applications of Confocal Fluorescence Microscopy

Application	Process
Fluorochrome staining and immunostaining	• Single, dual or multiple-labeled samples • Simultaneous detection of multiple fluorochromes • Fluorescence recovery after photobleaching (FRAP) • Fluorescence resonance energy transfer (FRET)
Green fluorescent protein	• Studies on cellular trafficking, intracellular signaling, gene expression, morphology
Organelle function	• Studying organelles (endoplasmic reticulum, golgi, mitochondria, nucleus, plasma membrane) with organelle-specific fluorescent probes • Nuclear and tissue changes during cancer
Nuclear function	• Spatial order in the nucleus: relationship of chromosome domains, telomeres and kinetochores • Localization of several genes in nucleus using fluorescence *in situ* hybridization (FISH)
Cytoskeletal function	• Analysis of associated proteins and arrangement of cytoskeleton within microarchitecure of cells
Ion fluctuations	• Intracellular ion imaging of calcium, magnesium, zinc, sodium, potassium, hydrogen, chloride ions
Membrane potential	• Analysis of changes in membrane potential dyes
Cell function	• Analysis of signaling pathways, cell division, apoptosis
Photoactivation	• Uncaging photoactivatible caged compounds
Microcirculation	• Velocity of blood cells in microvessels • Microcirculation of various agents in tissues
Diagnostic tool	• Determination of cornea thickness • Diagnostic virology as a survey tool to identify areas of necrosis or cell enlargement due to viral infection, in preparation for subsequent ultrastructural analysis
Morphometry	• Measure structures in 2-D images and 3-D reconstructions
Microarrays	• Gene expression
Reconstruction	• Three-dimensional reconstruction • Four-dimensional imaging

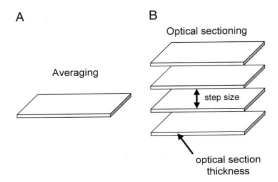

Fig. 33 Comparison between frame averaging (A) and optical sectioning (B). During frame averaging, the laser beam scans the specimen several times at the same focal plane. The number of scans depends on the set number of frames to integrate the image. During optical sectioning, the laser beam scans the sample at several focal planes, usually with the same increment in step size. The step size refers to the distance between optical sections. Optical section thickness is usually smaller than the step size.

Typically when one optical section is collected, the PMT detects signals that are converted into a digital image which is a matrix of numbers that are then placed in a storage region called the *digital frame memory*. Several digitized images are usually integrated and added together in the digital frame memory into one final image or *frame*. This greatly reduces image noise (from the PMT and laser) in the final image. For example, the laser in CLSMs may scan the sample 16 times (for 16 frames) at exactly the same focal plane. In the digital frame memory, the 16 frames are then added together and divided by the total number of frames (in this case, 16) to produce one integrated image with a reduction in image noise. This is called *averaging* (Fig. 33). Most confocal systems give a choice of the number of frames that can be averaged (e.g., 2, 4, 8, 16, 32, 64, 128, 256). To prevent photobleaching, it is desirable to choose the lowest possible frame average that will yield the best picture. Some confocal microscopes (Bio-Rad, Olympus) use weighted frame averaging, called *Kalman averaging* or Kalman filtering, in which the most recently acquired image contributes significantly more to the final image than previous frames. This allows ease of use while scanning for an area of interest. *Sectioning,* on the other hand, refers to a process where one optical section is taken at one focal plane, then another is taken at a different focal plane, and so on until a stack of optical sections or *Z-series* is generated (Fig. 33). The distance between images in the Z-series is usually identical and is referred to as the *step size*. The optical section thickness is usually smaller than the step size. Each image in the stack can be averaged to produce high-quality images ready for further applications (Sections VII.C–F).

To obtain the best images, the confocal microscopist works with several parameters. These include not only the type of objective lens, confocal aperture size, and laser intensity, but also imaging mode such as frame averaging and final image size. Often the researcher switches between different modes in one viewing session. For example, when changing from a smaller to a larger image size such as from 512×512 to 1024×1024, the images get dark when the same settings are used for the large image as for the smaller one because there are four times as many pixels to scan at the same scan speed. Therefore, adjustments need to be made to increase the scan speed, the laser intensity, or the aperture size when creating the high-resolution image. When working with unfamiliar samples, the microscopist should vary these parameters to determine which combination yields superior image quality.

A. Single-Label Imaging

Optical sections labeled with one fluorochrome provide much information, since the out-of-focus fluorescence is almost completely eliminated by the increase in contrast and apparent resolution (Fig. 34). This can be applied to single cells as well as thicker, multicellular specimens. Real-image colors can be observed confocally, directly by eye with the spinning disk confocal microscopes because the images are generated and viewed with white light. For example, when a tandem-scanning confocal microscope is used in the reflected-light mode, real colors of a single optical slice of cultured 3T3 cells display

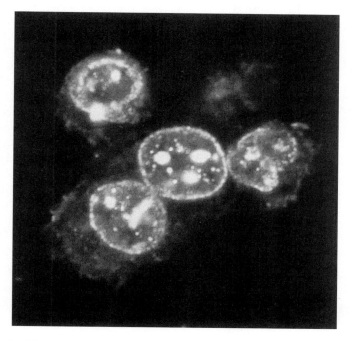

Fig. 34 Confocal fluorescence micrograph of tissue culture cells stained for topoisomerase II immunofluorescence revealing nuclear staining. Topoisomerase II was localized with a mouse monoclonal anti-topoisomerase II antibody and a FITC-conjugated goat anti-mouse IgG. The topoisomerase II antibody was generously donated by Dr. Thomas Rowe, University of Florida. A single *xy* optical section was taken with a Bio-Rad MRC-1024 confocal laser scanning microscope using a 60 × 1.4 NA objective on a Nikon Diaphot.

focal contacts at the cell substratum interface (Paddock *et al.,* 1989; Wright *et al.,* 1993). Although samples cannot be viewed directly by eye, an optical section acquired with the CLSM can be pseudocolored to highlight various structures, or colored to mimic the original fluorochrome (e.g., green for fluorescein). Many researchers superimpose the fluorescence images over a conventional brightfield (or Nomarski) image to provide contextual reference information. For example, to accurately identify *Drosophila* neuroblasts (embryonic neural precursor cells), Schmid and colleagues (1999) labeled the neuroblasts with the fluorescent dye DiI and used Nomarski optics in addition to confocal fluorescence microscopy to determine positions of the neuroblasts within the central nervous system. Confocal movies of this research can be viewed at Dr. Chris Doe's laboratory home page (see Section X.C.6 for URL).

B. Multiple-Label Imaging

With multiple labeling, two or more structures can be observed and traced simultaneously in the same cell or tissue (Fig. 35; see color plate). Although images for each channel can be placed side-by-side, a merged color image shows relationships better (Fig. 35). Care must be taken to prevent bleed-through from the different channels. In older confocal microscopes, the user had to select the correct filter set for each application. Often the 514-nm emission line of an argon ion laser was used to excite both fluorescein and rhodamine channels. Alternatively, a krypton–argon ion laser or two different lasers were used. In some newer CLSMs, the software-driven AOTF provides specific selection of excitation wavelengths for the various fluorochromes and the ability to individually attenuate the lines of a laser that emits multiple wavelengths. This feature coupled with better laser lines is essential in samples labeled with multiple fluorochromes since each dye has a different excitation wavelength and may photobleach at a different rate. Dyes with narrower absorbance and emission spectra better suited to the laser lines (Table II), such as BODIPY (Haugland, 1990), and digital image subtraction can also be used to reduce the bleed-through data.

An important feature of newer CLSMs is the ability to allow three different fluorochromes (e.g., red, green, and blue) to be imaged either simultaneously (where the emission from all dyes is collected at once) or sequentially. Each fluorochrome is illuminated by the appropriate laser lines, and filters or AOTFs are used to prevent bleed-through between the different channels. The images collected concurrently by three different PMTs can then be displayed as a colored image by merging the information from the three channels into a single RGB image.

C. Z-Series

A major application of confocal microscopes is to generate 2-D optical sections from fluorescently labeled samples without physically sectioning the sample. To obtain 3-D information from the specimen, it is common practice to acquire a series of optical

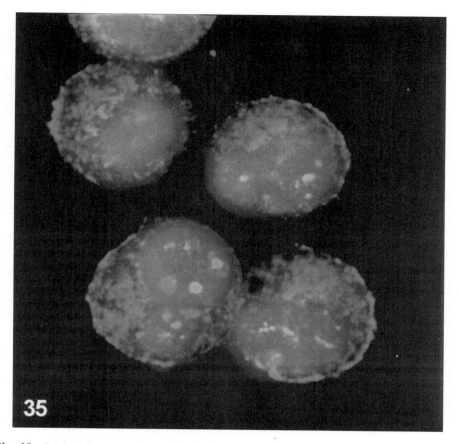

Fig. 35 Confocal fluorescence image of a single xy optical section of tissue culture cells double-labeled to show topoisomerase II (green) and nuclei (red). Topoisomerase II was stained with a rabbit polyclonal anti-topoisomerase II antibody followed by a FITC-conjugated goat anti-rabbit IgG. Positions of the nuclei were revealed using propidium iodide. The image was generated on a Bio-Rad MRC-1024 confocal laser scanning microscope using a 60×1.4 NA objective on a Nikon Diaphot. A separate channel was used to image each fluorochrome, and the images were later merged using Confocal Assistant 4.02. The image shows that topoisomerase II localizes to both the cytoplasm and nuclei. (See Color Plate.)

sections taken at successively higher or lower focal planes along the z axis. This stack of optical sections is called a *Z-series*. All the 2-D optical sections in a data set make up a volume data set. The principle of Z-series acquisition is demonstrated schematically in Fig. 36. The depth of field depends on the working distance of the objective, thickness of the sample, and penetration of the fluorochrome in the sample. To obtain 3-D information from the Z-series, individual slices from the stack may be viewed as a montage (Fig. 37). A Z-series can also be used to make stereo images, 3-D reconstructions, and 4-D images.

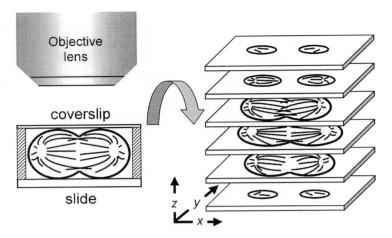

Fig. 36 Schematic diagram demonstrating the principle of optical sectioning. To preserve three-dimensional morphology, the mitotic cell is mounted with spacers to prevent the coverslip from flattening the cell. A through-focal series of optical slices, known as a *Z-series,* is taken to generate a volume data set which can then be rendered into a three-dimensional reconstruction. Both confocal laser scanning microscopes and spinning disk confocal microscopes can be used to generate the Z-series of optical sections. Modified from Wright, S. J., *et al.* (1993). *Methods Cell Biol.* **38,** 1–45, with permission of Academic Press, copyright © 1993.

D. Stereo Imaging

Although a Z-series montage provides some 3-D information, stereoscopic images produce a more 3-D perspective of the specimen (Fig. 38). CLSMs obtain stereoscopic images by software manipulation (pixel shifting) of a Z-series (White, 1995). Stereo pairs are also produced by reconstructing a Z-series three-dimensionally and observing two individual views of the rendered volume that are approximately 9° apart. In addition to pixel shifting, the TSM can produce stereo images by controlling objective movement in both the x and z axes to generate a left and right image at each focal plane (Wright *et al.,* 1993). This provides a separate stack of right and left images which, when compiled, produces a stereo image through two different volumes. Thus, each half of the stereo pair can be acquired directly from the specimen without digital manipulation (Boyde, 1985; Boyde *et al.,* 1990; Wright *et al.,* 1989, 1993). Stereoscopic images can be produced as red-green (RG) or red-blue (RB) anaglyphs which are transmitted to corresponding eyes by RG or RB glasses, respectively. Stereo pairs can also be displayed side-by-side as a right and left black-and-white image which can be viewed with polarizing eyeglasses. They can also be viewed directly by eye using either the parallel-eye method in which the binocular focus point is at infinity, or the cross-eye-method in which the eyes are converged (crossed) to a close point between the observer and the stereo-pair picture. Stereo images are especially useful for publication of 3-D reconstructions as 2-D projections on paper in journals.

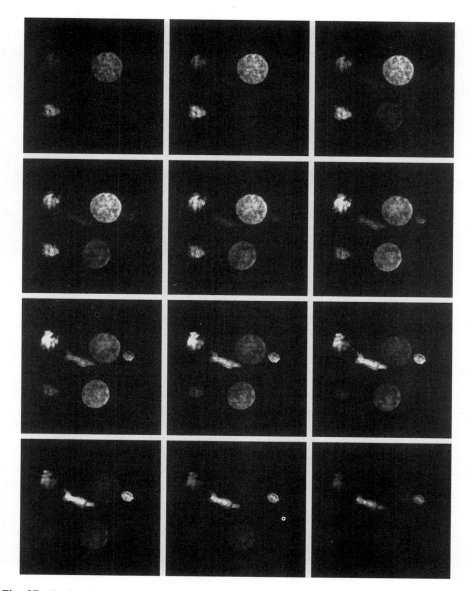

Fig. 37 Confocal fluorescence micrographs showing a Z-series montage of a 4-cell mouse embryo stained for histones. A mouse monoclonal anti-histone antibody (Roche Molecular Biochemicals) followed by a FITC-conjugated goat anti-mouse IgG was used to label histones. Serial, confocal xy optical sections were generated on a Bio-Rad MRC-1024 confocal laser scanning microscope using a 60 × 1.4 NA objective on a Nikon Diaphot. A Z-series of xy optical sections was taken at 0.5-μm increments using 3% laser power and a Kalman average of 2. The Z-series montage was generated using Confocal Assistant 4.02. As shown here, chromosomes in two blastomers are in interphase, whereas the others are in metaphase or anaphase. The small, oval bright spot next to the top interphase nucleus and metaphase chromosomes in the last six images of the Z-series is the second polar body. Cell division is asynchronous in mouse embryos.

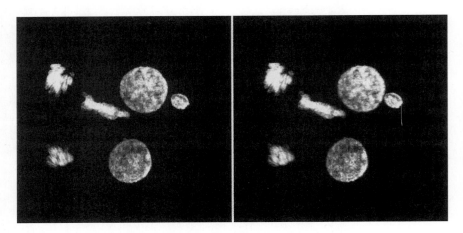

Fig. 38 Stereo pair confocal fluorescence micrographs of the same 4-cell mouse embryo shown in Fig. 37 reveals the three-dimensional organization of the histone-stained chromosomes and nuclei. A Z-series montage generated from 17 xy optical sections at 0.5-μm increments was three-dimensionally reconstructed using a 9° angle to produce the stereo pair with Confocal Assistant 4.02.

E. Three-Dimensional Reconstruction

Although volumetric information can be obtained from stereo images, it provides only a static view of the specimen from a single direction. An improved means of analyzing complex spatial relationships can be achieved with 3-D reconstruction, because the data volume can be viewed interactively from any angle (e.g., top, bottom, sides, obliquely). A 3-D reconstruction can be generated by using computer software to precisely stack the 2-D optical sections on top of each other into one single image. Confocal microscopes must be able to repeatedly and accurately align pixels during 3-D and 4-D reconstructions. Advances in computer graphics, computer memory, software, and digital image storage have made generating 3-D reconstructions relatively easy and fast as compared to early generation systems. One requirement, in addition to the computer hardware, is specialized software that can handle the intensive processing. These developments offer the investigator the ability to dissect the data interactively to develop powerful insights into complex 3-D structures of biological specimens. However, a difficulty in presenting 3-D reconstructions lies in the lack of appropriate visualization tools because most displays (output devices) provide a 2-D view. To overcome this problem, different perspectives of the specimen are sequentially displayed on the computer monitor in a movie as a rotating volume which gives a 3-D perspective of the data, provides depth, and brings into view structures and relationships otherwise unnoticed. There still is a lack of appropriate visualization tools to present 3-D information on a 2-D printed journal page. Some journals come with CD-ROMs so that the animations can be played as a digital movie. A number of Web sites also display 3-D reconstructions as movies. As journals are increasingly turning to the Web as a means of publishing electronically, authors of journal articles will be able to routinely display raw Z-series image stacks, and 3-D and 4-D movies to demonstrate their research results. For example, 3-D movies of DiI-labeled neuroblasts together with engrailed-GFP landmarks (Schmid *et al.,* 1999) were used to identify the parental neuro-blasts for all cell types of the embryonic central nervous system in *Drosophila* embryos. These cell lineage movies can be observed at http://www.neuro.uoregon.edu/doelab/ doelab.html and http://www.biologists.com/Development/movies/lineages/overview/ grasshop.htm.

1. Definitions

To aid in understanding 3-D reconstruction, several terms must be defined. *Rendering* refers to processes, such as rotation, projection, coloring, and shading, that are required to display a specimen on the computer image monitor (Van Zandt and Argiro, 1989; Wright *et al.,* 1993). Volume rendering works by taking a Z-series known as a *volume* and assumes that the 3-D objects are composed of volumetric building blocks or cubes called voxels. A *voxel* is defined as a volume element just as a *pixel* is defined as a picture element. A voxel can be considered to be a 3-D pixel. Each voxel of the volume is assigned a number representing a characteristic of the 3-D object such as density or luminosity. The 3-D reconstruction software summarizes the specimen as a 3-D array of numbers. To generate an image, the 3-D array of voxels or numbers is transferred into computer memory, the volume data is processed, and then displayed on the screen of the image monitor. One way to show the data in a journal article is as a *projection* which displays the entire stack of optical sections in the Z-series. This can be done for each channel in multiple-labeled specimens (Fig. 39) which are then displayed adjacent to each other. The Z-series can also be displayed as a merged color stereo projection (Fig. 40; see color plate).

2. Considerations

Several parameters must be considered before acquiring a Z-series suitable for 3-D reconstruction (Wright *et al.,* 1993). Careful attention must be paid to each parameter so the 3-D reconstructions are produced with minimal distortions and artifacts.

a. Specimen Preparation

Improper specimen preparation can produce distorted 3-D images (Blatter, 1999; Wright and Schatten, 1991; Wright *et al.,* 1993). Fixation and extraction protocols opti-mized for conventional fluorescence microscopy often result in flattening of specimens especially by the weight of the coverslip (Wright *et al.,* 1990; Wright and Schatten, 1991). Fixation and dehydration protocols also introduce distortional artifacts. Hence it is necessary to employ the greatest care when preparing a sample for 3-D reconstruction. It is important to keep in mind that the further one focuses in the z axis of the sample, the more prominent loss of contrast and shadowing become so that upper focal planes appear bright, while lower ones are dim. Increasing the illumination intensity can overcome some of the problem; however, photobleaching may become severe. Hence, antioxidants should

A   B

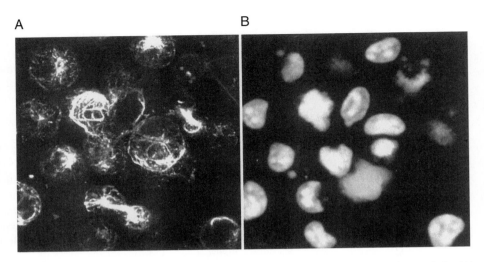

Fig. 39 Confocal fluorescence micrographs of tissue culture cells double-labeled for microtubules (A) and nuclei (B). Microtubules were stained with a mouse monoclonal E7 anti-tubulin antibody followed by a FITC-conjugated goat anti-mouse IgG. Positions of the nuclei were revealed using propidium iodide. The images were generated on a Bio-Rad MRC-1024 confocal laser scanning microscope using a 60 × 1.4 NA objective on a Nikon Diaphot. A separate channel was used to image each fluorochrome. A Z-series of xy optical sections was taken at 0.5-μm increments using 3% laser power and a Kalman average of 3 for each channel. A projection (volume data set) of each Z-series was generated using Confocal Assistant 4.02.

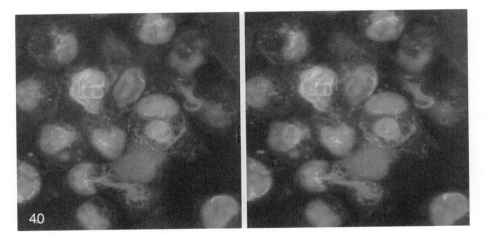

Fig. 40 Stereo pair confocal fluorescence image of the same tissue culture cells in Fig. 39 double-labeled for microtubules (green) and nuclei (red). The stereo pair was generated by merging the two projections using Confocal Assistant 4.02. The three-dimensional relationship of microtubules and nuclei is easier to observe in the merged color stereo image. (See Color Plate.)

be used to reduce photobleaching during acquisition of large data sets. Heterogeneity inherent in specimens can also create problems in acquiring a Z-series (Cheng and Kriete, 1995). The intensity of the excitation light and emitted fluorescence can be attenuated by structures situated between the focal plane of interest and the objective lens. This causes self-shadowing of structures and can significantly reduce the image contrast in some areas of the specimen.

b. Image Registration

A second requirement for effective 3-D reconstruction is image registration. Each optical section of the Z-series must be in register with its neighbors during image acquisition; otherwise distortion is introduced into the 3-D reconstruction (Shuman et al., 1989; Wright and Schatten, 1991; Wright et al., 1993). Hence care must be taken to prevent the specimen from moving during the acquisition of the Z-series. Specimen movement can be avoided by affixing cells to the coverslip or slide prior to mounting and by reducing vibrations of the microscope stage (Wright et al., 1993). Care must also be taken to make sure that the filter sets are not misaligned in confocal microscopes where the emissions from the fluorochromes in multiple-labeled samples are collected with separate filters and laser lines. In many newer CLSMs, filter changing is automated and this greatly reduces alignment problems for multiple-labeled specimens. Misalignment problems can be digitally corrected in some newer CLSMs. A quick way to check alignment in the microscope is through the use of dual- or triple-labeled fluorescent microspheres such as FocalCheck fluorescent microspheres (Molecular Probes, Eugene, OR) (Fig. 41).

c. Pixel Resolution

A third important consideration before obtaining a Z-series is the pixel resolution for the conditions used (objective, magnification, electronic zoom, image size, optical section thickness) so that the proportional dimensions of the specimen are maintained during 3-D reconstruction. If relatively too few or too many optical sections are collected, the rendered volume will appear distorted by being either squashed or stretched, respectively, as shown in Fig. 42 (Wright et al., 1990; Wright and Schatten, 1991). Because the computer bases its calculations on cube-shaped voxels during 3-D reconstruction, the z depth of each voxel must be of the same dimension as the pixels in the x and y dimensions of the optical section (Paddock et al., 1990; Wright et al., 1993). If too few optical sections are acquired, the computer software can synthesize an image by averaging the sections above and below the desired plane of interest, and insert the interpolated image between the acquired sections. Likewise, if too many sections are acquired, the program can remove sections from the Z-series.

d. Image Storage

A fourth consideration for collecting images for 3-D reconstruction revolves around image storage (Chen et al., 1995; Cox, 1999; Shotton, 1989; Shuman et al., 1989; van Zandt and Argiro, 1989; Wright and Schatten, 1991; Wright et al., 1990, 1993). Many 3-D reconstructions are displayed as digital movies and take up considerably

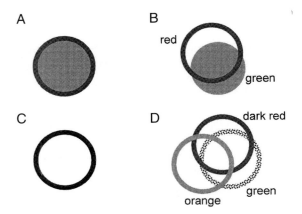

Fig. 41 Fluorescent microspheres aid in the optical alignment of the confocal laser scanning microscope (CLSM). Visible and ultraviolet wavelengths emitted by fluorescent samples have separate light paths in the CLSM and are detected with different photodetectors. Proper optical alignment of these paths is critical for subsequent data interpretation especially in multiple labeled specimens. In (A) and (B), 15-μm microspheres were labeled with two fluorochromes: the center was labeled with a green fluorochrome requiring excitation at 505 nm, whereas the periphery was labeled with a dark red fluorochrome requiring excitation at 660 nm. When the CLSM is aligned properly, the red ring surrounds the green sphere when two images collected in separate green and red channels are merged together (A). The CLSM is misaligned in (B), since the red ring and green sphere are out of register and areas of overlap will appear yellow. In (C) and (D), the periphery of each microsphere is labeled with three fluorochromes: green, orange, and dark red, which emit at 515 nm, 580 nm, and 680 nm, respectively. These microspheres are used to align three different channels simultaneously, by aligning the green, orange, and dark red rings on top of each other. The CLSM is aligned in (C) and misaligned in (D). Modified from Haugland, R. (1999). "Handbook of Fluorescent Probes and Research Chemicals," 7th ed., with permission from Molecular Probes, Inc., copyright © 1999.

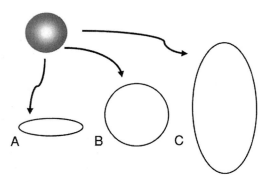

Fig. 42 Three-dimensional reconstruction of a spherical fluorescent microbead. In (A) too few optical sections were taken resulting in a squashed, three-dimensional reconstruction. In (B), the pixel resolution in the xy plane was used to generate the correct distance to obtain optical sections in the z axis. A near-perfect sphere results, as expected. Too many optical sections were taken in (C), and the sphere appears stretched into an elongated structure in the three-dimensional reconstruction.

more space than single optical sections. One Z-series containing 50 optical sections can easily require 300 MB of storage space (Cox, 1999). Compression of images aids in storage capabilities with little if any loss of information from the image. As a safety precaution, data files must be backed up, and they must be removed to make room for the next project. Although significant advances have been made in the availability of storage media with greater capacity such as rewritable CD-ROMs, Zip disks, and Jaz disks, the confocal microscopist is faced with higher resolution images and therefore image storage will remain an important issue in the future.

3. Image Processing

Once the Z-series has been collected, it may be processed to improve image quality prior to (or after) 3-D reconstruction. For example, mathematical algorithms may be applied on the entire stack of images to remove background noise and artifacts (background subtraction), smooth or sharpen the images, or correct for any brightness and contrast problems (thresholding or intensity segmentation) (Blatter, 1999). The images can also be masked prior to 3-D reconstruction to enclose an ROI for further processing. Color can also be added to the Z-series prior to 3-D reconstruction to highlight regions of interest, to selectively differentiate between various regions of interest, or to label structures collected with each channel in multiple-labeled samples (Fig. 40).

4. Volume Visualization

Two major methods are available for 3-D reconstruction and display of a Z-series of confocal data (Brotchie *et al.,* 1999; Chen *et al.,* 1995; Van Zandt and Argiro, 1989; Wright *et al.,* 1993). These systems use either geometric surface rendering (Harris and Stevens, 1989) or volume rendering (Van Zandt and Argiro, 1989). In geometric surface rendering, triangles and 2-D polygons are fused together to approximate the surfaces of 3-D objects. Surface rendering provides extensive information on the shape and surface of a specimen (Brotchie *et al.,* 1999). Volume rendering provides useful information on internal as well as surface structures of interest. In volume-rendered 3-D reconstructions, several manipulations can be performed to reveal particular features of interest in the volume. These include changing the view of the volume (from top, bottom, obliquely), viewing the entire volume or a single optical section from different angles, adjusting the relative transparency or opacity of the volume to reveal structures on the surface or deep within the specimen, obtaining maximal contrast, choosing a background color, pseudocoloring the image within the volume by voxel or gradient values, turning on and positioning a light source to enhance surface features of the specimen, activating antialiasing to prevent formation of linear artifacts, and compressing the data to eliminate nonrelevant voxels and decrease rendering time (Blatter, 1999; Webb and Dorey, 1995; Wright *et al.,* 1993). Once these parameters have been set, an animation of the 3-D reconstruction is generated by choosing the number of angles (e.g., every 9° of a total of 360°), total rotation, and the axis of rotation. The animation is produced by showing in rapid succession the views taken at different angles of the already rendered *Z*-series. The animated 3-D reconstruction can then be viewed interactively. Cutaway views can be generated to reveal hidden information. Both single and multiple-labeled samples can be rendered into a 3-D reconstruction.

A sample can be repeatedly imaged over the same single focal plane of interest at selected time intervals to generate a 3-D time-lapse image (Lui *et al.,* 1997; Stricker *et al.,* 1992; Summers *et al.,* 1993; Williams *et al.,* 1999). This generates one stack of optical sections over time. This stack can then be subjected to 3-D reconstruction into a single image. For example, intracellular Ca^{2+} levels have been imaged in ventricular myocytes as they become permeable to Ca^{2+} after exposure to the calcium ionophore ionomycin (Bkaily *et al.,* 1999). In another example, the pattern of calcium ion fluxes can be observed in the cytoplasm and nucleus during fertilization of a single sea urchin egg (Stricker *et al.,* 1992; Summers *et al.,* 1993; Wright *et al.,* 1993). Calcium waves can also be observed in astrocytes and Müller cells using the calcium indicator dye calcium green-1 and a video-rate CLSM (Newman and Zahs, 1997) by accessing http://physiology.med.umn.edu/faculty/krz1a.htm.

F. Four-Dimensional Imaging

The technique of generating 3-D reconstructions becomes even more powerful when *Z*-series of the same living sample are taken at periodic intervals over time and reconstructed three-dimensionally (Fig. 43). This method is referred to as *four-dimensional imaging* and produces a consecutive series of 3-D reconstructions over time (Mohler and White, 1998; Summers *et al.,* 1993; Wright *et al.,* 1993). The 4-D data sets can be observed as 3-D stereo movies or projections over time. For example, in Fig. 44, a

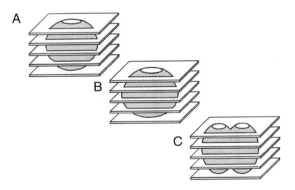

Fig. 43 Four-dimensional imaging. Stacks of optical sections through the same living sample can be generated over time. A single cell, such as a recently fertilized egg, is followed during interphase (A), mitosis (B), and the two-cell stage (C). Three-dimensional reconstructions can be generated from each stack using the same parameters. A four-dimensional movie can then be generated showing changes throughout the depth of the specimen over time.

Fig. 44 Confocal fluorescence micrographs showing a montage of four-dimensional imaging. Three-dimensional reconstructions over time of the same developing sea urchin (*Lytechinus pictus*) embryo stained with $DiOC_6(3)$. The embryo was imaged with a Bio-Rad MRC-600 equipped with an argon ion laser. As the embryo developed, a Z-series of ~30 optical sections at 5-μm steps was taken every 8 min. The Z-series was rendered with interpolated sections to obtain the appropriate pixel resolution. Developmental stages: (A) zygote; (B) 2-cell stage; (C) 4-cell stage; (D) 8-cell stage; (E) ~16-cell stage, with elongated blastomeres in the process of cell division; (F) early morula; (G) morula; (H) morula; (I) late morula/early blastula stage, prior to hatching. Modified from Wright, S. J., *et al.* (1993). *Methods Cell Biol.* **38,** 1–45, with permission from Academic Press, copyright © 1993.

sea urchin zygote labeled with $DiOC_6$ was successfully imaged every 8 min up to the early blastocyst stage (Stricker *et al.*, 1992; Wright *et al.*, 1993). Time-lapse, 4-D stereo movies can be observed of cell fusions in nematode (*Caenorhabditis elegans*) embryos labeled with the vital plasma membrane probe FM™ 4-64 (Molecular Probes) and apical protein MH27-GFP constructs (Mohler and White, 1998; Mohler *et al.*, 1998) by visiting http://www.loci.wisc.edu/stereo4d/Stereo-4D_demo.mov. With this technique, analysis of embryonic development reveals positions and lineages of various cells which can be traced back to their origins. This technique is especially powerful for imaging rapid signal pathways and ion fluctuations in cells and tissues with video-rate CLSMs. For example, each 3-D data set can be rendered, rotated to a predefined angle, and captured as a still image. The collection of 3-D still images taken over time can be combined into a movie to reveal changes in all four axes, x, y, z, and time (t).

VIII. Alternatives to Confocal Microscopy

A. Deconvolution

When an image is acquired with a conventional microscope, the microscope slightly degrades or blurs the image in a predictable way. This is exaggerated when one captures an image of a rapidly moving cell which may appear as a blurred streak when an image is collected with a video camera. This blurring is referred to as *convolution*. In convolution, every point in the image is replaced by a slightly blurred point referred to as the point-spread function (PSF) (Fig. 10B), and the final image is the sum of all the blurred points. *Deconvolution,* which is a technique that uses algorithms to mathematically calculate and remove the blurring from images, essentially reverses convolution. Deconvolution has been used as an alternative method to confocal microscopy since it uses software to remove out-of-focus information from images, rather than a confocal aperture as in confocal microscopes (Agard and Sedat, 1983; Agard *et al.,* 1989; Holmes *et al.,* 1995; Shaw, 1995; Wang, 1998). Therefore deconvolution is a technique to extract useful information from blurry images.

Deconvolution usually employs a conventional light microscope hooked up to a computerized camera system which is used to capture and digitize the images for later storage in the computer. The camera can be a cooled CCD or video camera. High-quality images work best for the this technique. Deconvolution algorithms, of which there are several to choose from, are then used to remove out-of-focus information from the images. The nearest-neighbor algorithm is commonly used and requires a minimum of three different focal planes or slices (one above, one below, and one at the focal plane of interest). The nearest-neighbor algorithm calculates a characteristic PSF for the lens in the microscope that was used to collect the images. The PSF is then used to deconvolve each image using the images above and below the image being processed. This step identifies the haze which is subtracted from the slice of interest, resulting in a clear, sharp deconvolved image. The PSF is a mathematical term for the point response or impulse of an optical system to a point input. With deconvolution programs, the PSF is calculated using diffraction theory. The shape of the PSF depends on the wavelength of light, NA, and aberrations of the objective lens used during image acquisition, as well as the distance between each pixel within a plane and the distance between acquired image planes, which usually ranges between 0.1 to 10 μm. By knowing the shape of the PSF, the deconvolution program removes excess light from the image plane, resulting in a high-resolution image. In this way, the out-of-focus fluorescence is mathematically calculated and subtracted from the in-focus image. The result is stunning images which can be observed at the Web sites in Section X.C.4.

Deconvolution technique has several advantages. The cost is less than that of a CLSM and the system can be mounted on a conventional light microscope. It uses less light than CLSMs which leads to less photodamage and photobleaching. When one stack of images is generated, it can be deconvolved several different ways without the need to acquire a new stack of images, unlike the CLSM. A disadvantage is that deconvolution is very computer intensive. In the past, inefficient computers and software made the generation of images very time consuming. Now very fast computers and better software make deconvolved image acquisition time rival that of slow-scan CLSMs. Many confocal microscope systems also have a digital deconvolution software package as part of the system to further enhance confocal images, since confocal imaging does not remove all of the out-of-focus information from an image. Several deconvolution programs are commercially available.

B. Multiphoton Microscopy

Multiphoton excitation is a new, laser-based microscopy technique that is an alternative to confocal microscopy (Denk *et al.,* 1990, 1995; Masters *et al.,* 1999; McCarthy, 1999; Piston, 1999; Straub and Hell, 1998). It is based on the principle that two or more photons are better than one to image a fluorescent sample (Potter, 1996). Most fluorochromes used for fluorescence microscopy can be excited by simultaneous absorption of multiple, low-energy photons. This process is based on nonlinear optic effects and results in shorter wavelength emission of the fluorochrome than the excitation wavelengths (Fig. 45).

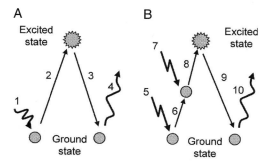

Fig. 45   Diagram of electron events during fluorochrome excitation by one or two photons. One electron in the ground state is shown for two identical fluorochrome molecules. In one-photon fluorescence excitation as performed with a conventional fluorescence or confocal microscope (A), a single photon in the UV or visible range (1) has sufficient energy to excite an electron in the fluorescent molecule from the ground state to an excited state (2). The excited electron decays back to the ground state (3) with emission of a longer wavelength photon (4). In two-photon fluorescence excitation with a multiphoton microscope (B), two photons which individually have insufficient energy are used to cooperatively excite the fluorochrome. As shown in (B), one longer wavelength red or infrared photon (5) partially excites a ground state electron (6). Within femtoseconds to picoseconds, a second red photon (7) excites the electron to a higher energy level (8). The high-energy electron returns to the ground state (9) and emits a shorter wavelength photon (10) than the excitation photons. Thus with two-photon imaging, two low-energy, longer wavelength photons result in emission of one shorter wavelength photon, whereas conventional and confocal fluorescence microscopy use one higher energy, short-wavelength photon to cause a fluorochrome to emit one longer wavelength photon.

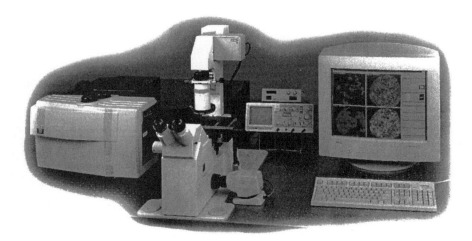

Fig. 46 The Bio-Rad MRC-1024MP multiphoton system. The scan head (left) is mounted on the video port of an inverted microscope and can easily be coupled to other models and types of microscopes. (Courtesy of Bio-Rad Microscopy Division, Hemel Hempstead, UK.)

The multiphoton microscope is similar to a CLSM in that it uses a laser to excite fluorescence in a sample. In fact, some CLSMs (PCM 2000, MRC-1024, Radiance2000, RTS2000) can be upgraded to a multiphoton microscope (Fig. 46). However, the multiphoton microscope does have important differences as compared to the CLSM. First, multiphoton fluorescence is excited only at a spot within the plane of focus where the photon density is high enough for the nonlinear process to occur (Fig. 47). Two-photon excitation occurs when two photons (approximately twice the wavelength required for

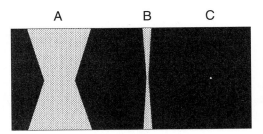

Fig. 47 A comparison of the illumination experienced by the specimen in a conventional fluorescence microscope (A), confocal laser scanning microscope (B), and multiphoton microscope (C). Each was taken at one instant in time using the same oil-immersion objective lens with a high numerical aperture. The whole depth of the fluorescent sample is continuously illuminated with the conventional light microscope (A), whereas a confocal microscope scans the sample with one or more finely focused spots of light that illuminate only a portion of the sample at one time (B). In (A) and much less in (B), photobleaching occurs above and below the plane of focus. In (C) the fluorochrome is excited only at a spot in the thin focal plane by two pulses of near-red or infrared light from a titanium–sapphire pulsed laser.

single-photon excitation) are absorbed almost simultaneously by the fluorochrome. Since the probability of multiphoton excitation depends on the density of the excitation photons, the effective excitation falls off rapidly above and below a defined point in the plane of focus. Thus, out-of-focus light is not produced above and below the plane of focus (Fig. 47). This eliminates the need for a confocal aperture to obtain optical sections. This prevents other focal planes in the sample from being photobleached or photodamaged before they can be imaged (Fig. 48). Since many cellular structures are

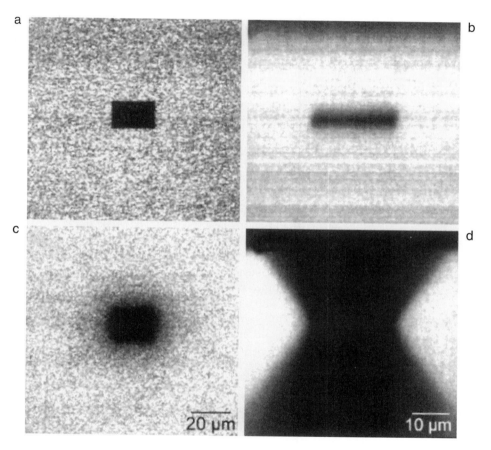

Fig. 48 Photobleaching several microns below the surface of a three-dimensional gel of FITC–dextran induced by two-photon absorption of light at 760 nm from a pulsed titanium–sapphire laser (a and b) or by single-photon absorption at 488 nm (c and d). (a and c) xy optical sections; (b and d) xz optical sections. In two-photon excitation, photobleaching is limited to the illuminated area in the focal plane (b), whereas with single-photon excitation the photobleaching extends in a double cone throughout the specimen (d). Reprinted by permission of the Royal Microscopical Society and Blackwell Science Ltd., from Kubitscheck, U., Tschödrich-Rotter, M., Wedeking, P., and Peters, R. (1996). *J. Microsc.* **182,** 225–233.

less absorbent in the red and infrared part of the spectrum, living samples are relatively unaffected. For example, a multiphoton microscope was used to collect a total of 2160 optical sections of a developing *C. elegans* embryo with no apparent affects on development (Mohler *et al.,* 1998; Mohler and White, 1998). Multiphoton microscopy has also been successfully used to image living hamster embryos over prolonged periods (Squirrell *et al.,* 1999). The result of multiphoton imaging is clear images with improved background discrimination and improved signal strength since the confocal aperture is absent.

A second difference in the multiphoton microscope is that a pulsed, long-wavelength (700–1000 nm) laser is used to excite fluorochromes in the sample that would have been excited by a single photon of a visible or UV laser. Therefore, another advantage of multiphoton microscopy is the ability to excite UV fluorochromes without the need of a UV laser (McCarthy, 1999; Piston, 1999). For example, a fluorochrome that requires UV light (350 nm) to fluoresce could also be excited by two red photons (700 nm) if they reach the fluorochrome at nearly the same time—within about 10^{-12} s (1 picosecond, ps) to 10^{-18} s (1 femtosecond, fs). To obtain the two-photon effect, the density of the long-wavelength photons must be at least 100,000 times greater than in one-photon excitation (Piston, 1999). To achieve this task, a fiber-coupled, mode-locked (pulsed) laser is used to generate rapid pulses of near-infrared or infrared light (for example, 1.2 ps at 82 MHz or 100 fs at 100 MHz). Lasers commonly used are a tunable, titanium–sapphire (Ti–sapphire) laser or a solid state, diode-pumped neodymium–yttrium aluminum garnet (Nd:YAG) laser because of their peak power, pulse width, range of wavelength tunability, cabling, and power requirements. Most lasers used for multiphoton research have 30- to 200-fs pulses, repetition frequencies of 80–100 MHz, and a power range of 30–300 mW. The infrared laser connects directly to the scan head via a fiber optic cable. A tunable Ti–sapphire laser which operates in the range of 690–1000 nm can provide excitation wavelengths of 350–500 nm when used in a multiphoton microscope. This covers a wide range of fluorochromes commonly used in microscopy. Four different fluorochromes have been excited with a multiphoton microscope equipped with one 700-nm laser and the fluorescence emissions were collected into four different channels (Xu *et al.,* 1996). Bleed-through can still be a problem in multiphoton microscopes, so careful selection of filters is needed.

The host microscope for the infrared laser is often modified to efficiently accept the infrared signals. For example, an infrared-enhanced version of a Nikon microscope (TE300DV) has been developed specifically for use with multiphoton systems such as the Radiance2000 MP from Bio-Rad Laboratories. In addition, Olympus has developed a series of long-wavelength transmitting objective lenses for the multiphoton microscope (McCarthy, 1999).

Using an infrared laser coupled to a multiphoton microscope has several advantages especially since infrared light scatters less than visible wavelengths. This allows deeper penetration into thick samples that normally scatter visible light. In addition, because there is no out-of-focus absorption, there is less photodamage and photobleaching (Kubitscheck *et al.,* 1996). This leads to extended sample viability and fewer cytotoxic effects. The excitation and fluorescence emission wavelengths are also well separated. Multiphoton microscopy does have some disadvantages. It cannot be used for reflected light imaging. It is also unsuitable for pigmented cells and tissues that absorb near-infrared light. The laser has a relatively high cost—at least $100,000 per infrared laser alone (McCarthy, 1999). Multiphoton microscopy does not offer better resolution than a confocal microscope (Piston, 1999).

Three-photon excitation works like two-photon excitation except that three photons must interact with the fluorochrome at the same time. Photon density must be 10 times more than two-photon excitation (Piston, 1999). Three-photon lasers can be used to simultaneously excite both UV-absorbing (350 nm) fluorochromes and green-absorbing fluorochromes (525 nm) (Piston, 1999). Four photon-excitation is also possible with the multiphoton microscope (Piston, 1999).

Because of these features, the multiphoton microscope has become a powerful tool to investigate living and very thick specimens such as brain slices and embryos. The increased spatial resolution is important for FRAP, fluorescence resonant energy transfer (FRET), and other micromanipulations. A powerful application of two-photon microscopy is 3-D resolved photorelease of caged compounds (uncaging). In pharmacological studies, multiphoton microscopy can be used to analyze drug uptake without UV-induced autofluorescence. A new real-time multiphoton microscope (Bio-Rad Laboratories, RTS2000) makes these applications easier to perform at video rates (Fan *et al.,* 1999). Multiphoton microscopes are available from Bio-Rad Laboratories [Radiance2000 MP, MRC-1024 MP (Fig. 46), RTS2000], Leica (TCS NT2 MP, TCS SP2 MP), and Carl Zeiss (LSM 510 NLO). See Chapter 3 for an in-depth discussion of two-photon microscopy and its applications.

IX. Conclusions and Future Directions

There is a wide range of confocal applications with fluorescently labeled samples. Many different confocal microscopes have been designed to meet the needs of these applications. Recent improvements in optics, lasers, and computer technology have yielded a new generation of confocal microscopes. What will future confocal microscopes be like? One hope is that as technology improves, the price for the systems will be reduced so that confocal microscopes with many features will be accessible to more users rather than just core facilities. Additional technological advances may also make confocal microscopes better tools in the future. For example, the use of adaptive optics in which a deformable mirror is bent to "de-blur" light (Schilling, 1999a,b) may benefit confocal microscopy. In addition, scientists at the Fraunhofer Institute for Silicon Technology in Itzehoe, Germany, have built a CLSM the size of a fountain pen which can fit into the tip of an endoscope (Hogan, 2000). The aim is to then use the instrument as a diagnostic tool during surgery. Future confocal microscopes may achieve better axial resolution. Improved ways to display 3-D and 4-D data sets especially for publication will greatly enhance our understanding of biological function. Miniaturizing the CLSM may lead to new applications.

X. Resources for Confocal Microscopy

At the time of manuscript submission, all information in Section X was valid; however due to the ever-changing nature of the web, addresses are subject to change without notice.

A. Major Suppliers of Confocal Microscopes

Atto Bioscience
15010 Broschart Road
Rockville, MD 20850
Phone: 800-245-2614
Fax: 301-340-9775
URL: www.atto.com

Bio-Rad Laboratories
Life Science Group
2000 Alfred Nobel Drive
Hercules, CA 94547
Phone: 800-424-6723
Fax: 510-741-5800
URL: www.microscopy.bio-rad.com

Bio-Rad Microscopy Division
Bio-Rad House
Maylands Avenue
Hemel Hempstead
Hertfordshire
HP2 7TD
United Kingdom
Phone: +44 (0)208 328 2000
Fax: +44 (0)208 328 2500
URL: www.microscopy.bio-rad.com

Carl Zeiss, Inc.
One Zeiss Drive
Thornwood, NY 10594
Phone: 800-233-2343
Fax: 914-681-7446
URL: www.zeiss.com

Leica Microsystems Inc.
410 Eagleview Blvd Suite 107
Exton, PA 19341
Phone: 610-321-0460
Fax: 610-321-0426
URL: www.leica.com

Leica Microsystems Heidelberg GmbH
Im Neuenheimer Feld 518
D-69120 Heidelberg, Germany
Phone: +49 6221/41 48 0
Fax: +49 6221/41 48 33
URL: www.leica.com

Nikon Inc.
1300 Walt Whitman Road
Melville, NY 11747-3064
Phone: 631-547-4200
Fax: 631-547-0299
URL: www.nikonusa.com

NORAN Instruments (discontinued confocal sales)
2551 West Beltline Highway
Middleton, WI 53562-2697
Phone: 608-831-6511
Fax: 608-836-7224
URL: www.noran.com

Olympus America Inc.
Two Corporate Center Drive
Melville, NY 11747-3157
Phone: 800-446-5967
Fax: 631-844-5112
URL: www.olympus.com

Optiscan
P.O. Box 1066
Mount Waverly MDC
Victoria, Australia 3149
Phone: 61-3-9538 3333
Fax: 61-3-9562 7742
URL: www.optiscan.com.au

PerkinElmer Life Sciences
9238 Gather Road
Gaithersburg, MD 20877
Phone: 800-638-6692
Fax: 301-963-3200
URL: www.perkinelmer.com

Solamere Technology Group
1427 Perry Avenue
Salt Lake City, UT 84103
Phone: (801) 322-2645
Fax: (801) 322-2645
URL: www.solameretech.com

Technical Instrument San Francisco
1826 Rollins Road
Burlingame, CA 94010
Phone: 650-651-3000
Fax: 650-651-3001
URL: www.techinst.com

B. Suppliers of Antibodies, Fluorescent Dyes, Fluorochrome-Conjugated Antibodies, and Antifade Reagents (Used in This Chapter)

Developmental Studies Hybridoma Bank
Department of Biological Sciences
The University of Iowa
007 Biology Building East
Iowa City, IA 52242
Phone: 319-335-3826
Fax: 319-335-2077
URL: www.uiowa.edu/~dshbwww/index.html

Jackson ImmunoResearch Laboratories, Inc.
P.O. Box 9
872 West Baltimore Pike
West Grove, PA 19390
Phone: 800-367-5296
Fax: 610-869-0171
URL: www.jacksonimmuno.com

Molecular Probes, Inc.
P.O. Box 22010
Eugene, OR 97402-0469
Phone: 800-438-2209
Fax: 800-438-0228
URL: www.probes.com

Roche Molecular Biochemicals
9115 Hague Road
P.O. Box 50414
Indianapolis, IN 46250
Phone: 800-428-5433
Fax: 800-428-2883
URL: www.biochem.roche.com

Sigma Chemical Co.
P.O. Box 14508
St. Louis, MO 63178
Phone: 800-325-3010
Fax: 800-325-5052
URL: www.sigma-aldrich.com

Vector Laboratories, Inc.
30 Inbold Road
Burlingame, CA 94010
Phone: 800-227-6666
Fax: 650-697-0339
URL: www.vectorlabs.com

Zymed Laboratories, Inc.
458 Carlton Court
South San Francisco, CA 94080
Phone: 800-874-4494
Fax: 650-871-4499
URL: www.zymed.com

C. Web Sites for Confocal Microscopy and Related Topics

1. Commercial

- www.mdyn.com/
  Confocal introduction and application notes prepared by Molecular Dynamics (now Amersham Pharmacia Biotech). See technical notes.

- http://fluorescence.bio-rad.com/
  Bio-Rad's interactive fluorochrome database.
- www.science.uwaterloo.ca/physics/research/confocal/main.html
  Scanning Laser Microscopy Lab at the University of Waterloo, Canada features
  a confocal transmission microscope and the confocal MACROscope. See also
  Biomedical Photometrics Inc. Web site (www.confocal.com).
- www.lasertec.co.jp
  Lasertec Corporation is a supplier of video-rate confocal laser scanning reflected
  light microscopes (1LM21, 2LM31) that detect only reflected light from the sample
  and are not configured for fluorescence microscopy.

2. Societies

- msa.microscopy.com
  Web site of the Microscopy Society of America with many useful resources and
  links.
- www.rms.org.uk/
  Web site for the Royal Microscopical Society with many useful resources and
  links.

3. Three-Dimensional Reconstruction

- www.vitalimages.com
  Vital Images, Inc. 3-D reconstruction software, formerly VoxelView; now Vitrea 2.
- rsb.info.nih.gov/nih-image/
  NIH Image, 3-D image processing and analysis software for the Macintosh. A PC
  version is also available.
- http://biocomp.stanford.edu/3dreconstruction/index.html
  National Biocomputation Center's 3-D reconstruction home page features an
  introduction, software, movies, and references.

4. Deconvolution

- www.vaytek.com
  Vaytek's deconvolution software, VoxBlast and MicroTome.
- www.aqi.com
  AutoQuant Imaging, Inc., 3-D blind deconvolution software, AutoDeblur, and
  image processing/3-D rendering software, AutoVisualize-3D.
- www.scanalytics.com
  Scanalytics, Inc. software for the acquisition, processing and rendering of 3-D
  images; includes 3-D deconvolution and visualization software.

- www.api.com
  Applied Precision, Inc. deconvolution software and DeltaVision Restoration Microscope.
- www.msg.ucsf.edu/sedat/sedat.html
  Dr. John Sedat's laboratory home page features many movie clips aimed at understanding nuclear architecture and chromosome organization. Techniques include 3-D microscopy and computerized wide-field deconvolution microscopy for multiwavelength fluorescence imaging of living cells.

5. Multiphoton Microscopy

See confocal company Web sites in Section X.A.

6. Researchers and Applications

- www.cs.ubc.ca/spider/ladic/confocal.html
  Dr. Lance Ladic's confocal microscopy Web site with extensive resources from specimen preparation to volume visualization. Has many links including one to Confocal Assistant software.
- www.ou.edu/research/electron/www-vl/
  The WWW-Virtual Library: Microscopy Web site sponsored by the Samuel Roberts Electron Microscopy Laboratory and University of Oklahoma covers many aspects and resources of light and electron microscopy including confocal microscopy.
- www.nrcam.uchc.edu
  The National Resource for Cell Analysis and Modeling couples computational cell biology with high-resolution light microscopy to simulate specific cellular processes. Includes the Virtual Cell program and movies which model cell biological processes.
- www.cyto.purdue.edu
  Purdue University Cytometry Laboratories' site for many microscopy resources including the "Microscopy Image Analysis & 3-D Reconstruction" CD-ROM, and confocal fluorescence and reflection imaging.
- http://swehsc.pharmacy.arizona.edu/exppath/index.html
  Dr. Douglas Cromey's confocal microscopy resources on the World Wide Web.
- http://froglab.biology.utah.edu
  Dr. David Gard's home page at the University of Utah shows confocal imaging of amphibian oocytes and embryos.
- http://web.uvic.ca/ail/
  Advanced Imaging Laboratory of the Biology Department at the University of Victoria, British Columbia.

- http://chem.csustan.edu/confocal
  CSUPERB Confocal Microscope Core Facility of California State University which provides on-site and remote access to a CLSM for faculty and students.
- www.loci.wisc.edu
  Laboratory of Optical and Computational Instrumentation developed by Drs. John White and Jeff Walker at the University of Wisconsin at Madison has information on biophotonics, optical instrumentation, confocal and multiphoton imaging, and 4-D microscopy.
- www.biology.ucsc.edu/people/sullivan/
  Dr. Bill Sullivan's Lab movie page at the University of California, Santa Cruz, shows confocal movie clips of living *Drosophila* embryos injected with either fluorescently labeled proteins or GFP to show cellular dynamics during development.
- http://155.37.3.143/panda
  Dr. Mark Terasaki's home page at the University of Connecticut Health Center shows confocal movie clips of cellular dynamics during fertilization and development.
- www.neuro.uoregon.edu/doelab/doelab.html
  Dr. Chris Doe's laboratory home page at the University of Oregon shows confocal cell lineage movies of parental neuroblasts in *Drosophila* embryos.

D. Abbreviations Used in This Chapter

2-D	Two-dimensional
3-D	Three-dimensional
4-D	Four-dimensional
AOD	Acoustooptical deflector
AOTF	Acoustooptical tunable filter
CCD	Charge-coupled device
CLSM	Confocal laser scanning microscope
DABCO	1,4-Diazabicyclo[2.2.2]octane
FRAP	Fluorescence recovery after photobleaching
FRET	Fluorescence resonant energy transfer
fs	Femtosecond
GFP	Green fluorescent protein
ICU	Instrument control unit
MB	Megabyte
NA	Numerical aperture
PBS	Phosphate-buffered saline, pH 7.2
PMT	Photomultiplier tube
ps	Picosecond
PSF	Point-spread function
RB	Red, blue
RG	Red, green
RGB	Red, green, blue
RISC	Reduced instruction set computing
ROI	Region of interest
TSM	Tandem scanning confocal microscope

Acknowledgments

The authors thank Dr. Brian Matsumoto for his technical assistance in obtaining optical sections with the MRC-1024 and his many stimulating discussions, as well as for critically reading the manuscript. The authors thank Dr. Gerald Schatten for introducing us to confocal microscopy. The authors gratefully acknowledge the figures and helpful discussions by Bio-Rad, Carl Zeiss, Leica, Nikon, NORAN, Olympus, and PerkinElmer.

References

Agard, D. A., and Sedat, J. W. (1983). Three-dimensional architecture of a polytene nucleus. *Nature (London)* **302,** 676–681.

Agard, D. A., Hiraoka, Y., Shaw, P. J., and Sedat, J. W. (1989). Fluorescence microscopy in three dimensions. *In* "Methods in Cell Biology" (D. L. Taylor and Y.-L. Wang, eds.), Vol. 30, pp. 353–378. Academic Press, San Diego.

Amos, W. B., White, J. G., and Fordham, M. (1987). Use of confocal imaging in the study of biological structures. *Appl. Opt.* **26,** 3239–3243.

Art, J. J., and Goodman, M. B. (1993). Rapid scanning confocal microscopy. *In* "Methods in Cell Biology" (B. Matsumoto, ed.), Vol. 38, pp. 47–77. Academic Press, San Diego.

Berrios, M., Conlon, K. A., and Colflesh, D. E. (1999). Antifading agents for confocal fluorescence microscopy. *In* "Methods in Enzymology" (P. M. Conn, ed.), Vol. 307, pp. 55–79. Academic Press, San Diego.

Bkaily, G., Jacques, D., and Pothier, P. (1999). Use of confocal microscopy to investigate cell structure and function. *In* "Methods in Enzymology" (P. M. Conn, ed.), Vol. 307, pp. 119–135. Academic Press, San Diego.

Blatter, L. A. (1999). Cell volume measurements by fluorescence confocal microscopy: Theoretical and practical aspects. *In* "Methods in Enzymology" (P. M. Conn, ed.), Vol. 307, pp. 274–295. Academic Press, San Diego.

Bowtell, D. D. L. (1999). Options available—from start to finish—for obtaining expression data by microarray. *Nat. Genet. Suppl.* **21,** 25–32.

Boyde, A. (1985). Stereoscopic images in confocal (tandem scanning) microscopy. *Science* **230,** 1270–1272.

Boyde, A. (1990). Real-time direct-view confocal light microscopy. *In* "Electronic Light Microscopy: Techniques in Modern Biomedical Microscopy" (D. M. Shotton, ed.), pp. 289–314. Wiley-Liss, New York.

Boyde, A., Jones, S. J., Taylor, M. L., Wolfe, A., and Watson, T. F. (1990). Fluorescence in the tandem scanning microscope. *J. Microsc. (Oxford)* **157,** 39–49.

Brakenhoff, G. J. (1979). Imaging modes in confocal scanning light microscopy (CSLM). *J. Microsc. (Oxford)* **117,** 233–242.

Brelje, T. C., Wessendorf, M. W., and Sorenson, R. L. (1993). Multicolor laser scanning confocal immunofluorescence microscopy: Practical application and limitations. *In* "Methods in Cell Biology" (B. Matsumoto, ed.), Vol. 38, pp. 97–181. Academic Press, San Diego.

Brotchie, D., Roberts, N., Birch, M., Hogg, P., Howard, C. V., and Grierson, I. (1999). Characterization of ocular cellular and extracellular structures using confocal microscopy and computerized three-dimensional reconstruction. *In* "Methods in Enzymology" (P. M. Conn, ed.), Vol. 307, pp. 496–513. Academic Press, San Diego.

Calloway, C. B. (1999). A confocal microscope with spectrophotometric detection. *Microsc. Microanal.* **5 (suppl. 2: Proceedings),** 460–461.

Carlsson, K., Danielsson, P. E., Lenz, R., Liljeborg, A., Majlöf, L., and Aslund, N. (1985). Three-dimensional microscopy using a confocal scanning laser microscope. *Optics Lett.* **10,** 53–55.

Chen, H., Swedlow, J., Grote, M., Sedat, J. W., and Agard, D. A. (1995). The collection, processing and display of digital three-dimensional images of biological specimens. *In* "Handbook of Biological Confocal Microscopy" (J. B. Pawley, ed.), 2nd ed., pp. 197–210. Plenum Press, New York.

Cheng, P. C., and Kriete, A. (1995). Image contrast in confocal light microscopy. *In* "Handbook of Biological Confocal Microscopy." (J. B. Pawley, ed.), 2nd ed., pp. 281–310. Plenum Press, New York.

Cheung, V. G., Morley, M., Aguilar, F., Massimi, A., Kucherlapati, R., and Childs, G. (1999). Making and reading microarrays. *Nat. Genet. Suppl.* **21,** 15–19.

Cogswell, C. J., and Sheppard, C. J. R. (1990). Confocal brightfield imaging techniques using an on-axis scanning optical microscope. *In* "Confocal Microscopy" (T. Wilson, ed.), pp. 213–243. Academic Press, London.

Cox, G. (1999). Equipment for mass storage and processing of data. *In* "Methods in Enzymology" (P. M. Conn, ed.), Vol. 307, pp. 29–55. Academic Press, San Diego.

Davidovits, P., and Egger, M. D. (1971). Scanning laser microscope for biological investigations. *Appl. Optics* **10,** 1615–1619.

Davidovits, P., and Egger, M. D. (1972). Scanning Optical Microscope. U.S. Patent No. 3,643,015

Davidovits, P., and Egger, M. D. (1973). Photomicrography of corneal endothelial cells in vivo. *Nature* **244,** 366–367.

Denk, W., Strickler, J. H., and Webb, W. W. (1990). Two-photon laser-scanning fluorescence microscopy. *Science* **248,** 73–76.

Denk, W., Piston, D. W., and Webb, W. W. (1995). Two-photon molecular excitation in laser-scanning microscopy. *In* "Handbook of Biological Confocal Microscopy" (J. B. Pawley, ed.), 2nd ed., pp. 445–458. Plenum Press, New York.

DeRisi, J., Penland, L., Brown, P. O., Bittner, M. L., Meltzer, P. S., Ray, M., Chen, Y., Su, Y. A., and Trent, J. M. (1996). Use of a cDNA microarray to analyze gene expression patterns in human cancer. *Nat. Genet.* **14,** 457–460.

Draaijer, A., and Houpt, P. M. (1988). A standard video-rate confocal laser-scanning reflection and fluorescence microscope. *Scanning* **10,** 139–145.

Draaijer, A., and Houpt, P. M. (1993). High scan-rate confocal laser scanning microscopy. *In* "Electronic Light Microscopy: Techniques in Modern Biomedical Microscopy" (D. M. Shotton, ed.), pp. 273–288. Wiley-Liss, New York.

Dunn, K. W., and Wang, E. (2000). Optical aberrations and objective choice in multicolor confocal microscopy. *BioTechniques* **28,** 542–550.

Egger, M. D. (1989). The development of confocal microscopy. *Trends Biochem. Sci.* **12,** 11.

Egger, M. D., and Petráň, M. (1967). New reflected light microscope for viewing unstained brain and ganglion cells. *Science* **157,** 305–307.

Fan, G. Y., Fujisaki, H., Miyawaki, A., Tsay, R. K., Tsien, R. Y., and Ellisman, M. H. (1999). Video-rate scanning two-photon excitation fluorescence microscopy and ratio imaging with cameleons. *Biophysic. J.* **76,** 2412–2420.

Genka, C., Ishida, H., Ichimori, K., Hirota, Y., Tanaami, T., and Nakazawa, H. (1999). Visualization of biphasic Ca^{2+} diffusion from cytosol to nucleus in contracting adult rat cardiac myocytes with an ultra-fast confocal imaging system. *Cell Calcium* **25,** 199–208.

Giloh, H., and Sedat, J. W. (1982). Fluorescence microscopy: Reduced photobleaching of rhodamine and fluorescein protein conjugates by *n*-propyl gallate. *Science* **217,** 1252–1255.

Golub, T. R., Slonim, D. K., Tamayo, P., Huard, C., Gaasenbeek, M., Mesirov, J. P., Coller, J., Loh, M. L., Downing, J. R., Caligiuri, M. A., Bloomfield, C. D., and Lander, E. S. (1999). Molecular classification of cancer: Class discovery and prediction by gene expression monitoring. *Science* **286,** 531–537.

Gratton, E., and vandeVen, M. J. (1995). Laser sources for confocal microscopy. *In* "Handbook of Biological Confocal Microscopy" (J. B. Pawley, ed.), 2nd ed., pp. 69–109. Plenum Press, New York.

Hale, I. L., and Matsumoto, B. (1993). Resolution of subcellular detail in thick tissue sections: Immunohistochemical preparation and fluorescence confocal microscopy. *In* "Methods in Cell Biology" (B. Matsumoto, ed.), Vol. 38, pp. 289–324. Academic Press, San Diego.

Hama, T., Takahashi, A., Ichihara, A., and Takamatsu, T. (1998). Real time *in situ* confocal imaging of calcium wave in the perfused whole heart of the rat. *Cell. Signal.* **10,** 331–337.

Harris, K. M., and Stevens, J. K. (1989). Dendritic spines of CA 1 pyramidal cells in the rat hippocampus: Serial electron microscopy with reference to their biophysical characteristics. *J. Neurosci.* **9,** 2982–2997.

Haugland, R. P. (1990). Fluorescein substitutes for microscopy and imaging. *In* "Optical Microscopy for Biology" (B. Herman and K. Jacobson, eds.), pp. 143–157. Alan R. Liss, New York.

Haugland, R. P. (1999). "Handbook of Fluorescent Probes and Research Chemicals," 7th ed. Molecular Probes, Inc., Eugene, OR.

Hogan, H. (2000). Miniature microscope clears things up. *Biophotonics Int.* **April/May**, 20.

Holmes, T. J., Bhattacharya, S., Cooper, J. A., Hanzel, D., Krishnamurthi, V., Lin, W-C, Roysam, B., Szarowski, D. H., and Turner, J. N. (1995). Light microscopic images reconstructed by maximum likelihood deconvolution. *In* "Handbook of Biological Confocal Microscopy" (J. B. Pawley, ed.), 2nd ed., pp. 389–402. Plenum Press, New York.

Hoyt, C. (1996). Liquid crystal filters clear the way for imaging multiprobe fluorescence. *Biophotonics Int.* **July/Aug,** 49–51.

Ichihara, A., Tanaami, T., Isozaki, K., Sugiyama, Y., Kosugi, Y., Mikuriya, K., Abe, M., and Uemura, I. (1996). High-speed confocal fluorescence microscopy using a Nipkow scanner with microlenses for 3-D imaging of single fluorescent molecule in real-time. *Bioimages* **4,** 57–62.

Inoué, S. (1995). Foundations of confocal scanned imaging in light microscopy. *In* "Handbook of Biological Confocal Microscopy" (J. B. Pawley, ed.), 2nd ed., pp. 1–17.

Inoué, S., and Spring, K. R. (1997). "Video Microscopy, the Fundamentals," 2nd ed. Plenum Press, New York.

Jester, J. V., Petroll, W. M., and Cavanagh, H. D. (1999). Measurement of tissue thickness using confocal microscopy. *In* "Methods in Enzymology" (P. M. Conn, ed.), Vol. 307, pp. 230–245. Academic Press, San Diego.

Johnson, G. D., and de C Nogueira Arajo, G. M. (1981). A simple method of reducing the fading of immunofluorescence during microscopy. *J. Immunol. Methods* **43,** 349–350.

Johnson, G. D., Davidson, R. S., McNamee, K. C., Russel, G., Goodwin, D., and Holbrow, E. J. (1982). Fading of immunofluorescence during microscopy: A study of the phenomenon and its remedy. *J. Immunol. Methods* **55,** 231–242.

Kawasaki, E., Schermer, M., and Zeleny, R. (1999). Confocal laser scanners for gene analysis. *Biophotonics Int.* **March/April,** 47–49.

Keller, H. E. (1995). Objective lenses for confocal microscopy. *In* "Handbook of Biological Confocal Microscopy" (J. B. Pawley, ed.), 2nd ed., pp. 111–126. Plenum Press, New York.

Kino, G. S. (1989). Efficiency in Nipkow disk microscopes. *In* "The Handbook of Biological Confocal Microscopy" (J. Pawley, ed.), pp. 93–97. IMR Press, Madison, WI.

Kino, G. S. (1995). Intermediate optics in Nipkow disk microscopes. *In* "Handbook of Biological Confocal Microscopy" (J. B. Pawley, ed.), 2nd ed., pp. 155–165. Plenum Press, New York.

Kino, G. S., and Xiao, G. Q. (1990). Real-time scanning optical microscopes. *In* "Confocal Microscopy" (T. Wilson, ed.), pp. 361–387. Academic Press, London.

Kong, S. K., Ko, S., Lee, C. Y., and Lui, P. Y. (1999). Practical considerations in acquiring biological signals from confocal microscope. *In* "Methods in Enzymology" (P. M. Conn, ed.), Vol. 307, pp. 20–26. Academic Press, San Diego.

Kubitscheck, U., Tschödrich-Rotter, M., Wedekind, P., and Peters, R. (1996). Two-photon scanning microphotolysis for three-dimensional data storage and biological transport measurements. *J. Microsc.* **182,** 225–233.

Langanger, G., De Mey, J., and Adam, H. (1983). 1,4-Diazobicycle-(2,2,2)-octane (DABCO) retards the fading of immunofluorescence preparations. *Mikroskopie* **40,** 237–241.

Lichtman, J. W., Sunderland, W. J., and Wilkinson, R. S. (1989). High-resolution imaging of synaptic structure with a simple confocal microscope. *New Biol.* **1,** 75–82.

Longin, A., Souchier, C., French, M., and Byron, P. A. (1993). Comparison of anti-fading agents used in fluorescence microscopy: Image analysis and laser confocal microscopy study. *J. Histochem. Cytochem.* **41,** 1833–1840.

Lui, P. P., Lee, M. M., Ko, S., Lee, C. Y., and Kong, S. K. (1997). Practical considerations in acquiring biological signals from confocal microscope. II. Laser-induced rise of fluorescence and effect of agonist droplet application. *Biol. Signals* **6,** 45–51.

Majlof, L., and Forsgren, P.-O. (1993). Confocal microscopy: Important considerations for accurate imaging. *In* "Methods in Cell Biology" (B. Matsumoto, ed.), Vol. 38, pp. 79–95. Academic Press, San Diego.

Masters, B. R., So, P. T. C., Kim, K. H., Buehler, C., and Gratton, E. (1999). Multiphoton excitation microscopy, confocal microscopy, and spectroscopy of living cells and tissues; functional metabolic imaging of human skin *in vivo*. *In* "Methods in Enzymology" (P. M. Conn, ed.), Vol. 307, pp. 513–536. Academic Press, San Diego.

McCarthy, D. C. (1999). A start on parts for multiphoton microscopy. *Photonics Spectra* **November,** 80–88.

Minsky, M. (1957). Microscopy Apparatus. U.S. Patent No. 3,013,467

Minsky, M. (1988). Memoir on inventing the confocal scanning microscope. *Scanning* **10,** 128–138.

Mohler, W. A., and White, J. G. (1998). Stereo-3-D reconstruction and animation from living fluorescent specimens. *BioTechniques* **24,** 1006–1012.

Mohler, W. A., Simske, J. S., Williams-Masson, E. M., Hardin, J. D., and White, J. G. (1998). Dynamics and ultrastructure of developmental cell fusions in *Caenorhabditis elegans* hypodermis. *Curr. Biol.* **8,** 1087–1090.

Newman, E. A., and Zahs, K. R. (1997). Calcium waves in retinal glial cells. *Science* **275,** 844–849.

Ohta, H., Yamamoto, M., Ujike, Y., Rie, G., and Momose, K. (1999). Confocal imaging analysis of intracellular ions in mixed cellular systems or *in situ* using two types of confocal microscopic systems. *In* "Methods in Enzymology" (P. M. Conn, ed.), Vol. 307, pp. 425–441. Academic Press, San Diego.

Paddock, S. W. (1989). Tandem scanning reflected-light microscopy of cell–substratum adhesions and stress fibers in Swiss 3T3 cells. *J. Cell Sci.* **93,** 143–146.

Paddock, S., DeVries, P., Holy, J., and Schatten, G. (1990). On laser scanning confocal microscopy and three dimensional volume rendering of biological structures. *Proc. Soc. Photo-Opt. Instrum. Eng.* **1205,** 20–28.

Petráň, M., Hadravský, M., Egger, D., and Galambos, R. (1968). Tandem-scanning reflected-light microscope. *J. Opt. Soc. Am.* **58,** 661–664.

Petráň, M., Hadravský, M., Benes, J., Kucera, R., and Boyde, A. (1985). The tandem scanning reflected light microscope (TSRLM). *Proc. R. Microsc. Soc.* **20,** 125–139.

Petráň, M., Boyde, A., and Hadravský, M. (1990). Direct view confocal microscopy. *In* "Confocal Microscopy" (T. Wilson, ed.), pp. 245–283. Academic Press, London.

Piston, D. W. (1999). Imaging living cells and tissues by two-photon excitation microscopy. *Trends Cell Biol.* **9,** 66–69.

Potter, S. M. (1996). Vital imaging: Two photons are better than one. *Curr. Biol.* **6,** 1595–1598.

Schilling, G. (1999a). Giant eyes on the sky. *Astronomy* **December,** 49–51.

Schilling, G. (1999b). Techniques for unblurring the stars come of age. *Science* **286,** 1504–1506.

Schmid, A., Chiba, A., and Doe, C. Q. (1999). Clonal analysis of *Drosophila* embryonic neuroblasts: neural cell types, axon projections and muscle targets. *Development* **126,** 4653–4689.

Shaw, P. J. (1995). Comparison of wide-field/deconvolution and confocal microscopy for 3D imaging. *In* "Handbook of Biological Confocal Microscopy" (J. B. Pawley, ed.), 2nd ed., pp. 373–387. Plenum Press, New York.

Sheppard, C. J. R., and Choudhury, A. (1977). Image formation in the scanning microscope. *Opt. Acta* **24,** 1051–1073.

Sheppard, C. J. R., and Shotton, D. M. (1997). "Confocal Laser Scanning Microscopy." Bios Scientific Publishers, Oxford, UK.

Shotton, D. M. (1989). Confocal scanning optical microscopy and its applications for biological specimens. *J. Cell Sci.* **94,** 175–206.

Shuman, H., Murray, J. M., and DiLullo, C. (1989). Confocal microscopy: An overview. *BioTechniques* **7,** 154–163.

Snyman, C. J., Raidoo, D. M., and Bhoola, K. D. (1999). Localization of proteases and peptide receptors by confocal microscopy. *In* "Methods in Enzymology" (P. M. Conn, ed.), Vol. 307, pp. 368–394. Academic Press, San Diego.

Squirrell, J. M., Wokosin, D. L., White, J. G., and Bavister, B. D. (1999). Long-term two-photon fluorescence imaging of mammalian embryos without compromising viability. *Nat. Biotechnol.* **17,** 763–767.

Straub, M., and Hell, S. W. (1998). Multifocal multiphoton microscopy: A fast and efficient tool for 3-D fluorescence imaging. *Bioimaging* **6,** 177–185.

Stricker, S. A., Centonze, V. E., Paddock, S. W., and Schatten, G. (1992). Confocal microscopy of fertilization-induced calcium dynamics in sea urchin eggs. *Dev. Biol.* **149,** 370–380.

Summers, R. G., Stricker, S. A., and Cameron, R. A. (1993). Applications of confocal microscopy to studies of sea urchin embryogenesis. *In* "Methods in Cell Biology" (B. Matsumoto, ed.), Vol. 38, pp. 265–287. Academic Press, San Diego.

Tsien, R. Y., and Bacskai, B. J. (1995). Video-rate confocal microscopy. *In* "Handbook of Biological Confocal Microscopy" (J. B. Pawley, ed.), 2nd ed., pp. 449–478. Plenum Press, New York.

Van Zandt, W., and Argiro, V. (1989). A new "inlook" on life. *Unix Rev.* **7,** 52–57.

Wang, Y.-L. (1998). Digital deconvolution and fluorescence images for biologists. *In* "Methods in Cell Biology" (G. Sluder and D. E. Wolf, eds.), Vol. 56, pp. 305–315. Academic Press, San Diego.

Webb, R. H., and Dorey, C. K. (1995). The pixilated image. *In* "Handbook of Biological Confocal Microscopy" (J. B. Pawley, ed.), 2nd ed., pp. 55–67. Plenum Press, New York.

Wei, X., Somanathan, S., Samarabandu, J., and Berezney, R. (1999). Three-dimensional visualization of transcription sites and their association with splicing factor-rich nuclear speckles. *J. Cell Biol.* **146,** 543–558.

White, J. G., Amos, W. B., and Fordham, M. (1987). An evaluation of confocal *versus* conventional imaging of biological structures by fluorescence light microscopy. *J. Cell Biol.* **105,** 41–48.

White, N. S. (1995). Visualization systems for multidimensional CLSM images. *In* "Handbook of Biological Confocal Microscopy" (J. B. Pawley, ed.), 2nd ed., pp. 211–254. Plenum Press, New York.

Wijnaendts van Resandt, R. W., Marsman, H. J. B., Kaplan, R., Davoust, J., Stelzer, E. H. K., and Stricker, R. (1985). Optical fluorescence microscopy in 3 dimensions: Microtomoscopy. *J. Microsc. (Oxford)* **138,** 29–34.

Wilke, V. (1985). Optical scanning microscopy—the laser scan microscope. *Scanning* **7,** 88–96.

Williams, D. A., Bowser, D. N., and Petrou, S. (1999). Confocal Ca^{2+} imaging of organelles, cells, tissues, and organs. *In* "Methods in Enzymology" (P. M. Conn, ed.), Vol. 307, pp. 441–469. Academic Press, San Diego.

Wilson, T., and Sheppard, C. J. R. (1984). "Theory and Practice of Scanning Optical Microscopy." Academic Press, New York.

Wright, S. J., and Schatten, G. (1991). Confocal fluorescence microscopy and three-dimensional reconstruction. *J. Electron Microsc. Tech.* **18,** 2–10.

Wright, S. J., Walker, J. S., Schatten, H., Simerly, C., McCarthy, J. J., and Schatten, G. (1989). Confocal fluorescence microscopy with the tandem scanning light microscope. *J. Cell Sci.* **94,** 617–624.

Wright, S. J., Schatten, H., Simerly, C., and Schatten, G. (1990). Three-dimensional fluorescence imaging with the tandem scanning confocal microscope. *In* "Optical Microscopy for Biology" (B. Herman and K. Jacobson, eds.), pp. 29–43. Alan R. Liss, New York.

Wright, S. J., Centonze, V. E., Stricker, S. A., DeVries, P. J., Paddock, S. W., and Schatten, G. (1993). Introduction to confocal microscopy and three-dimensional reconstruction. *In* "Methods in Cell Biology" (B. Matsumoto, ed.), Vol. 38, pp. 1–45. Academic Press, San Diego.

Xiao, G. Q., and Kino, G. S. (1987). A real-time confocal scanning optical microscope. *Proc. Soc. Photo-Opt. Instrum. Eng.* **809,** 107–113.

Xiao, G. Q., Corle, T. T., and Kino, G. S. (1988). Real-time confocal scanning optical microscope. *Appl. Phys. Lett.* **53,** 716–718.

Xu, C., Zipfel, W., Shear, J. B., Williams, R. M., and Webb, W. W. (1996). Multiphoton fluorescence excitation: New spectral windows for biological nonlinear microscopy. *Proc. Natl. Acad. Sci. USA* **93,** 10763–10768.

Yamaguchi, K. (1999). Observation of microcirculatory kinetics by real-time confocal laser scanning microscopy. *In* "Methods in Enzymology" (P. M. Conn, ed.), Vol. 307, pp. 394–422. Academic Press, San Diego.

Yamaguchi, K., Nishio, K., Sato, N., Tsumura, H., Ichihara, A., Kudo, H., Aoki, T., Naoki, K., Suzuki, K., Miyata, A., Suzuki, Y., and Morooka, S. (1997). Leukocyte kinetics in the pulmonary microcirculation: Observations using real-time confocal luminescence microscopy coupled with high-speed video analysis. *Lab. Invest.* **76,** 809–822.

Zalensky, A. O., Allen, M. J., Kobayashi, A., Zalenskaya, A., Balhorn, R., and Bradbury, E. M. (1995). Well-defined genome architecture in the human sperm nucleus. *Chromosoma* **103,** 577–590.

CHAPTER 2

Direct-View High-Speed Confocal Scanner: The CSU-10

Shinya Inoué* and Ted Inoué

Marine Biological Laboratory
Woods Hole, Massachusetts

Universal Imaging Corporation
West Chester, Pennsylvania

*This author is a consultant to Yokogawa Electric Corporation.

METHODS IN CELL BIOLOGY, VOL. 70
Copyright 2002, Elsevier Science (USA). All rights reserved.
0091-679X/02 $35.00

I. Introduction

Over the past decade, confocal microscopes have dramatically improved our ability to examine the structural and functional detail of biological tissues and cells. The same characteristics that made these advances possible, namely the ability of confocal microscopes to provide exceptionally clean serial optical sections that are free of out-of-focus flare, have also made it possible to directly generate tilted sections, as well as to generate striking three-dimensional (3-D) images.

Alternative methods, such as the application of computationally intensive deconvolution algorithms to serial sections (obtained with nonconfocal, or wide-field, fluorescence microscopes) can also yield clean optical sections and reconstructed 3-D images (e.g., Agard, 1984; Holmes *et al.*, 1995).

More recently, two- (and multi-) photon optics have been used to further improve optical sectioning, especially for deep, living tissues. With only the fluorophores lying in the focal plane being exposed to the coincident action of the double- (multiple-) wavelength excitation wave, fluorescence is strictly limited to that plane. Meanwhile, the longer wavelength excitation wave dramatically reduces light scattering in general and photobleaching of the fluorophore that lie outside of the focal plane.

Although many of these confocal and deconvolution approaches yield exquisite optical sections and their composites, it may take from many seconds to considerably longer for each two-dimensional image covering a reasonable area to become available.

The CSU-10 described in this chapter is a disk-scanning, direct-view confocal scanning unit.[1] As detailed in the next section, the CSU-10 incorporates, in addition to the main Nipkow disk, a second disk with some 20,000 microlenses that are each aligned with a corresponding pinhole on the main Nipkow disk, thus substantially improving the transmission of the confocal illuminating beam. Thus, one can view confocal fluorescence images in real time, i.e., at video rate or faster, through the eyepiece or captured through a video camera.

The compact unit, attached to an upright or inverted microscope, transforms a research microscope into an exceptionally easy-to-use and effective, direct-view epifluorescence confocal microscope (Fig. 1). Thus, optical sections of fluorescent specimens show with high resolution, and in true color, directly through the ocular as one focuses through the specimen. Sparsely distributed, weakly fluorescent objects are readily brought to view.

In addition to viewing the image through the ocular in real time, the confocal image generated by the CSU-10 can be directly captured by a photographic camera or a video or CCD camera attached to the C-mount on top of the unit.

Since images of the microscope field are scanned at 360 frames/s by the multiple arrays of pinholes on its Nipkow disk, the CSU-10 not only provides very clean, full-frame video-rate images but, by use of high-speed intensified cameras, can even capture frames in as short an interval as 3 ms or less.

In this paper prepared in 1999, we describe the basic design of the CSU-10; some sample applications including video-rate and higher speed confocal imaging; the low

---

[1] See Chapter 1 for an overview of Nipkow-disk-type confocal microscopes and other video-rate, direct-view confocal systems. See also Amos and White (1995), Juškaitis *et al.* (1996), Lanni and Wilson (1999), and Yuste *et al.* (1999).

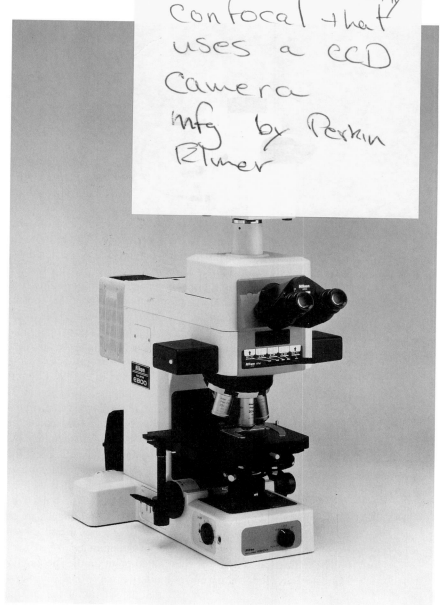

Fig. 1 The CSU-10 confocal scanning unit attached to the C-mount video port on top of an upright microscope (Nikon E-800). The scanner can also be attached to the video port of an inverted microscope. Illumination is conducted through a polarization-maintaining optical fiber from the laser source (not shown). (Figure courtesy of Yokogawa Electric Corporation.)

rate of fluorescence bleaching observed using the CSU-10; rapid digital processing for further haze removal and image sharpening and for generation of dynamic stereo images; mechanical and optical performance of the CSU-10; and potential future applications. Addendum A provides some recent updates.

II. Overview of Confocal Microscopes

In a point-scanning confocal microscope, the specimen is illuminated through a well-corrected objective lens by an intense, reduced image of a point source that is located in the image plane (or its conjugate) of the objective lens. Light emitted by the focused point on the specimen traces the light path back through the objective lens and (after deviation through a dichromatic mirror) progresses to an exit pinhole. The small exit pinhole, which is also placed in the image plane of the objective lens, effectively excludes light emanating from planes in the specimen other than those in focus. Thus, confocal imaging selectively collects signals from the focused spot in the specimen with dramatic reduction of signals from out-of-focus planes.

To achieve an image of the specimen with a point-scanning confocal system, either the specimen itself is moved in a raster pattern or the confocal spot and exit pinhole together are made to scan the specimen. In other words, either the specimen is precisely raster scanned in the $x-y$ plane, or the illuminating and return imaging beam are made to raster scan a stationary specimen by tilting these beams at the aperture plane of the objective lens (commonly with galvanometer-driven mirrors) (see, e.g., this volume, Chapter 1, and Pawley, 1995). The very rapidly changing intensity of light passing the exit pinhole is detected by a photomultiplier tube and captured in a digital frame buffer. The output of the frame buffer, the confocal image, is displayed on a computer monitor.

For a number of reasons, it generally requires a few seconds to generate a low-noise, full-screen confocal fluorescent image with a point-scanning confocal system (see, e.g., Amos and White, 1995). One way of overcoming this speed limitation is to use many points to scan the specimen in parallel, such as by the use of a Nipkow disk.

In Nipkow-disk-type confocal microscopes, the disk with multiple sets of spirally arranged pinholes is placed in the image plane of the objective lens. The pinholes are illuminated from the rear, and their highly reduced images are focused by the objective lens onto the specimen. As the Nipkow disk spins, the specimen is thus raster scanned by successive sets of reduced images of the pinholes.

The light emitted by each illuminated point on the specimen is focused back, by the same objective lens, onto a corresponding pinhole on the Nipkow disk. This exit pinhole may be the same pinhole that provided the scanning spot, as in the Kino-type confocal system (Kino, 1995), or may be one located on the diametrically opposite side of the Nipkow disk as in the Petráň-type confocal microscope (Petráň et al., 1968; see this volume, Chapter 1, Figs. 8 and 9).

With any Nipkow-type confocal system, one needs to maintain a moderately large separation between adjacent pinholes relative to their diameters in order to minimize crosstalk (i.e., leakage) of the return beam through neighboring pinholes. On the other hand, with the pinholes separated by, say, 10 times their diameter, only 1% of the incoming beam is transmitted by the Nipkow disk since the pinholes occupy only that fraction of the disk area. Nipkow disk systems, therefore, generally tend to suffer from low levels of transmission of the beam that illuminates the specimen.

In addition, a major fraction of the illuminating beam can be backscattered by the Nipkow disk and contribute to unwanted background light in the Kino-type arrangement. The Kino system thus includes several design features to minimize this source of unwanted light (see, e.g., Kino, 1995). The Petráň, tandem-scanning-type does not suffer from backscatter of the illuminating beam but does require very high precision of pinhole placement on the Nipkow disk as well as stability of rotational axis since the entrance and exit pinholes are located on opposite sides of the axis of rotation of the disk.

III. Design of the CSU–10

In order to overcome these difficulties encountered with the past Nipkow disk systems, the Yokogawa CSU-10 uses the same pinholes for the entrance and exit beams but is equipped with a second, coaxially aligned Nipkow disk that contains some 20,000 microlenses. Each microlens is precisely aligned with its corresponding pinhole on the main Nipkow disk onto which the illuminating beam is focused. Thus, instead of the 1% or so found with conventional Nipkow disk systems, some 40% to 60% of the light impinging on the disk containing the microlenses becomes transmitted through the pinholes to illuminate the specimen (Fig. 2).

In the CSU-10, each fluorescent beam emitted by the specimen is focused by the objective lens onto the same pinhole that acted as the entrance pinhole. Light passing the pinhole is then reflected by a dichromatic filter placed between the two Nipkow disks, behind the pinhole-containing disk, but before the one with the microlenses (Figs. 2 and 3). Thus, much of the backscattered light that could contaminate the image-forming beam is eliminated before the illumination beam reaches the main Nipkow disk. The image-forming beam reflected from the dichromatic mirror is projected by a collector lens onto the camera faceplate or into the eyepiece.

In the CSU-10, the microlenses and pinholes and arranged into 12 sets of a unique geometrical pattern (constant-pitch helical; see Section VIII). This unique pattern provides image frames that are homogeneously illuminated and traced by uniformly spaced scan lines. The image is thus free from nonuniform distribution of light intensity across the field or from scan-line inhomogeneity (scanning stripes) that can detract from Nipkow disk confocal systems using other pinhole patterns (Ichihara et al., 1996).

With the (standard) CSU-10, the two Nipkow disks revolve together on the axis of an electric motor spinning at 1800 rpm, or 30 rps. Since there are 12 sets of helically arranged pinholes on the Nipkow scanners of the CSU-10, the scanned image is averaged 12 times per video frame (NTSC video repeats 30 frames per second). At any instant,

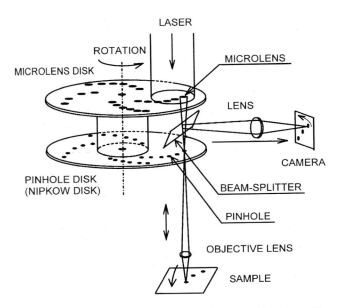

Fig. 2 Schematic of optics in the CSU-10. The expanded and collimated laser beam illuminates the upper Nipkow disk containing some 20,000 microlenses. Each microlens focuses the laser beam onto its corresponding pinhole, thus significantly raising the fraction of the illuminating beam that is transmitted by the main Nipkow disk containing the pinhole array. From the pinholes, the beams progress down to fill the aperture of the objective lens. The objective lens generates a reduced image of the pinholes into the specimen focal plane. Fluorescence given off by the illuminated regions in the specimen is captured by the objective lens and focused back onto the Nipkow disk containing the pinhole array. Each pinhole now acts as the confocal exit pinhole and eliminates fluorescence from out-of-focus regions, thus selectively transmitting fluorescence that originated from the specimen region illuminated by that particular pinhole. (However, for specimens with fluorescence distributed over large depths, some out-of-focus fluorescence can leak through adjacent pinholes in multiple pinhole systems such as the CSU-10.) The rays transmitted by the exit pinholes are deflected by the dichromatic beam splitter, located between the two Nipkow disks, and proceed to the image plane. (Figure courtesy of Yokogawa Electric Corporation.)

some 1200 pinholes scan the field in succession (Fig. 4). This produces a very clean, 12-frame-averaged image for every video frame[2] or as directly viewed through the eyepiece. Image noise due to low number of photons can, if necessary, be further reduced by on-chip averaging on a low-noise CCD camera, or by digital image averaging as described in Section VII.

---

[2]Whether with the standard speed CSU-10 or those equipped with high-speed motors for faster-than-video-rate recording, the video signal and pinhole locations are synchronized through the "Video Sync" BNC connector located (above the DC power input plug) on the CSU-10. Lack of sync between the video camera V-sweep (or shutter intervals, if present, in CCD cameras) and the pinhole locations can introduce uneven spacing and/or illumination of scan lines that show up as "streaks" or scanning stripes in the video record.

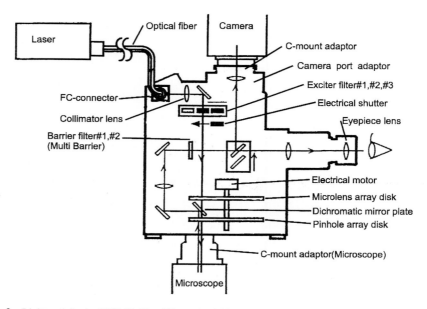

Fig. 3 Light path in the CSU-10. The 488-nm (and 568-nm) light emitted by the argon laser (or argon-krypton gas laser) is conducted through a polarization-maintaining, shielded optical fiber to the FC connector on the CSU-10. The collimated expanded beam, after passing the selected barrier and/or neutral density filters and electrically controlled shutter, impinges on the Nipkow disk containing the microlenses. Each microlens focuses the beam onto its corresponding pinhole and thence onto the specimen and back to the same pinhole, which now acts as the exit pinhole as explained in Fig. 2. The fluorescent light transmitted by the exit pinholes is deviated by the beam-splitting dichromatic filter located between the two spinning disks. Thereafter it passes through the barrier filter and is focused into the image plane in the eyepiece or camera faceplate. In the standard CSU-10, the modified Nipkow disks containing the pinholes and the microlenses spin at 1800 rpm on the shaft of a DC motor. The whole compact scanning unit is coupled to the microscope, as well as the video, CCD, or photographic camera, through female and male C-mount connectors. (Figure courtesy of Yokogawa Electric Corporation.)

IV. Sample Biological Applications

Figure 5 (right column) shows a series of optical sections of the autofluorescence of a dandelion pollen grain viewed through the CSU-10. The images were obtained with 488-nm excitation through a 40/0.8 Fluor objective lens and captured with a monochrome CCD camera. The left column shows optical sections recorded through the same microscope but by switching to standard epi-fluorescence.

Figure 6 shows the autofluorescence of another species of pollen grain in natural color, again excited with a 488-nm laser beam but captured with a chilled three-chip color CCD camera attached to the CSU-10. (See color plate.)

Figure 7 illustrates a high-resolution optical section taken through a rat heart muscle cell showing the distribution of ryanodine receptors in the sarcoplasmic reticulum.

As these figures illustrate, the CSU-10 provides uniformly lit, well-resolved, optical sections of 3-D fluorescent specimens. Without resorting to the use of a

Fig. 4 Pinholes in a single visual field of the CSU-10. The 10-mm wide by 7-mm high visual field of the pinhole arrays on the main Nipkow disk was recorded with a SIT camera while the disks were not spinning. As seen, some 1200 pinholes cover the field at any one instant. When the disks spin, the spirally arrayed pinholes successively scan the specimen. Since the radical scan pitch between the adjacent spirals is arranged in a pattern that offsets the adjoining pinholes by a fraction of the pinhole diameter, the specimen is fully raster-scanned by partially overlapping images of the pinholes.

vibration-isolation bench, the image is free from vibration-induced waviness that sometimes plagues a point scanner system. The image is also free from scan line "streaks" (nonuniform spacing of scan lines) that are common in other disk scanning systems (see footnote 2).

Through the eyepiece of the CSU-10, these detailed images can be seen in natural fluorescence color, one optical section after another, in real time, as one focuses the microscope fine adjustment.

The direct images seen through the CSU-10 eyepiece and the video records impressively represent the specimens' fluorescence distribution in sequential optical sections. Nevertheless, with thick fluorescent specimens, such as those illustrated, they are somewhat contaminated by haze from out-of-focus fluorescence that is quite absent in confocal images generated by point scanners and especially multiphoton systems. In multipinhole systems such as the CSU-10, some out-of-focus light inevitably leaks through adjacent exit pinholes. The haze can be removed and the image sharpened considerably further by fairly simple and rapid digital processing of the direct or stored video signal as described in Section VII.

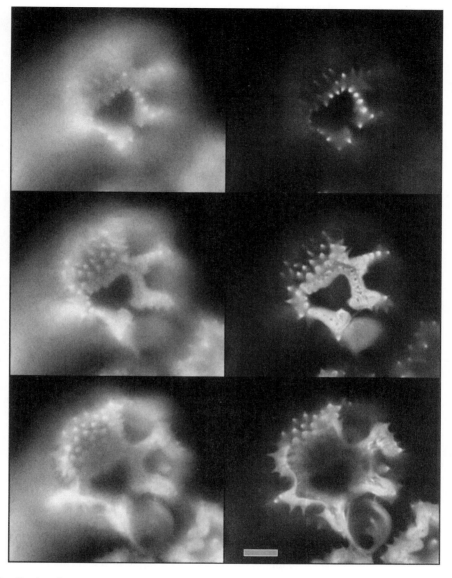

Fig. 5 Autofluorescence of dandelion pollen grain. Specimen slide of dandelion pollen grains boiled in mineral acid courtesy of Dr. Brad Amos, MRC, Cambridge, UK. Right column: Selected frames from serial optical sections spaced 0.5 μm apart were taken with a chilled interline CCD camera (Hamamatsu C5985) through the 488-nm illuminated CSU-10 mounted on a custom-made microscope (Fig. 3-13 in Inoué and Spring, 1997). An Olympus FLA epifluorescence unit was inserted between the calcite analyzer and the trinocular tube, and the specimen was imaged with a Nikon 40/0.85 Fluor objective lens equipped with a correction collar. Left column: Images taken at the same focal levels in normal epifluorescence through the same lens. See Figures 14 to 17 for improvement of the same video signals from the CSU-10 after digital image processing. Scale bar = 10 μm.

Fig. 6 Autofluorescence of pollen grain of mallow. This multicolor autofluorescence was captured in ca. 0.6 s with a chilled, 3-CCD color video camera (Dage-MTI Model 330, Michigan City, IN) mounted on the CSU-10 and illuminated with 488-nm laser light. The video signal, after conversion from RGB to standard (Y/C) format through a scan converter (VIDI/O Box, Truevision Inc., Indianapolis, IN), was captured onto a Sony ED-Beta VCR. For this illustration, a single frame of the VCR output was captured and printed with a Sony Mavigraph video printer. The same full-color image is seen through the eyepiece of the CSU-10 in real time. (See Color Plate.)

Turning to dynamic images of active living cells, Fig. 8 illustrates the process of macropincytosis in a GFP-coronin-labeled slime mold (*Dictyostelium discoideum*) ameba. Time-lapse video sequences of optical sections recorded through the CSU-10 were postprocessed as described in the figure legend. The three panels clearly depict the sequence by which the coronin-rich cortex of the ameba engulfs the fluid medium and then forms the pinocytotic vesicle (Fukui *et al.,* 1999a). Other observations of dynamic changes, seen by optical sections using the CSU-10 in this small ameba expressing GFP-coronin and GFP-actin, are illustrated in the paper just cited for cytokinesis, and in Fukui *et al.* (1999b) for the dynamics of eupodia (true feet) formation and resorption in a crawling ameba.

Figure 9 illustrates mitochondria in a ciliate protozoan *Tetrahymena,* swimming in the confocal field of the CSU-10. The images in the whole panel were taken in 1 s.

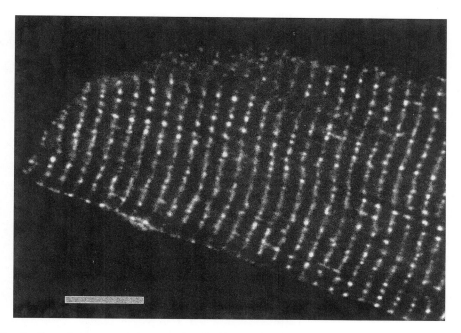

Fig. 7 Ryanodine receptors in heart muscle cell. This optical section of a rat myocyte shows the distribution of type 2 ryanodine receptors. The receptor proteins were stained with a polyclonal antibody against ryanodine 2 and visualized with an FITC-labeled secondary antibody. (The tetrametric receptor proteins form channels in the membranes of the sarcoplasmic reticulum and control the rapid release of Ca^{2+} to the myofibrils.) Scale bar $= 10\ \mu$m. The confocal optical section of the rat ventricular myocyte was captured by Drs. Peter Lipp, M. Laine, and M. D. Batman of the Babraham Institute, UK, using a modified CSU-10 ("UltraView" by PerkinElmer Wallac; see footnote 7) on a Wallac chilled-CCD camera equipped with a Sony 1300 × 1000 pixel interline CCD chip. Data courtesy of Dr. Baggi Somasundaram (bsomasundaram@wallacus.egginc.com).

As the panels illustrate, one can obtain very sharp, full-frame video images of selected organelles and their dynamics, in high spatial and temporal resolution as the protist swims about (motion slowed down with Protoslo). We believe such detailed fluorescence images of swimming protozoa have never before been observed or recorded.

Figure 10 illustrates the dynamic growth, shortening, motility, and binding to the cell wall of microtubules in the fission yeast of *Schizosaccharomycetes pombe*. The confocal time-lapse frames of these small, living cells clearly show, in high spatial and temporal resolution, the distribution and changes of the microtubules (expressing GFP-tubulin) in optical section. In the optical sections, the intensity of fluorescence reflects the number of microtubules (one to a few) in the microtubule bundles, which coupled with their dynamic behavior reveals the sites of microtubule organizing centers.

Figure 11 illustrates, in natural color, the rapidly changing fluorescence in a clam sperm stained with Eosin-B. (See color plate.) Live cells are commonly thought to be unstained by Eosin-B, and entrance of this dye is considered an indicator of cell death. We nevertheless find that membrane regions of the swimming sperm are stained and exhibit

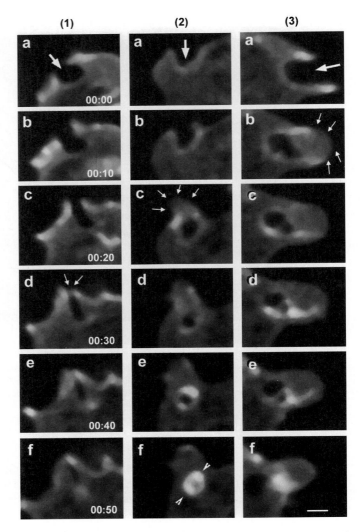

Fig. 8 Three representative sequences of macropinocytosis in amoeba of the slime mold *Dictyostelium*. Fluorescence in these time-lapsed (10-s interval) CSU-10 confocal images is due to GFP-coronin, which accumulates around the pinocytotic crowns and remains surrounding the internalized pinocytotic vesicles as shown. Scale bar = 1 μm. *Technical details:* Vegetative-stage *Dictyostelium discoideum* expressing GFP-coronin fusion protein were overlaid by an agarose sheet and observed in a microchamber using 1.4 NA Plan Apo objective lenses on a Nikon E-800 microscope. The shutter in the CSU-10 was opened for 0.75 s while the signal was integrated on chip in a chilled CCD camera (Hamamatsu C-5985) for 0.5 s once every 10 s. Regions of interest were selected in the image stacks captured on hard disk (in the same image-processing and system-control computer operated by the UIC MetaMorph program that controlled the shutter operation), low-pass filtered, and unsharp masked, and a median filter was applied. (From Fukui, Y., Engler, S., lnoué, S., and de Hostos, E. L. Architectural dynamics and gene replacement of coronin suggest its role in cytokinesis. *Cell Motil. Cytoskel.* **42,** 204–217. Reprinted with permission of Wiley-Liss, Inc., a subsidiary of John Wiley & Sons, Inc.)

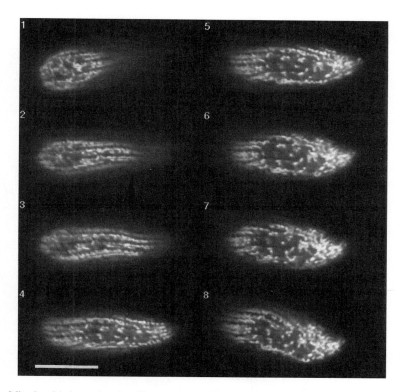

Fig. 9 Mitochondria in a swimming ciliate protozoan. The MitoTracker-G-stained *Tetrahymena pyriformis* generated these sequential series of optical sections as it tumbled through the confocal field viewed with the CSU-10. This whole sequence covers a 1-s period. In panels 1–3, the head (to the right) of the *Tetrahymena* is bent below the confocal plane. In panels 6–8, the head is lifted further up. Where the confocal optical section lies just below the cell surface, mitochondria (that sandwich the linearly arranged basal bodies of cilia, not seen here) show in paired rows. Where a region further inside the cell comes to lie in the optical section, one sees randomly oriented mitochondria. Scale bar = 20 μm^3. *Technical details:* The mitochondria were selectively stained with the cell-permanent fluorescent dye MitoTracker-G (Molecular Probes, Eugene, OR). Optical sections of the swimming *Tetrahymena,* captured through a Leica 100/1.3 Plan Fluotar oil-immersion objective lens, were recorded directly to a VCR through a low-light-level silicon intensifier target (SIT) camera at video rate. Frame intervals ca 0.1 s except 0.26 s between panels 1 and 2. The *Tetrahymena,* whose swimming motion was slowed down to between 10 and 20 μm/s with Protoslo (Carolina Science, Burlington, NC), was viewed through a Leica 100/1.3 Plan Fluotar oil-immersion objective lens. The CSU-10 confocal unit mounted on top of the Leica DMRX microscope was illuminated with a 488-nm laser, and the image captured with a SIT camera (Dage-MTI 66, Michigan City, IN) was recorded at video rate onto an ED-Beta VCR (Sony EDW 30F, San Jose, CA). Several seconds of the video played back from the VCR was passed through a time-base corrector (For-A FA-400, Newton, MA) and captured into RAM as a "stack" through an InstruTech DVP-32 Digital Video Processor (Port Washington, NY) using the "Adjust Analog Contrast" and "Acquire Stream" functions in MetaMorph (Version 4.0, UIC, West Chester, PA). Using MetaMorph, a sequence of contiguous frames was selected from the stack held in the RAM, from which eight partial frames (region of interest) were chosen to generate the montage. (S. Inoué and B. Matsumoto, 1999, original images.)

[3] The confocal fluorescence images seen through the CSU-10 can easily be superimposed in real time with the brightfield or DIC images of the specimen. All one has to do is to turn on the transilluminating light source to an appropriate brightness. Then the fluorescing mitochondria, for example, that come to lie in the confocal plane shine in succession within the (nonconfocal) brightfield image of the swimming *Tetrahymena* that show the ciliary rows and other cell structures.

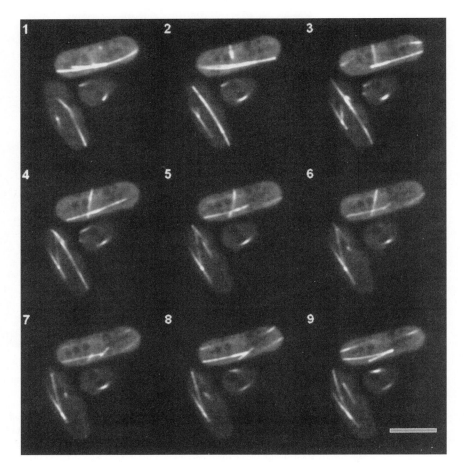

Fig. 10 Optical sections of microtubules in the fission yeast, *Schizosaccharomycetes pombe*. The growth and bending, bright overlap region, mutual gliding, and shortening of the GFP-expressing microtubules are clearly visible in these selected time-lapsed frames imaged with the CSU-10. Same microscope optical system as used for Fig. 9, but captured with a Hamamatsu Orca-1, chilled high-resolution CCD camera, whose digital output was directly captured in the MetaMorph digital image processor. The processor synchronously controlled the shutter on the CSU-10 to prevent excess exposure of the live yeast cell to the 488-nm illumination. The interval for these selected, FFT-blurred frames is ca 30 s. Scale bar = 5 μm. (From Tran, P. T., Maddox, P., Chang, F., and Inoué, S. (1999). Dynamic confocal imaging of interphase and mitotic microtubules in fission yeast, *S. pombe. Biol. Bull.* **197,** 262–263. Reprinted with permission of The Biological Bulletin.)

a faint pink fluorescence when excited with the 488-nm beam of the CSU-10. However, after excitation with the 488-nm beam for about 10 s, a striking series of progressive changes were observed within the sperm cell. The four spherical mitochondria in the mid-piece suddenly swelled as their fluorescence turned to a bright yellow. Over the next second, the nucleus swelled and took on a bright red fluorescence, progressively from the mid-piece region forward. Concurrently, the nuclear envelope took on a bright yellow fluorescence. After several more seconds, a strong pink fluorescence suddenly appeared at the tip of the sperm inside the acrosomal vesicle. Up to that point, the vesicle contents showed no fluorescence except for the very thin acrosome itself. Meanwhile the faint pink fluorescence of the sperm tail turned red and then yellow, again progressively away from the mid-piece region.

Thus, we observe, in optical sections of the minuscule sperm cell, rapid sequential changes in morphology, fluorescence color, and dye penetration into organelle compartments. These sudden, stepwise changes to the membranes of the cell compartments were no doubt induced by photodynamic damage by Eosin-B in the presence of the 488-nm excitation beam.[4] They nevertheless illustrate how changes in small cellular compartments and their membranes can be followed dynamically in true fluorescent color and with high spatial and temporal resolution, with the real-time confocal scanning unit.

V. Low Bleaching of Fluorescence Observed with the CSU-10

Bleaching of specimen fluorescence is a major roadblock to long-term observations and experiments using wide-field fluorescence or point-scanning confocal microscopy. This is especially noticeable when observations are attempted on living cells, e.g., to follow dynamic changes in microtubules incorporating GFP-tubulin or when one needs to detect very low levels of fluorescence.

One approach to reduce the bleaching is to use a two-photon scanning microscope where the presence of the short-wavelength excitation photons is confined to the focal plane (see, e.g., Denk *et al.,* 1995). Indeed, Squirrell *et al.* (1999) have recorded time-lapse images for 24 h (recording five optical sections every 15 min) of developing mammalian embryos from which they could recover viable fetuses. Where applicable, use of scavengers for oxygen-free radicals can also be effective.

On the other hand, several of us have been impressed by the low degree of fluorescence bleaching in live specimens observed with the CSU-10. Examples include MitoTracker-stained organelles in swimming protozoa, GFP-tubulin in living yeast cells, and X-rhodamine-conjugated microtubules exhibiting fluorescence speckles in *Xenopus* egg extract. Several images of these specimens are illustrated in the present article.

In particular, the dynamics of GFP-tubulin in dividing yeast cells (Fig. 10) could be recorded for up to 300 time-lapsed frames with the CSU-10 without appreciable bleaching. In comparison, many fewer images with the same signal-to-noise quality could be captured with a regular point-scanning confocal microscope before bleaching became unacceptable or the cell was killed (Dr. Phong Tran, Columbia University, Dr. Ted Salmon, Paul Maddox, University of North Carolina, personal communication).

We believe the surprisingly low degree of fluorescence bleaching with the CSU-10 stems from a low dose rate of the fluorescence excitation beam, coupled with the high rate of return beam transmission and the better performance of chilled CCDs compared to the photomultipliers used in point-scanning confocal microscopes as explained later.

---

[4] In the absence of Eosin-B, the clam sperm receiving the 488-nm irradiation under the CSU-10 continued to swim for many minutes.

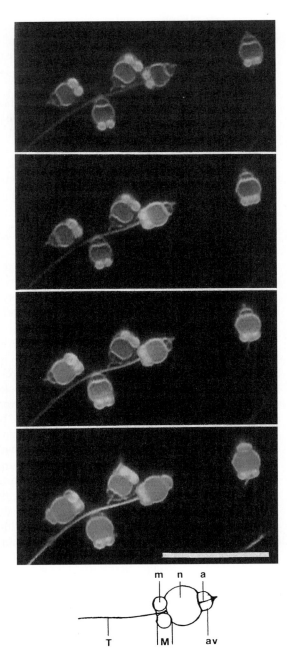

Experts do not seem to agree on whether fluorescence bleaching is linearly dependent on the dose rates of the excitation beam seen by the specimen, especially at the very high doses [brief illumination for near saturation of fluorescence that each specimen part experiences in a point-scanning confocal system (see, e.g., the several articles that discuss bleaching in Pawley, 1989)]. At lower dose rates, it appears that bleaching is, in fact, linearly dependent on dose rates, i.e., the degree of bleaching is determined by (energy absorbed per unit of time) × (duration of exposure) and is independent of whether a higher dose of energy is given over a brief interval of time or a lower dose is given over a longer period of time (Fred Lanni, Carnegie Mellon University, personal communication).

In the CSU-10, the power of the excitation beam that reaches the specimen is quite low. In fact, we find that it is several times lower, using the same objective lens, than in the wide-field fluorescence mode illuminated by an HBO-100 short-arc mercury lamp.

As to the efficiency of transmission of the imaging system, we measured the throughput of the imaging beam through the CSU-10 to be 70% of that compared to the same microscope with the CSU-10 removed.[5]

Coupled with the high throughput of the return beam, the higher quantum efficiency (perhaps by a factor of 2 to 4) and lower noise level of quality chilled-CCD cameras compared to photomultipliers (as used in current point-scanning confocal systems) may explain the greater photon-capture efficiency and better S/N ratio for systems such as the

---

[5] We measured the peak pixel brightness in the image of the pinholes (with the disk stopped) to be 70% of that compared to the CSU-10 removed from the microscope (Inoué and Knudson, unpublished data). These measurements were made (using monochromatic green and red trans-illumination respectively for the 488- and 532-nm excitation filter cubes) by checking the integration time required to generate the same pixel gray values in the middle of the stationary pinhole image compared to the pixel gray values in the same region of the blank, trans-illuminated field (i.e., with the CSU-10 removed but with the microscope optics and illumination otherwise unchanged). The image magnification was confirmed to be unchanged when the CSU-10 was removed from the microscope (by imaging the same object onto the CCD camera). This increased throughput should not be confused with the improved throughput for the entrance pinhole that is brought about by the use of microlenses. The imaging beam from the specimen is diverted to the camera (and eyepiece) before it reaches the microlens-containing disk (see Figs. 2 and 3).

Fig. 11 Selected frames from CSU-10 confocal video sequences showing striking photodynamic changes in Eosin-B-stained *Spisula* (clam) sperm exposed to the 488-nm (fluorescence exciting) laser beam. Top to bottom panels: 13, 14, 15, and 31s after exposure to the laser beam (0.01 mW per μm^2 at the specimen). Except for the fourth sperm from the left, much of the tail is invisible owing to the shallow depth of field of the confocal optics. Scale bar = 10 μm. Bottom: Schematic diagram of *Spisula* sperm. a; Acrosome: av; acrosomal vesicle; M; midpiece; m; mitochondrion; n; nucleus; T; sperm tail (part only shown). *Technical details:* Sperm were suspended in seawater containing 0.1% Eosin B and observed through a Nikon E-800 upright microscope equipped with a 100 × /1.4 NA Plan Apo oil-immersion lens and a CSU-10. The weak fluorescence of the sperm parts excited by the 488-nm laser beam was integrated on chip for 20 frames (0.66 s) on a chilled, 3-chip color camera (Dage-MTI Model 330, Michigan City, IN) attached to the CSU-10 and observed continuously on an RGB color monitor. The RGB output through the monitor was converted by a scan converter (VIDI/O Box, Truevision Inc., Indianapolis, IN) to Y/C format, color balance and background levels adjusted with a video processor (Elite Video BVP-4 Plus, www.elitevideo.com), and recorded to an ED-Beta VCR. (From Inoué *et al.*, 1997, with permission) (See Color Plate.)

CSU-10 compared to point-scanning systems. Those factors would also allow for use of lower levels of excitation light (and hence lower rates of fluorescence bleaching)[6] to obtain comparable low noise images with the CSU-10 using a chilled-CCD camera compared to a point scanner that captures the image signal with a photomultiplier tube.

With wide-field fluorescence using chilled-CCD cameras, comparably low levels of illumination could be used to reduce bleaching as with multipinhole scanning systems such as the CSU-10. However, background fluorescence, e.g., from the dyes used in the culture medium when observing GFP-microtubules in the yeast cells, obscures the desired images of the weakly fluorescent microtubules (Paul Maddox, University of North Carolina, personal communication).

Although we do not have a precise measure as to how much each of these factors contributes to the low fluorescence bleaching, there is no question, empirically, that the CSU-10 can provide excellent quality, optically sectioned, time-lapse images of sensitive, living specimens over extended periods of time (See also Addendum B).

VI. Very-High-Speed Full-Frame Confocal Imaging with the CSU-10

With point-scanning confocal systems one can scan along the horizontal axis at very high speeds. Thus, single-to-a-few sequential horizontal scans can be used to record rapid physiological transients and to generate kymographs (slit images versus time plot) depicting those transient events (see, e.g., Tsien and Bacskai, 1995). However, as mentioned earlier, a few to several seconds are generally required to display or record each low-noise full frame with a point-scanning confocal system. In contrast, with real-time, disk-scanning confocal systems, one can record sequential full-frame images at intervals of 33 ms, i.e., at video rate as already illustrated. Using a different motor to spin the disks in the CSU-10 at higher speeds, one can even generate full-frame signals at millisecond intervals (See also Addendum A).

Thus, Genka *et al.* (1999) modified the CSU-10 to drive the Nipkow disk motor at 5000 instead of 1800 rpm. At this higher speed, each full-frame optical section can be obtained in 1 ms. Combined with a Gen II image intensifier (ILS-3, Gen II, Imco, Essex, UK) and a high-speed CCD camera with controller (model 1000HR, Eastman Kodak, San Diego, CA), the 512×384 pixel 8-bit images were stored in the controller memory at 1–4 ms/frame. Such data were used to measure the 3-D diffusion pattern of

---

[6] When somewhat greater noise levels can be tolerated, the light-capturing efficiency of intensifier video tubes, such as SIT, or Gen-III or Gen IV intensifiers coupled with CCD cameras, also allows the use of low fluorescence excitation combined with high rates of image capture (as illustrated, e.g., in Figs. 9 and 12).

[7] In addition to the striking calcium waves in optical section seen in these excerpted frames, many sparks and more waves can be seen by playing back the full sequence of the 290 frames available. Contact Dr. Baggi Somasundaram at Perkin Elmer Wallac, or see http:\\www.wallac.com, to view this and many other impressive dynamic scenes taken with the Wallac UltraView system (mitosis in Drosophila embryo expressing histone-GFP, endoplasmic reticulum and organelles in onion cells, flow of cortical actin in fibroblasts, etc.). Data courtesy of Dr. Baggi Somasundaram.

Fluo-3-bound calcium ions surrounding the nucleus of contracting myocytes (see also Ishida *et al.,* 1999).

Figure 12 shows vivid pseudocolor images of Ca^{2+} transients in myocytes taken through a standard-speed CSU-10 equipped with an image intensifier. (See color plate.) The images in Ishida *et al.* (1999) and Genka *et al.* (1999), recorded with considerably greater temporal and spatial resolution than in this illustration, display hitherto unknown changes in Ca^{2+} localization and dynamic distribution that take place within the myocytes.

Yamaguchi *et al.* (1997) describe a system for recording confocal images of fluorescein-tagged blood cells at 250 and 500 frames/s flowing in pulmonary capillaries (which were also fluorescently labeled for selected adhesion molecules; see also Aoki *et al.,* 1997). They use the CSU-10 equipped with a 5000-rpm motor, coupled with an image intensifier CCD camera (EktaPro Intensified Imager VSC, Kodak, San Diego, CA), connected with a high-speed video analysis system (EktaPro 1000 Processor, Kodak) to record the high-speed images.

Before continuing our discussion on the performance of the CSU-10 in Section VIII, in the next section we shall examine the advantage, means, and rationale for applying digital image processing to multibeam confocal and related image data. With appropriate digital processing, one can very effectively and nearly instantaneously improve the quality of optical sections by enhancing details in the in-focus image while suppressing the background fluorescence. Other forms of image noise can also be reduced or removed. Such approaches are especially useful in cell biology, for bringing out subtle or fine image details, for preparing a stack of images for 3-D projection, and for processing large quantities of image data, e.g., when many time points are involved and/or when each time point is associated with a substantial stack of serial optical sections.

VII. Haze Removal, Image Sharpening, and Dynamic Stereo Image Generation by Digital Signal Processing

Once digital computers became widely available for image processing, wide-field microscope images were effectively enhanced using simple, neighborhood-based digital convolution operations such as sharpening and unsharp masking. While such simple convolution operations typically may not be as effective at removing haze from images as are iterative, constrained deconvolution algorithms, they can nevertheless significantly improve images to the extent necessary to obtain excellent 3-D reconstructions and stereo pairs. Moreover, their speed makes these operations well suited to 4-D microscopy (3-D imaging over time) where large data sets are common.

The CSU-10, by virtue of its ability to collect images at high speed with low rates of fluorescence bleaching and its ease of use, makes it possible to collect massive quantities of 3-D image data. This allows it to be used quite effectively in 4-D microscopy. As noted earlier, the CSU-10 is a multiple-pinhole, high-light-throughput scanning confocal

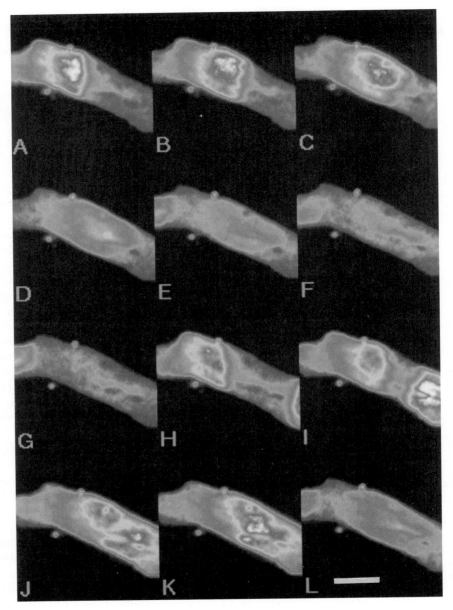

Fig. 12 Mouse cardiac myocyte. Waves and sparks of elevated cytosolic Ca^{2+} are presented in pseudocolor in this heart muscle cell injected with Fluo-3. (Ca^{2+} concentrations rise in order of purple, light-blue, dark-blue, green, yellow, orange, red, and white.) Frame intervals each 200 ms between panels A–F. Then after 1.4s, panel G-L show the approach and annihilation of two calcium waves. Frame intervals: G–H = 360 ms, H–I = 160 ms, I–J = 120 ms, J–K = 40 ms. Scale bar = 5 μm.[7] *Technical details*. The image from the CSU-10 (modified by addition of motorized filter wheel as a part of the "UltraView" system distributed by PerkinElmer Wallac), viewed through an Olympus 60/1.4 oil-immersion objective lens, was intensified with a Videoscope Intensifier. The images, each exposed for 25 ms, were captured at 25 frames per second onto a PerkinElmer FKI-300 cooled-CCD camera equipped with a Kodak Olympics interline chip. (See Color Plate.)

system. The fixed size of its pinhole apertures is necessarily a compromise between light-gathering capacity and confocal rejection. Because of these design choices, images from the CSU-10 can benefit from digital image processing to further suppress residual background fluorescence and to sharpen image detail.

Ideally, one might use image deconvolution operations for precise image restoration. However, the speed of such operations, typically on the order of hours per image stack,[8] is too slow for routine processing of 4-D data sets. Fortunately, when used properly, image processing operations such as sharpening and unsharp masking are adequate, rapidly generating significantly improved results.

A. Neighborhood-Based Processing

Many of the most commonly used image processing programs, such as Universal Imaging's MetaMorph and Adobe's PhotoShop, use *neighborhood*-based convolution algorithms for operations such as image sharpening. Here, neighborhood refers to the fact that these algorithms modify every pixel in an image by replacing it with the weighted sum of the pixel and those surrounding it. The matrix containing these weighing factors is called the *convolution kernel* or *kernel* for short. Typically, the kernel covers a 3 × 3-pixel area, although other sizes, such as 5 × 5 pixels, are also common.

The results of convolution operations depend on the values used in the kernel. For example, the convolution blurs the image if all kernel elements have the same value, e.g., 1. To convolve with this kernel, the brightness of the central pixel and of the surrounding 8 pixels in a 3 × 3-pixel area are each multiplied by 1 and summed together. The result is divided by 9. Then the central pixel is replaced by this average brightness of the 3 × 3-pixel area, and thus the image is blurred. Such operation is also referred to as a low-pass filter or neighborhood average.

The sharpening kernel, on the other hand, subtracts the average brightness of the neighboring pixels from the weighted brightness of the central pixel, resulting in the enhancement of brightness gradients in the image. A typical sharpening kernel, called the Laplacian sharpening kernel (Fig. 13, panel 1), subtracts the sum of the 8 neighboring pixels from the central pixel multiplied by 9. The result is an image containing the same values as the original in areas of uniform brightness, and enhanced details where there are brightness gradients. A variant of this operation only subtracts the sum of the 4 nearest pixels from the central pixel multiplied by 5 (Fig. 13, panel 2) (Castleman, 1979; Inoué and Spring, 1997, Section 12.7; Russ, 1995).

Neighborhood-based computations are well suited for processing stacks of images because current microprocessors, such as the Intel Pentium, AMD Athlon, and Motorola PowerPC, perform the computations required for neighborhood operations very rapidly. The system in the authors' laboratory uses the MetaMorph software under the Windows NT operating system. The computer has 384 megabytes of RAM and an Intel Pentium II running at 400 MHz. This system sharpens 100 images, each 640 × 480 pixels, using

[8] With the advent of the new Gigaflop MacIntosh and other microprocessors, we should see this number drop by a factor of 2 to 5 during the next few years. However, even with this performance boost, deconvolution will be too slow for routine processing of 4-D data.

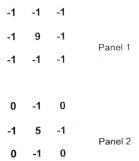

Fig. 13 The Laplacian sharpening kernel. See text.

a 3×3 sharpening kernel, in approximately 5 s, i.e., each frame nearly at video rate. Newer computers using the Intel Pentium 4 running at over two GHz are several times faster, reducing the computation time to approximately one second, or several times faster than video rate!

A typical 4-D imaging experiment might include 200 time points each with 30 images collected through the depth of the sample collected. If the reference system takes 3 s to process each image stack, the entire experiment would be processed in about 10 min, much shorter than the time in which a typical deconvolution system processes a *single* image stack! It is this efficiency that makes neighborhood operations so useful.

B. Sharpening and Unsharp Masking

Whereas sharpening often substantially improves images from the CSU-10, *unsharp masking* provides superior haze removal with only a modest increase in computational complexity.

In unsharp masking the image is enhanced by masking it with an unsharp (blurred) version of the same image. The result of this masking is a brightness reduction in areas that lack fine details. Conversely, unsharp masking increases the contrast of areas that do contain fine details. The resulting image contains enhanced fine details, much like a sharpened image, but also has a reduced background level, very useful for fluorescence applications containing background haze (compare Fig. 14 with Fig. 5 right).

The MetaMorph software provides controls for fine-tuning the unsharp masking process, a feature not typically found in other sharpening methods. One option sets the amount of blurring to use, thereby selecting the size of the features to enhance. This is very useful as it permits control of the degree of sharpening of image details without significantly increasing image noise.

A second option sets the amount of the blurred image to be removed from the original. This parameter adjusts the magnitude of the enhancement that is applied to the image or, to think of it in another way, this sets how much haze is removed. For example, if 100%

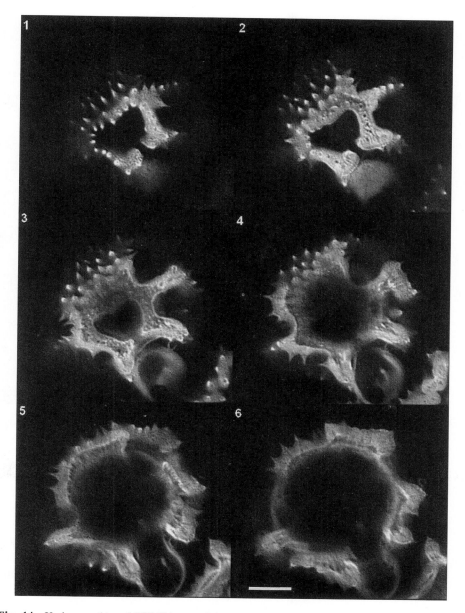

Fig. 14 Unsharp masking of CSU-10 images. Selected images (ca 2 μm apart) of the serial optical sections taken through the CSU-10 were digitally unsharp masked (using a 2 × 2 kernel and 0.90 scaling factor after first applying a 2 × 3 median filter to prevent the appearance of white noise spots). The images in panels 2 and 4 resulted from processing of the two bottom right images in Fig. 5. Scale bar = 10 μm.

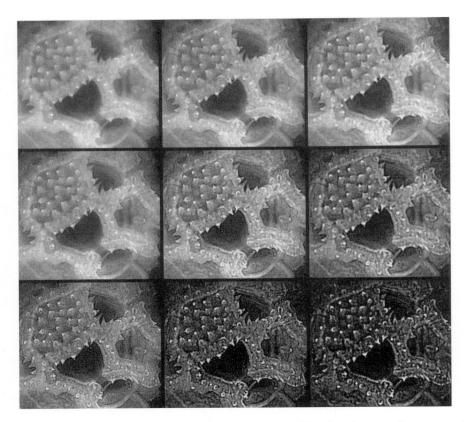

Fig. 15 Influence of kernel size and scaling factor on unsharp masking. Central portion of max-projected images [sum of all optical sections taking the brightest pixels projected through the stack of images (see Section VII.F.)] of the pollen grain shown in Fig. 14. The kernels used to produce the unsharp masks for the three columns were, from left to right, 3×3, 7×7, and 11×11. The scaling factors used to determine the fraction of contrast given the masks were, from top to bottom rows, 50%, 80%, and 90%. The MetaMorph software automatically enhances contrast to fill the gray value ranges.

of the blurred image is subtracted, then only fine details will remain in the resultant image; all regions of uniform brightness, such as background haze, are reduced to black. On the other hand, if only 50% is subtracted, then the resultant image will look much like the original with only a slight enhancement.

Figure 15 shows the results of applying a variety of blur sizes and subtraction percentages. The upper left panel has been processed the least (and is virtually indistinguishable from the original image) whereas the lower right panel represents the highest amount of processing, large blur and large subtraction percentage. For this particular sample, a small amount of blur and a moderate subtraction percentage results in the most esthetically pleasing image.

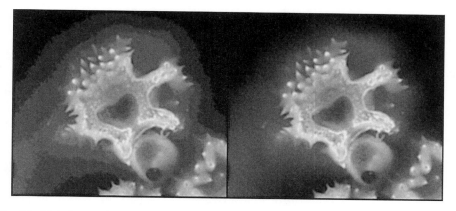

Fig. 16 Digital processing of 8-versus 16-bit images. The left panel shows quantizing error (discrete steps of gray values, especially noticeable in the background) by digital contrast enhancement of an 8-bit image (see text). Right panel: Even though the original image was taken with a CCD camera capable of capturing no more than 8 bits of image gray values, the quantizing error does not appear when the same image is processed after first converting to a 16-bit image.

C. Arithmetic Precision

Because unsharp masking is a subtractive process, the dynamic range of the processed image is necessarily less than that of the original. For example, if 95% of the haze is removed from an original pixel value of 100, then the processed pixel will have a pixel value of roughly 5. With an 8-bit image containing 255 gray values, the best resultant image will have integer gray values ranging only between 0 and 12 (0 to 5% of 255).

Although these values may be scaled through contrast enhancement to fill the full black-to-white dynamic range of the video system, there will never be more than 12 distinct gray values. In most cases this results in unsatisfactory quantization effects (Fig. 16, left panel). Fortunately, we can minimize this undesirable effect by scaling the original image.

If the original image values were first multiplied by 100, then the range of gray values would span the range of 0 to 25,500. Even though there are still only 255 unique gray values, the convolution operation is more accurate when starting with larger values since the computed values retain more precision. In other words, scaling allows integer-based operations to compute results that would have contained fractional values, which are lost in integer computations.

After scaling, even with a 95% haze removal factor, the resultant image could contain pixel values ranging from 0 to 1275, more than enough for excellent quality visualization (Fig. 16, right panel). Most importantly, a scaled image after processing contains a greater number of distinct gray values than an unscaled image after processing, resulting in an image free from quantization artifacts.

It should be noted that such improvements by scaling can be obtained for most convolution operations and many arithmetic operations (image averaging, division, etc.), since such operations typically involve steps that would generate fractional values that are rounded off to integer values in the resultant image data.

D. Significance of Unsharp Masking

Unsharp masking selectively increases the contrast of fine image details, including edges (without the ringing that accompanies conventional high-pass filtering or convolution), while reducing the contrast of low spatial frequency components of the image, suppressing haze and background fluorescence, etc.

Some argue that such sharpening generates an image that is artificial and even artifactual. Indeed, images convolved with any kernel could be viewed with such criticism, especially when one is concerned with exact distribution of intensity in the image. However, one should recognize that the generation of an image through *any* optical instrument, such as a microscope, or the recording or transmission of image data through analog video recorders or image processors, involves a series of hidden convolutions that can differentially attenuate or accentuate the spatial and temporal frequencies present. Not only microscopes using any contrast-generating mode, but our own visual system, including the optics in our eye, the retina, and other parts of our visual neural system, processes an image in such a way as to extract or emphasize certain significant features. Extreme examples would be the loss of contrast of high-frequency components (by the drop in modulation function) as one approaches the resolution limit in any diffraction-limited optical instrument including microscopes; selective enhancement of the higher frequency components in darkfield, phase contrast, DIC, etc.; loss and compensation of losses for certain frequency signals in video format signal transmission and especially recording; and edge and motion detection and image feature filtering, or accentuation, in our own visual system (see, e.g., Inoué and Spring, 1997, Chapters 2, 4, and 11).

In interpreting the significance of any image or image feature, we naturally need to be cognizant of these series of filtering or convolving and deconvolving steps. Yet, as the listed examples show, selection or emphasis of particular image features at the cost of suppressing other features is not only often an advantage, but also a necessity for generating data or information that is meaningful from a particular perspective. The question is, from what perspective or for seeing what image attributes?

Unsharp masking, in part, compensates for the high-frequency image components that were reduced or lost in the optical or electronic transmission and recording process. Thus, e.g., it can significantly restore image details whose contrast was reduced and that *appeared to be* lost in video recording. On the other hand, as with any operation that enhances the high spatial frequency components, it could bring forth undesirable high-frequency image noise. Unlike convolving with simple high-pass kernels, however, there is less tendency to introduce ringing or extraneous high-frequency boundary lines, etc., as described earlier.

If excessive noise, such as random bright spots, is seen after unsharp masking, the noise can often be reduced or eliminated by first appropriately filtering the original image, e.g., by applying a median filter (an operation that replaces each pixel with the median value of the local neighborhood's pixel values). Similarly, periodic noise, such as video scan lines and motion artifacts, is eliminated by Fourier filtering to blur the image before unsharp masking, e.g., by applying a "blur" fast Fourier transform (FFT).

Unsharp masking effectively brings out fine details in the image and suppresses haze, background, and contributions from any large image regions with uniform brightness. Thus, it is most effective for aiding in the visualization or measurement of locations or distances between finer image features. It is clearly not intended for photometric measurements.

E. Image Processing in 3-D Projection and Stereo Image Generation

Unsharp masking is especially effective when building up 3-D or 4-D (3-D with time) image projections, for example, for stereoscopic presentation of complex objects as seen in Fig. 17.

In contrast to unsharp masking and other convolution operations that are applied to images taken at single focal planes, deconvolution operations use images in a whole stack of focal planes to compute each theoretical, unadulterated, in-focus image. Thus, they eliminate, or minimize, haze and other undesirable image features that were present in the original microscope image that arose by superimposition of light waves that originated from out-of-focus planes. Although it is computationally extremely intensive, involving many rounds of iterative operations using the full stack of images obtained at many foci, deconvolution can produce nearly perfect optical sections free from out-of-focus contributions. It is desirable for obtaining very clean optical sections, for photometry, and for 3-D reconstruction of relatively uncomplicated or isolated systems with high image contrast and low noise in the initial image stack.

For generating time lapse, and especially 4-D (time-lapsed 3-D) confocal images, or for preparation for morphometric measurements of specimens with complex 3-D structures, unsharp masking and *nearest-neighbor haze-removal* or *inverse filtering* deconvolution algorithms provide major advantages. Although limited by the caveats already described, data stacks for many time points can be processed with simpler computers and in a much shorter time. Indeed, today one can carry out such operations as fast as the images can be acquired using the faster, commonly available personal computers.

F. From Image Stacks to Projections

Presenting 3-D image stacks is a significant challenge for microscopists. As illustrated in Fig. 14, one can display a few representative images from the stack. Although this results in the clearest display of fine details in each focal plane, it can be difficult to interpret the 3-D structure of the specimen from serial optical sections.

A second, common approach is the *maximum projection* (e.g., Shotton, 1993, Chapter 2). This method compresses a stack into a single image based on the maximum brightness for each pixel through all the planes in the stack. For example, the resulting images illustrated in each panel in Fig. 17 contain information on the most prominent features through the sample. Maximum projection also gives some degree of transparency, as bright objects within the sample have precedence over darker objects in front of them.

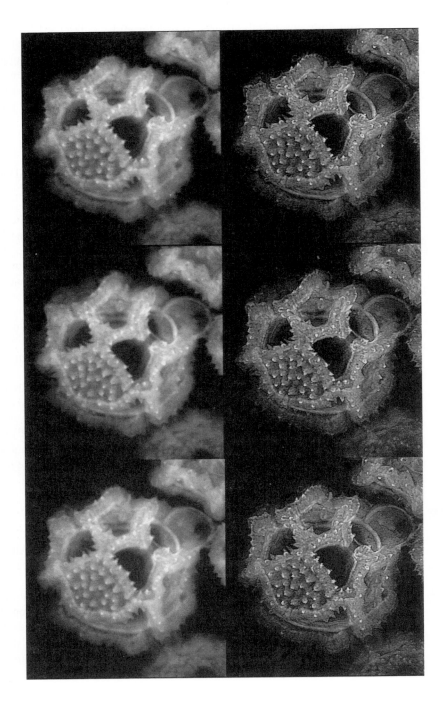

However, as with all two-dimensional representations of real objects, the image appears flat, containing few cues as to the true 3-D structure of the object being observed.

In order to obtain more *depth cues* to assist in the interpretation of the image, the maximum projection may be computed at two or more angles. When viewed as an animation, the projected stack appears to rotate, resulting in a compelling 3-D display. Alternatively, pairs of projections may be displayed as stereo pairs, allowing truly remarkable views of the specimen (Fig. 17).

Rotation of stereo pairs adds even more visual cues. These dynamic stereo images take advantage of our visual system's ability to interpret 3-D structures through relative motion in addition to stereopsis resulting in a clear 3-D view.

G. Viewing of 3-D and 4-D Images

Just as microscopists migrated from static photography to motion pictures, today video microscopists are moving from single image stacks to series of stacks collected over time, popularly called *4-D microscopy*.

However, outside the laboratory, while 2-D movies are the norm, 3-D movies are rare, still only seen at places like Disney World. Why is this?

Importantly, there have been few breakthroughs in 3-D image visualization, even for static 3-D images. Although some promising technologies exist for creating holograms or other *objective space 3-D images* (see, e.g., Inoué, 1986, Section 11.5), most practical methods depend on special glasses that allow the light from only one perspective to reach each eye.

Fig. 17 Stereo pairs of dandelion pollen grain. Fig. 17 is presented as a novel "three-panel stereogram" in which the two end panels are identical. With such an arrangement, one can view the stereo images with the unaided eyes by "cross-eyed" or "parallel-eyed" viewing, or with the aid of stereo viewing glasses. Either way, upon fusion, one sees a set of four panels with the two inner panels being an ortho- and pseudo-stereoscopic (distance cues reversed) view. For cross-eyed viewing, the panels here generate the pseudo-stereo image as the left and the ortho-stereo image as the right of the middle two panels. (In the pseudo-stereo view, the pollen grain appears concave, with the external ridges and spines pointing away, and the pollen grain appears in the ortho-view mound-shaped with the appendages pointing towards and to the side of the viewer.) Observed with stereopticons or wall-eyed viewing, the locations of the ortho- and pseudo-views are reversed. The three-panel stereogram makes the stereoscopic images directly available to those able to fuse the images with their unaided eye (whether cross- or wall-eyed), as well as for those needing to use viewing glasses. In addition to the convenience of access, side-by-side presentation of ortho- and pseudo-stereograms helps one perceive the true three-dimensional structure of the specimen (with the closer and distant features, respectively, pointing towards the observer) without being misled by shadowing, perspectives, and other cues that can otherwise confuse our senses. Technical details. Each panel in Fig. 17 was generated by projecting (at inclinations of ± 5 degrees) the brightest pixel in the line of sight (max-projection) through a stack of 31 optical sections covering the proximal half of the pollen grain. Top pair without, and bottom pair with, unsharp masking (some individual optical sections shown in Fig. 5 and 14). Before unsharp masking, the original stack was converted to 16 bit and Fast Fourier filtered with the "low blur" filter in order to reduce final image noise (see Section VII.D.). Stereo images of the lower panels, generated after unsharp masking, exquisitely display details of the thin, fin-like ridges and the spines with thin pores. Those of the unprocessed upper panels better show some of the larger features, such as the balloon-shaped vellum lying outside of the lower right pore.

For many years, *anaglyphs* (red–green or red–blue stereo pairs) have been used for the presentation of stereo images. These work by printing or otherwise presenting each perspective image in a unique color, usually red and green. The viewer then wears glasses containing green and red filters so that each eye sees only a single image.

Although anaglyphs are convenient for certain types of presentation, they do not work well with analog video equipment. Because anaglyphs depend on displaying pure colors that match colored filters in glasses, they are generally not compatible with video taping and other analog recording systems. Nor do they work well with image compression techniques that can also result in image color shift. These systems encode color information for recording and transmission in such a way that the colors shift, resulting in bleed-through that destroys the stereo effect. However, with the prevalence of purely digital storage and display, anaglyphs are again becoming a feasible stereo display technology.

Far better, however, are 3-D viewing systems based on polarizing glasses (see, e.g., Inoué and Spring, 1997, Section 10.8). There exist at least two methods for displaying and viewing stereo images using polarizing systems.

The first method, used for years in movie theaters, uses two projectors, one for each viewing perspective. The two projectors contain polarizing filters that are oriented so as to polarize the light in exactly opposite orientation from one another. For example, one might contain a horizontal polarizer while the other would contain a vertical polarizer.

The viewer wears glasses with corresponding polarizers so that each eye sees only light from the desired projector. The crossed polarizer blocks the other image. A slight variant of this technique uses left and right circularly polarized light rather than linearly polarized light. This version does not require exact alignment of the axes of the viewing glasses relative to those of the projector.

A second method that produces even better results uses special glasses that have electronically controllable liquid crystal shutters. In this method, the images for the left and right eye are displayed alternately (at twice the normal frame rate to avoid image flicker) and synchronized with the shutters in the glasses. When the left eye's image is displayed, the right eye's shutter is closed and vice versa (Lipton, 1991).

This last method of stereo image presentation has been the most effective of all that we have evaluated, as the shutters force each eye to see the appropriate image. Even people who have difficulty seeing stereoscopic displays using other methods typically see 3-D images with ease when using shuttered stereo goggles.

H. Storage Requirements

In the late 1980s, when we started to acquire stacks of 3-D images digitally, computer memory and hard disks were quite expensive. A typical image stack consisting of 50 images uses 12 megabytes per 3-D data set. At that time, hard drive capacity was only large enough to store approximately 10 of these images, making it impractical to store time-lapsed 3-D images on a hard drive. It was, therefore, often necessary to save image stacks using analog video media such as videotapes and laser disks.

Today in 2002, the tiny Zip disk holds as much data as the previous state-of-the-art hard drives, and current hard drive capacities have swollen to beyond 250 gigabytes, over a 1000-fold increase in less than 12 years. Image sizes have also increased, in many cases to approximately 2 megabytes per image and 100 megabytes per stack. Even with the vastly increased storage capacities, the largest disks still only store approximately 2500 3-D time points. Although this is an impressive amount of storage, one can imagine the difficulties that might be encountered when working with such large data sets.

A primary consideration is data backup. If a single image stack consumes 100 megabytes of storage space, even CD-R disks, with a capacity of 650 megabytes, are inadequate. Newly released devices, such as the writeable DVD technologies, with capacities of approximately 5 gigabytes are considerably better, though still too small for anything but relatively short experiments.

VIII. Mechanical and Optical Performance of the CSU–10

In order to generate a compact scanner that provides an excellent quality image, the designers of the CSU-10 chose a Nipkow disk diameter of 55 mm, on which some 20.000 pinholes were placed in a novel arrangement (Fig. 18; Ichihara *et al.,* 1996). They avoided placing the pinholes in a "fixed-angle helical pattern," which produces reduced illumination towards the periphery of the spinning disk, or a "tetragonal pattern," which while providing uniform illumination, generates an uneven scanning pattern, i.e., scanning stripes or "streaks." Instead they placed the pinholes in a "constant-pitch helical pattern" that gives both uniform illumination and even scanning that is free from streaks (assuming that the Nipkow disk rotation and video camera vertical scanning rates are synchronized; see footnote 2).

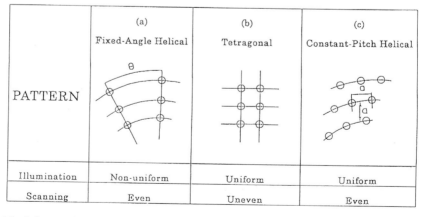

PATTERN	(a) Fixed-Angle Helical	(b) Tetragonal	(c) Constant-Pitch Helical
Illumination	Non-uniform	Uniform	Uniform
Scanning	Even	Uneven	Even

Fig. 18   Influence of pinhole pattern in the Nipkow disk on the uniformity of field illumination and evenness of scanning pattern and patented arrangement of pinholes (constant-pitch helical) used in the CSU-10. (From Ichihara *et al.,* 1996, with permission.)

Indeed, as shown in the illustrations in this paper, the CSU-10 provides a uniformly illuminated confocal field free of scanning stripes.

The diameter of each pinhole in the primary Nipkow disk is 50 μm, and they are spaced 250 μm apart, yielding a pinhole-fill factor of 4% $[(50/250)^2]$. The disk containing the microlenses is mechanically fixed to the second Nipkow disk so that each microlens is well aligned and focused onto the corresponding pinhole. Since the same pinhole acts both as the entrance and exit pinhole, a slight run-out of the axis of disk rotation would have minimum impact on the confocality of the system.

Nevertheless, any mechanical vibration introduced by the disks or the electric motor driving the disks could result in significant image deterioration, especially with the scanner perched on top of a microscope using a high-power objective lens. In fact, we were impressed to find no such sign of image vibration, or deterioration, on any of the three brands of research microscopes that we tested. Indeed, the lateral image resolution through the CSU-10 seemed to be fully determined by the NA of the objective lens used and the wavelength involved (for expected lateral resolution, see, e.g., Inoué and Oldenbourg, 1995).

The depth of field provided by the CSU-10, measured on fluorescent microspheres and a first surface mirror, has been reported to lie between 1.0 and 1.3 μm using $\times 100$, 1.3 to 1.4 NA objective lenses and 488-nm excitation (Yokogawa Research Center, personal communication). Such measurements made on extremely thin test specimen, and with confocal pinholes somewhat larger than the Airy disk produced by the objective lens, should yield a "depth of field" (i.e., the depth, projected into specimen space, where the intensity of the diffraction pattern falls to a half of its peak value) that is more characteristic of the objective lens performance than the confocal system performance. Rather than the depth of field defined by very thin objects, for multiple pinhole systems such as in the CSU-10, our concern is more in terms of the ability, or efficiency, of the system to eliminate or reduce the contribution from out-of-focus regions of the specimen.

The authors so far have no numerical value to ascribe to such efficiency for the CSU-10. Instead, the reader needs to judge the efficacy of the scanner from the illustrations presented here, through copies of dynamic images available on CD disks (see footnote 7), and through direct examination of their own specimens through a microscope equipped with a scanner unit. It may well be that for specimens or scenes in which the specimen is not moving or changing rapidly, or subject to bleaching, a laser point-scanning microscope would be more suited for obtaining better discrimination of out-of-focus regions.

Nevertheless, as illustrated, images obtained with the CSU-10 can rapidly be post processed to reduce out-of-focus contributions, to bring out greater image detail, and/or to generate impressive rotating, depth-stacked, or stereoscopic images.

IX. Potentials of the CSU-10

As stressed in this chapter, a major advantage of real-time, direct-view confocal microscopy is the ability to very easily observe, display, or record moving specimens or scenes that are changing under our eyes.

Another advantage, which may not be as obvious, is the ability to readily find sparsely distributed, weakly fluorescent objects. Such objects are quite difficult to find with point-scanning systems since point scanners yield only one low-noise, full-frame optical section every few seconds.

In contrast, for example, with the CSU-10 equipped with a video-rate intensifier CCD camera, one can focus through the specimen and rapidly find sparsely distributed sources of very weak fluorescence. Such images can be somewhat noisy as characteristic of highly intensified low-light-level video images. Nevertheless, we were able to find images of very sparsely distributed, 20-nm-diameter fluorescent spheres (which are most difficult to find with a point scanner) without any trouble on the video monitor as we focused through the specimen (Smith, S., Stanford University, and Inoué, S., 1998, unpublished observation).

Such ability to rapidly locate and follow weak fluorescence becomes especially important for observing the behavior of a few fluorophores. They may be sparsely distributed in space, and their fluorescence emission may last only very briefly.

Indeed, Tadakuma *et al.* (2001) confirms the possibility of visualizing fluorescence from single molecules of tetramethyl rhodamine in cultured cells of *Xenopus* excited with 514.5-nm argon laser emission in the CSU-10 (see also Funatsu, 1995). Funatsu *et al.* found the intensity of fluorescence (which lasted for several seconds before suddenly disappearing) to be comparable to that established earlier for single fluorophores using total internal reflection (evanescent wave) fluorescence microscopy (Funatsu *et al.*, 1995; Vale *et al.*, 1996). Thus, the CSU-10 can be expected even to serve as a tool for studying, in a living cell, the behavior or activity of single molecules, provided they are not undergoing rapid Brownian motion.

Waterman-Storer and Salmon (1998) have devised a method for visualizing specific regions on a single microtubule where the tubulin molecules were assembling or disassembling. This was accomplished by fluorescent speckle microscopy (FSM) in which the polymer, labeled with a very low concentration of X-rhodamine-conjugated tubulin, was observed with a low-noise CCD camera. The nonuniform statistical distribution of the sparse label then gave rise to a punctate, or speckly, appearance to the individual microtubules when the fluorescence was observed with a high-NA objective lens and recorded digitally with high resolution. The speckle pattern, originating from the small, random numbers of fluorophores that were incorporated per unit length of the polymer (i.e., per length corresponding to the resolution limit of the objective lens), remained sufficiently stable during the observation to serve as unique markers for the particular stretch of polymer. Thus, one can identify which region of the single microtubule had disassembled or is being newly assembled and which remains stable as shown in Fig. 19.[9]

---

[9] FSM reveals, among other attributes, three events related to the motility and exchange of subunits in microtubules: a) the microtubules may move without exchange of subunits (the whole speckle pattern moves without adding or losing speckles); b) free ends of microtubules may grow or shorten by addition or subtraction of subunits (as in Fig. 19, the existing speckle pattern remains stationary, new speckles are added at growing ends, and old speckles disappear from shortening ends); or c) the subunits may treadmill through the microtubules (as in Fig. 20, the same pattern flows through the microtubule).

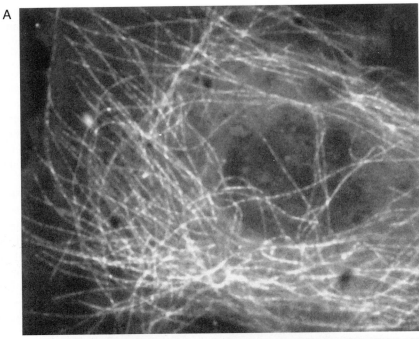

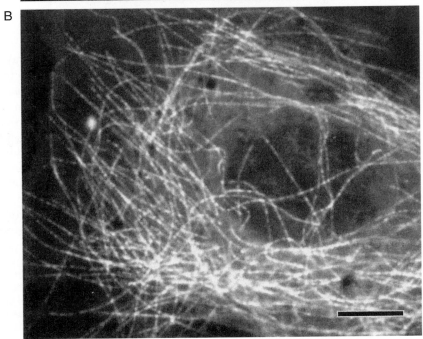

Using FSM, Maddox and co-workers (1999) recorded the poleward flux of tubulin taking place in spindles formed in *Xenopus* egg extracts (see also Waterman-Storer *et al.,* 1998). In order to visualize the flux of speckle pattern through these relatively thick (ca 5 μm) spindles with wide-field fluorescence microscopy (Fig. 20C), it was necessary to lable the microtubules with very low fractions (0.05%) of fluorescent tubulin (Fig. 20D). With such a low fraction of tubulin contributing to the fluorescent image, one could not readily discern the detailed structure of the spindle or discern the individual microtubules.

On the other hand, by observing a more heavily (1%) labeled *Xenopus* spindle (Fig. 20A) through the CSU-10, one could clearly record the poleward tubulin flux in the spindle fibers (Fig. 20B). In dynamic playback of the recorded time-lapsed scenes, such flux, reflecting the distribution of very few molecules of labeled tubulin, could even be observed in single microtubules growing from the spindle pole outside of the central spindle region (Fig. 20A, arrows).

As discussed in Section VI, the capability of the CSU-10 for capturing full-frame optical-sectioned images at considerably greater than video rate (1 to 4 ms/frame), albeit with the use of special high-speed cameras and recording devices, opens up further possibilities for monitoring rapid phenomena hitherto unobservable.

Very recently the CSU-10 reportedly has also been used for fluorescence resonance energy transfer experiments between modified GFP molecules (M. Shimizu, Yokogawa Electric Corporation, personal communication).

In addition to such advanced applications, the real-time, direct-view capability of the CSU-10, the ability to directly view confocal fluorescence images in true color, its adaptability to virtually any upright or inverted research microscope, its compact size and mechanical stability, the high quality of images obtained with brief exposures, its low rate of fluorescence bleaching, and its exceptional ease of use should open up many ready applications in cellular, developmental, and neurobiology, medical screening, etc.

Fig. 19 Confocal fluorescence speckle microscopy of microtubules (MTs) in tissue cultured cell. Panels A and B show two selected frames from a time-lapsed sequence of an interphase PTk2 cell injected with a low concentration of porcine tubulin labeled with X-rhodamine. The cells in panels A and B are the 32nd and 151st of the 152 frames that were recorded once every 4 s. Scale bar = 10 μm. As seen in the two panels, the speckle patterns on the *persistent* interphase MTs (seen in the middle of the cell) are stationary and remain remarkably constant over this long period of time. In contrast, the *free ends* of many of the MTs (pointing toward the cell periphery to the left of the panels) grow and shorten actively. In panel B, several new or longer individual MTs can be seen, while many others present in panel A have shortened or disappeared. These changes are dramatically displayed when the full sequence is played back dynamically. *Technical details:* Using the 568-nm illumination from a Kr/Ar laser, portions of the cell lying next to the coverslip were observed through the CSU-10 with an Olympus 100 × /1.3 NA objective lens. The confocal image (each exposed for 1700 ms) was captured with a Hamamatsu chilled-CCD camera (Orca II) in the 14-bit mode and stored as a stack file in MetaMorph. The somewhat-less sharp signal for panel A was enhanced to match panel B by first applying a median filter (horizontal = vertical size = 3, subsample ratio = 2) to prevent the rise of background noise, and then by applying an unsharp masking operation (low-pass kernel size = 3, scaling factor = 0.5, with auto result scale). (For the meaning of these operations, see Section VII.B of this article.) Data courtesy of Dr. Clare Waterman-Storer, Scripps Research Institute.

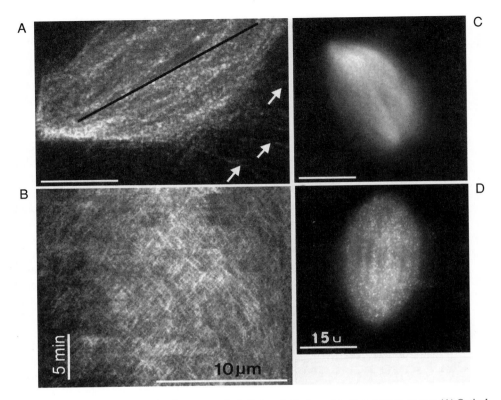

Fig. 20   Fluorescence speckle microscopy of tubulin in spindle formed in *Xenopus* egg extract. (A) Optical section recorded on a low-noise, chilled-CCD camera (Orca I, Hamamatsu Photonics) through CSU-10. Speckles are seen along spindle fibrils as well as in microtubules fanning out (arrows) from the spindle. Approximately 1% of tubulin was labeled with rhodamine-X. (B) Kimograph along the dark line in panel A. The speckles (revealing tubulin flux in the microtubules) move poleward at ca 2.0 µm/min. (C,D) Wide-field epifluorescence of similar spindle. Fluorescence speckles are visible in D in a spindle incorporating only ca 0.05% fluorescent tubulin, but not in C with ca 2% labeled tubulin. Scale bars for A, B = 10 µm, for C, D = 15 µm. (From Maddox *et al.,* 1999, with permission.)

The capability for video-rate and faster recording and display of dynamic events in optical section or as through-focused images, digital processing to rapidly obtain exceptionally sharp and clean optical sections or projections at any tilt angle, and striking depth-stacked or stereoscopic views that are generated using commercially available digital processors open up various possibilities for additional experiments and observations.

X. Addendum A (April, 2002)

Although three years have elapsed since this article was prepared, we believe that much of its content stands as originally presented. Nevertheless, we have appended the following brief descriptions to bring the article up to date.

A. Yokogawa has now introduced the CSU-21, which does not completely supercede but adds functions not present in the CSU-10. In brief, instead of a fixed speed of 1,800 rpm, the CSU-21 incorporates a variable speed motor that allows choice of Nipkow disk rotational speed of up to 5,000 rpm under computer control. Thus up to 1,000 full frame images can be acquired every second with the CSU-21. In addition, the fluorescence excitation filters, emission filters and dichromatic mirrors (3 each), neutral density filters, and illumination beam shutter can all be controlled either using the panel switches located below the eyepiece on the CSU-21 or remotely via computer. Remarkably, these new functions have been introduced without altering the external dimensions or the number of pinholes and microlenses (20,000 each) on the spinning disks. As before, the unit can be attached to any research-grade upright or inverted microscope, and images in full color can be viewed through the eyepiece in real time, or monitored or recorded through appropriate cameras mounted through C-mount threads.

B. The speed and capacity of computers and storage devices continue to improve by leaps and bounds. For example, newer computers using the Intel Pentium 4 running at over two GHz are several times faster than our reference computer, a 400 MHz Pentium II machine, reducing the computation time for a 3×3 neighborhood processing operation on 100 images from five to approximately one second, or several times faster than video rate (Section VII.A.). We have updated some numbers in Section VII.H. to reflect these changes.

C. Meanwhile, CCD cameras with considerably improved sensitivity, noise level, effective wavelength range, and general capabilities have become available (see, e.g., Maddox *et al.*, 2002, and articles in Pawley Handbook of Confocal Microscopy, 3rd edition, 2002, in preparation).

D. In an article on a Biorad confocal microscope incorporating SELS (signal-enhancing lens system), Reichelt and Amos compare the performance of their system with the CSU-10 (Microscopy and Microanalysis, Nov. 2002, pp. 9–11). As with the CSU-10, their SELS system operates with high photon capture efficiency, and hence can be used for observations of living cells with low rates of fluorescence bleaching and photon damage. The SELS point-scanning confocal system has the advantage that it can be switched to standard point scanning confocal mode with its greater depth discrimination. However, in either mode with the point scanner, extended scan time is required to capture images with any appreciable frame size. In contrast, the CSU provides full-frame capture at video- and higher-rate-imaging, an essential feature whenever the specimen is changing or moving moderately rapidly. While the authors of the SELS system argue that the SELS provides a much better z-discrimination compared to the CSU, their comparison is based on normalized z-axis intensity profiles obtained on a large thin film of a fluorescence dye (Nile Red). In fact, for cell biology applications where the imaging of fine specimen detail (rather than the location of an "infinite-sized fluorescent layer") is important, we believe that the CSU provides a considerably better depth discrimination than the SELS. For example, for yeast cells (*S. cervisiae*) the authors of the SELS article report that "the depth of focus became so great that most of the depth of each cell was included." In contrast, with the CSU-10 the focal depth was some one fifth of the thickness of the yeast cell, *S. pombe*, whose diameter is only ca. 3 μm (Fig. 10 of our article.) With both the SELS and CSU systems, depth discrimination and specimen detail are improved further by post processing, such as with unsharp masking (Fig. 14) or by applying deconvolution algorithms (Maddox *et al.,* 2002). The CSU is clearly the system of choice whenever full frame images of rapidly changing or moving specimens are to be recorded or viewed (e.g., Figs. 9 to 12), or when serial optical sections need to be acquired in rapid succession (e.g., Maddox *et al.,* 2002).

E. Selected additional references: (i) Members of the Salmon laboratory at the University of North Carolina provide details on the practical use of the CSU-10 for achieving very high resolution images in GFP imaging of a variety of living cells undergoing mitosis, including methods for visualizing the dynamic flow of tubulin molecules within the microtubules (Maddox *et al.,* 2002; see also Cimini *et al.,* 2001; Grego *et al.,* 2001; Waterman-Storer and Salmon, 1998). Maddox *et al.* (2002) also provide important data on the performance of the CSU-10 used in conjunction with selected hardware and software. (ii) Nakano and coworkers have made extensive use of the CSU-10 for analyzing endocytic and exocytic pathways of selected molecules in plant cells, including video-rate observations of the movement of Cop1 molecules between Golgi and ER in yeast cells (Saito *et al.,* 2002; Sato *et al.,* 2001; Ueda *et al.,* 2001; Yahara *et al.,* 2001). (iii) Tadakuma *et al.,* (2001) from the Funatsu group present data showing the real-time imaging capability of the CSU-10 for observing the behavior of fluorescence-labeled single protein molecules.

XI. Addendum B (personal communication from Dr. Kenneth R. Spring, June, 2002)

The extent of bleaching of fluorescence with the CSU-10 during illumination of a test specimen with 488-nm laser light was determined by K. R. Spring. A 1-μm-thick film of eosin-stained polymer was used to determine the fluorescence-bleaching rate caused by continuous illumination through a $40 \times /0.75$ objective lens at an intensity at the specimen of 45 microwatts/cm^2. The resulting fluorescence intensity decreased exponentially, as expected from photo bleaching with a half time of 445 seconds. When the experiment was repeated under identical conditions with another video-rate confocal attachment, the Noran Odyssey, the half time was not significantly different, 420 seconds.

Since the rate of bleaching is that expected for the illumination intensity, it is worthwhile to consider why users of the CSU-10 have reported less bleaching. We have identified two factors that contribute to the improved S/N of the CSU-10 that others and we have noted. First, because the excitation light is spread over 1200 pinholes in the field of view, each point in the image is irradiated with 1/1200 of the intensity of that in a video-rate point-scanning confocal system. Second, because each point is illuminated 12 times in a video frame, the collected image has already undergone some averaging.

When a point in the specimen is illuminated using the CSU-10, the fluorophores are probably all in their ground state because of the comparatively long time between each period of illumination (about 3 msec), the relatively long illumination period (about 30 μsec), and the relatively low intensity of the exciting light. In video-rate point-scanning of 512 points per horizontal line, each point is excited only once during each video frame for about 80 nsec with an intensity that is 1200 times higher than in the CSU-10. Under these circumstances, most of the fluorophores in the illuminated volume will be excited and cycle as rapidly as possible-between the ground and excited states (a fluorescein molecule could undergo only about 15–20 cycles of absorption and emission during the 8-nsec interval). The total fluorescence emission that a single fluorophore can produce with saturating excitation light in the 80-nsec interval is far less than that emanating from the same molecule illuminated for 12–30-μsec intervals with three orders of magnitude lower excitation light intensity.

The S/N achieved in the CSU-10 is improved to such a great extent that much lower excitation levels can be employed to produce visually acceptable images. The reduced bleaching that is observed with the CSU-10 most probably stems form the need for less excitation light on the specimen rather than form anything special about the manner in which that light is delivered to the specimen through the objective lens.

Acknowledgments

The authors thank Dr. Ken Spring of NHLBI, NIH, for contributing Addendum B, Dr. Brad Amos of Cambridge University for the gift of the dandelion pollen grain slide, Dr. John Murray of the University of Pennsylvania and Rieko Arimoto of Washington University for their efforts to prepare isolated sea urchin spindles, Dr. Brian Matsumoto of the University of California at Santa Barbara for participating in recording the swimming *Tetrahymena,* Dr. Phong Tran of Columbia University for providing records of yeast cells, Dr. Clare Waterman-Storer of Scripps Research Institute and Paul Maddox and co-workers of the University of North Carolina for providing the fluorescence speckle images, Dr. Yoshio Fukui of Northwestern University for providing the panels of *Dictyostelium,* Dr. Baggi Somasundaram of PerkinElmer Wallac for providing extensive image data, and Dr. Hideyuki Ishida of Tokai University and Dr. Akihiko Nakano of Riken Institute for providing information their recent work. We are grateful to members of the Yokogawa Electric Corporation for use of the CSU-10 confocal scanning unit, as well as for providing several original illustrations, information on the CS4-21 (see Addendum A), and list of publications; to members of Leica, Nikon Inc., and Carl Zeiss Inc. for use of their microscopes; and to members of Dage-MTI and Hamamatsu Photonics Systems for making available their chilled-CCD cameras. The authors also thank Dr. Edward D. Salmon and Paul Maddox of the University of North Carolina, Dr. Fred Lanni of the Carnegie Mellon University, Dr. Rudolf Oldenbourg of the Marine Biological Laboratory, Dr. Clare Waterman-Storer of Scripps Research Institute, and Dr. Phong Tran of Columbia University for thoughtful discussion on problems related to fluorescence fading and image-capture efficiency of the CSU-10. Finally we thank David Szent-Gyorgyi of Universal Imaging Corporation for providing extensive help in acquiring high-quality printouts from the MetaMorph imaging system, and Jane MacNeil and Bob Knudson of the MBL for invaluable help throughout the preparation of the many revisions of this article.

References

Agard, D. A. (1984). Optical sectioning microscopy: Cellular architecture in three dimensions. *Annu. Rev. Biophys. Bioeng.* **13,** 191–219.

Amos, W. B., and White, J. G. (1995). Direct view confocal imaging systems using a slit aperture. *In* "Handbook of Biological Confocal Microscopy" (J. B. Pawley, Ed.), pp. 403–415. Plenum Press, New York.

Aoki, T., Suzuki, Y., Nishino, K., Suzuki, K., Miyata, A., Iigou, Y., Serizawa, H., Tsumura, H., Ishimura, Y., Suematsu, M., and Yamaguchi, K. (1997). Role of CD18-ICAM-1 in the entrapment of stimulated leukocytes in alveolar capillaries of perfused rat lungs. *Am. J. Physiol.* **273,** H2361–H2371.

Castleman, K. R. (1979). "Digital Image Processing." Prentice-Hall, Englewood Cliffs, NJ.

Cimini, D., Howell, B., Maddox, P., Khodjakov, A., Degrassi, F., and Salmon, E. D. (2001). Merotelic kine-tochore orientation is a major mechanism of aneuploidy in mitotic mammalian tissue cells. *J. Cell Biol.* **153**(3), 517–527.

Denk, W., Piston, D. W., and Webb, W. W. (1995). Two-photon molecular excitation in laser-scanning microscopy. *In* "Handbook of Biological Confocal Microscopy" (J. B. Pawley, ed.), pp. 445–458. Plenum Press, New York.

Fukui, Y., Engler, S., Inoué, S., and de Hostos, E. L. (1999a). Architectural dynamics and gene replacement of coronin suggest its role in cytokinesis *Cell Motil. Cytoskel.* **42**, 204–217.

Fukui, Y., de Hostos, E., Yumura, S., Kitanishi-Yumura, T., and Inoué, S. (1999b). Architectural dynamics of F-actin in eupodia suggests their role in invasive locomotion in *Dictyostelium. Exp. Cell Res.* **249**, 33–45.

Funatsu, T., Harada, Y., Tokunaga, M., Saito, K., and Yanagida, T. (1995). Imaging of single fluorescent molecules and individual ATP turnovers by single myosin molecules in aqueous solution. *Nature* **374**, 555–559.

Genka, C., Ishida, H., Ichinori, K., Hirota, Y., Tanaami, T., and Nakazawa, H. (1999). Visualization of biphasic Ca^{2+} diffusion from cytosol to nucleus in contracting adult rat cardiac myocytes with an ultra-fast confocal imaging system. *Cell Calcium* **25**, 199–208.

Grego, S., Cantillana, V., and Salmon, E. D. (2001). Microtubule treadmilling *in vitro* investigated by fluores-cence speckle and confocal microscopy. *Biophys. J.* **81**, 66–78.

Holmes, T. J., Bhattacharyya, S., Cooper, J. A., Hanzel, D., Krishnamurthi, V., Lin, W.-C., Roysam, B., Szarowski, D. H., and Turner, J. N. (1995). Light microscopic images reconstructed by maximum likelihood deconvolution. *In* "Handbook of Biological Confocal Microscopy" (J. B. Pawley, ed.), pp. 389–402. Plenum Press, New York.

Ichihara, A, Tanaami, T., Isozaki, K., Sugiyama, Y., Kosugi, Y., Mikuriya, K, Abe, M., and Uemura, I. (1996). High-speed confocal fluorescence microscopy using a Nipkow scanner with microlenses for 3-D imaging of single fluorescent molecule in real time. *Bioimages* **4**, 57–62.

Inoué, S. (1986). "Video Microscopy." Plenum Press, New York.

Inoué, S., and Oldenbourg, R. (1995). Optical instruments. Microscopes. *In* Optical Society of America, ed., "Handbook of Optics," 2nd ed., Vol. 2. McGraw-Hill, Inc., pp. 17.1–17.52.

Inoué, S., and Spring, K. R. (1997). "Video Microscopy," 2nd ed. Plenum Press, New York.

Inoué, S., Tran, P., and Burgos, M. (1997). Photodynamic effect of 488-nm light on Eosin-B-stained *Spisula* sperm. *Biol. Bull.* **193**, 225–226.

Ishida, H., Genka, C., Hirota, Y., Nakazawa, H., and Barry, W. H. (1999). Formation of planar and spiral Ca^{2+} waves in isolated cardiac myocytes. *Biophys. J.* **77**, 2114–2122.

Juškaitis, R., Wilson, T., Neil, M. A. A., and Kozubek, M. (1996). Efficient real-time confocal microscopy with white light sources. *Nature* **383**, 804–806. See also News and Views by Dixon, T., on pp. 760–761.

Kino, G. S. (1995). Intermediate optics in Nipkow disk microscopes. *In* "Handbook of Biological Confocal Microscopy" (J. B. Pawley, Ed.), pp. 155–165. Plenum Press, New York.

Lanni, F., and Wilson, T. (1999). Grating image systems for optical sectioning fluorescence microscopy of cells, tissues, and small organisms. *In* "Imaging Neurons—A Laboratory Manual" (R. Yuste, F. Lanni, and A. Konnerth, Eds.), pp. 8.1–8.9. Cold Spring Harbor Laboratory Press, Cold Spring Harbor, NY.

Lipton, L. (1991). "The Crystal Eyes Handbook." StereoGraphics Corp., San Rafael, CA.

Maddox, P., Desai, A., Salmon, E. D., Mitchison, T. J., Oogema, K., Kapoor, T., Matsumoto, B., and Inoué, S. (1999). Dynamic confocal imaging of mitochondria in swimming *Tetrahymena* and of microtubule poleward flux in *Xenopus* extract spindles. *Biol. Bull.* **197**, 263–265.

Maddox, P. S., Moree, B., Canman, J., and Salmon, E. D. (2002). A spinning disk confocal microscopy system for rapid high resolution, multimode, fluorescence speckle microscopy and GFP imaging in living cells. *Methods in Enzymology,* in press.

Pawley, J. B., Ed. (1995). "Handbook of Biological Confocal Microscopy." Plenum Press, New York.

Petráň, M., Hadravsky, M., Egger, M. D., and Galambos, R. (1968). Tandem scanning reflected light microscope. *J. Opt. Soc. Am.* **58**, 661–664.

Russ, J. C. (1995). "The Image Processing Handbook," 2nd ed. CRC Press, Boca Raton, FL.

Saito, C., Ueda, T., Abe, H., Wada, Y., Kuroiwa, T., Hisada, A., Furuya, M., and Nakano, A. (2002). A complex and mobile structure forms a distinct subregion within the continuous vacuolar membrane in young cotyledons of *Arabidopsis. Plant J.* **29,** 245–255.

Sato, K., Sato, M., and Nakano, A. (2001). Rer1p, a retrieval receptor for ER membrane proteins, is dynamically localized to Golgi by coatomer. *J. Cell Biol.* **152,** 935–944.

Shotton, D. Ed. (1993). "Electronic Light Microscopy." Wiley-Liss, New York.

Squirrell, J. M., Wokosin, D. L., White, J. G., and Bavister, B. D. (1999). Long-term two-photon fluorescence imaging of mammalian embryos without compromising viability *Nat. Biotechnol.* **17,** 763–767.

Tadakuma, H., Yamaguchi, J., Ishihama, Y., and Funatsu, T. (2001). Imaging of single fluorescent molecules using video-rate confocal microscopy. *Biochem. Biophys. Res. Comm.* **287,** 323–327.

Tran, P. T., Maddox, P., Chang, F., and Inoué, S. (1999). Dynamic confocal imaging of interphase and mitotic microtubules in fission yeast, *S. pombe. Biol. Bull.* **197,** 262–263.

Tsien, R., and Bacskai, B. J. (1995). Video-rate confocal microscopy. *In* "Handbook of Biological Confocal Microscopy" (J. B. Pawley, Ed.), pp. 459–478. Plenum Press, New York.

Ueda, T., Yamaguchi, M., Uchimiya, H., and Nakano, A. (2001). Ara6, a plant-unique novel-type Rab GTPase, functions in the endocytic pathway of *Arabidopsis thaliana. EMBO J.* **20,** 4730–4741.

Vale, R. D., Funatsu, T., Pierce, D. W., Romberg, L., Harada, Y., and Yanagida, T. (1996). Direct observation of single kinesin molecules moving along microtubules. *Nature* **380,** 451–453.

Waterman-Storer, C. M., and Salmon, E. D. (1998). How microtubules get fluorescent speckles. *Biophys. J.* **75,** 2059–2069.

Waterman-Storer, C. M., Desai, A., Bulinski, J. C., and Salmon, E. D. (1998). Fluorescent speckle microscopy, a method to visualize the dynamics of protein assemblies in living cells. *Curr. Biol.* **8** 22, 1227–1230.

Yahara, N., Ueda, T., Sato, K., and Nakano, A. (2001). Multiple roles of Arf1 GTPase in the yeast exocytic and endocytic pathways. *Mol. Biol. Cell* **12,** 221–238.

Yamaguchi, K., Nishio, K., Sato, N., Tsumura, H., Ichihara, A., Kudo, H., Aoki, T., Naoki, K., Suzuki, K., Miyata, A., Suzuki, Y., and Morooka, S. (1997). Leukocyte kinetics in the pulmonary microcirculation: observations using real-time confocal luminescence microscopy coupled with high-seed video analysis. *Lab. Invest.* **76**(6), 809–822.

Yuste, R., Lanni, F., and Konnerth, A. Eds. (1999). "Imaging Neurons—A Laboratory Manual." Cold Spring Harbor Laboratory Press, Cold Spring Harbor, NY.

CHAPTER 3

Introduction to Multiphoton Excitation Imaging for the Biological Sciences

Victoria E. Centonze

Department of Cellular and Structural Biology
University of Texas Health Science Center at San Antonio
San Antonio, Texas 78229-3900

I. Introduction

Since its inception more than 350 years ago with van Leeuwenhoek's single lens magnifier, the light microscope has undergone a steady evolution punctuated by significant and sometime revolutionary leaps in technology. One such revolution was the development of the laser scanning confocal microscope in the late 1980s (Carlsson *et al.,*

1985; White *et al.,* 1987; Brakenhoff *et al.,* 1989) and more recently the development of multiphoton excitation microscopy (Denk *et al.,* 1990).

For many decades epifluorescence microscopy was accepted as a powerful tool for biological research and medical diagnostics. Since the signal emitted from a fluorescent probe is detected against a dark background it provides a sensitive method for detection of structures within cells and tissues. Over the years significant effort was devoted to the development of reagents (antibodies and ligands) to probe for practically any desired biological molecule. Although fluorescence techniques are more commonly applied to the inspection of fixed samples, it has become possible to use fluorescent techniques for observation of living material. Initially, purified proteins were derivatized and subsequently introduced into living cells where their dynamic behavior could be visualized. Labeled proteins can now be expressed within cells by genetically engineering a chimeric protein that has a naturally fluorescent protein, such as GFP (Chalfie *et al.,* 1994), as an endogenous probe. Indicator molecules that change fluorescence characteristics as they "sense" changes in ion concentration (Ca^{2+}, Na^+, K^+, Mg^{2+}, Zn^{2+}), pH, and membrane potential have also been introduced into cells, thus allowing the physiological state of individual cells to be monitored optically.

For fluorescence microscopy to be useful one must be able to successfully excite the fluorophore and subsequently detect its emission. And in the case of live cell observation, one must do so under conditions that minimize the production of phototoxic compounds that can be detrimental to the biological material. Fluorescence microscopy is problematic when attempting to make observations within thick or light-scattering material. The image collected from a wide-field fluorescence microscope consists of signal from fluorophore in the plane of focus over- and underlaid by blurred signal emanating from all other regions throughout the volume of the specimen. As the specimen thickness increases more fluorescent material contributes to the out-of-focus haze. This increase in background can be great enough to completely obscure the signal from the plane of focus. Optical sectioning fluorescence microscopy provides elegant solutions to the problem of imaging thicker specimens.

Deconvolution microscopy is a wide-field-based imaging system that computationally removes out-of-focus blur (Castleman, 1979; Agard *et al.,* 1989). A sequential stack of images through the volume of a specimen is digitally acquired from a conventional epifluorescence microscope. Each image contains focused signal from a particular plane as well as defocused signal from other planes. Algorithms are then applied to the data that account for the behavior of light through the biological material and the optical system through which it is observed. Out-of-focus light is deblurred and reassigned to its origin, thus leaving in the image of each plane only signal that is calculated to have originated from that plane. This method has several disadvantages: it is computationally intensive, an optical section is not immediately viewable, it is less successful when samples are so blurry that they appear virtually featureless, and repeated wide-field illumination required to acquire a stack of images could result in significant photodamage (Periasamy *et al.,* 1999).

Laser scanning confocal microscopy is an alternative optical sectioning fluorescence technique for imaging thick specimens. For confocal imaging, a specimen is illuminated point-by-point in a raster scan. Signal is generated at the point of focus, collected by the objective, and reimaged as a spot of light (an Airy disk) at a conjugate plane immediately before the detector. Signal is also generated within a cone of light both above and below the point of focus. This out-of-focus light appears at the conjugate plane as concentric rings surrounding the Airy disk of in-focus light. A pinhole at the conjugate plane is positioned to pass in-focus light to the detector but block out-of-focus light from reaching the detector. The finest optical section is obtained when the pinhole diameter matches the diameter of the Airy disk. A high-contrast image of fluorescent signal in an optical section is digitally acquired and displayed on a computer monitor as the specimen is being scanned. In this way a high signal-to-noise image can be generated as long as a focused point of light reaches the fluorophore within the depths of a specimen and as long as the emitted light can travel back out of the specimen with sufficient integrity to be reimaged at the pinhole. It is difficult to meet these criteria as the sample becomes excessively thick, if the specimen is heavily stained, or if the specimen has significant light-scattering properties. To generate a single optical section photodamage is less of a concern in confocal imaging than in wide-field deconvolution. However, if multiple optical sections are to be acquired (XYT, XYZ, and especially XYZT), then photodamage to the specimen may be significant. It must be remembered that even though out-of-focus fluorescence is not detected, it is generated. This represents a comparatively large volume in which phototoxic products may be produced (Centonze and Pawley, 1995).

Multiphoton excitation imaging represents a further refinement of optical sectioning fluorescence microscopy techniques. Although it suffers from being a slightly lower resolution imaging technique, it is superior to any other imaging technique for thick tissue observation and in some instances for preservation of live specimen viability.

II. Principles of Multiphoton Imaging

Multiphoton excitation imaging is based upon the theoretical work of Maria Göppert-Mayer (Göppert-Mayer, 1931). In her doctoral thesis she predicted that fluorophore excitation could occur if two photons were absorbed simultaneously in a single quantized event. A fluorophore will reach its excited state if it absorbs sufficient energy. This energy can be delivered in packets of one, two, or three photons, as long as those packets are incident on the fluorophore simultaneously. Regardless of the mode of excitation, the decay is the same, resulting in the release of a photon of the expected wavelength from the fluorophore. Since the energy of a photon is inversely proportional to its wavelength, excitation of a fluorophore will require two photons of twice the wavelength (but half the energy) of conventional one-photon excitation. Demonstration Göppert-Mayer's theory was not possible until the development of laser technology in the 1960s when two-photon fluorescence excitation was achieved by Kaiser and Garrett (1961) and three-photon excitation by Singh and Bradley (1964).

It was another 30 years before multiphoton excitation was recognized as a practical imaging method for optical sectioning fluorescence microscopy. Denk, Strickler, and Webb (1990) were the first to apply two-photon excitation fluorescence to laser scanning microscopy. Multiple photon absorption events are extremely rare at typical photon densities used for epifluorescence microscopy (Denk and Svoboda, 1997). Therefore, to initiate enough excitation events for practical imaging it was essential to illuminate

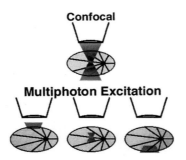

Fig. 1 Diagrammatic representation of the volume of excitation for confocal and multiphoton imaging. Upper panel depicts illumination used in confocal imaging: all fluorescent structures in the beam path are excited; light emanating from above and below the plane of focus must be eliminated by the confocal blocking aperture. Lower three panels depict passage of short pulse of infrared light through a sample. Only at the point of focus is the photon density sufficient to elicit multiphoton excitation. Fluorescence continues to be emitted for the duration of the excited-state lifetime of the fluorophore after the passage of the short-pulse excitation. (Reproduced from Fig. 97.1, "Cells: A Laboratory Manual," Vol. 2, Spector, Goldman, and Leinwand, eds. Cold Spring Harbor Laboratory Press, New York, 1998.) (See Color Plate.)

with a very high photon density. Delivering the required photon density in a continuous wave would require excessively high laser powers resulting in damage to the specimen being observed. The high photon density required for multiphoton excitation is attained by designing the laser to deliver light in short pulses of high amplitude that is then focused into a confined region by the objective lens of the microscope (Fig. 1; see color plate). The relatively low duty cycle of the pulses keeps the mean power incident on the specimen relatively low to avoid specimen damage due to heating. Repetition rates of around 100 MHz ensure that pulses are delivered at a rate that is longer than the average fluorescent lifetime (a few nanoseconds) of a fluorophore. In this way the fluorophore has sufficient time between excitation events to decay to its ground state, thus avoiding fluorophore saturation that can decrease the signal obtained from a sample.

Once focused by the objective the local beam intensity is high enough to make multiple photon absorption events sufficiently frequent to generate detectable signal while imaging. Excitation is spatially confined because the photon density required for excitation dissipates rapidly away from the plane of focus. For two-photon excitation the local beam intensity sufficient for excitation falls off as the square of the distance from the plane of focus. Three-photon excitation gives improved spatial resolution compared to two-photon excitation at the same wavelength because of the cubic dependence of excitation on incident illumination power density. Thus, for each point of illumination, the excitation of the fluorophore occurs only in a small, three-dimensionally confined volume and for a given fluorophore the axial resolution is greater for three-photon excitation (Gryczynski *et al.,* 1995; Gu, 1996; Hell *et al.,* 1996; Wokosin *et al.,* 1996a). No excitation of fluorophore occurs in regions immediately adjacent to the plane of focus so that as the beam is scanned a thin optical section is generated by virtue of excitation without the need to introduce a pinhole (Denk *et al.,* 1994; Williams *et al.,* 1994; Denk, 1996; Denk and Svoboda, 1997). The focal spot of the longer wavelength laser light

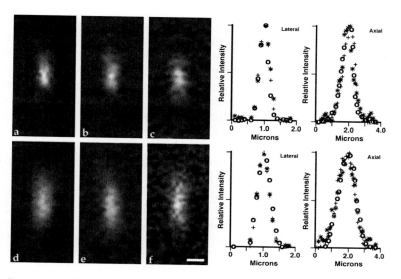

Fig. 2 Comparisons of point spread functions between confocal (a–c) and multiphoton (d–f) microscopy. Z-Series were taken of subresolution beads within a model sample at three depths: shallow, with 10 μm of the surface (a and d); middle, between 30 and 40 μm (b and e); and deep, approx. 70 μm (c and f). Visually, the dimensions of the point-spread functions do not appear to increase with depth of imaging. This result is illustrated by graphs of normalized intensity versus axial position in the bead imaged by confocal (upper graph) or multiphoton (lower graph) microscopy at three levels within the sample (○, shallow; +, middle; ∗, deep). The lateral resolutions for confocal and multiphoton microscopy were 0.35 μm and 0.45 μm, respectively. The axial resolutions for the confocal and multiphoton microscopy were 0.85 μm and 1.25 μm, respectively. Images were collected at a pixel resolution of 0.1 μm with a Kalman 1 collection filter. Scale bar, 0.5 μm. (Reproduced from Centonze and White, © 1998, *Biophysical Journal* **75,** 2015–2024.)

used for multiphoton excitation is larger than that produced by shorter wavelength lasers used for conventional excitation. As a result there is a slight degradation of resolution (Fig. 2; Gu and Sheppard, 1995; Centonze and White, 1998). Introduction of the confocal pinhole can improve the resolution slightly; however, this advantage is offset by a significant loss in signal (Gauderon *et al.,* 1999). Since the confocal pinhole is not a requirement for imaging it is no longer necessary to descan the fluorescent signal. Rather, emission from the fluorophore can be collected directly from the objective lens by a photon detector, thus simplifying the detection pathway and increasing the sensitivity of detection (Williams *et al.,* 1994; Wokosin and White, 1997; Soeller and Cannell, 1999).

III. Advantages of Multiphoton Imaging

A. Increased Depth of Imaging Penetration

As one attempts to visualize fluorescent structures deep within light-scattering samples with confocal microscopy there are a number of problems that can arise which cause significant deterioration of the signal. First, excitation light may be absorbed by fluorophore above the plane of focus thus attenuating the light available for excitation in lower planes. A bright fluorescent structure in upper planes can effectively shadow the lower regions. Second, the excitation beam may be scattered by the biological material thereby reducing the amount of light that reaches the focal plane. Third, detectable signal may be reduced as emitted light is scattered from its predicted path and is therefore not reimaged at the pinhole. Essentially scattered emission from the point of focus gets mixed up with the out-of-focus signal and is blocked by the pinhole. Fourth, light generated out of the plane of focus is also scattered and can "sneak" in through the pinhole thus increasing the background. The overall result is that signal is reduced and background increases. Attempts to penetrate deeper into tissue by increasing incident laser power results in a brighter, blurry image.

Multiphoton excitation imaging does not suffer from the limitations described above. First, photons are absorbed only in the plane of focus. Overlying material is essentially transparent to long wavelengths of light and does not attenuate the incident beam. Second, longer wavelengths are scattered less than shorter wavelengths (Stelzer *et al.*, 1994; Potter *et al.*, 1996). Since multiphoton excitation starts with light that is about twice as long as required for conventional excitation, there is a better chance for the beam to reach deeper in and still be focused. Third, the emitted light is still scattered but this becomes less of an issue since a pinhole is not needed to create the optical section. Multiphoton excitation generates emitted photons from a spatially defined point in a sample, so to contribute to the image all they must do is reach the detector; the path taken through the biological material is irrelevant. As such, the detection system for multiphoton imaging can be simplified by placing the detectors external to the scanhead for whole-area detection (Denk *et al.*, 1994). This results in a threefold increase in detection sensitivity compared to confocal imaging with the same photomultiplier tube. This was measured by David Wokosin at the Integrated Microscopy Resource, Madison, WI, and by Brad Amos at the Medical Research Council (MRC), Cambridge, UK. Fourth, multiphoton excitation does not generate signal out of the plane of focus so the issue of increased background is eliminated. Considering these factors one can increase the depth of imaging penetration through light scattering material by increasing the laser power. The result is a high-contrast image that shows significant structural detail (Fig. 3).

In general, multiphoton imaging increases the depth of imaging penetration into a thick sample by at least two to three times compared with confocal imaging (Centonze and White, 1998; Gerritsen and De Grauw, 1999). As one images deeper into thick material it becomes particularly important to match the index of refraction of the sample medium to the immersion medium. Mismatch between the specimen and the objective results in a progressive degradation of the point-spread function (PSF) as depth into the specimen increases. Configuring the multiphoton microscope with wide-field detection makes it insensitive to aberrations in the detection pathway, but aberrations in the illumination pathway still cause degradation of the PSF. Degrading the PSF reduces the axial resolution giving rise to a severe loss of detectable fluorescence signal (Jacobsen *et al.*, 1994; Gerritsen and De Grauw, 1999). Therefore, selection of mounting medium and choice of lenses is particularly important when using multiphoton excitation to image deep within tissue.

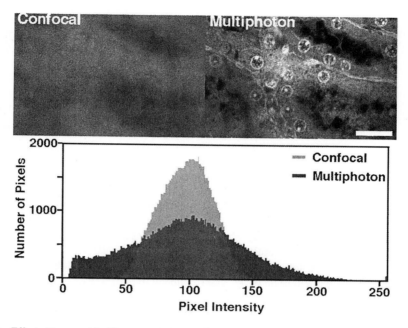

Fig. 3 Effect of increased incident power on generation of signal. Samples of acid-fucsin-stained monkey kidney were imaged at a depth of 60 μm into the sample by confocal (550 μW of 532-nm light) and by multiphoton (12 mW of 1047-nm light) microscopy. Laser intensities were adjusted to produce the same mean number of photons per pixel. The confocal image exhibits a significantly narrower spread of pixel intensities compared to the multiphoton image indicating a lower signal-to-background ratio. Multiphoton imaging therefore provides a high-contrast image even at significant depths within a light-scattering sample. Images were collected at a pixel resolution of 0.27 μm with a Kalman 3 collection filter. Scale bar = 20 μm. (Reproduced from Centonze and White, © 1998, *Biophysical Journal* **75**, 2015–2024.)

B. Improved Specimen Viability

Since nonlinear excitation of a fluorophore is restricted to the volume close to the geometrical focus of the light, it therefore follows that for each scan of a sample the processes of photobleaching and photodamage are also restricted to the same limited volume (Denk *et al.*, 1990; Stelzer *et al.*, 1994). Using multiphoton excitation imaging it is no longer possible to photobleach signal in the depths of the sample while focusing on signal at the surface. The reduced volume of photobleaching means that a three-dimensional data set may be collected for the same integrated excitation dose as a single focal plane in a confocal microscope. As such the total volume of fluorophore bleached in a sample in one confocal Z-series of 10 sections is comparable to the total volume of fluorophore bleached in a multiphoton Z-series of 10 sections repeated 10 times. Additionally, limiting the volume of excitation using multiphoton excitation also limits the concentration of phototoxic by-products of excitation such as singlet oxygens and free radicals. This may bring the rate of production of these toxic products down to levels that can be eliminated by the intrinsic scavenging mechanisms of living tissue, such as catalase and superoxide dismutase (Halliwell, 1996). Thus physiological experiments can be performed with less perturbation due to imaging (Denk *et al.,* 1994, 1995b; Yuste and Denk, 1995; Svoboda *et al.,* 1996, 1997).

The benign nature of multiphoton imaging compared with confocal imaging has been promoted from the beginning (Denk *et al.,* 1990). However, a demonstration of this phenomenon was not published until 1999 (Squirrell *et al.,* 1999). Hamster embryos are extremely sensitive to light, making imaging of these living embryos with any form of microscopy, in particular fluorescence microscopy, technically difficult. Time-lapse imaging of living embryos was a goal to monitor responses of the embryos to various environmental conditions and ultimately assess developmental competence (Barnett *et al.,* 1997). Two-cell embryos were stained and imaged in timelapse for 8 h. They were subsequently cultured to an age that should produce a blastocyst. Virtually all of the embryos imaged by confocal microscopy using 514-nm, 532-nm, or 568-nm light were developmentally arrested at the two-cell stage. However, similarly prepared embryos imaged by multiphoton excitation using a 1047-nm femtosecond pulsed Nd:YLF laser were able to divide during the period of imaging and continued to blastocyst stage subsequent to imaging. The frequency of data collection and duration of time-lapse imaging were increased without affecting developmental progression. In fact, blastocysts that had undergone long-term multiphoton imaging were implanted in surrogate females, were born, and proved to be reproductively competent adults. These data show unequivocally the benign nature of multiphoton imaging with 1047-nm light compared with visible light confocal imaging. Heating due to absorption of long-wavelength light by water is of concern when using the high photon flux demanded by multiphoton imaging (Williams *et al.,* 1994); however these data and others (Wokosin *et al.,* 1996b; Schönle and Hell, 1998) show that it is not the major mechanism causing photodamage of living specimens. An intriguing result of the studies with hamster embryos was that developmental progression was halted by visible light illumination alone and did not require the presence of fluorophore, so by-products of fluorophore excitation were not the culprits. Photoinduced production of H_2O_2 was demonstrated to occur during wide-field and confocal imaging but not multiphoton imaging at 1047 nm. Production of H_2O_2 is suspected to result from light-initiated photoreduction of flavins that activate flavin-containing oxidases in mitochondria and peroxisomes resulting in H_2O_2 production (Hockberger *et al.,* 1999; Squirrell *et al.,* 1999).

The dramatic improvement in viability of a developing embryo imaged by multiphoton excitation was documented for 1047-nm light from a Nd:YLF laser (Squirrell *et al.,* 1999). In this study and others (Wokosin *et al.,* 1996b; Centonze, unpublished data) it was difficult to elicit observable damage in living specimens due to irradiation with pulsed light at this wavelength even at maximum power from the laser (in excess of 10 mW average power at the specimen plane using high-NA objectives or 100 mW average power at the specimen plane using low-NA lenses). The susceptibility of a specimen to photodamage by multiphoton excitation will likely vary with the specimen and with the wavelength of light used for excitation. Cloning efficiency of cells has been used as a benchmark for cell viability under various multiphoton excitation regimes. Unstained

Chinese hamster ovary (CHO) cells exposed to pulsed light ($\leq$800 nm, 150 fs) were unaffected when the average power was $\leq$1 mW; however, at $\geq$6 mW cells died or were significantly impaired and complete cell destruction occurred at >10 mW (König et al., 1997). In a subsequent study (König et al., 1999) CHO cells were irradiated with either 780-nm or 920-nm light with a pulse width of approx. 120 fs in order to assess the effect of different wavelengths on cell viability. Endogenous materials can absorb each of these wavelengths by either a two- or three-photon process thus having the potential to affect cell viability. In addition, the absorption coefficient of water is seven times greater at 920 nm than at 780 nm. Changing the wavelength to 920 nm reduced destructive effects, indicating that photothermal effects are not responsible for photodamage (König et al., 1999).

IV. Imaging Parameters

A. Pulse–Width Considerations

Since damage due to water heating is not the major culprit compromising photodamage, destructive nonlinear effects must be due to one-photon absorption of the primary beam, two-photon absorption by materials such as the coenzymes NAD(P)H, porphyrins, and flavins, or three-photon absorption by nucleic acids and amino acids. If linear absorption is the dominant mechanism for photodamage then damage depends on average intensity and not on peak intensity of excitation light. In this case minimizing the pulse width is essential. However, the efficacy of using increased average power to perform two-photon imaging with picosecond pulse (Bewersdorf and Hell, 1998) or CW excitation (Hell et al., 1998) indicates that nonlinear optical effects are the major limitation on live cell imaging. If two-photon absorption is the dominant mechanism, then photodamage and the two-photon excitation rate will be proportional to each other. If three-photon (or higher) absorption is the dominant mechanism for photodamage to occur then increasing the peak intensity by decreasing the pulse width would be a detriment to live cell imaging. Two studies have each demonstrate the quadratic dependence of absorption and photodamage. In one study, cell cloning was assessed for CHO cells irradiated with either 240-fs or 2.2-ps pulses of 780-nm light. A ninefold increase in pulse length required a threefold greater mean power to generate the same damage effect (König et al., 1999). The second study monitored the relative change in basal fluorescence of basal dendrites deep within slices of rat neocortex. The rate of photodamage measured as a function of average laser power for 75 fs to 0.8–3.2 ps pulses of laser light showed a power order of 2. This indicates that two-photon absorption is the predominant mechanism for photodamage. Therefore, photodamage is proportional to the signal and signal can only be generated at the expense of photodamage (Koester et al., 1999). The results of these studies prove that for low excitation rates the pulse width is irrelevant for imaging of live cells. It should be noted that at higher excitation rates the power order can be greater than 2, indicating that higher order effects may contribute to photodamage. Choice of pulse width for multiphoton excitation of most specimens can now be considered moot.

Femtosecond pulses are preferred when one wants to perform three-photon excitation and when one needs to image so deep into thick tissue that the maximum power output of the picosecond pulsed laser is inadequate.

B. Three-Photon Excitation

Use of short-wavelength dyes for imaging is problematic for both technical and biological reasons. Specially corrected optics are required to be able to pass wavelengths ranging from the UV to the visible (Webb, 1990; Piston et al., 1995). The short wavelengths needed for excitation are often damaging to living specimens because of the presence of endogenous UV-absorbing proteins. Multiphoton excitation imaging provides a mechanism for circumventing these limitations; short-wavelength dyes are excited by relatively benign long-wavelength light, and no UV wavelengths need be used (Williams et al., 1994). UV-excitable dyes can be excited at shorter NIR wavelengths by two-photon absorption or at longer NIR wavelengths by three-photon absorption (Wokosin et al., 1996a,b; Hell et al., 1996; Xu et al., 1996). Three-photon excitation may be the preferred mode for better resolution or for limiting the volume of photobleaching or uncaging. The complement of molecules that can be used for optical imaging may now include fluorophores that excite deep in the UV (<300 nm), a region that is very problematic for conventional microscope optics (Maiti et al., 1997).

The facility to excite some fluorophores by either two- or three-photon excitation is advantageous for imaging specimens containing multiple labels; the "red" and "green" dyes can be excited by two-photon absorption while the "blue" dyes can be excited by three-photon absorption. In practice, this means that multiphoton excitation can be used to image multiple labels simultaneously (Wokosin et al., 1996a, 1997). Use of a single wavelength for excitation ensures that the signals from spectrally different fluorophores originated from the same volume. If different wavelengths are used to excite each fluorophore a correction may be needed to compensate for differences in focal depth for each wavelength.

C. Probes

The option to excite fluorophores with either two- or three-photon absorption greatly expands the range of molecules that can be useful for multiphoton imaging. However, it is difficult to predict the multiphoton excitation spectrum from the single-photon spectrum. Some fluorophores exhibit differences in excitation characteristics depending on whether the mode of excitation is from single- or two-photon absorption. Knowing this, it will be important to determine the multiphoton excitation spectra in order to select optimum excitation wavelengths and to select appropriate combinations of fluorophores for multilabeling. Multiphoton spectral data exists for only a handful of dyes commonly used in biological research (Xu et al., 1996). Therefore, it falls to the user to determine empirically whether the fluorophore will be appropriate for a particular biological application. Measurements that have been done show that generally the peak of excitation is either the same or blue-shifted from that predicted from the single-photon excitation data

(Xu *et al.,* 1996). Another characteristic of two-photon absorption is that the excitation spectrum is often broader than for single-photon excitation. As a result, irradiation of a fluorophore with wavelengths off of the peak can still produce sufficient excitation for imaging. Broadening of the excitation spectra increases the likelihood of being able to excite simultaneously by multiphoton absorption dyes that have dissimilar single-photon absorption spectra (e.g., rhodamine and fluorescein; Wokosin *et al.,* 1996a, 1997). Three-photon cross-sections of fluorophores have been documented (Davey *et al.,* 1995; He *et al.,* 1995). Measurement of three-photon excitation spectra reveals that they parallel the single-photon spectra (Xu *et al.,* 1996). Most known known molecules have relatively small two-photon cross sections. Based on structure–properties studies for two-photon absorption, design strategies have been developed resulting in the synthesis of molecules with dramatically larger two-photon cross sections (Albota *et al.,* 1998). In addition, caging groups, azido (Adams *et al.,* 1997) or bromo-7-hydroxycoumarin-4-ylmethyl (Furuta *et al.,* 1999), have higher two-photon cross-sections than the conventional CNB group. Development of such probes will facilitate a variety of applications of multiphoton excitation thus enabling the full utility of multiphoton excitation imaging to be realized.

While the excitation spectra of fluorophores are sensitive to the mode of excitation, the emission spectra of these fluorophores are not. There appears to be no difference between the emission spectra of one- and two-photon excited fluorescence (Curley *et al.,* 1992; Xu *et al.,* 1995; Xu and Webb, 1996), nor is there a difference between two- and three-photon excited fluorescence (Xu *et al.,* 1996). This suggests that the same fluorescence states are reached by either linear or nonlinear excitation. Therefore, conventional barrier filters may be used to selectively detect the emitted signal.

D. Instrumentation

1. Optimization of a Multiphoton Microscope

In principle, any beam-scanning confocal microscope can be converted to a multi-photon microscope by replacing the continuous wave laser source with a pulse laser (Denk *et al.,* 1990, 1995a). The different excitation wavelengths require replacement of a dichroic beam splitter that separates fluorescence from excitation. The scanning mirrors should also have an enhanced silver coating to optimize reflectivity of the infrared wavelengths (Denk and Svoboda, 1997; Wokosin and White, 1997). Signal may be descanned back into the PMTs inside the scanhead or whole-area detection may be employed. Whole area detection may be achieved by placing a PMT directly beneath the objective (Williams *et al.,* 1994; Wokosin and White, 1997; Koester *et al.,* 1999; Soeller and Cannell, 1999) or by placing a PMT behind a high-numerical-aperture condenser (Koester *et al.,* 1999).

If the signal is descanned then the option of inserting the imaging aperture is retained. This may be desirable if it is necessary to obtain the maximum possible resolution. However, the increase in resolution is obtained at the expense of loss of signal, the pinhole will block scattered light emanating from the sample which could otherwise contribute to the formation of the image (Gauderon *et al.,* 1999). For this reason there will also be a decrease the depth from which signal can be collected. With the pinhole in place the signal falloff for multiphoton imaging at increasing depths within the sample becomes comparable to the falloff seen for confocal imaging.

The option of using external PMTs for whole-area signal detection avoids sending the emission back through the scanhead. Since all of the signal generated in multiple-photon excitation originates from the plane of focus, it is no longer necessary to reimage the diffraction-limited spot of emitted light in order to impose an imaging aperture to eliminate out-of-focus information. Therefore a PMT can be placed immediately beneath the objective or above the condenser to detect all the light collected by the lens. Having a shorter emission path facilitates the collection of photons from a greater solid angle. A demagnification lens may be placed in the detection path to direct emitted photons to the face of the detector (Koester *et al.,* 1999). Thus, more scattered photons can be detected, thereby increasing the signal detected very deep in light scattering samples. Whole-area detection results in a significant increase in detection sensitivity (Wokosin and White, 1997; Koester *et al.,* 1999).

2. Laser Choices

A suitable laser source for multiphoton excitation must be capable of generating short pulse width (picoseconds or femtoseconds; see section IV.A, Pulse-Width Considerations) at repetition frequencies of around 100 MHz with sufficient peak power to generate detectable amounts of multiphoton fluorescence. Ideally, to take advantage of any available fluorophore, the laser should also be widely tunable. The implementation of all-solid-state crystals and passive mode-locking for pulse formation has enabled the development of ultrafast, pulsed laser sources that are more affordable, robust, and easier to use than the original pulsed laser systems. The new lasers conventionally have average power outputs in excess of 500 mW and can therefore deliver sufficient power at the specimen to make multiphoton imaging possible.

Tunability in a laser may be an attractive option for a laboratory in which a variety of fluorophores are used. In practice, though, it is more likely that the prospective imaging applications will define a range of required wavelengths. Once this range is determined then a choice can be made from the available laser alternatives (Wokosin *et al.,* 1997). However, the inconvenience of tuning the laser to different wavelengths may in fact make tunability a liability. Manufacturers have developed extended range optics, thus making laser tuning a less invasive though still bothersome activity. For investigators requiring only a limited tuning range (<100 nm), the newest model lasers may be a viable option. These lasers have "one-knob" tuning that makes changing between wavelengths even less difficult.

Another option is to select a long-wavelength laser and take full advantage of multiple photon excitation. A fixed-wavelength Nd:YLF laser emitting only 1047-nm light has been used to good effect as a source for a combination of two- and three-photon excitation. By illuminating a specimen with only a single wavelength it is possible to excite three different fluorophores simultaneously, e.g., rhodamine and fluorescein excited by two-photon absorption and DAPI excited by three-photon absorption (Wokosin *et al.,* 1996a, 1997).

V. Applications of Multiphoton Imaging

For most of the applications requiring fluorescence optical sectioning, confocal imaging is the method of choice for acquiring images. The data are collected at high resolution and high signal-to-noise thus producing crisp images of fluorescent structures within an optical section. However, this technology is limited by problems of photobleaching, phototoxicity, and image degradation due to light scatter. In situations when these phenomena prevent the acquisition of data, multiphoton becomes the better alternative.

A. *In Vivo* Imaging of Cancerous Tissue

Multiphoton excitation can penetrate through the stratum corneum to image fluorescence generated deep within the multiple cell layers of human skin. Autofluorescence in skin from connective tissue (e.g., collagen and elastin) in the dermis and from reduced pyridine nucleotides [NAD(P)H] in the epidermis can be viewed to determine structural integrity and to monitor cellular oxidative metabolism (Piston *et al.,* 1995; Masters *et al.,* 1998a,b). One can imagine that in the future multiphoton imaging will be utilized in clinical settings as a method for performing immediate, noninvasive biopsy of accessible tissue such as skin or cornea (Masters and So, 1999).

The improved depth of imaging penetration afforded by multiphoton excitation can also be used to visualize the dynamics of cells within tumors. Intravital imaging of metastatic cells in a primary tumor can provide useful information on behavioral phenotype of metastatic and nonmetastatic cells (Farina *et al.,* 1998). Confocal imaging is severely limited by fatty tissue attached to the tumors. Multiphoton excitation at 900-nm wavelengths penetrates through the fatty tissue to depths in excess of 60 μm to generate images of GFP-expressing tumor cells. Autofluorescence of connective tissue emitting at wavelengths less than 500 nm provides information about the gross architecture of the tumor (Fig. 4; see color plate). Using multiphoton technology it is now possible to observe the metastatic process *in vivo.*

B. Visualizing Calcium Dynamics

The light scattering properties of brain tissue and phototoxicity resulting from fluorophore excitation have confounded efforts to obtaining high-resolution images using conventional microscopies. The improved depth of imaging obtained with multiphoton excitation imaging makes it possible to resolve small structures such as dendritic spines (<1 μm) at depths of 150 μm beneath the surface of the tissue section (Yuste and Denk, 1995). It is now feasible to conduct physiological experiments that require imaging deep within brain tissue (Denk *et al.,* 1994, 1995; Yuste and Denk, 1995b; Svoboda *et al.,* 1996).

Ratiometric images of calcium dynamics can be acquired by multiphoton excitation of Indo-1 via either two-photon (Piston *et al.,* 1994) or three-photon excitation (Szmacinski *et al.,* 1996). Alternatively, protein-based "cameleons" can be genetically encoded and targeted to cells and organelles of interest (Miyawaki *et al.,* 1997). The efficacy of using such reporters for observation of calcium transients *in vivo* has been documented using

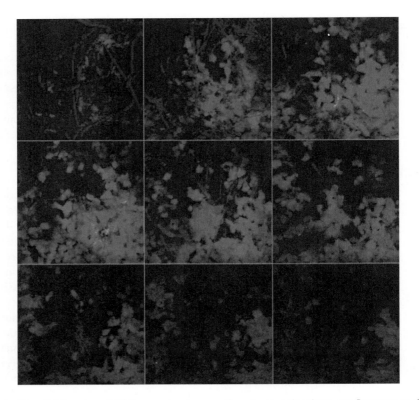

Fig. 4 Intravital imaging of GFP-expressing tumor cells and connective tissue autofluorescence. GFP-expressing MTLn3 cells were injected in the mammary fat pads of Fischer 344 rats. Tumors which formed after 2 weeks postinjection were viewed intravitally with multiphoton microscopy using 900-nm femtosecond pulses. The GFP images were collected in one channel that selected for 500–550 nm emission. The images of connective tissue were collected simultaneously in the second channel that selected for 350–480 nm emission. The images were collected at 10-μm intervals through the volume of the sample. Image degradation was significant at depths beyond 60 μm. (Specimen courtesy of J. B. Wyckoff and J. S. Condeelis, Albert Einstein College of Medicine.) (See Color Plate.)

either "slow-scanning" (Fig. 5; Centonze *et al.*, 1999; see color plate) or video-rate (Fan *et al.*, 1999) multiphoton excitation imaging systems.

C. Long-Term Imaging of Embryo Development

Cytoskeletal dynamics and organellar distribution in early-stage embryos is of interest for understanding the intricacies of the developmental process and to provide information about the overall health of embryos. Confocal imaging of structures within early embryos is difficult for two reasons: signal falloff inside large embryos with yolky cytoplasm and disruption of the developmental regime. Imaging by multiphoton excitation provides a

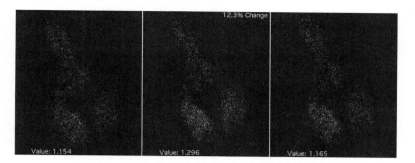

Fig. 5 Ratioed images of cameleon FRET detection using multiphoton excitation imaging. ycam2 cameleon (provided by R. Tsien) was expressed in pharyngeal muscles cells of *C. elegans* worms using the myo-2 promoter. Illumination with 800-nm light primarily excited the CFP moiety of the chameleon. Emission was collected between 380 and 485 nm for CFP and between 535 and 550 nm for YFP. Intensity measurements in regions of interest of the images were ratioed to determine the binding of free calcium by chameleon. At least a 10% change in YFP/CFP emission ratio from baseline to peak was correlated with opening of the terminal bulb lumen. (Specimen courtesy of E. Maryon, B. Saari, and P. Anderson, University of Wisconsin.) (See Color Plate.)

method by which a more comprehensive view of large cells may be obtained by increasing the depth from which useful images can be obtained and by perturbing the cell less during repeated illuminations required for three- (XYZ or XYT) and four (XYZT)-dimensional imaging. Living samples can withstand extended imaging by multiphoton excitation presumably because there is an overall lower concentration of phototoxic byproducts produced that falls within the buffering capacity of the cell (Squirrell *et al.*, 1999).

The study of morphogenetic movements of cells in the developing embryo requires the collection of four-dimensional data sets to document changes in cell shape and position in a dynamic, highly three-dimensional assembly. Multiphoton techniques can be applied to great advantage. Four-dimensional data sets are generated as a series of optical sections through an embryo acquired at regular time intervals. The deeper imaging penetration of multiphoton excitation allows for the collection of a complete series of images through the various layers of the embryo, thus creating a comprehensive representation of its organization. To maintain the impression of continuity of events throughout the developmental process, three-dimensional data must be acquired within short intervals, thus requiring the embryo to be scanned thousands of times. The lessened photodamage experienced by the living embryo during multiphoton excitation imaging is essential to maintain viability. This imaging regime has been used to observe morphogenetic movements of cells in the forming archenteron of sea-urchin embryos (Laxson and Hardin, 1998) and to observe cell fusion events associated with formation of the body wall muscle in *Caenorhabditis elegans* embryos (Mohler *et al.*, 1998; Mohler and White, 1998). In both of these studies widefield and confocal imaging photobleached the fluorophores used to track the cells and caused development to arrest or be abnormal. With the use of multiphoton imaging, however, the signal from the dyes remained strong throughout the term of observation and the embryos appeared unaffected: development proceeded normally.

D. Photoactivation

Optical sectioning capabilities of multiphoton excitation may prove very valuable for photolysis of bioactive materials. A wide range of bioactive molecules such as second messengers or neurotransmitters can be conjugated with photolabile protecting groups. Multiphoton photolysis has been used to photorelease calcium (Brown *et al.,* 1999; Soeller and Cannell, 1999) and EGTA (DelPrincipe *et al.,* 1999); however, the full potential of this technique has not been realized. A major limitation is that only a handful of good probes for multiphoton photolysis are readily available. The commonly used caging groups (e.g., 2-nitrobenzyl, 3-nitrophenyls, benzoins, and phenacyls) have been reported to be sensitive to photolysis by multiphoton irradiation without accompanying tissue damage. However, these molecules have relatively low photosensitivity at wavelengths around 700 nm. Carboxylate, carbamate and phosphate esters of brominated 6-hydroxycoumarin-4-ylmethanol have significantly better action cross-sections (Furuta *et al.,* 1999). The existence of compounds such as these will make feasible the localized release of bioactive molecules in such applications as photodynamic therapy and mapping of local responses to neurotransmitters and messengers.

VI. New Development in Nonlinear Optics

Third-harmonic-generation microscopy is an emerging technology based on nonlinear optical effects. It is an entirely coherent process that will enable visualization and characterization of transparent specimens (Barad *et al.,* 1997; Squier *et al.,* 1998).

Light is generated when three photons of the primary beam are directly converted to one photon at the third harmonic wavelength. These events occur only when the axial focal symmetry is broken by a change in properties of the material, such as a change in refractive index or nonlinear susceptibility. This imaging method is so sensitive to inhomogeneities that it can actually detect boundaries between index-matched materials, such as glass and immersion oil. Considering this, third-harmonic-generation imaging should be very useful for imaging intracellular structures such as vesicles and cytoskeletal elements without the need for exogenously added probes (Yelin and Silberberg, 1999). As the third-harmonic light is generated without production of heat associated with a Stokes shift it is likely that this form of microscopy will be even less damaging to biological specimens than multiphoton excitation imaging. Because of the nonlinear nature of third-harmonic signal generation it is inherently an optical sectioning technique. The images are not sensitive to lateral shift and are therefore well suited to three-dimensional reconstruction.

As this field develops it will be important to study the contrast mechanisms involved in creating third-harmonic-generation images. A thorough understanding of these contrast mechanisms will enable the quantitative measure of material properties of the specimen, thus helping to distinguish biological structure from artifact. It will also allow for the determination of optimal resolution.

Third-harmonic-generation imaging has already been used to observe biological specimens over extended periods of time (Squier *et al.*, 1998). It is therefore probable that third-harmonic-generation imaging will be a useful, noninvasive tool for future studies of dynamics of intracellular structures.

VII. Conclusion

Considering its potential impact on research and possibly diagnostic methods, the development of multiphoton excitation imaging may be viewed as a revolution in light microscopy. The advantages of deep optical sectioning and virtually benign observation of living systems provide the researcher with powerful tools to pursue projects previously thought impossible. In the near future, technological advancements in laser technology and imaging system design should result in less expensive, simpler to operate instruments that should make the technology more accessible. This, along with the versatility afforded by proposed advancements in probe development and signal detection design, promises to make multiphoton microscopy a mainstay of the modern biological research environment.

References

Adams, S. R., Lev-Ram, V., and Tsien, R. Y. (1997). A new caged Ca^{2+}, azid-1, is far more photosensitive than nitrobenzyl-based chelators. *Chem. Biol.* **4**, 867–878.

Agard, D. A., Hiraoka, Y., Shaw, P. J., and Sedat, J. W. (1989). Fluorescence microscopy in three dimensions. *Methods Cell Biol.* **30**, 353–378.

Albota, M., Beljonne, D., Bredas, J.-L., Ehrlich, J. E., Fu, J.-Y., Heikal, A. A., Hess, S. E., Kogej, T., Levin, M. D., Marder, S. R., McCord-Maughon, D., Perry, J. W., Rockel, H., Rumi, M., Subramaniam, G., Webb, W. W., Wu, X.-L., and Xu, C. (1998). Design of organic molecules with large two-photon absorption cross sections. *Science* **281**, 1653–1656.

Barad, Y., Eizenber, H., Horowitz, M., and Silberberg, Y. (1997). Nonlinear scanning laser microscopy by third-harmonic generation. *Appl. Phys. Lett.* **70**, 922–924.

Barnett, D. K., Clayton, M. K., Kimura, J., and Bavister, B. D. (1997). Glucose and phosphate toxicity in hamster preimplantation embryos involves disruption of cellular organization, including distribution of active mitochondria. *Mol. Reprod. Dev.* **48**, 227–237.

Bewersdorf, J., and Hell, S. W. (1998). Picosecond pulsed two-photon imaging with repetition rates of 200 and 400 MHz. *J. Microsc.* **19**, 28–38.

Brakenhoff, G. J., van Spronsen, E. A., van der Voort, H. T., and Nanninga, N. (1989). Three-dimensional confocal fluorescence microscopy. *Methods Cell Biol.* **30**, 379–398.

Brown, E. B., Shear, J. B., Adams, S. R., Tsien, R. Y., and Webb, W. W. (1999). Photolysis of caged calcium in femtoliter volumes using two-photon excitation. *Biophys. J.* **76**, 489–499.

Carlsson, K., Danielsson, P. E., Lenz, R., Liljeborg, A., Majlof, L., and Aslund, N. (1985). Three-dimensional microscopy using a confocal scanning laser microscope. *Optics Lett.* **10**, 53–55.

Castleman, K. R. (1979). "Digital Image Processing." Prentice-Hall, Englewood Cliffs, NJ.

Centonze, V. E., and Pawley, J. B. (1995). Tutorial on practical confocal microscopy and use of the confocal test specimen. *In* "Handbook of Biological Confocal Microscopy," (J. Pawley, Ed.), pp. 549–569. Plenum Press, New York.

Centonze, V. E., and White, J. G. (1998). Multiphoton excitation provides optical sections from deeper within scattering specimens than confocal imaging. *Biophys. J.* **75,** 2015–2024.

Centonze, V. E., Maryon, E., Wokosin, D., Saari, B., and Anderson, P. (1999). Excitation–contraction coupling in *C. elegans* muscle cells as monitored by multiphoton excitation FRET imaging of calcium indicator proteins. *1999 International Worm Meeting Abstract # 226.*

Chalfie, M., Tu, Y., Euskirchen, G., Ward, W. W., and Prasher, D. C. (1994). Green fluorescent protein as a marker for gene expression. *Science* **263,** 802–805.

Curley, P. F., Ferguson, A. I., White, J. G., and Amos, W. B. (1992). Applications of a femtosecond self-sustaining mode-locked Ti:sapphire laser to the field of laser scanning confocal microscopy. *Opt. Quantum Electron.* **24,** 851–859.

Davey, A. P., Bourdin, E., Henari, F., and Blau, W. (1995). Three-photon induced fluorescence from a conjugated organic polymer for infrared frequency upconversion. *Appl. Phys. Lett.* **67,** 884–885.

DelPrincipe, F., Egger, M., Ellis-Davies, G. C., and Niggli, E. (1999). Two-photon and UV-laser flash photolysis of the Ca^{2+} cage, dimethoxynitrophenyl-EGTA-4. *Cell Calcium* **25,** 85–91.

Denk, W. (1996). Two-photon excitation in functional biological imaging. *J. Biomed. Optics* **1,** 296–304.

Denk, W., and Svoboda, K. (1997). Photon upmanship: Why multiphoton imaging is more than a gimmick. *Neuron* **18,** 351–357.

Denk, W., Delaney, K. R., Gelperin, A. L., Kleinfield, D., Stowbridge, B. W., Tank, D. W., and Yuste, R. (1994). Anatomical and functional imaging of neurons using 2-photon laser scanning microscopy. *J. Neurosci. Methods* **54,** 151–162.

Denk, W., Strickler, J. H., and Webb, W. W. (1990). Two-photon laser scanning fluorescence microscopy. *Science* **248,** 73–76.

Denk, W., Piston, D. W., and Webb, W. W. (1995a). Two-photon molecular excitation in laser scanning microscopy. *In* "The Handbook of Biological Confocal Microscopy" (J. Pawley, Ed.), pp. 445–458. Plenum Press, New York.

Denk, W., Sugimori, M., and Llinas, R. (1995b). Two types of calcium response limited to single spines in cerebellar Purkinje cells. *Proc. Nat. Acad. Sci. USA* **92,** 8279–8282.

Fan, G. Y., Fujisaki, H., Miyawaki, A., Tsay, R.-K., Tsien, R. Y., and Ellisman, M. H. (1999). Video-rate scanning two-photon excitation fluorescence microscopy and ratio imaging with cameleons. *Biophys. J.* **76,** 2412–2420.

Farina, K. L., Wyckoff, J. B., Rivera, J., Lee, H., Segall, J. E., Condeelis, J. S., and Jones, J. G. (1998). Cell motility of tumor cells visualized in living intact primary tumors using green fluorescent protein. *Cancer Res.* **58,** 2528–2532.

Furuta, T., Wang, S. S.-H. Dantzker, J. L., Dore, T. M., Bybee, W. J., Callaway, E. M., Denk, W., and Tsien, R. Y. (1999). Brominated 7-hydroxycoumarin-4-ylmethyls: Photolabile protecting groups with biologically useful cross-sections for two photon photolysis. *Proc. Natl. Acad. Sci. USA* **96,** 1193–1200.

Gauderon, R., Lukins, P. B., and Sheppard, C. J. R. (1999). Effect of a confocal pinhole in two-photon microscopy. *Micros. Res. Tech.* **47,** 210–214.

Gerritsen, H. C., and De Grauw, C. J. (1999). Imaging of optically thick specimen using two-photon excitation microscopy. *Microsc. Res. Tech.* **47,** 206–209.

Göppert-Mayer, M. (1931). Über Elementarakte mit zwei Quantensprüngen. *Ann. Phys.* **9,** 273–295.

Gryczynski, I., Szmacinski, H., and Lakowicz, J. R. (1995). On the possibility of calcium imaging using indo-1 with three-photon excitation. *Photochem. Photobiol.* **62,** 804–8.

Gu, M. (1996). Resolution in three-photon fluorescence scanning microscopy. *Opt. Lett.* **21,** 988–990 (erratum **21,** 1414).

Gu, M., and Sheppard, C. J. R. (1995). Comparison of three-dimensional imaging properties between two-photon and single-photon fluorescence microscopy. *J. Microsc.* **177,** 128–137.

Halliwell, B. (1996). Mechanisms involved in the generation of free radicals. *Path. Biol.* **44,** 6–13.

He, G. S., Bhawalkar, J. D., Prasad, P. N., and Reinhardt, B. A. (1995). Three-photon-absorption-induced fluorescence and optical limiting effects in an organic compound. *Opt. Lett.* **20,** 1524–1526.

Hell, S. W., Bahlmann, K., Schrader, M., Soini, A., Malak, H., Gryczynski, I., and Lakowicz, J. R. (1996). Three-photon excitation in fluorescence microscopy. *J. Biomed. Opt.* **1,** 71–74.

Hell, S. W., Booth, M., Wilms, S., Schnetter, J. C. M., Kirsch, A. K., Arndt-Jovin, D. J., and Jovin, T. (1998). Two-photon near and far-field fluorescence microscopy with continuous-wave exciation. *Opt. Lett.* **23**, 1238–1240.

Hockberger, P. E., Skimina, T. A., Centonze, V. E., Lavin, C., Chu, S., Dadras, S., Reddy, J. K., and White, J. G. (1999). Activation of flavin-containing oxidases underlies light-induced production of H_2O_2 in mammalian cells. *Proc. Natl. Acad. Sci. USA* **96**, 6255–6260.

Jacobsen, H., Hanninen, P. E., Soini, E., and Hell, S. W. (1994). Refractive-index-induced aberrations in two-photon confocal microscopy. *J. Microsc.* **176**, 226–230.

Kaiser, W., and Garrett, C. G. B. (1961). Two-photon excitation in $CaF_2:Eu^{2+}$. *Phys. Rev. Lett.* **7**, 229–231.

Koester, H. J., Baur, D., Uhl, R., and Hell, S. W. (1999). Ca^{2+} fluorescence imaging with pico- and femtosecond two-photon excitation: Signal and photodamage. *Biophys. J.* **77**, 2226–2236.

König, K., Becker, T. W., Fischer, P., Riemann, I., and Halbhuber, K. J. (1999). Pulse-length dependence of cellular response to intense near-infrared laser pulses in multiphoton microscopes. *Opt. Lett.* **24**, 113–115.

König, K., So, P. T. C., Mantulin, W. W., and Gratton, E. (1997). Cellular response to near-infrared femtosecond laser pulses in two-photon microscopes. *Opt. Lett.* **22**, 135–136.

Laxson, E., and Hardin, J. (1998). Bottle cells are required for the initiation of primary invagination in the sea urchin embryo. *Dev. Biol.* **204**, 235–250.

Maiti, S., Shear, J. B., Williams, R. M., Zipfel, W. R., and Webb, W. W. (1997). Measuring serotonin distribution in live cells with three-photon excitation. *Science* **275**, 530–532.

Masters, B. R., So, P. T. C., and Gratton, E. (1998a). Multi-photon excitation microscopy of in vivo human skin. *In* "Advances in Optical Biopsy and Optical Mammography" (R. R. Alfano, Ed.), *Ann. NY Acad. Sci.* **838**, 58–67.

Masters, B. R., So, P. T. C., and Gratton, E. (1998b). Multi-photon excitation microscopy of in vivo human skin: Functional and morphological optical biopsy based on three-dimensional imaging, lifetime measurements and fluorescence spectroscopy. *Laser Med. Sci.* **13**, 196–203.

Masters, B. R., and So, P. T. C. (1999). Multi-photon excitation microscopy and confocal microscopy imaging of in vivo human skin: A comparison. *Microsc. Microanal.* **5**, 282–289.

Miyawaki, A., Llopis, J., Heim, R., McCaffery, J. M., Adams, J. A., Ikura, M., and Tsein, R. Y. (1997). Fluorescent indicators for Ca^{2+} based on green fluorescent proteins and calmodulin. *Nature* **388**, 882–887.

Mohler, W. A., and White, J. G. (1998). Multiphoton laser scanning microscopy for four-dimensional analysis of *C. elegans* embryonic development. *Optics Express* **3**, 325–331.

Mohler, W., Simske, J., Williams-Masson, E., Hardin, J., and White, J. (1998). Dynamics and ultrastructure of developmental cell fusions in the *Caenorhabditis elegans* hypodermis. *Curr. Biol.* **8**, 1087–1090.

Periasamy, A., Skoglund, P., Noakes, C., and Keller, R. (1999). An evaluation of two-photon excitation versus confocal and digital deconvolution fluorescence microscopy imaging in *Xenopus* morphogenesis. *Microsc. Res. Tech.* **47**, 172–181.

Piston, D. W., Kirby, M. S., Cheng, H., Lederer, W. J., and Webb, W. W. (1994). Two-photon excitation fluorescence imaging of three-dimensional calcium-ion activity. *Appl. Optics* **33**, 662–669.

Piston, D. W., Masters, B. R., and Webb, W. W. (1995). Three-dimensionally resolved NAD(P)H cellular metabolic redox imaging of the *in situ* cornea with two-photon excitation laser scanning microscopy. *J. Microsc.* **178**, 20–27.

Potter, S. M., Wwang, C. M., Garrity, P. A., and Fraser, S. E. (1996). Intravital imaging of green fluorescent protein using 2-photon laser-scanning microscopy. *Gene* **173**, 25–31.

Schönle, A., and Hell, S. W. (1998). Heating by absorption in the focus of an objective lens. *Opt. Lett.* **23**, 325.

Singh, S., and Bradley, L. T. (1964). Three-photon absorption in napthalene crystals by laser excitation. *Phys. Rev. Lett.* **12**, 612–614.

Soeller, C., and Cannell, M. B. (1999). Two-photon microscopy: Imaging in scattering samples and three-dimensionally resolved flash photolysis. *Microsc. Res. Tech.* **47**, 182–195.

Squier, J. A., Muller, M., Brakenhoff, G. J., and Wilson, K. R. (1998). Third harmonic generation microscopy. *Opt. Express* **3**, 315–324.

Squirrell, J. M., Wokosin, D. M., White, J. G., and Bavister, B. D. (1999). Long-term two-photon fluorescence imaging of mammalian embryos without compromising viability. *Nat. Biotechnol.* **17**, 763–767.

Stelzer, E. H. K., Hell, S., and Lindek, S. (1994). Nonlinear absorption extends confocal fluorescence microscopy into the ultra-violet regime and confines the illumination volume. *Optics Commun.* **104,** 223–228.

Svoboda, K., Denk, W., Kleinfeld, D., and Tank, D. W. (1997). In vivo dendritic calcium dynamics in neocortical pyramidal neurons. *Nature* **385,** 161–165.

Svoboda, K., Tank, D. W., and Denk, W. (1996). Direct measurement of coupling between dendritic spindes and shafts. *Science* **272,** 716–719.

Szmacinski, H., Gryczynski, I., and Lakowicz, J. (1996). Three-photon induced fluorescence of the calcium probe Indo-1. *Biophys. J.* **70,** 547–555.

Webb, W. W. (1990). Two photon excitation in laser scanning fluorescence microscopy. *MICRO* **90,** 445–449.

White, J. G, Amos, W. B., and Fordham, M. (1987). An evaluation of confocal versus conventional imaging of biological structures by fluorescence light microscopy. *J. Cell Biol.* **105,** 41–48.

Willams, R. M., Piston, D. W., and Webb, W. W. (1994). Two-photon molecular excitation provides intrinsic 3-dimensional resolution for laser-based microscopy and microphotochemistry. *FASEB J.* **8,** 804–813.

Wokosin, D. L., and White, J. G. (1997). Optimization of the design of a multiple-photon excitation laser scanning fluorescence imaging system. *In* "Three Dimensional Microscopy: Image Acquisition and Processing IV," *Proc. SPIE* **2984,** 25–29.

Wokosin, D. W., Centonze, V. E., Crittenden, S., and White, J. G. (1996a). Three-photon excitation fluroescence imaging of biological specimens using an all-solid-state laser. *Bioimaging* **4,** 208–214.

Wokosin, D. L., Centonze, V. E., White, J. G., Hird, S. N., Sepsenwol, S., Malcolm, G. P. A. Maker, G. T., and Ferguson, A. I. (1996b). Multiple-photon excitation imaging with an all-solid-state laser. *In* "Optical Diagnostics of Living Cells and Biofluids," *Proc. SPIE* **2678,** 38–49.

Wokosin, D. L., Centonze, V. E., White, J. G., Armstrong, D., Robertson, G., and Ferguson, A. I. (1997). All-solid-state ultrafast lasers facilitate multiphoton excitation fluorescence imaging. *IEEE J. Select. Top. Quant. Electron.* **2,** 1051–1065.

Xu, C., and Webb, W. W. (1996). Measurement of two-photon excitation cross sections of molecular fluorophores with data from 690 to 1050 nm. *J. Opt. Soc. Am.* **13,** 481–491.

Xu, C., Guild, J., Webb, W. W., and Denk, W. (1995). Determination of absolute two-photon excitation cross-sections by in situ second-order autocorrelation. *Opt. Lett.* **20,** 2372–2374.

Xu, C., Zipfel, W., Shear, J. B., Williams, R. M., and Webb, W. W. (1996). Multi-photon fluorescence excitation: new spectral windows for biological nonlinear microscopy. *Proc. Natl. Acad. Sci. USA* **93,** 10763–10768.

Yelin, D., and Silberberg, Y. (1999). Laser scanning third-harmonic-generation microscopy in biology. *Optics Express* **5,** 169–175.

Yuste, R., and Denk, W. (1995). Dendritic spines as basic functional units of neuronal integration. *Nature* **375,** 682–684.

CHAPTER 4

Confocal Microscopy: Important Considerations for Accurate Imaging

Lars Majlof and Per-Ola Forsgren

Molecular Dynamics
Sunnyvale, California 94086

I. Introduction

Confocal microscopy is a powerful tool for visualization and quantification of three-dimensional structures that were previously impossible to capture. During the first few years of the existence of confocal microscopes the emphasis was indeed on visualization, and many of the issues we discuss here were of minor importance. Now, as demands for quantitative measurements are commonplace, it is becoming more and more important to understand the limitations of the confocal microscope as a measurement tool.

The commercially available confocal microscopes can essentially be divided into two categories: spinning disk type and laser scanning. We will not specifically address any of these types of instruments, but will rather deal with general issues encountered when using a confocal microscope for fluorescence imaging. However, because the laser-scanning instrument type is dominating this field, we have also included a section on detector characteristics for photomultiplier-type detectors. (Spinning disk instruments typically use conventional video-type cameras to digitize the images.)

We investigate the influence of refractive index in immersion and mount media on depth measurements and show that one can easily make systematic errors of 30% or more if care is not taken. This article also addresses lens performance and selection criteria and presents depth resolution measurements for a selection of objectives. The conversion of a continuous optical image into a digital image is a process known as spatial sampling. Information may be lost in this process, unless appropriate measures are taken. We present basic rules of thumb for correct spatial sampling.

II. Factors Affecting Confocal Imaging

A. Refractive Index Differences and Their Effect on Focus

The refractive index of a medium determines the speed of light in it. At the interface between two media with different indices of refraction, light adheres to Snell's law of refraction, that is, rays that come in at an angle other than perpendicular change their direction in the new medium. A rather nonintuitive consequence, that has been known at least since the 1950s (Galbraith, 1955), is that the movement of the focal plane in a microscopic specimen does not always correspond to the movement of the specimen stage. A full treatment of the phenomenon is complicated but one can use ray optics and Snell's law of refraction to derive an approximate formula for correction of focus movement. Figure 1 illustrates the ray paths in a microscope objective.

The reason that the plane of focus moves a different distance than the specimen stage is that the immersion medium between objective and coverslip has a different index of refraction than the medium in which the specimen is mounted. The distance traveled by

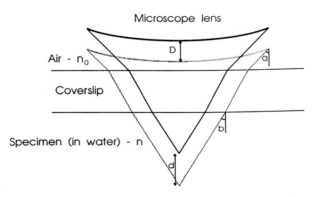

Fig. 1   The optical rays of a microscope lens in two positions above the specimen. D, The physical movement of the stage; d, the movement of the plane of focus. The rays in the figure are drawn for air as immersion medium and water as mount medium. Water has a higher refractive index than air, therefore the distance traveled by the focal plane is greater than the distance traveled by the objective.

the plane of focus is larger than the stage movement if the mount medium has a higher refractive index than the immersion medium. A higher refractive index in the immersion medium than in the mount medium gives the opposite result.

The following formula is derived from geometry and from Snell's law of refraction:

$$d/D = (n/n_0)(\cos a/\cos b) \tag{1}$$

where d is the movement of the focus plane, D is the movement of the lens, n is the refractive index of the mounting medium, n_0 is the refractive index of the immersion medium, and a and b are the angles of incident light. The equation shows that the correction of focus movement d is dependent both on n/n_0 and the incident angles of the rays from the objective. (The angles a and b are related by Snell's law and thus also depend on the refractive indices n_0 and n.) Microscope objectives for biological use are optimized for a certain combination of immersion and mount media. They can, however, if the refractive indices are different, give only a single optimally thin plane of focus for one depth in the specimen. Typically, well-corrected objectives are constructed to trade optimal resolution at a single depth for less optimal but good resolution over a range of depths. A rigorous treatment of the depth correction would call for an integration of the above equation over all angles a, with a weight function for different light intensities at different angles. It would also require knowledge about the particular optimizations for the objective, data that are generally known only by the manufacturer. Furthermore, the analysis so far has been performed on axis (points in the middle of the field); if off-axis imaging points are considered, the treatment becomes even more complicated.

Restricting the analysis to close-to-axis rays and making a zero-order approximation in which both angles a and b are small, the ratio of the cosines can be assumed to be 1. Under these limitations, the ratio of distances becomes

$$d/D = n/n_0 \tag{2}$$

In Table I a number of different combinations of immersion and mount media are listed, showing how large the discrepancies may be between stage and focal plane movements.

The two images in Fig. 2 illustrate the depth correction effect. The sample contains fluorescent beads, 6.5 μm in diameter. They were dried onto a coverslip and then placed on an object glass with glycerol in between. This produced a monolayer of beads at an

Table I

Examples of Corrections for Some Combinations of Immersion and Mount Media

Immersion medium	Mount medium	Correction
Air	Water	1.33
Air	Glycerol	1.47
Oil	Air	0.66
Oil	Water	0.88
Oil	Glycerol	0.97

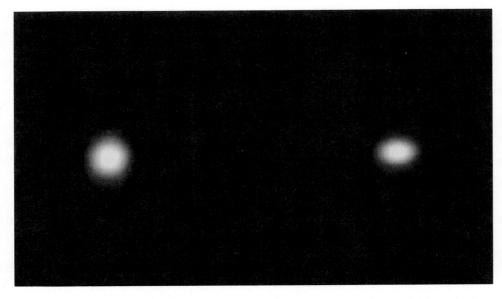

Fig. 2 Two spherical beads in glycerol, imaged with a ×40 NA 0.95 objective. The left image has been corrected for refractive index differences, giving the correct round shape of the bead, whereas the right image shows the distortion when no correction is used.

optimal imaging distance for the lens. The images were scanned in the vertical direction with a × 40 NA 0.95 air objective. To the right is a bead scanned without taking refractive index differences into account and to the left the correction has been performed. Large beads were used to reduce the influence of the point-spread function of the lens.

The conclusion derived from Table I, Fig. 2, and the preceding discussion is that to make any type of accurate measurement that includes the depth dimension, it is imperative to compensate for the refractive index variations. Otherwise, errors of up to 50% are possible.

The previous discussion serves as a good starting point to analyze spherical aberrations. As indicated by Eq. (1), paraxial (close to the center axis) and marginal (furthest away from the axis) rays will have focuses at different depths. Let us consider a theoretical objective with one optimal focus plane at the top of the specimen, where all rays intersect at the same point. As indicated in Fig. 3, the paraxial and marginal rays do not coincide as we go further into the specimen. If we assume that paraxial and marginal rays have the same plane of focus at the top of the specimen, the difference in focal depth between the two rays is (Carlsson, 1990)

$$\delta = z \left(\frac{n}{n_0} - \frac{\sqrt{n^2 - NA^2}}{\sqrt{n_0^2 - NA^2}} \right) \tag{3}$$

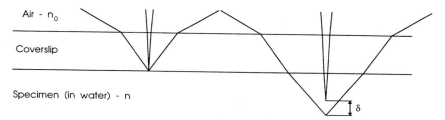

Fig. 3 The optical rays in a theoretical microscope objective. Both paraxial and marginal rays are shown for two different locations of the objective with respect to the specimen. The spherical aberration is evident in the difference of focal depth of the two rays for the lower position of the objective.

where δ is the difference in focal point for paraxial and marginal rays, z is the distance from the optimal focus plane, n is the refractive index of the mount medium, n_0 is the refractive index of the immersion medium, and NA is the numerical aperture of the lens.

One can thus see that the width of the point-spread function in the depth direction depends both on the numerical aperture of the lens and the distance into the specimen.

Figure 4 shows vertical cross-sections of 0.5 μm beads at 5, 30, and 115 μm from the coverslip. A common way to describe the resolution of an optical system is to measure the FWHM (full width at half-maximum) values of a point source. Figure 5 shows a diagram of measured vertical FWHM for 0.5-μm fluorescent latex beads as a function of depth below the coverslip for a $\times 40$/NA 0.95 Plan Apo objective. The results reveal the strong widening of the point-spread function caused by the spherical aberration.

Another consequence of the spherical aberration is that the measured intensity in the image is reduced. The reason is that the effective gathering angle for fluorescent light from a particular point becomes smaller with increasing spherical aberration. Marginal rays collect fluorescence only from points excited by marginal rays of incident light whereas in a nonaberrated situation they would also collect fluorescence excited by paraxial incident light. In Fig. 6 we show the maximum intensity in the same beads that

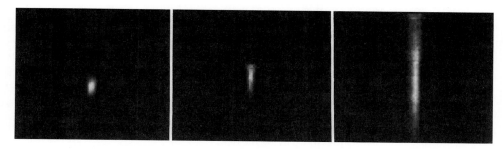

Fig. 4 Projections from the side of 0.5-μm beads at different distances from the coverslip. The beads are in water and were scanned with a $\times 40$ NA 0.95 lens. *Left:* 5 μm from coverslip. *Middle:* 30 μm from coverslip. *Right:* 115 μm from coverslip.

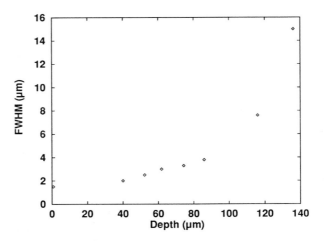

Fig. 5 Vertical full width at half-maximum (FWHM) measurement of 0.5-μm beads as a function of distance (in water) below the coverslip. A $\times$40 NA 0.95 lens was used in the experiment and the correction collar was set for actual coverslip thickness.

were the basis for the FWHM measurement (see Fig. 5) as a function of distance into the specimen. The specimen consisted of a sparse distribution of beads in gelatin.

Large variations in index of refraction within a specimen will make it difficult to image it correctly. To illustrate this effect we prepared two specimens with 6.5-μm latex beads. In one the beads were in water and in the second they were in glycerol. The difference in

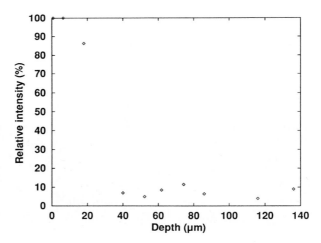

Fig. 6 Measured intensity of the brightest section of 0.5-μm beads as a function of distance (in water) below the coverslip. A $\times$40 NA 0.95 lens was used in the experiment and the correction collar was set for actual coverslip thickness.

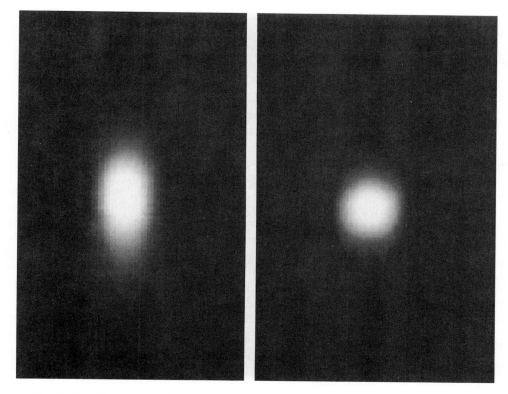

Fig. 7 Two 6.5-μm latex beads, the left in water and the right in glycerol, which has a refractive index more similar to that of the latex bead. The bead measured in water acts as a lens and changes the image of itself.

refractive index between bead and medium is larger in water than in glycerol. Figure 7 shows the difference between the imaged vertical cross section for a single bead in the two samples.

B. Lens Selection

The selection of the right lens has always been crucial to high quality-microscopic imaging. The partially new requirements imposed by confocal microscopy have made the choice even more critical. Now the proper choice of lens may well make the difference between a successfully collected data set or no data at all. In this section we highlight some of the most important factors to consider when making the choice.

The classic criterion for lens selection (Taylor and Salmon, 1989) (after having determined the needed magnification) is the numerical aperture (NA) of the lens. The NA is a measure of the light-gathering power of the lens and is defined by

$$NA = n \sin \theta \qquad (4)$$

where n is the index of refraction for the immersion medium and θ is the angle between the optical axis and the greatest marginal ray entering the lens. The NA is also a measure of the resolving power of the lens, as shown below.

The image brightness for conventional epifluorescence microscopy is related to the NA and the magnification according to

$$\text{Brightness} \cong \text{NA}^4/\text{magnification}^2 \qquad (5)$$

This, and the fact that the in-plane, two-point resolution (according to the Raleigh criterion) can be expressed as

$$R_f = 0.61\lambda/\text{NA} \qquad (6)$$

clearly show why the practical rule for conventional microscopy is to select the lens with the highest numerical aperture. (Even a small increase in NA has a large impact on light collection and makes it possible to resolve smaller details in the specimen.)

In confocal microscopy similar relationships apply. Here, because of the optical properties of the scanning system, the resolution criteria are more critically dependent on the NA. In the confocal fluorescence case the theoretical two-point resolution is defined by

$$R_f = 0.46\lambda/\text{NA} \qquad (7)$$

which is a 30% improvement over the conventional case. Figure 8 shows how the NA affects the intensity and in-plane resolution when imaging close to the coverslip, well within the design parameters for the lenses used.

C. Depth Discrimination

The discussion above relates to lens performance in two-dimensional imaging. In confocal microscopy the depth dimension is, however, also important. One can theoretically derive an expression for the thickness of the confocal sections, at least for the limiting case of an infinitely small confocal aperture (Carlsson and Aslund, 1987). This expression,

$$R_d = 1.4n_0\lambda/\text{NA}^2 \qquad (8)$$

shows that a high NA may be favorable; the higher the NA, the thinner the slice that can be cut from the specimen. Note, however, that the index of refraction of the immersion medium is part of the expression. This means that the nominal vertical resolution of an NA 0.95 dry objective is better than that for an NA 1.0 oil immersion objective (although the resolution of the latter is better in the planar image)!

It is also important to remember that this expression applies close to the coverslip (and close to the optical axis). The discussions above about spherical aberrations make it clear that performance can be expected to decrease farther from the coverslip. Nevertheless, it is interesting to calculate the theoretical resolution to obtain an understanding of the limitations of the technique. Table II shows data for some typical lenses (assuming that the wavelength is 500 nm, i.e., green fluorescence).

Fig. 8 The four images show the same monolayer of fluorescent latex beads (1.6-μm diameter) imaged through four different ×40 lenses. The lenses used were as follows: (*lower left*) ×40 NA 0.70, (*lower right*) ×40 NA 0.85, (*upper left*) ×40 NA 0.95, and (*upper right*) ×40 NA 1.0 oil. The images represent the best confocal section for each lens. The laser power and detector sensitivity used were the same for each image to allow direct comparisons of intensities.

Although high-NA lenses seem to be the preferred choice, they may not be practical in all experimental situations. A basic limitation is the short working distance they possess (one cannot image past the point where the lens runs into the coverslip). A thinner than normal coverslip may provide a way to add several tens of microns of useful depth range. This may seem harmless, especially when used with a oil immersion objective, but will in fact always lead to reduced depth resolution due to spherical aberration (Keller, 1990).

A potentially valuable option is the use of water and glycerol immersion lenses. Because the entire optical path between the specimen and the front lens of the objective

Table II
Theoretical Lateral and Depth Resolutions
(at 500 nm) of Typical High-Quality
Microscopy Objectives[a]

Objective	$R_f(\mu m)$	$R_d(\mu m)$	R_d/R_f
$\times 40$ NA 0.85 dry	0.27	0.97	3.5
$\times 40$ NA 0.95 dry	0.24	0.78	3.2
$\times 40$ NA 1.0 oil	0.23	1.06	4.6
$\times 60$ NA 1.4 oil	0.17	0.54	3.3
$\times 100$ NA 1.4 oil	0.17	0.54	3.3

[a] R_f, Lateral two-point resolution; R_d, full-width half-maximum (FWHM) of the point-spread function in depth. The last column shows the ratio between lateral and axial resolving power.

is composed of the same medium (no coverslip is used) no depth correction issues arise and the spherical aberration with depth is minimized.

D. Behavior of Real Lenses

To investigate the three-dimensional imaging properties of different types of microscope lenses, we created slides with fluorescent latex beads suspended in a water-based gel. The bead concentration was kept low to minimize the influence of laser absorption on the measurement. (Using a $\times 40$ lens there would be only a few beads in focus within the full field of view of the microscope.) In the preparation of the slides we also made sure to have a few beads adhere to the coverslip (thus serving as markers for the position of the coverslip), making for reliable depth measurements with respect to the coverslip position.

We have chosen to investigate $\times 40$ lenses, basically because there is a wide range of choice for this magnification, but also because it is a representative magnification for many types of biological work. Table III lists the lenses used in our study.

For each of these lenses, we scanned series of sections that covered the entire useful depth range (as limited by the thickness of the sample) of the lenses. In the resulting data sets (200–250 sections, 512×512 pixels) we then measured the vertical diameter of the beads, as well as their intensities as a function of depth below the coverslip.

It is interesting to note that the $\times 40$/NA 0.95 Plan Apo that theoretically has the best vertical resolution only does so within a reasonably short distance from the coverslip. The extremely rapid decay in measured fluorescence intensity for this lens is even more astonishing. This does not correlate with the normal fluorescence image; beads are perceived as quite bright throughout the entire working range of the lens. We can only speculate as to the reason for this, but it is possible that this particular lens design sacrifices axial chromatic aberration to achieve other corrections.

Table III
Working Parameters of ×40
Objectives Investigated[a]

Type	NA	Working distance (mm)	n/NA^2
Plan Achro	0.70	0.50	2.04
Fluor	0.85	0.39	1.38
Plan Apo	0.95	0.13	1.11
Plan Apo-oil	1.0	0.1	1.52

[a]Listed are their numerical apertures (NA), their working distance, and the ratio between intended immersion medium and the square of the numerical aperture.

E. Detector Characteristics

The detector system in all current laser-scanning confocal microscopes is based on the photomultiplier tube. This device is a vacuum tube with a photosensitive cathode and a set of secondary electrodes (dynodes). It is capable of high amplification with excellent signal-to-noise characteristics, and can be used to measure a large dynamic range (several orders of magnitude) of photon flux.

Unfortunately not every incident photon is detected. The wavelength of the light (i.e., its energy) strongly affects the probability of detection. Figure 12 shows the quantum

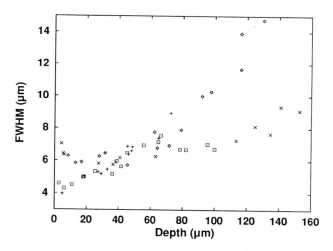

Fig. 9 Diagram showing the full-width half-maximum vertical diameter of fluorescent latex beads as a function of the depth below the coverslip for four different objectives. ◇, ×40 NA 0.70; +, ×40 NA 0.85; □, ×40 NA 0.95; ×, ×40 NA 1.0.

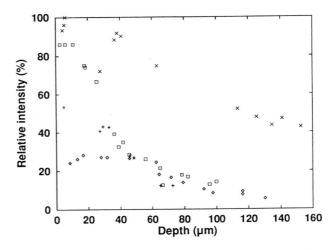

Fig. 10 Diagram of relative measured fluorescence intensity as a function of the depth below the coverslip for each lens. All intensity values are scaled with respect to the brightest average bead image (right under the coverlip) for the Plan Apo 1.0/oil lens. ◇, ×40 NA 0.70; +, ×40 NA 0.85; □, ×40 Na 0.95; ×, ×40 NA 1.0.

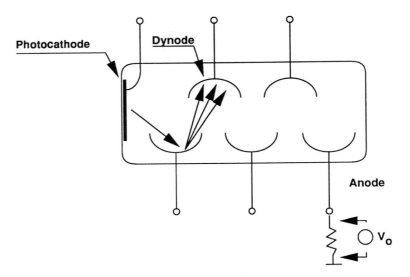

Fig. 11 Photomultiplier tube. Photons with energy above a certain threshold (which depends on the material in the photocathode) may excite electrons enough to let them escape from the cathode. An electric field between the cathode and the following dynode accelerates the free electrons from the cathode toward the dynode. The electrons accumulate enough energy before striking the dynode to create several free secondary electrons in the dynode. This process is repeated throughout the dynode chain, resulting in high multiplication factors, that is, for every photon that results in a primary electron, the photomultiplier tube can generate a pulse of electrons that may easily be measured and quantified.

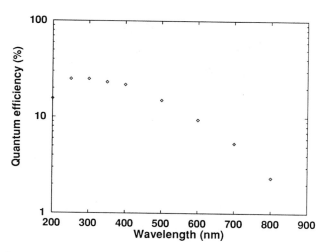

Fig. 12 Quantum efficiency for a "red-enhanced" photocathode as a function of wavelength.

efficiency (i.e., the probability that an incident photon will generate an electron) as a function of the wavelength for a typical photomultiplier tube.

Two observations can immediately be made: the response is highly nonlinear over the visible area, and the overall sensitivity is poor (typically less than 10% of the photons are detected). More specifically, the sensitivity to green fluorescence is two to three times larger than the sensitivity to red. The most important consequence of this is that direct comparisons of measured fluorescent intensities are difficult. This is especially true when comparing intensities from multiple dyes (the fluorescence of which are from different parts of the spectrum and thus measured with different sensitivities). Caution may also be in order when comparing intensities from samples in which the sample environment (pH, etc.) may affect the fluorescence distribution of the dye during the course of the experiment. If the fluorescence is in the area where the quantum sensitivity changes rapidly, small shifts in the peak fluorescence wavelength may result in comparatively large changes in the measured value.

In light of the poor sensitivity for the red part of the spectrum, optical filters and beam splitters for the detection system become more critical. This is especially true in a dual-labeling situation, in which long-pass filters must be applied to suppress the shorter wavelength in the "red channel." The cut-on wavelength must be well matched to the actual fluorescence spectrum of the dye to accomplish good imaging in the red channel.

F. Effects of Spatial Sampling

All digital imaging systems are affected by the fact that they reproduce image information only in discrete points with discrete values. The conversion is done by sampling the intensity in the original continuous image in points laid out on a rectangular grid.

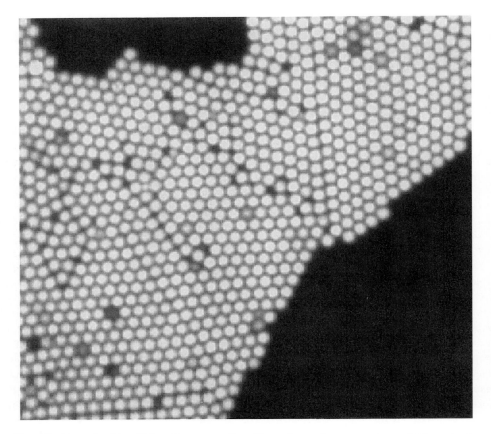

Fig. 13 Example of insufficient discrete sampling. *Above:* Image shows a specimen of 1.6-μm beads sampled every 0.25 μm. *On the following page:* The same specimen sampled every 1.0 μm, which is not enough to resolve the beads accurately.

This process may introduce some surprising artifacts if it is not performed properly, as demonstrated in Fig. 13. The problem encountered is that the sampling points are too sparse to represent the information in the original image properly. Although the illustration here appears to indicate that the problem is two dimensional, it affects confocal imaging in all three dimensions. Horizontal sampling artifacts are often easy to detect during data capture simply by comparing the optical image through the microscope binocular to the digital image. Artifacts due to vertical undersampling are much more difficult to see because no such direct comparison can be made.

Theoretical work (Shannon, 1948) shows that the minimum sampling distance that can be used without loss of information is half of the smallest feature in the input. In practice the rule of thumb is that the sampling rate should be about 2.3 times the largest spatial frequency in the input. Because the resolution of the optical system in the microscope

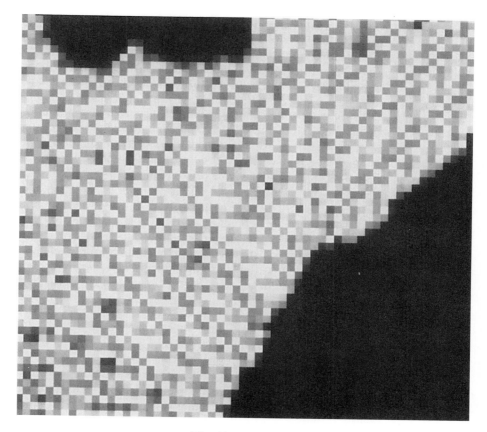

Fig. 13 (*continued*)

sets the limit on information presented to the scanning system, the ultimate limit on pixel and section spacing is set by the point-spread function of the lens. The specimen itself may, however, allow a less strict condition depending on the size and spacing of the features present in the imaged area. If the sample is believed to be isotropic one should select the section-to-section distance equal to the horizontal pixel size to ensure that no vertical sampling artifacts are introduced. It is, however, never useful to scan sections that are closer together than the vertical resolution limit of the lens divided by 2.3.

III. Conclusions

We have described artifacts that the interaction between the optics in the microscope and the specimen can cause. The most important is that the focal plane can move more or less than the microscope stage, depending on the index of refraction of the media above and below the coverslip. This effect can cause vertical distance errors in excess of 40%, when using a dry objective on a specimen in glycerol. It is therefore imperative that one either designs the experiment properly (e.g., a glycerol immersion objective could be used) or uses software that is cognizant of this phenomenon to collect and measure the data.

The lenses that are currently used for confocal microscopy are designed based on a set of optimization criteria that is not necessarily appropriate for confocal applications. We have shown that highly corrected lenses that theoretically ought to be ideal may behave erratically when used in confocal systems. It is also evident that spherical aberrations limit the performance of all microscope objectives when used for imaging deep below the coverslip.

References

Carlsson, C. (1990). *Proc. Soc. Photo-Opt. Instrum. Eng.* **1245,** 68–80.

Carlsson, C., and Aslund, N. (1987). *Appl. Op.* **26,** 3232–3238.

Galbraith, W. Q. (1955). *Q. J. Microsc.* **96,** 285–289.

Keller, H. E. (1990). *In* "Handbook of Biological Confocal Microscopy" (J. B. Pawley, ed.), Rev. Ed., pp. 77–86. Plenum, New York.

Shannon, C. E. (1948). *Bell Syst. Tech. J.* **27,** 379–423; 623–656.

Taylor, D. L., and Salmon, E. D. (1989). *Methods Cell Biol.* **29,** 207–237.

CHAPTER 5

Multicolor Laser Scanning Confocal Immunofluorescence Microscopy: Practical Application and Limitations

T. Clark Brelje, Martin W. Wessendorf, and Robert L. Sorenson

Department of Cell Biology and Neuroanatomy
University of Minnesota Medical School
Minneapolis, Minnesota 55455

METHODS IN CELL BIOLOGY, VOL. 70
Copyright 1993, Elsevier Science (USA). All rights reserved
0091-679X/93 $35.00

I. Introduction

Fluorescence is presently the most important imaging mode in biological confocal microscopy (Schotten, 1989; Tsien and Waggoner, 1990). In conventional microscopy, the illumination stimulates fluorescence throughout the entire depth of a specimen, rather than only in the focal plane. The contrast and resolution observed for structures within the focal plane can be severely reduced by the background fluorescence from out-of-focus structures. In contrast, confocal microscopy uses the combination of a focused illumination spot and a detection pinhole to restrict excitation and detection to a small, diffraction-limited volume within the focal plane (Brakenhoff *et al.,* 1979; Wilson and Sheppard, 1984). Out-of-focus structures do not contribute to the background because they receive little or no illumination, and any signal derived from them is rejected by the detector aperture. The illumination point can be scanned across the specimen to build an image of structures within the focal plane in which the out-of-focus background is virtually absent (White *et al.,* 1987; Amos, 1988; Carlsson, 1990). Over and above the improvement in image contrast and resolution, it is the capability to optically section intact or thick specimens that makes confocal microscopy extremely attractive for studies using fluorescent probes. Application of this optical sectioning capability has found widespread use within biology for the examination of the three-dimensional distribution of fluorescent probes within intact, fixed, or living specimens (Schotten, 1989; Paddock, 1991).

Because many biological problems cannot be unambiguously characterized by the examination of a single parameter, techniques for the independent detection of signals from multiple fluorescent probes are required. At present, this can most easily be done by using fluorescent probes that can be viewed selectively based on differences in their excitation and emission spectra. Until recently, the capacity of confocal microscopes to examine multiple fluorescent probes in a single specimen has been extremely limited compared to conventional fluorescence microscopy (DeBiasio *et al.,* 1987; Waggoner *et al.,* 1989; Wessendorf, 1990; Galbraith *et al.,* 1991) and flow cytometry (Hoffman, 1988; Shapiro, 1988; Lanier and Recktenwald, 1991).

Advances in fluorescent probe chemistry, economical laser availability, and confocal microscope instrumentation are making the enormous potential of multicolor laser scanning confocal microscopy (LSCM) available to a wider range of biologists. This article outlines the requirements for performing multicolor immunofluorescence studies with LSCM. First, the principles of immunofluorescence histochemistry necessary for the preparation of multilabeled specimens are summarized. Second, the technical aspects of confocal microscope instrumentation that affect its application to multicolor studies are examined. Third, the practical application and limitations of this technology are demonstrated for multicolor LSCM, using the inexpensive, air-cooled argon ion and krypton–argon ion lasers as light sources. Fourth, aspects of confocal microscopy that affect the comparison of images acquired with different excitation and/or emission wavelengths are examined. With this information, users should have a better understanding of the compromises needed to effect the accurate imaging of specimens stained with multiple fluorophores. Although this article concentrates on studies of fixed biological specimens, much of the information concerning the application and limitations of multicolor LSCM is applicable to the viewing of living specimens.

II. Immunofluorescence Histochemistry

Immunofluorescence histochemistry involves the use of antibodies labeled with fluorophores to detect substances within a specimen. Several excellent discussions on various aspects of immunofluorescence histochemistry have been published (Pearse, 1980; Larsson, 1983; Sternberger, 1986; Wessendorf, 1990). Therefore this section focuses on issues of special concern in the use of multicolor immunofluorescence for the detection of multiple substances in a single specimen.

Immunofluorescence can be performed by either direct or indirect methods. Direct immunofluorescence involves the conjugation of the primary antibody with a fluorophore, such as fluorescein or rhodamine. For indirect immunofluorescence, the primary antibody is visualized by using a fluorophore-conjugated secondary antibody raised against the immunoglobulins of the species in which the primary antibody was raised. Indirect immunofluorescence is more commonly used because the labeling of each primary antibody is laborious and can decrease its affinity or specificity for its antigen. Moreover, direct methods may be less sensitive than indirect methods because theoretically more than one molecule of the secondary antibody can bind to a given molecule of the primary antibody. With the widespread use of indirect immunofluorescence for biological studies, a large number of secondary antibodies conjugated to various fluorophores have become commercially available.

Before accepting the localization of an antigen by indirect immunofluorescence, the specificity of the staining and visualization must be established. This characterization of a staining protocol is especially important for multicolor immunofluorescence because of the greatly expanded opportunities for cross-reactivity and artifactual staining.

A. Fluorophores and Labeling Reagents

The most important factors in determining the limitations and capabilities of fluorescence microscopy are the physical characteristics of the fluorescent dyes used in biological studies (Tsien and Waggoner, 1990; Wells and Johnson, 1993). In particular, the extent to which a fluorophore absorbs light (i.e., the extinction coefficient) and the likelihood that an excited fluorophore will emit fluorescence (i.e., the quantum yield) or spontaneously decompose (i.e., photobleaching) imposes strict limits on this method (Wells *et al.,* 1990). Because specimens can be stained with a finite amount of fluorescent dye, it is important that the microscope efficiently excite and detect the limited number of photons emitted before its photodestruction. Although numerous fluorescent dyes exist, relatively few have been found suitable for immunofluorescence studies (Fig. 1). Because the selection of appropriate fluorophores is critical to the success of multicolor immunofluorescence studies, the important physical and spectral properties concerning their use will be briefly described.

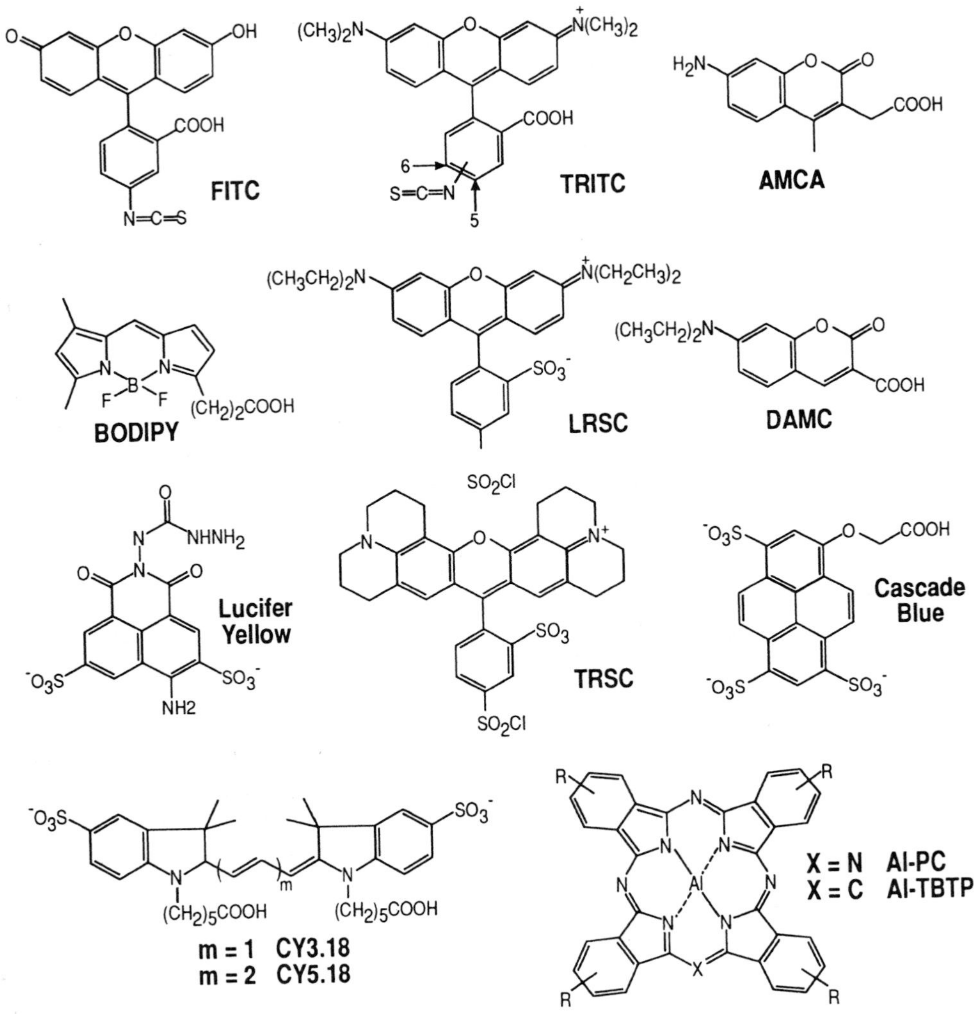

Fig. 1 Chemical structures of common fluorophores for immunofluorescence studies. See Table I for the abbreviations used for each fluorophore and its spectral properties. Note that most commercial preparations of tetramethylrhodamine isothiocyanate (TRITC) are a mixture of the 5′- and 6′-isothiocyanate isomers.

The most important consideration in the selection of fluorescent dyes for the covalent labeling of antibodies, or other relevant biological molecules, is their intensity of fluorescence. This is particularly important for LSCM, in which exceedingly small quantities of fluorescent dyes must be detected over short intervals of time. With nonsaturating excitation rates, the relative brightness of a fluorescent dye is proportional to the product of its extinction coefficient at a given excitation wavelength and its quantum yield (Table I).

Table I
Spectroscopic Properties of Common Probes for Immunofluorescence[a]

Fluorophore or labeling reagent	Abbreviation	Molecular weight	Excitation maximum (nm)	Emission maximum (nm)	Extinction coefficient ($\times 10^3$ M^{-1}cm^{-1})	Quantum yield	Reference
7-Amino-4-methylcoumarin-3-acetic acid	AMCA	233	347	445	15	—	Khalfan et al. (1986); Wessendorf et al. (1990b,c)
Pyrenyloxytrisulfonic acid	Cascade Blue	486	376, 399	423	23, 28	—	Whitaker et al. (1991a); MP[b]
7-Diethylaminocoumarin-3-carboxylic acid	DAMC	261	391	474	—	—	Staines et al. (1988)
Lucifer yellow	LY	457	428	533	12	—	MP
Fluorescein isothiocyanate	FITC	389	496	518	67	0.20–0.35	Waggoner et al. (1989)
4,4-Difluora-5,7-dimethyl-4-bora-3a,4a-diazaindacene 3-propionic acid	BODIPY	292	503	511	80	0.40	Haugland (1990); MP
Cyanine 3.18	Cy3.18	718	554	565	150	0.15	Mujumdar et al. (1993); BDS[c]
Tetramethylrhodamine isothiocyanate	TRITC	444	554	576	67	—	
Lissamine rhodamine sulfonyl chloride	LRSC	577	572	590	83	0.04	Chen (1969)
R-Phycoerythrin	R-PE	240,000	480, 565	578	2,000	0.85	Oi et al. (1982)
B-Phycoerythrin	B-PE	240,000	546, 565	575	2,410	0.59	Oi et al. (1982)
Texas Red sulfonyl chloride	TRSC	625	592	610	87	0.01	Titus et al. (1982)
Allophycocyanin	APC	110,000	650	661	690	0.68	Oi et al. (1982)
Cyanine 5.18	Cy5.18	734	649	667	240	0.28	Mujumdar et al. (1993); BDS
Aluminum tetrabenztriazaporphyrin	Al-TBTP	~1,000	395, 656, 675	678	189	0.24	Renzoni et al. (1992); BTG USA[d]
Aluminum phthalocyanine	Al-PC	876	350, 675	680	160	0.50	Renzoni et al. (1992)
Peridinin chlorophyll a-binding protein	PerCP	35,000	470	680	380	1.00	Recktenwald (1989)

[a] These fluorescent probes are listed in order of increasing excitation and emission wavelengths. Except for LY and PerCP, all values are given for the individual fluorophores conjugated to IgG antibodies. All conjugates had dye/protein ratios of 2 to 4, except for the biological pigments (i.e., R-PE, B-PE, APC, and PerCP), which had dye/protein ratios of 1. The values for AMCA, Cy3.18, and Cy5.18 are for conjugates prepared from the corresponding succinimidyl esters. The extinction maxima for R-PE, B-PE, and APC are for the phycobiliproteins, which contain 34, 34, and 6 pigment molecules, respectively.

[b] MP, Molecular Probes, Inc. (Eugene, OR).

[c] BDS, Biological Detection Systems, Inc. (Pittsburgh, PA).

[d] BTG USA, British Technology Group USA, Inc. (Gulph Mills, PA).

Because both of these properties are sensitive to the local molecular environment (for example, pH of aqueous media, solvent polarity, and the proximity of quenching species), they must be determined for the actual conjugates of the fluorophores rather than for the unconjugated fluorescent dyes. For example, the high quantum yield for fluorescein in aqueous solution (~0.70) decreases to 0.20–0.35 when four fluorophores are bound to each antibody molecule (Tsien and Waggoner, 1990). This reduction in fluorescence on conjugation is due to the quenching by other bound dye molecules or through interactions with the labeled protein. Moreover, these interactions limit the increase in fluorescence intensity that can be obtained by conjugating higher numbers of fluorophores to each protein molecule. Besides these intrinsic properties of the conjugated fluorophore, their observed brightness is also directly influenced by the efficiency of excitation by the available light sources and detection by the commonly used detectors (such as the human eye, photographic film, or photomultiplier tubes).

Although the total fluorescence signal can be increased with higher illumination intensities or by integrating for a longer time, photochemical side effects such as photobleaching of the fluorophore or damage to the specimen limits the allowable light exposure. However, the high illumination intensities necessary for LSCM are already near the optical saturation limit of many fluorescent probes (Wells *et al.*, 1990; Wells and Johnson, 1993). Further increases in the illumination intensity will actually decrease the observed signal-to-background ratio because much higher levels of light are needed to saturate the background autofluorescence (Tsien and Waggoner, 1990). It is important to recognize that the average number of photons emitted by a fluorophore before photodestruction is determined by the ratio of the quantum yields for fluorescence and photobleaching (Mathies and Stryer, 1986). This suggests that longer observation times can be achieved only by increasing the concentration of the fluorophore, using lower illumination intensities to reduce the rate of fluorescence, or decreasing the rate of photobleaching. Therefore, the lowest illumination intensities that give images with an adequate signal to noise should be used for the collection of images by LSCM. The rate of photobleaching for some fluorophores can be reduced by the addition of chemical antioxidants, such as *p*-phenylenediamine (Johnson and de C. Nogueira Araujo, 1981; Johnson *et al.*, 1982) or *n*-propyl gallate (Giloh and Sedat, 1982), to the mounting media for fixed specimens. However, these antifade reagents are incompatible with living cells.

Although the intensity of fluorescence is important, the utility of a fluorophore for the labeling of antibodies is determined by several additional properties. An important practical consideration is the availability of suitable labeling reagents for the preparation of conjugates with biological molecules (Haugland, 1983). Isothiocyanates are widely used to attach fluorescent dyes to neutral amino groups of proteins in buffers at pH 8.5–9.5. Other alternatives for the labeling of amino groups include succinimidyl esters, chlorotriazinyl groups, aziridines, and anhydrides. The fluorophore must not adversely affect the properties of the biological molecule after conjugation. For example, the fluorophore-to-protein ratio for antibodies must be carefully controlled to prevent their denaturation, precipitation, or loss of binding activity. Finally, the fluorophore should have low toxicity and biological activity so that studies with living tissue can be done.

1. Fluorescein

Fluorescein isothiocyanate (FITC) is the most widely used fluorescent probe for the preparation of conjugates with biological molecules (Hansen, 1967; Haugland, 1990). This xanthene dye (Fig. 1) is particularly useful for several reasons: conjugates are easily prepared because of the water solubility of FITC; it is brightly fluorescent because of its reasonably large extinction coefficients and high quantum yields after conjugation (Table I); and it has low nonspecific binding with most biological tissues. The preparation of fluorescein conjugates from the closely related (4,6-dichlorotriazinyl) aminofluorescein (DTAF) has been recommended because of its higher purity and stability (Blakeslee and Baines, 1976). Typically, three to five fluoresceins can be conjugated to each IgG antibody before self-quenching and altered binding affinities are observed. Fluorescein is maximally excited by blue light and emits primarily green to yellow fluorescence (Fig. 2). Although the excitation spectrum of fluorescein does not overlap with any of the intense emission peaks of mercury arc lamps (313, 334, 365, 405, 435, 546, and 578 nm), the output intensity in the range of 450 to 500 nm is sufficient for the excitation of fluorescein by conventional fluorescence microscopy. The 488-nm line of the argon ion laser used in flow cytometry and LSCM is ideally suited for near-maximal excitation of fluorescein.

In spite of its general usefulness, fluorescein does have several unfavorable properties. It is not particularly photostable, it is sensitive to changes in pH and solvent polarity, and its emission spectrum overlaps extensively with cellular autofluorescence. For fixed specimens, the photobleaching of fluorescein is effectively retarded by the addition of an antifade reagent to the mounting medium (see above). Structurally, the pH sensitivity of fluorescein is due to the presence of an ionizable carboxylate group conjugated to the π electron system of the fluorophore, and has been employed in pH indicators such as 2',7'-bis-(2-carboxyethyl)-5(and 6)-carboxyfluorescein (BCECF), seminaphthorhodafluors (SNARFs), and seminaphthofluoresceins (SNAFLs) (Whitaker et al., 1991b). Therefore specimens stained with fluorescein should be mounted in aqueous media with a pH of at least 8 for the maximum intensity of fluorescence (Hiramoto et al., 1964).

2. BODIPY

Boron dipyrromethene difluoride (BODIPY) has been proposed as a substitute for fluorescein (Worries et al., 1985; Haugland, 1990). BODIPY conjugates are efficiently excited by blue light and emit primarily green fluorescence like fluorescein, but with a smaller Stokes shift and narrower excitation and emission peaks (Fig. 2). This characteristic usually requires the excitation or detection of BODIPY staining at suboptimal wavelengths. Major advantages of BODIPY compared to fluorescein include its insensitivity to changes in pH and solvent polarity. In addition, BODIPY has been

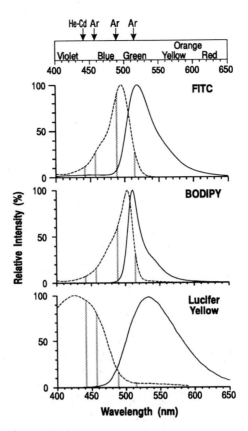

Fig. 2 Excitation (dashed) and emission (solid) spectra of green fluorophores. *Top:* Fluorescein isothiocyanate (FITC)-conjugated donkey anti-mouse IgG (Jackson ImmunoResearch Laboratories, West Grove, PA). The excitation spectrum was scanned while measuring emission at 570 nm, and the emission spectrum was scanned while exciting at 370 nm. *Middle:* BODIPY-conjugated goat anti-mouse IgG (Molecular Probes, Eugene, OR). The excitation spectrum was scanned while measuring emission at 600 nm, and the emission spectrum was scanned while exciting at 370 nm. *Bottom:* Lucifer yellow (Sigma, St. Louis, MO). The excitation spectrum was scanned while measuring emission at 600 nm, and the emission spectrum was scanned while exciting at 370 nm. The emission lines for argon ion (Ar) and helium–cadmium (He–Cd) lasers are shown as vertical lines beneath the excitation spectra.

reported to be more photostable than fluorescein conjugates (Haugland, 1990), but less photostable than the rhodamine Texas Red (Robitaille *et al.,* 1990). Unfortunately, at high concentrations of BODIPY the green fluorescence emission peak at ~515 nm decreases and a second peak of red fluorescence appears at ~620 nm (Pagano *et al.,* 1991; Haugland, 1992). Although this phenomenon probably requires interaction of multiple fluorophores, a similar relatively weak, red fluorescence can be observed from specimens brightly stained with BODIPY-conjugated secondary antibodies (T. C. Brelje, unpublished observations).

The development of additional BODIPY fluorophores with excitation and emission spectra shifted to longer wavelengths has been described by Molecular Probes (Eugene, OR) (Haugland, 1992). Unfortunately, the fluorescence from many of these derivatives is quenched upon conjugation to proteins. The development of red-shifted derivatives which are suitable for conjugation to proteins is being actively pursed and should be available in the near future (R. Haugland, personal communication).

3. Lucifer Yellow

Although infrequently used as a covalent label for biological molecules, Lucifer yellow is often used in immunofluorescence studies for the intracellular filling of cells (Stewart, 1978). It is intensely fluorescent, relatively photostable, and the presence of a free amino group makes it fixable in specimens. However, the extremely broad excitation and emission spectra of Lucifer yellow (Fig. 2) can complicate its independent detection in the presence of other fluorophores. In these cases, the use of alternate intracellular markers, such as biocytin (Horikawa and Armstrong, 1988) or N-(2-aminoethyl)biotinamide (Neurobiotin; Kita and Armstrong, 1991), that can be visualized with more specific fluorophores conjugated to avidin or anti-biotin antibodies is preferred.

4. Rhodamine Fluorophores

The rhodamines are xanthene derivatives structurally related to fluorescein, but with additional chemical substitutions that shift their excitation and emission spectra to longer wavelengths (Fig. 1). The most widely used rhodamines in order of increasing excitation and emission wavelengths are tetramethylrhodamine, Lissamine rhodamine, and Texas Red (Fig. 3). Although the lower quantum yields of rhodamine conjugates make them significantly dimmer than comparable fluorescein conjugates (Table I), they are generally more photostable and are pH insensitive (McKay *et al.*, 1981). Rhodamine conjugates must be carefully prepared because they are particularly susceptible to quenching when more than two or three dye molecules are covalently attached to each antibody molecule. Nonetheless, rhodamine staining can often appear quite bright by conventional fluorescence microscopy because their excitation spectra coincide with the strong 546-nm emission peak of mercury arc lamps. The hydrophobic nature of the rhodamines, especially for Texas Red and less so for tetramethylrhodamine, requires careful control of the number of fluorophores conjugated to each antibody. Higher fluorophore-to-protein ratios result in fluorescent quenching, denaturation and precipitation of the antibodies, and higher background staining. Although the excitation source available influences which rhodamine to use, the larger spectral overlap between fluorescein and tetramethylrhodamine has been observed to be a problem in the specific visualization of these fluorophores in multiple-labeled specimens (Wessendorf *et al.*, 1990a).

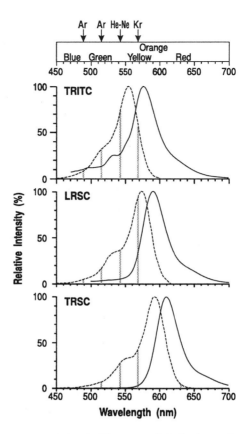

Fig. 3 Excitation (dashed) and emission (solid) spectra of rhodamine-conjugated secondary antibodies (Jackson ImmunoResearch Laboratories). *Top:* Tetramethylrhodamine isothiocyanate (TRITC)-conjugated donkey anti-mouse IgG. The excitation spectrum was scanned while measuring emission at 610 nm, and the emission spectrum was scanned while exciting at 450 nm. *Middle:* Lissamine rhodamine sulfonyl chloride (LRSS)-conjugated donkey anti-mouse IgG. The excitation spectrum was scanned while measuring emission at 630 nm, and the emission spectrum was scanned while exciting at 480 nm. *Bottom:* Texas Red sulfonyl chloride (TRSC)-conjugated donkey anti-mouse IgG. The excitation spectrum was scanned while measuring emission at 650 nm, and the emission spectrum was scanned while exciting at 500 nm. The emission lines for argon ion (Ar), krypton ion (Kr), and helium–neon (He–Ne) lasers are shown as vertical lines beneath the excitation spectra.

5. Phycobiliproteins

Phycobiliproteins are naturally occurring components of the light-collecting complexes of certain cyanobacteria and algae, and have considerable potential as fluorescent probes (Oi *et al.,* 1982; Kronick, 1986). These proteins have been engineered by natural selection for large extinction coefficients, high quantum yields, and to protect the covalently linked tetrapyrrole chromophores from quenching processes (Table I) (Glazer,

1989). Although designed to efficiently transfer energy from blue light to the chlorophyll photosynthetic system, the purified phycobiliproteins are highly fluorescent because the molecules no longer have any nearby acceptors to which to transfer the absorbed energy.

The most widely used phycobiliproteins for fluorescence studies are R-phycoerythrin, B-phycoerythrin, and allophycocyanin (Haugland, 1992). The phycoerythrins are efficiently excited by blue–green light and emit primarily yellow–orange fluorescence (Fig. 4). Because each contains a large number of chromophores, their fluorescence

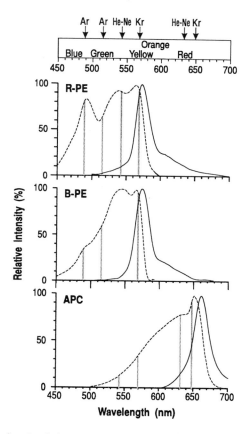

Fig. 4 Excitation (dashed) and emission (solid) spectra of phycobiliprotein-conjugated secondary antibodies. *Top:* R-Phycoerythrin (R-PE)-conjugated goat anti-mouse IgG (Jackson ImmunoResearch Laboratories). The excitation spectrum was scanned while measuring emission at 625 nm, and the emission spectrum was scanned while exciting at 475 nm. *Middle:* B-Phycoerythrin (B-PE)-conjugated goat anti-mouse IgG (Jackson ImmunoResearch Laboratories). The excitation spectrum was scanned while measuring emission at 625 nm, and the emission spectrum was scanned while exciting at 475 nm. *Bottom:* Allophycocyanin (APC)-conjugated goat anti-mouse IgG (Biomeda, Foster City, CA). The excitation spectrum was scanned while measuring emission at 710 nm, and the emission spectrum was scanned while exciting at 500 nm. The emission lines for argon ion (Ar), krypton ion (Kr), and helium–neon (He–Ne) lasers are shown as vertical lines beneath the excitation spectra.

yield is equivalent to at least 30 fluorescein or 100 rhodamine fluorophores at comparable wavelengths (Oi *et al.*, 1982). The more efficient excitation of R-phycoerythrin by the 488-nm emission line of an argon ion laser has led to its widespread use with fluorescein for two-color studies using flow cytometry (Hoffman, 1988; Shapiro, 1988; Lanier and Recktenwald, 1991). The photostability of R-phycoerythrin is slightly less than that of fluorescein, whereas B-phycoerythrin is slightly more stable (White and Stryer, 1987). Although not as bright as the phycoerythrins, the longer wavelength allophycocyanin is efficiently excited by red light and emits far red fluorescence (Fig. 4).

Although these phycobiliproteins are brightly fluorescent, their unique properties must be considered when used as fluorescent labels. Their intensity of fluorescence is greatly reduced by denaturation of their protein component. R-Phycoerythrin fades rather rapidly with the high-intensity illumination used with LSCM (Schubert, 1991). Furthermore, the high molecular weights of the phycobiliproteins (Table I) restrict their penetration into the denser regions of fixed cells and thick specimens.

6. Cyanine Fluorophores

Cyanine 3.18 and 5.18 are sulfoindocyanine dyes from a family of fluorophores developed by A. Waggoner and colleagues at Carnegie-Mellon University (Southwick *et al.*, 1990; Mujumdar *et al.*, 1993). These sulfoindocyanine dyes are highly water soluble, pH insensitive, and exhibit low nonspecific binding to biological specimens. Because of their large extinction coefficients and moderate quantum yields (Table I), their antibody conjugates are typically brighter and have greater photostability than those with fluorescein (Yu *et al.*, 1992; Mujumdar *et al.*, 1993). Because these cyanine fluorophores are as bright, if not brighter, in organic solvents, they are particularly useful with thick specimens that must be dehydrated and cleared before examination by LSCM (Mesce *et al.*, 1993).

With increasing length of the polymethine chain $[(-C=)_n]$ separating the indolenine nuclei (Fig. 1), the absorption and emission wavelengths of the cyanine chromophore shift to longer wavelengths (Southwick *et al.*, 1990). The excitation and emission spectra of cyanine 3.18 conjugates are similar to tetra-methylrhodamine (Fig. 5). When used with conventional fluorescence microscopy and a mercury arc lamp, cyanine 3.18 has been found to give significantly brighter specific staining than either fluorescein or any of the rhodamines (Wessendorf and Brelje, 1992). The longer wavelength cyanine 5.18 is excited by red light and emits far red fluorescence, and can be used as a low molecular weight alternative to allophycocyanin (Fig. 5). Like other far red fluorophores, cyanine 5.18 is extremely difficult to observe by eye in conventional fluorescence microscopes because of the low sensitivity of the eye to red light above 650 nm.

7. Phthalocyanine Derivatives

Phthalocyanine derivatives suitable for conjugation to biological molecules were developed by Ultra Diagnostics Corp. (Schindele and Renzoni, 1990; Renzoni *et al.*, 1992) and licensed by British Technology Group USA (Gulph Mills, PA). Structurally related

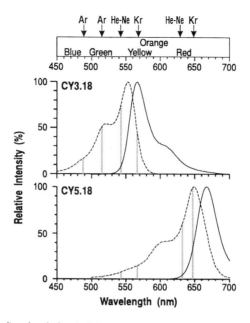

Fig. 5 Excitation (dashed) and emission (solid) spectra of cyanine-conjugated secondary antibodies (Jackson ImmunoResearch Laboratories). *Top:* Cyanine 3.18 (CY3.18)-conjugated donkey anti-mouse IgGs. The excitation spectrum was scanned while measuring emission at 610 nm, and the emission spectrum was scanned while exciting at 490 nm. *Bottom:* Cyanine 5.18 (CY5.18)-conjugated donkey anti-mouse IgGs. The excitation spectrum was scanned while measuring emission at 710 nm, and the emission spectrum was scanned while exciting at 500 nm. The emission lines for argon ion (Ar), krypton ion (Kr), and helium–neon (He–Ne) lasers are shown as vertical lines beneath the excitation spectra.

to the linking of tetrapyrroles by methine ($-C=$) groups in porphyrins, the phthalocyanines are composed of four isoindole units linked by aza nitrogen atoms ($-N=$; Fig. 1). As with the porphyrins, metal atoms such as aluminum are usually inserted in the central ring of phthalocyanines. These metal-phthalocyanines are excited by both ultraviolet and far red light, and emit far red fluorescence (Fig. 6). Replacing one of the aza nitrogens by a methine carbon gives a tetrabenztriazaporphyrin that is similar to the corresponding phthalocyanine except that an additional red excitation peak occurs (Fig. 6). Although phthalocyanine conjugates are currently not commercially available, the increased use of far red fluorophores for all types of fluorescence studies should encourage their production.

8. Blue Fluorophores

Several ultraviolet-excitable, blue fluorophores have been used for immunofluorescence studies: 7-amino-4-methylcoumarin-3-acetic acid (AMCA; Khalfan *et al.,* 1986; Wessendorf *et al.,* 1990a), diethylaminocoumarin (DAMC; Staines *et al.,* 1988), and

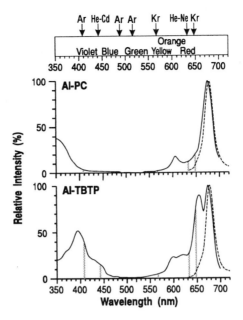

Fig. 6 Excitation (dashed) and emission (solid) spectra of phthalocyanine derivatives conjugated to secondary antibodies (British Technology Group USA, Gulph Mills, PA). *Top:* Aluminum phthalocyanine (AI-PC)-conjugated goat anti-mouse IgG (Ultralight 680). *Bottom:* Aluminum tetrabenztriazaporphyrin (Al-TBTP)-conjugated goat anti-mouse IgG (Ultralight T680). The excitation spectra were scanned while measuring emission at 725 nm, and the emission spectra were scanned while exciting at 340 nm. The emission lines for argon ion (Ar), krypton ion (Kr), helium–cadmium (He–Cd), and helium–neon (He–Ne) lasers are shown as vertical lines beneath the excitation spectra.

Cascade Blue (pyrenyloxytrisulfonic acid; Whitaker *et al.,* 1991a; Fritschy *et al.,* 1992). All of these blue fluorophores are excited by ultraviolet light and emit blue fluorescence (Fig. 7). However, the greater overlap between the emission spectra of DAMC and fluorescein makes it less attractive for multicolor studies. AMCA is the most widely available of these blue fluorophores and has been successfully used for triple-labeling studies by conventional fluorescence microscopy (Wessendorf *et al.,* 1990a), flow cytometry (Delia *et al.,* 1991), and LSCM (Schubert, 1991; Ulfhake *et al.,* 1991).

B. Immunofluorescence Staining Techniques

Staining methods using multicolor immunofluorescence techniques require careful characterization if the observed staining is to be validly interpreted. To prevent the occurrence of false positives (i.e., the appearance of staining in instances in which there is none), it is necessary to demonstrate for the staining protocol that (1) the fluorophores can be distinguished from each other by using the microscope filter sets, (2) the secondary

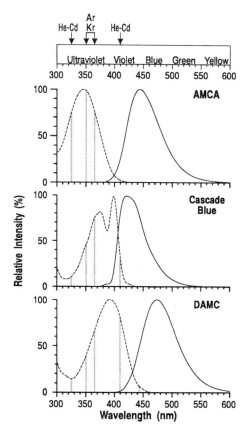

Fig. 7 Excitation (dashed) and emission (solid) spectra of blue fluorophore-conjugated secondary antibodies. *Top:* 7-Amino-4-methylcoumarni-3-acetic acid (AMCA)-conjugated donkey anti-rabbit IgG (Jackson ImmunoReseah Laboratories). The excitation spectrum was scanned while measuring emission at 475 nm, and the emission spectrum was scanned while exciting at 370 nm. *Middle:* Pyrenyloxytrisulfonic acid (Cascade Blue)-conjugated goat anti-mouse IgG (Molecular Probes). The excitation spectrum was scanned while measuring emission at 500 nm, and the emission spectrum was scanned while exciting at 340 nm. *Bottom:* 7-Diethylaminocoumarin-3-carboxylic acid (DAMC)-conjugated avidin (Organon Teknika-Cappel, Durham, NC). The excitation spectrum was scanned while measuring emission at 530 nm, and the emission spectrum was scanned while exciting at 370 nm. The emission lines for argon ion (Ar), krypton ion (Kr), and helium–cadmium (He–Cd) lasers are shown as vertical lines beneath the excitation spectra.

antibodies specifically recognize only one of the primary antibodies, and (3) the primary antibodies are specific in their recognition of the substance against which they were raised (Fig. 8). This characterization can be most efficiently performed by addressing these issues sequentially. First, the specificity of visualization of the fluorophores is tested. Having established the specificity of the fluorophores, the specificity of the

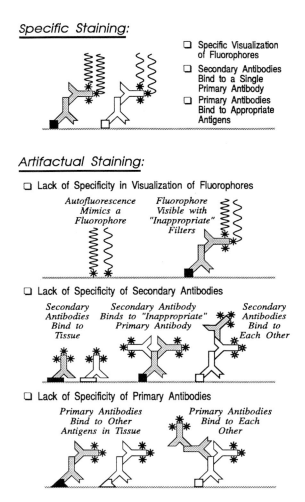

Fig. 8 Diagrammatic representation of specific vs artifactual staining observed with two-color indirect immunofluorescence. The diagram of specific staining depicts the detection of individual antigens (solid and open squares) with separate combinations of primary and fluorophore-conjugated secondary antibodies (shaded and open). The visibility of fluorophores, using only the appropriate filters, is shown as the emission of a single wavelength of light from each secondary antibody. Other substances intrinsic to the tissue that may have an affinity for the primary or secondary antibodies is shown as rectangular and triangular binding sites.

fluorophore-conjugated secondary antibodies can be tested. Once the specificity of the secondary antibodies is established, it is possible to use these reagents to examine the specificity of the primary antibodies. Although the tests for the occurrence of false positives will be addressed, it should also be noted that the combination of multiple detection schemes may obscure potential occurrences of multiple labeling and result in false negatives (Wessendorf, 1990; Wessendorf *et al.*, 1990a).

1. Visualization Specificity of Fluorophores

The choice of fluorophores used for immunofluorescence studies is critical. First, it is necessary to determine whether autofluorescence from substances intrinsic to the tissue can mimic the appearance of the fluorophores. This can be done by examining unstained specimens with the same set of excitation and emission filters intended for use with a fluorophore. If no fluorescence is detected, then no substances in the tissue would appear to mimic the emission of the fluorophore. Although the exact nature of the autofluorescent substances in biological tissues is unclear, it has been suggested that endogenous fluorophores include flavins, flavoproteins, reduced pyridine nucleotides, and lipofuscin pigments (Pearse, 1980; Lang *et al.*, 1991). Autofluorescence is usually the highest with excitation by violet–blue to blue light (400 to 450 nm) and decreases with increasing wavelength (Aubin, 1979; Benson *et al.*, 1979). However, cells containing significant amounts of porphyrins or chlorophyll will exhibit considerable far-red autofluorescence.

When more than one fluorophore is used for the simultaneous detection of antibodies, it is also important to know whether one of the fluorophores can mimic the appearance of any other fluorophores used. The specificity of visualization of the fluorophores can be tested by staining individual specimens with one of the primary antibodies followed by the appropriate secondary antibody. When these specimens are then examined by using the filter sets for each of the fluorophores, each specimen should be visible only with the appropriate filter set. Observation of staining with any of the other filter sets indicates that the protocol is not specific. This control should be performed in areas of brightest staining so as to maximize the likelihood of observing any lack of specificity.

2. Secondary Antibody Specificity

The simultaneous detection of more than one primary antibody depends on the availability of secondary antibodies that (1) do not cross-react with proteins intrinsic to the tissues being examined, (2) recognize only one of the primary antibodies, and (3) do not recognize each other. Without negative secondary antibody controls, reliable immunofluorescence is impossible. However, demonstrating secondary antibody specificity does not replace the need for primary antibody controls.

Whether the secondary antibody recognizes any substances intrinsic to a specimen can be examined by staining the specimen with only the secondary antibody in the absence of the primary antibody. The presence of staining could suggest that the secondary antibody is not specific. Although such staining may result from a specific interaction of the secondary antibody with an IgG-like epitope in the tissue, there may be other causes. These include the presence of naturally occurring antibodies in the serum against tissue epitopes, the presence of antibodies to contaminants in the immunizing antigen, and the presence of other serum proteins that can bind to constituents within the tissue. For this reason, the use of secondary antibodies that were affinity isolated before conjugation with the fluorophore may be beneficial. Although this should reduce the problem, it is not totally eliminated because the IgG coupled to the affinity column may also contain contaminants.

To prove that a secondary antibody recognizes only the primary antibody and not other constituents in the primary antibody solution is more difficult. First, it is necessary to demonstrate whether the secondary antibody is capable of staining the primary antibody. If staining results from sequential application of the primary and secondary antibodies, but not from the application of the secondary antibody alone, it would suggest that the secondary antibody recognizes something in the primary antibody solution that binds to the specimen. Second, it is necessary to demonstrate that the secondary antibody specifically recognizes the immunoglobulins of the primary antibody and not additional constituents in the antibody solution. This possibility can be tested by preincubating the primary antibody with the antigen against which it is directed. If this absorption control blocks all staining, it would appear that the staining is due to the secondary antibody recognizing the immunoglobulins of the primary antibody. If an absorption control is not possible, a similar control could be performed by examining the staining with preimmune serum from the same animal.

The simultaneous detection of more than one primary antibody with the corresponding secondary antibodies requires several more controls. It must be established that no secondary antibody has affinity for a primary antibody other than the one against which it is directed. This can be tested by staining specimens with only one of the primary antibodies and each of the secondary antibodies to be used. Assuming that the fluorophores can be specifically visualized, staining should be detected only with the filter set for the appropriate secondary antibody.

Once it is established that none of the secondary antibodies cross-react with the "inappropriate" primary antibodies, it is necessary to establish that no secondary antibody has an affinity for any of the other secondary antibodies being used. If multiple labeling is observed after incubating tissue with one primary antibody followed by all of the other secondary antibodies, this would suggest that one or more of the secondary antibodies have an affinity for another. This is most likely to occur when one of the secondary antibodies has been raised in the same or a closely related species to that in which one of the primary antibodies has been raised. Therefore it is essential to avoid using secondary antibodies raised in such species. The best choice is to use secondary antibodies raised in the same or similar species.

Secondary antibodies prepared specifically to avoid the above problems have become commercially available (e.g., Jackson ImmunoResearch Laboratories, West Grove, PA). These secondary antibodies are extensively adsorbed against solid phase-immobilized immunoglobulins and serum proteins of other species to minimize cross-reactivity with them. For example, the specific detection of mouse and rat monoclonal antibodies when used together is difficult because similar epitopes in both species will be recognized by most secondary antibodies to one of these species. However, these monoclonal antibodies can be distinguished by using fluorescein-conjugated donkey anti-mouse IgG that has been stripped of cross-reactivity to rat serum proteins, and rhodamine-conjugated donkey anti-rat IgG that has been stripped of cross-reactivity to mouse serum proteins. (It should be noted, however, that secondary antibodies adsorbed against closely related species should be used only when necessary because the IgG that is removed reduces epitope recognition of the "appropriate" IgG. As a result, the intensity of staining may suffer with some subclasses of IgGs.) An additional benefit in using reagents stripped of antibodies against serum proteins from the appropriate species is the reduction or elimination of cross-reactions to traces of these proteins present in normal tissues or absorbed from the culture medium by cells *in vitro* (Houser *et al.,* 1984).

Another issue to consider when using multiple secondary antibodies is whether the specimen is stained sequentially or simultaneously with each of the antibodies. The presence of small amounts of isoantibodies in serum from different individuals of the same species and/or heterophile antibodies in serum from different species may preclude the mixing of the sera into a single staining solution. If this occurs, the microprecipitates can frequently be observed on the surface of the specimen as small fluorescent particles. In this case, it would be best to stain the specimen by sequential application of the secondary antibodies.

3. Primary Antibody Specificity

Specific simultaneous detection of more than one substance depends on the availability of primary antibodies that (1) do not recognize other substances in the tissue being examined, (2) can be recognized specifically by secondary antibodies (see above), and (3) do not recognize each other. Even with well-characterized primary antibodies, it is important to demonstrate the specificity of staining with the primary antibodies whenever a new tissue is examined.

Before using a primary antibody for multicolor immunofluorescence studies, it is important to establish that each primary antibody recognizes the substance against which it was raised. This is typically done by testing whether staining can be blocked by absorption of the antibody with the relevant antigen. However, if the antiserum contains additional antibodies to contaminants in the antigen preparation, the absorption of the antiserum with an impure antigen may block staining but no distinction has been made between the desired and contaminating antibodies. In addition, testing whether staining can be blocked in an absorption control does not guarantee an antibody will recognize the substance as it occurs within the specimen, or after it has been altered by fixation. It is more difficult to test for cross-reactivity by a primary antibody with unknown substances in the specimen that contain epitopes similar to those against which the antibody had been raised. Because it is unclear how many tissue components are recognized by an antibody, it is preferable to refer to staining for a substance by immunocytochemistry as "immunoreactivity" in recognition of the shortcomings of the technique.

An additional concern with the simultaneous detection of more than one antigen is that the primary antibodies do not have an affinity for each other. This can be checked by simultaneously staining a region of a specimen known to contain immunoreactivity for only one of the antigens, using all of the primary and secondary antibodies for the multicolor immunofluorescence protocol. Assuming the secondary antibodies and fluorophores have been shown to be specific as described above, the presence of multiple labeling would strongly suggest that one of the other primary antibodies has affinity for the primary antibody being tested.

4. Background Staining

Because the detection of specific staining requires adequate contrast from the background staining, it is important to minimize the background observed when using immunofluorescent techniques. This is particularly important when image processing is employed (such as LSCM) because it is often possible to obtain acceptable images from the low-contrast background staining observed with control specimens. More important for LSCM is the effect of high background staining on the depth to which optical sections of acceptable quality can be acquired within thick specimens (see Section VI,A,4). There have been several suggestions to reduce the nonspecific staining observed with primary and secondary antibodies.

The presence of other antibodies in low amounts within the primary and secondary antisera can result in high background staining when the reagents are used at high concentrations. The first remedy for such "nonspecific staining" is to increase the dilution of the antisera. However, it can not easily be determined whether the lower background staining results from the decreased nonspecific or specific staining. For this reason, it might be expected that the greater capability of the microscope to visualize less intense staining would permit higher dilutions of the antisera to be used for increased specificity and low background staining.

Another widely used method to reduce background staining is to incubate the specimen with normal serum from the same species as the secondary antibody before applying the fluorophore-conjugated secondary antibody. The incubation with normal serum will presumably saturate nonspecific antibody-binding sites in the specimen. Although it may be possible to compete for the nonspecific binding by including normal serum in the dilution of the secondary antibody, in some cases the presence of small amounts of isoantibodies in serum from different individuals of the same species may result in the formation of microprecipitates.

Nonspecific background staining may also be increased by the presence of overconjugated secondary antibodies, unconjugated fluorophore molecules, or impurities from the labeling reagent bound noncovalently to the secondary antibodies. These fluorescent compounds may then bind nonspecifically to tissue components when the secondary antibody is used. Early on, it was recognized that this type of background staining could be substantially reduced by absorption of conjugated antibodies with tissue powders (Coons and Kaplan, 1950). Today, this problem is reduced by using solid-phase adsorption with either hydrophobic beads (Spack *et al.,* 1986) or normal serum from the same species as that of the specimen to be stained (Gailbraith *et al.,* 1978). However, repeating this adsorption before use of the fluorophore-conjugated antibody can still be helpful to remove free dye released during storage. Similarly, we routinely reuse a secondary antibody solution because background staining appears to be reduced as the nonspecific staining is adsorbed with each use of the antibody. Typically, the reuse of antibody is limited only by its stability when stored in a diluted state.

Besides pretreating thick specimens with detergents to increase penetration into thick specimens, the addition of detergents to the various incubation solutions can help reduce background staining. Typically, we add Triton X-100 to the antisera (0.3%) and washing solutions (1%). At these concentrations, the Triton X-100 reduces nonspecific, low-affinity protein interactions with little or no affects on antigen-antibody binding. For especially thick specimens, relatively long wash times (24–48 hr) may be necessary to observe a reduction in background staining.

Although it is widely recognized that repeated freezing and thawing can affect the reactivity of antibodies, this is especially important for fluorophore-conjugated secondary antibodies. Besides altering the affinity of the secondary antibody, freezing and thawing of the conjugates may denature or aggregate the antibody molecules. These changes can make the conjugates more "sticky," with a marked increase in background staining. Therefore it may be preferable to store the antisera in an unfrozen state as a 50% glycerol solution in buffer at −20°C. Alternatively, the undiluted antisera can be stored for prolonged periods at 4°C if kept as sterile solutions.

III. Instrumentation

The basic information describing the instrumentation necessary for fluorescence LSCM has been thoroughly presented in many excellent reviews and books (Wilson and Shepherd, 1984; Brakenhoff et al., 1989b; Schotten, 1989; Pawley, 1990). In this section, only the aspects of confocal microscope design that are important for the adaptation of LSCM to multicolor immunofluorescence studies is examined. These issues include the ability to excite the common fluorophores with the widely available lasers, the selection of filters for the spectral isolation of individual fluorescence signals, and the limits imposed by the use of photomultiplier tubes as detectors.

A. Laser Light Sources

Lasers have a number of unique properties, compared to other available light sources, that make them ideal for use with confocal microscopy (Gratton and van de Ven, 1990). Lasers have extremely high brightness, spatial coherence, low beam divergence, monochromatic output, low noise (i.e., beam intensity fluctuations), and can be focused into extremely small regions. Because the power requirements for fluorescence LSCM are minimal, usually only 1–3 m W (Gratton and van de Ven, 1990; Tsien and Waggoner, 1990; Wells et al., 1990), most lasers with even minor emission lines in the ultraviolet and visible part of the light spectrum could potentially be used as light sources for LSCM. More important for biological applications is how efficiently the emission lines of a given laser can excite the fluorophores one wishes to use (Fig. 9). Unfortunately, the monochromatic output of lasers is also a weakness. None of the widely available air-cooled lasers can match the broad emission of the mercury or xenon arc lamps used with conventional fluorescence microscopy to excite fluorophores ranging from the ultraviolet to red regions of the light spectrum. The output of the laser must

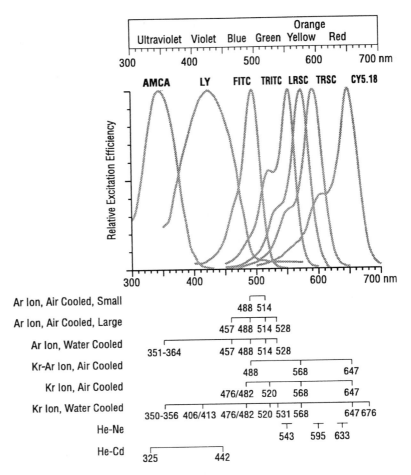

Fig. 9 Comparison of the emission wavelengths from common lasers and the excitation spectra of fluorescent probes. The emission wavelength(s) available from various argon (Ar), krypton (Kr), krypton–argon (Kr–Ar), helium–neon (He–Ne), and helium–cadmium (He–Cd) lasers are shown. Note that the broad range of emission lines may not be simultaneously available from the larger, water-cooled ion lasers. The intense emission peaks from high-pressure mercury (Hg) arc lamps commonly used as a light source for conventional fluorescence microscopy are shown for comparison. The excitation spectra of representative fluorophores used in immunofluorescence studies are shown: 7-amino-methylcoumarin-3-acetic acid (AMCA), fluorescein (FITC), tetramethylrhodamine (TRITC), Lissamine rhodamine (LRSC), Texas Red (TRSC), and cyanine 5.18 (CY5.18). Lucifer yellow (LY) is commonly used as an intracellular label for microinjection experiments (Stewart, 1978).

be more stable than the relative differences in the intensity of fluorescence to be detected during the acquisition of an image. Otherwise, differences in excitation intensity may be confused for actual variations in staining intensity. The output beam of the laser must be laterally stable, with the individual emission lines being colinear and parallel to lessen registration problems. Other important concerns regarding laser use with

LSCM are cost, power requirements, method of waste heat removal, and reliability of operation.

1. Argon Ion Lasers

The most common type of laser for LSCM is a small, air-cooled argon ion laser with 25–50 mW of output power. This can be attributed to their low cost, stable output with low noise levels, minimal maintenance requirements, and long operational lifetimes (3000–5000 hr). These argon ion lasers have two major emission lines at 488 nm (blue) and 514 nm (green). The 488-nm line efficiently excites fluorescein (87% of its excitation maximum at 496 nm), but the 514-nm line is suboptimal for the various rhodamines (only 5 to 30% of their excitation maxima). Because 514-nm light also excites fluorescein (30% of its excitation maximum), other fluorophores (e.g., rhodamines) cannot be specifically excited in the presence of fluorescein with the 514-nm line. Typically, the larger, air-cooled argon ion lasers with 100–500 mW of output power are equipped with optics that allow a broader selection of emission lines ranging from 457 nm (violet–blue) to 528 nm (green). Lucifer yellow is more efficiently excited at 457 nm compared to 488 nm (81% vs 17% of its excitation maximum at 428 nm). Hence these larger lasers should be superior for the imaging of the small processes of Lucifer yellow-injected neurons (Mossberg and Ericsson, 1990). However, the increased autofluorescence observed with this shorter wavelength excitation line may be a problem with some specimens. The 528-nm line is the longest wavelength emission available from either air- or water-cooled argon ion lasers, but is infrequently used because it is difficult to maintain during the operational lifetimes of the laser. However, it does permit slightly more efficient (10–40% of their excitation maxima) excitation of the red fluorophores (Fox *et al.*, 1991).

The water-cooled argon ion lasers of 5–20 W of output power have also been used as a reliable source of ultraviolet emissions (35–363 nm) for LSCM (Amdt-Jovin *et al.*, 1990; Montag *et al.*, 1991; Schubert, 1991; Ulfhake *et al.*, 1991; Bliton *et al.*, 1993). Because of the higher ionization states of argon required for the stimulation of ultraviolet emissions, there are higher electrical and cooling requirements for these lasers. This reduces their operational lifetimes (1000 to 2000 hr) compared to the air-cooled argon ion lasers. Although water-cooled argon ion lasers have been used for live-cell imaging of ultraviolet excitable ion-sensitive probes (Lechleiter and Clapham, 1992), the high cost and maintenance requirements of these lasers have so far limited their use with LSCM. Nonetheless, the ultraviolet output of these lasers can also be used for immunofluorescence studies to efficiently excite blue fluorophores such as AMCA (Schubert, 1991; Ulfhake *et al.*, 1991). Unfortunately, the chromatic aberration present in most microscope lenses in the ultraviolet can severely compromise imaging (Ulfhake *et al.*, 1991; Bliton *et al.*, 1993). However, at least one commerical instrument has become available with the necessary optical modifications for ultraviolet confocal imaging (Bio-Rad Microscience, Cambridge, MA).

Although water-cooled argon ion lasers can be equipped with optics that permit simultaneous emission at both ultraviolet and visible wavelengths, emission of multiple lines makes it more difficult to stabilize the output power of these lasers when operating in a multiline mode. Typically, both air- and water-cooled ion lasers are operated in a light-control mode to minimize intensity fluctuations. The power supply output is regulated by a feedback circuit that samples the laser beam and adjusts the laser current to maintain constant light output. Although this works well for lasers emitting at a single wavelength, it is more difficult when the simultaneous emission from several lines occurs. For example, if the detector for the light control circuitry responds primarily to 488- and 514-nm light, there may be considerable fluctuations in the ultraviolet power output. In contrast, if the light control circuitry is used to stabilze the ultraviolet emissions, large fluctuations in the power output at 488 nm may result. Therefore it may be preferable to use separate lasers as sources of ultraviolet and visible wavelengths to minimize intensity fluctuations.

2. Krypton and Krypton–Argon Ion Lasers

Although krypton ion lasers are similar in operation to argon ion lasers, the krypton ion has a wider range of visible emission wavelengths. The air-cooled krypton ion lasers of 15- to 200-mW output power have blue emission lines at 468, 476, and 482 nm, green lines at 520 and 531 nm, a yellow line at 568 nm, and red lines at 647 and 676 nm. This broad selection of wavelengths has the potential to efficiently excite most fluorophores requiring excitation by visible light (Fig. 9). Although this capability is enticing, krypton ion lasers are known for less reliable operation than similar argon ion lasers because of their lower output powers, higher noise levels, shorter operational lifetimes (<2000 hr), and greater sensitivity to changes in operation conditions (e.g., misaligned or dirty optics, gas pressure of the plasma tube, and power supply stability). Typically, these lasers have a limited lifetime due to depletion of the gas in their reservoirs that is used to replenish the plasma tube during aging. As if this were not bad enough, competition occurs between several of the krypton emission lines. This means that although the total output power of a krypton ion laser may stay constant when operating in a multiline mode, wide fluctuations in the intensity of the individual emission lines may occur. This excessive noise can make some of the wavelengths unusable for LSCM. Although it is less common, the use of the larger, water-cooled krypton ion lasers for LSCM has also been described (Brakenhoff *et al.,* 1989a; Van Dekken *et al.,* 1990).

Because of the promising characteristics of krypton–argon ion lasers, we attempted to address some of the shortcomings. We developed in collaboration with Bio-Rad Microscience and Ion Laser Technologies (Salt Lake City, UT) an air-cooled krypton–argon ion laser of 15-m W output power for use with LSCM. This laser operates in a multiline mode with a 4- to 5-m W output at 488 nm (blue), 568 nm (yellow), and 647 nm (red). To ensure the availability of more than enough light for the excitation of blue fluorophores such as fluorescein, a mixed gas krypton–argon ion laser was designed with the strong 488-nm emission of the argon ion instead of the weak blue emissions from the krypton ion. Similar to the argon ion lasers, the 488-nm line efficiently excites fluorescein, but the 568-nm krypton line is considerably more efficient than the 514-nm line for the excitation of the various rhodamines, especially Lissamine rhodamine (92% of its excitation maximum at 572 nm). In addition, the 647-nm line can efficiently excite far-red fluorophores such as cyanine 5.18 (98% of its excitation maximum at 649 nm).

3. Helium–Neon Lasers

As an alternative to krypton ion lasers for longer wavelength lines, helium–neon lasers are available with a 1- to 10-m W output at 543 nm (green), 595 nm (yellow), or 633 nm (red). These air-cooled lasers are inexpensive, have long lifetimes (~20,000 hr), and no adjustments of laser optics are needed because their emission wavelength cannot be changed. The 543-nm helium–neon laser can be used for more efficient excitation of rhodamines than is possible with an argon ion laser (Fig. 9) (Stelzer, 1990; Montag *et al.*, 1991; Schubert, 1991). As with the 647-nm line of the krypton ion lasers, the 633-nm helium–neon laser can be used for the excitation of far red fluorophores such as allophycocyanin or cyanine 5.18 (Fig. 9).

4. Helium–Cadmium Lasers

Helium–cadmium lasers can be used as an alternative ultraviolet light source instead of the water-cooled ion lasers. These lasers are air cooled and have 1–50 mW of output power at 325 nm (ultraviolet) or 442 nm (violet–blue). Because of their lower cost and operating requirements than the water-cooled argon lasers, helium–cadmium lasers have been used for many applications with flow cytometry that require ultraviolet excitation (Shapiro, 1988; Goller and Kubbies, 1992). A 442-nm helium–cadmium laser has been used in combination with an air-cooled argon ion laser for live-cell imaging by LSCM of changes in pH, using the ion-sensitive dye BCECF (Wang and Kurtz, 1990). Also, a true-color confocal reflection microscope with 442-nm helium–cadmium, 532-nm frequency-doubled neodymium–yttrium aluminum gamet (Nd-YAG), and 633-nm helium-neon lasers has been described (Cogswell *et al.*, 1992). The 325-nm line has less frequently been used for LSCM because it requires the use of special quartz optics (Brakenhoff *et al.*, 1979; Kuba *et al.*, 1991; Fricker and White, 1992).

5. Other Lasers

In addition to these lasers, a broad range of wavelengths is available from the use of tunable dye lasers and nonlinear wavelength expansion techniques (Gratton and van de Ven, 1990). The development of diode lasers emitting near 650 nm should also be excellent excitation sources for the excitation of far-red fluorophores. A particularly elegant application of high-power lasers is the two-photon excitation of ultraviolet-absorbing fluorophores by light with a wavelength twice that of the actual excitation wavelength (Denk *et al.*, 1990). However, these more exotic lasers are of limited availability to most biologists because of their high purchase and operating costs.

B. Optical Filters

Regardless of whether conventional or confocal optics are used, the selection of appropriate optical filters is critical to the successful operation of any fluorescent microscope. Two types of optical filters are commonly used for fluorescence microscopy. Colored glass filters consist of dye molecules suspended in glass or plastic that absorbs some wavelengths of light and transmits other. Although these colored glass filters are highly efficient blockers of light, their optical properties will vary with the thickness of the filter, the dyes may themselves fluoresce, and conversion of absorbed high-intensity light to heat may change the special characteristics of the filter. In contrast, interference filters are made of multiple layers of reflective materials deposited on glass substrates. These thin layers of metals or dielectrics work by an additive process whereby the amplitudes of two or more overlapping waves are attenuated or reinforced. The major advantage of interference filters are their intrinsic nonfluorescence, negligible light scatter, and that they allow almost any desired spectral response to be obtained. These factors allow the design of complicated interference filters with multiple reflectance and transmittance bands (see Section IV,B,3 and 4). Because optical filters of all types are subject to damage and degradation, filter performance needs to be periodically monitored to verify continued performance at their design specifications.

Optical filters are characterized by their transmission and reflection properties and their intended use. Bandpass interference filters transmit one particular region of the light spectrum. Typically, they are specified according to their peak wavelength of transmission and the full bandwidth at half-maximal transmission. Shortpass and longpass filters are specified according to the wavelength at which half-maximal transmission occurs. Dichroic mirrors are typically long-pass filters designed to be utilized at an angle of incidence of 45° to the light path. As such, they reflect light of wavelengths shorter than the specified cut-off wavelength and transmit light of longer wavelengths. When used at other angles of incidence, the cut-off wavelength of the dichroic mirror will be shifted and the reflectance/transmittance boundary broadened.

C. Detector Apertures

For multicolor LSCM with multiple detectors, it is beneficial to have separate, variable apertures for each detector. Although a single aperture could be positioned in the detection path before splitting the fluorescence into separate detectors, this configuration has several disadvantages. First, different aperture sizes are required to obtain the minimum depth of field for each fluorophore because the diameter of the Airy disk formed by their fluorescence is wavelength dependent (Brakenhoff *et al.,* 1990). Second, the use of separate, variable apertures provides the capability to balance the strength of the fluorescence signal against the rejection of the out-of-focus background fluorescence (i.e., depth of field) for the optimum overall performance for each fluorophore (Wilson, 1989; Wells *et al.,* 1990). This is particularly important for biological specimens, in which considerable differences in the intensity of staining of each fluorophore in multiple-labeled specimens are frequently observed.

D. Detectors

Although photomultiplier tubes (PMTs) are the most common detectors used in laser-scanning confocal microscopes (Art, 1990; Pawley, 1990), they are not necessarily the ideal detector. In an instrument with an efficient optical path, the detection of the fluorescence by the PMT can be the single least efficient step (Wells *et al.,* 1990). This is particularly important for multicolor LSCM because the response of PMTs is extremely sensitive to the wavelength of the detected photons. Although the quantum efficiency of the PMT may be as high as 25–30% for the detection of blue and green photons, this efficiency rapidly decreases as a nonlinear function with increasing wavelength. For example, the Bio-Rad MRC-600 uses PMTs with a prismatic S-20 photocathode surface (model 9828B; Thorn EMI, Rockaway, NJ) with a response typical of PMTs with extended red sensitivity. Nevertheless, the quantum efficiency for detection of photons decreases from ~11% at 500 nm, to ~5% at 600 nm, and <1% at 700 nm. Assuming the same fluorescence signal, this means that a fluorophore detected as red light will be detected with less than half of the efficiency of one detected as green light.

Several other issues must be considered when using PMTs with extended red sensitivity in confocal microscopes. Because the dark current of a PMT is primarily a function of the extent of red spectral response, the detection of weak fluorescence signals may be limited by their higher dark currents. In these red-sensitive PMTs, the lower work function of the photocathode results in more thermal events spontaneously ejecting electrons (Art, 1990). When imaging far red fluorophores, using the 647-nm line of a krypton–argon ion laser, we have observed a reduction in image contrast and increased dark current with prolonged use of the microscope. Presumably, this results from the heating of the PMTs by electrical components within the scan unit. Unfortunately, it is not possible to discriminate thermal events originating from the photocathode from photon-induced events. For optimal sensitivity it is necessary to use cooled PMTs to reduce the frequency of thermal events from the photocathode. Similarly, the higher dark current of PMTs induced by exposure to high light levels or room lighting can reduce the sensitivity of far red imaging for several hours to days.

IV. Approaches to Multicolor Laser Scanning Confocal Microscopy

Examination of multiple fluorescent probes in a single specimen requires the independent detection of signals from each probe. This requires differences in their excitation and emission spectra. Two approaches have been used to spectrally isolate each fluorophore in a multiple-labeled specimen: (1) Use a single wavelength to simultaneously excite all fluorophores and separate their emissions in the detection path; and (2) use different excitation wavelengths and emission filters for the detection of each fluorophore. In both cases, the fluorophores and microscope filters must be carefully chosen to allow differences in their excitation and emission spectra to be exploited (Wessendorf, 1990; Galbraith et al., 1989). Although not discussed here, separate detection by LSCM of fluorescent probes based on differences in fluorescence lifetimes is possible (Buurman et al., 1992; Morgan et al., 1992).

The advantages and disadvantages of these approaches for the spectral isolation of multiple fluorescence with LSCM will be examined below. Specific microscope configurations will be discussed to demonstrate the decisions and compromises necessary for the design of an instrument for multicolor LSCM. Because the use of lasers with ultraviolet emission lines with LSCM requires special optics and instrumentation, the following discussion will be limited to configurations with the widely available visible light lasers.

A. Single-Wavelength Excitation of Multiple Fluorophores

The simultaneous excitation by a single wavelength of multiple fluorophores and separation of their emissions in the detection path places severe restrictions on the fluorophores to be examined. The excitation spectra of the fluorophores must partially overlap to permit excitation by a single wavelength of light. (However, for most combinations of fluorophores one will be excited much more efficiently than the others.) In contrast, their emission spectra should not overlap significantly so the signal from each fluorophore can be spectrally isolated with carefully chosen bandpass filters. In addition, the emission peaks of the shorter wavelength fluorophores should overlap minimally with the excitation peaks of the longer wavelength fluorophores to reduce the possibility of energy transfer between fluorophores (Chapple *et al.*, 1988). Because of these strict requirements, a considerable compromise in the efficiency of excitation and degree of spectral isolation of each signal is usually required for most combinations of fluorophores.

These factors have not prevented this approach from being extensively utilized for two-color immunofluorescence studies using flow cytometry with a single excitation beam (Loken *et al.*, 1977; Hoffman, 1988; Shapiro, 1988). Similar to LSCM, flow cytometry with laser light sources has relatively few wavelengths available for the excitation of fluorophores. However, a significant advantage for flow cytometry is that the sampling volume is a single cell, not the considerably smaller diffraction-limited volume of confocal microscopy. As a result, there is more latitude for balancing the efficiency of excitation, spectral isolation of the fluorescence signals, and sensitivity of detection. Nevertheless, it is usually impossible to achieve total spectral isolation of the fluorophores because their emission spectra always overlap to some extent. In these cases, the cross-talk between multiple detectors can usually be reduced by using electronic compensation to subtract a fraction of the signal detected in each spectral band from the others (Loken *et al.*, 1977). With this approach, double-labeling studies have become widespread, using the combination of fluorescein and R-phycoerythrin excited by the 488-nm line of argon ion lasers (Oi *et al.*, 1982). For this combination of fluorophores, fluorescein is measured as green fluorescence around 530 nm, whereas R-phycoerythrin is detected as orange–red fluorescence above 570 nm.

As for flow cytometry, this approach can be adapted to LSCM for multicolor studies using a single excitation wavelength. For single-detector instruments, the images of each spectral region must be collected after switching emission filters (Brodin *et al.*, 1988). However, fading or movement of the specimen during the sequential collection of the images complicates the comparison of the images. These problems are avoided by sampling simultaneously several spectral regions with multiple detectors (Amos, 1988; Carlsson, 1990). In this case, the corresponding pixels within the images will be in spatial and temporal registration because the same illumination point is the source of all fluorescence signals.

1. Two–Color Laser Scanning Confocal Microscopy with 514–nm Line of an Argon Ion Laser

The 514-nm line of the widely available air-cooled argon ion lasers has been successfully used for the imaging of specimens double labeled with fluorescein and one of the rhodamines (Brelje *et al.,* 1989; Fehon *et al.,* 1990; Schweitzer and Paddock, 1990; Stamatoglou *et al.,* 1990; Brelje and Sorenson, 1991; Vincent *et al.,* 1991; Wetmore and Elde, 1991). Although neither fluorophore is efficiently excited by the 514-nm line, sufficient light is absorbed by each to allow their fluorescence signals to be detected from brightly stained specimens. In this configuration (Fig. 10), the 514-nm line

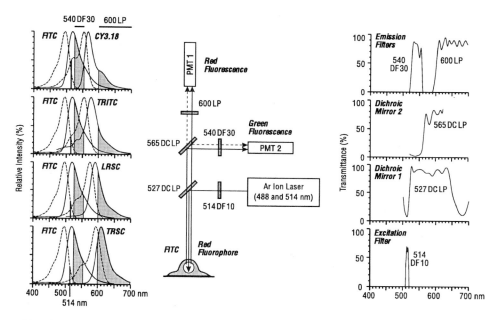

Fig. 10 Filter configuration for simultaneous two-color imaging of specimens double labeled with fluorescein and a red fluorophore, using the 514-nm line of an argon laser. *Left:* The excitation (dashed) and emission (solid) spectra of various combinations of fluorescein and a red fluorophore, either cyanine 3.18 (CY3.18), tetramethylrhodamine (TRITC), Lissamine rhodamine (LRSC), or Texas Red (TRSC). The 514-nm line is shown as a vertical line beneath the excitation spectra. The wavelengths transmitted by the bandpass emission filters for detecting green and red fluorescence are shown as shaded regions on the emission spectra. Although a substantial region of each emission spectrum is sampled, the amount of cross-talk between the detectors cannot be determined from the normalized emission spectra shown, because the intensity of observed signals will depend on their relative concentrations within the specimen. *Center:* The filter configuration for simultaneous excitation of both fluorophores by 514-nm light. A second dichroic mirror is used to direct the green and red fluorescence into separate photomultiplier tubes (PMTs). *Right:* The transmittance spectra of the various excitation filters, dichroic mirrors, and emission filters are shown. Filter abbreviations: DF, bandpass filter with the specified center wavelength (nm) and full bandwidth at half-maximal transmission; DC, dichroic mirror with half-maximal transmission at the specified wavelength (nm) when at an angle of incidence of 45°; LP, longpass barrier filter with the specified cut-off wavelength (nm) between the shorter reflecting and long transmitting wavelengths.

is selected with a narrow bandpass filter, reflected by a longpass dichroic mirror, and focused to a diffraction-limited volume within the specimen by the objective. The fluorescence signals are then transmitted by the same dichroic mirror into the detection path (Fig. 11). A second longpass dichroic mirror in the detection path reflects the shorter wavelength green fluorescence into one detector, and transmits the longer wavelength

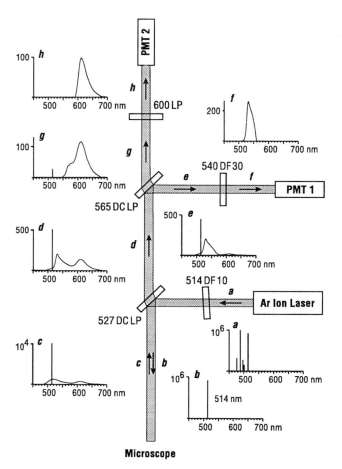

Fig. 11 Explanation of how the filter configuration for two-color imaging with the 514-nm line of an argon ion laser separates the green and red fluorescence signals from a specimen double labeled with fluorescein and Texas Red. The argon (Ar) ion laser operates in a multiline mode with output at several wavelengths (a). A narrow bandpass filter (514 DF 10) selects the 514-nm line (b), reflected by a longpass dichroic mirror (527 DC LP), and directed into the microscope. The fluorescence signals and a significant fraction of the excitation light return from the microscope (c), and are passed by the dichroic mirror (d). A second longpass dichroic mirror (565 DC LP) in the detection path reflects the shorter wavelength green fluorescence into one detector (e), and transmits the longer wavelength red fluorescence into another detector (g). Two emission filters (540 DF 30 and 600 LP) are used to restrict the detection to the regions of maximum separation of the fluorescence signals and to block the remaining reflected 514-nm light (f and h). Note that the spectra are shown so the total signal detected in both detectors will be approximately equal.

red fluorescence into another detector. The wavelength of light detected is further restricted by emission filters to 525–555 nm for the green fluorescence detector and longer than 600 nm for the red fluorescence detector. In addition, these filters must efficiently block the 514-nm light reflected by the specimen to prevent degradation of the observed signal-to-noise ratio.

From the emission spectra of fluorescein and the various rhodamines (Fig. 10), it appears that this configuration would require only minor corrections for the cross-talk between the green and red fluorescence detectors. Although excited less efficiently than tetramethylrhodamine, the longer wavelength Lissamine rhodamine and Texas Red should give better separation from the emission filter used for the green fluorescence detector. However, it should be noted that the emission spectra have been normalized as a percentage of the maximum fluorescence signal observed for each fluorophore. This comparison does not consider the differences in the efficiency of excitation and quantum yield of the fluorophores, the fluorophore/protein ratios of the secondary antibodies, and the relative intensity of the staining within a specimen. To demonstrate the magnitude of these issues, solutions of fluorescein and Texas Red conjugated to secondary antibodies were examined by spectrofluorometry (Fig. 12). When this is done, the previously small contribution of fluorescein to the red fluorescence signal detected at wavelengths longer than 600 nm is now as much as, if not more than, that from Texas Red. This is partially the result of the more efficient excitation of FITC by 514-nm light, but may also reflect the lower quantum yields of Texas Red conjugates (Table I). As a result, staining with the rhodamine will have to be especially bright and cross-talk compensation may still be necessary to remove the fluorescein signal observed in the red fluorescence detector (see Section V,A,3).

2. Cross-Talk Compensation

As routinely done for two-color studies with flow cytometry, the cross-talk between detectors in two-color LSCM with a single excitation wavelength can be reduced by subtracting a fraction of each signal from the others (Mossberg and Ericsson, 1990; Vassy *et al.,* 1990; Brelje and Sorenson, 1991; Lynch *et al.,* 1991; Carlsson and Mossberg, 1992). This cross-talk compensation is typically done after collection of the individual images by using linear image processing operations to subtract, with an appropriate coefficient, the image observed in one detector from the other image. The magnitude of this cross-talk correction can be estimated by imaging single-labeled specimens with the "inappropriate" detector(s), using the identical instrument configuration (i.e., intensity of excitation light, pinhole sizes, gain and black level of the photomultiplier tubes, and objective). Alternately, if the images contain spatially distinct structures stained only with one of the fluorophores, the correction factor can be estimated from the ratio between intensities of respective pixels in each image.

This approach to cross-talk compensation requires that the relationship between fluorophore concentration and the intensity of fluorescence be constant for all pixels within the images. Because photomultiplier tubes are normally used as detectors for LSCM, good linearity between the observed intensity of fluorescence and the measured signal is expected. More important is the influence of the local environment on the fluorophore

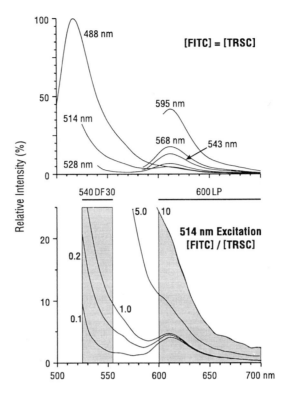

Fig. 12 Spectroscopic examination of the fluorescence observed with fluorescein (FITC)- and Texas Red (TRSC)-conjugated IgGs. *Top:* The emission spectra of an equimolar solution of the fluorophore-conjugated goat anti-mouse IgGs at the indicated excitation wavelengths. (Note that the concentrations are based on the fluorophore instead of protein because of the slight differences in the fluorophore/protein ratios of the conjugates.) For excitation by 514-nm light, the amount of red fluorescence observed at wavelengths longer than 600 nm is barely detectable in the presence of the longer wavelength fluorescence from FITC. Although more red fluorescence from TRSC is observed with increasing excitation wavelength, its relative intensity compared to that from FITC with 488-nm excitation is considerably less. *Bottom:* The red fluorescence from FITC increases as its relative concentration increases in comparison to TRSC. Even for an equimolar solution, the longer wavelength emissions from FITC still contribute one-third of the total red fluorescence signal observed at wavelengths longer than 600 nm.

and fading of the fluorescence from the region of the specimen being examined. If these effects only alter the observed intensity of fluorescence, the relative contribution of fluorescence signals collected in different spectral regions will not be altered. However, any shifts in the emission spectra of the fluorophores will affect the relative distribution of their fluorescence signals between the detectors. Although large changes in the emission spectra are unlikely for the widely used fluorophores, the occurrence of small shifts with the intense illumination conditions used for LSCM cannot be dismissed. A comparatively large change in the measured signals may be observed because of this shift due to the highly nonlinear relationship between quantum efficiency and wavelength of detected light for the photomultiplier tubes typically used as detectors (see Section III,D). For example, shifts in the emission spectra of fluorescein and tetramethylrhodamine at different concentrations have been observed with excitation intensities similar to those used with LSCM (Carlsson and Mossberg, 1992). Because of these difficulties, the relative relationship between the fluorescence signals observed in different spectral regions will not be constant over the region of the specimen examined.

Unlike flow cytometry, the additional demands imposed by the requirements of confocal imaging can further reduce the effectiveness of this cross-talk compensation technique. In particular, the wavelength-dependent imaging properties can alter the detection efficiency and the region of the specimen sampled for each pixel in images acquired in different spectral regions. Axial chromatic aberration within the optical components of the microscope will result in a noticeable shift in the focal plane at the different wavelengths (Fricker and White, 1992). Lateral chromatic aberration can also vary the off-axis collection efficiencies for light of different wavelengths (Wells *et al.,* 1990). Similarly, a small amount of spherical aberration is sufficient to cause severe reduction in collection efficiency and resolution as the focal plane is moved deeper within a specimen (Sheppard and Gu, 1991; see also Section VI,A,2). Even if these aberrations can be reduced or avoided, differences in resolution will alter the diffraction-limited volume of the specimen sampled for the same pixel in images acquired in different spectral regions (Table II). As the difference between the excitation and detection wavelengths for a fluorophore increases, the images will be of lower resolution and have less rejection of out-of-focus fluorescence (i.e., thicker optical sections; Wilson, 1989). However, this loss in resolution will be less than that observed with conventional microscopy because resolution is dependent on both the excitation and emission wavelengths for fluorescence confocal microscopy. As a result of these wavelength dependencies of imaging, the contribution of a stained structure to the same pixel will not be identical in images acquired in different spectral regions.

The difficulties in using this compensation technique for the correction of cross-talk between images acquired in different spectral regions can be easily demonstrated by examining specimens stained only with fluorescein. Using the previously discussed configuration (Fig. 10), images of a fluorescein-stained specimen were acquired in both the green and red fluorescence detectors (Fig. 13A and B, respectively). Although the images appear similar, the lower signals observed in the red fluorescence detector result in a lower signal-to-background ratio (i.e., higher background staining) and a loss of detail. Because both images were acquired with an identical peak pixel value, the red/green fluorescence ratio image should have a value of 1.0 for all pixels. Actually, a range of values from 0.8 to 6.0 is observed in the ratio image (Fig. 13C). The low-intensity background pixels apparently skew the histogram to larger ratio values (Fig. 14A). If pixels having an intensity below 10% of the maximum value are ignored in calculating the ratio image (Fig. 13D), a narrower range of ratio values centered on a peak value of 1.0 is observed (Fig. 14, *top*). However, this peak still represents a rather large range of ratio values (0.8 to 1.2). If the pixel value of the red fluorescence image is graphed against its ratio value, a relatively constant ratio (1.00 ± 0.2) is observed for the brighter pixels

Table II
Theoretical Microscope Resolutions[a]

Wavelength (nm)	Lateral resolution (μm)			Axial resolution (μm)		
	×10 NA 0.40 $n = 1.0$	×40 NA 0.85 $n = 1.0$	×60 NA 1.40 $n = 1.516$	×10 NA 0.40 $n = 1.0$	×40 NA 0.85 $n = 1.0$	×60 NA 1.40 $n = 1.516$
Confocal fluorescence microscope						
351/450	0.43	0.20	0.12	3.48	0.78	0.45
488/518	0.55	0.26	0.16	4.50	0.99	0.56
514/540	0.58	0.27	0.17	4.70	1.03	0.58
514/610	0.61	0.29	0.18	4.95	1.09	0.58
514/680	0.64	0.30	0.18	5.20	1.10	0.64
568/590	0.64	0.30	0.18	5.17	1.09	0.64
647/680	0.72	0.34	0.21	5.88	1.28	0.72
Confocal reflection microscope						
351	0.39	0.18	0.11	3.12	0.70	0.39
488	0.54	0.25	0.15	4.45	0.97	0.54
514	0.56	0.27	0.16	4.58	1.01	0.57
568	0.63	0.29	0.18	5.05	1.12	0.62
647	0.71	0.34	0.21	5.76	1.27	0.71
Conventional microscope						
351	0.54	0.25	0.17	4.39	0.97	0.63
450	0.69	0.32	0.21	5.63	1.25	0.81
488	0.74	0.35	0.23	6.10	1.35	0.88
518	0.79	0.37	0.24	6.48	1.43	0.93
568	0.87	0.41	0.27	7.10	1.57	1.02
590	0.90	0.42	0.28	7.38	1.63	1.06
647	0.99	0.46	0.30	8.09	1.79	1.16
680	1.04	0.49	0.32	8.50	1.88	1.22

[a] The conventional microscope resolutions were calculated with Eq. (1) for the lateral and Eq. (2) for the axial directions:

$$\Delta r = 0.61\lambda/\eta \sin \alpha \tag{1}$$

$$\Delta z = 2\lambda/\eta \sin^2 \alpha \tag{2}$$

where λ is the wavelength of light, $\eta \sin \alpha$ is the numerical aperture (NA) of the objective, and η is the refractive index of the immersion medium. A decrease in light intensity of ~27.7% laterally and ~19.3% axially is detectable between two point light sources placed this distance apart (Bliton *et al.,* 1993). The confocal resolutions are given as the distance between the peaks of two confocal point spread functions (PSFs) required to give a summed profile with the appropriate intensity decrease between the peaks. The confocal PSFs were calculated by multiplying together the PSF intensity profiles for the excitation and emission wavelengths as described by Eq. (3) in the lateral and Eq. (4) in the axial directions (Brakenhoff *et al.,* 1989):

$$I_{lateral}(v) = \text{const}[J_1(v)/v]^2 \qquad \text{with } v = 2\pi r\eta \sin \alpha \tag{3}$$

$$I_{axial}(\omega) = \text{const}[\sin(\omega/4)]^2/(\omega/4)^2 \qquad \text{with } \omega = 2\pi z\eta \sin^2 \alpha/\lambda \tag{4}$$

where $J_1(v)$ is the first-order Bessel function.

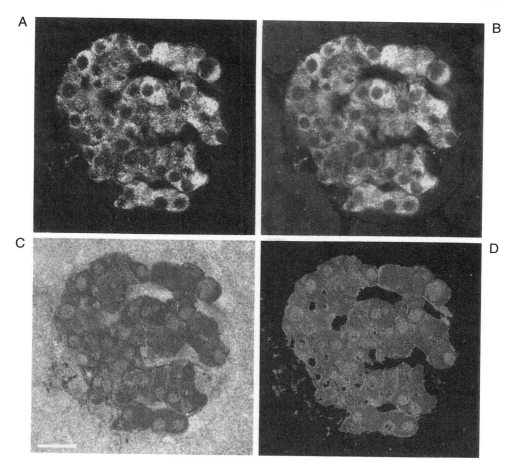

Fig. 13 Test of cross-talk compensation with image acquired by LSCM, using the 514-nm line of an argon ion laser for two-color imaging (Fig. 10). (A) Image acquired in the green fluorescence detector and (B) image acquired in the red fluorescence detector. Each image was collected with a similar peak pixel value of white (i.e., 255). (C) The ratio image obtained by dividing the red by the green fluorescence image. The ratio values are displayed so a value of 4.0 or greater is shown as white and a value of 1.0 is dark gray (i.e., 64). (D) The ratio image obtained by considering pixels having an intensity below 10% of the maximum value (i.e., 25) as background; these pixels are displayed in black. Scale bar = 25 μm. Nikon ×40, NA 0.85.

(Fig. 14, *bottom*). The use of a nonlinear function for the correction factor should give better results than choosing a single value based only on the brightest pixels. However, the large range of ratio values for pixels of the same intensity (note the large error bars in Fig. 14B) suggests the local environment within the specimen affected the relative distribution of the fluorescence signal between the detectors. This means that even pixels

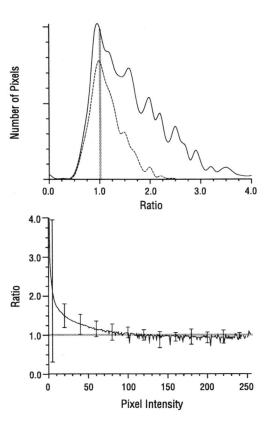

Fig. 14 Estimation of the ratio value necessary to compensate the cross-talk of the fluorescein signal into the red fluorescence detector. *Top:* Histogram of the ratio values observed for Fig. 13C (solid) and Fig. 13D (dashed). Even when the background pixels are ignored, a broad peak centered at 1.0 is still observed. *Bottom:* A graph of the observed ratio value (average ± SEM) vs pixel intensity of the red fluorescence image. Although a relatively constant ratio is observed for the brighter pixels, note the large error bars indicating a large range of values for pixels of the same intensity.

of identical intensity cannot be corrected to the same extent. Therefore it is impossible to avoid having pixels within the image that either under- or overcompensated irregardless of the correction factor used. The extent to which this increases the uncertainty in the corrected intensity values will have to be determined for each type of preparation because of the large number of factors involved.

Therefore this cross-talk compensation technique is only partially effective for images acquired in different spectral regions by LSCM. Unfortunately, this approach can only improve the separation of the fluorescence signals from brightly stained, double-labeled specimens. It does improve the efficiency of excitation or detection of the fluorophores. This makes it especially difficult, if not impossible, to determine if a structure is lightly stained with one fluorophore when substantial "bleed-through" from the other fluorophore occurs. Therefore this compensation technique is best suited for the correction of crosstalk between images in which each fluorophore stains spatially distinct structures.

B. Multiple-Wavelength Excitation of Multiple Fluorophores

The most powerful approach for the spectral isolation of multiple fluorophores is to use different excitation wavelengths and emission filters for the detection of each fluorophore. The preferential excitation of each fluorophore is possible because for every fluorophore, at some wavelength longer than its excitation maximum, the photons will not be sufficiently energic to stimulate the electrons of the fluorophore into the excited states from which fluorescent emissions occur. However, an emission filter is necessary to block the small amount of fluorescence inefficiently excited from any longer wavelength fluorophores also present in the specimen. Thus, the use of separate filter sets to image each fluorophore provides the requisite degrees of freedom necessary for the optimal excitation of each fluorophore and the spectral isolation of its fluorescence emission. Although this approach is widely used for conventional fluorescence microscopy (DeBiasio et al., 1987; Waggoner et al., 1989; Wessendorf, 1990), its application for multicolor LSCM has been limited by the lack of economical lasers with broadly spaced emission lines.

Besides the availability of appropriate lasers, a major difficulty in using multiple excitation wavelengths with LSCM is the need to use separate filter sets to obtain images of each fluorophore. When a set of images is taken of the same field of the microscope, in general the images are not accurately superimposable. For example, we frequently observed lateral shifts of at least 0.5–1.0 μm when changing filter sets and using a ×60 objective. More important is the misalignment between the illumination beam and the detector apertures required for optimal confocal imaging. Practically, this means that the microscope must be realigned after changing filter sets to ensure that the illumination source and detector aperture overlap in the focal plane (Stelzer, 1990).

Even minor changes in the orientation, thickness, or surface characteristics of filters in the illumination path can shift the images or require realignment of the confocal apertures when using separate filter sets. This is mostly due to differences in the dichroic mirror. Therefore the use of filter configurations in which images of each fluorophore are acquired with the same dichroic mirror is preferred. In this case, the images will be within registration to the maximum extent possible with the objective lens being used. The following configurations will illustrate several ways to use the optimal excitation and emission filters for each fluorophore without having to use separate dichroic mirrors.

1. Two-Color Laser Scanning Confocal Microscopy with 488/514-nm Lines of an Argon Ion Laser

A simple application of using multiple excitation wavelengths with LSCM is to use two of the emission lines from an air-cooled argon ion laser to improve the imaging

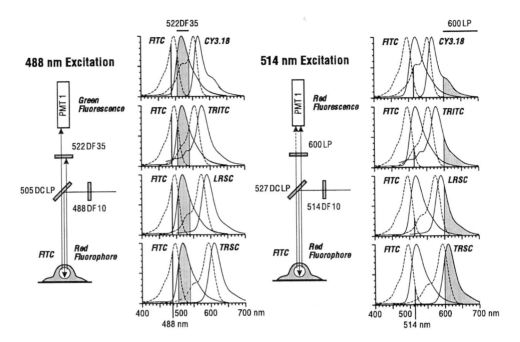

Fig. 15 Filter configurations for sequential imaging of specimens double labeled with fluorescein and a red fluorophore, using the 488- and 514-nm lines of an argon ion laser. *Left:* The filter set for detection of green fluorescence, using 488-nm excitation. *Right:* The filter set for detection of red fluorescence, using 514-nm excitation. The separate filter sets must be changed to allow sequential imaging of the green and red fluorescence from the specimen. The excitation (dashed) and emission (solid) spectra of various combinations of fluorescein and a red fluorophore, either cyanine 3.18 (CY3.18), tetramethylrhodamine (TRITC), Lissamine rhodamine (LRSC), or Texas Red (TRSC), are also shown. The excitation wavelength (solid line) and regions of the emission spectra transmitted by the bandpass emission filters (shaded regions) are also indicated. See Fig. 10 for filter abbreviations.

of specimens double labeled with fluorescein and rhodamine (Mossberg and Ericsson, 1990; Akner *et al.,* 1991). Although the imaging of the rhodamine is not improved, the 488-nm line excites fluorescein much more efficiently than the 514-nm line (Fig. 15). Similarly, the emission filter for the detection of the green fluorescence can be shifted to shorter wavelengths centered on the emission peak of fluorescein at 518 nm. Unlike the configuration using simultaneous 514-nm excitation, this filter set allows the imaging of fluorescein almost as efficiently as in single-labeled specimens. It will also reduce the amount of fluorescence from tetramethylrhodamine observed in the green fluorescence image. With a second filter set, the rhodamine is imaged as before, using the 514-nm line and the same red fluorescence emission filter. If the 528-nm line is available from the argon ion laser, it can be used to more efficiently excite the rhodamine and reduce cross-talk because fluorescein is only weakly excited by this wavelength (5% of its excitation maximum).

Unfortunately, these advantages for imaging of double-labeled specimens are partially offset by the use of separate filter sets for each excitation wavelength. The nonoverlapping images and misalignment of the microscope could be avoided if the filter configuration for simultaneous excitation of both fluorophores with the 514-nm line (Fig. 10) could be modified for use with both the 488- and 514-nm lines. One approach is to use the same filter configuration and use multiple excitation filters to select either the 488- or 514-nm line. This is possible because the longpass dichroic mirror will also reflect the shorter wavelength 488-nm light onto the specimen. If both excitation lines are used for simultaneous imaging, the more efficient excitation of fluorescein by the 488-nm line will increase the breakthrough of fluorescein into the red fluorescence detector. Another approach is to use a partially reflecting mirror (for example, a 10% reflectance/90% transmittance mirror) instead of a longpass dichroic mirror to reflect the illumination beam onto the specimen (Fig. 16). This permits the use of either excitation wavelength and the more efficient emission filter for the detection of fluorescein previously used with the separate filter sets (Fig. 15). When this shorter wavelength emission filter is used, the simultaneous imaging of both fluorophores is no longer possible because the reflected 514-nm light will be observed in the green fluorescence detector. The disadvantage of this mirror is that it requires higher output powers because only 10% of the available

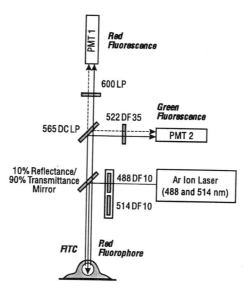

Fig. 16 Filter configuration for sequential imaging of specimens double labeled with fluorescein and a red fluorophore using the 488- and 514-nm lines of an argon ion laser with a single filter set. This filter set is identical to that shown for simultaneous excitation by 514-nm light (Fig. 10), except that a 10% reflectance/90% transmittance mirror is used instead of a dichroic mirror. In this case, 10% of the laser light will be reflected onto the specimen, and 90% of the emitted fluorescence transmitted into the detection path. Selection of the various laser wavelengths is done by positioning the appropriate excitation filter in the illumination path. See Fig. 10 for filter abbreviations.

light is actually reflected on the specimen. In addition, the other emission lines (i.e., 457 or 528 nm) available from argon ion lasers could be used with this configuration (Mossberg and Ericsson, 1990), but their lower output powers may be a problem.

2. Two-Color Laser Scanning Confocal Microscopy with Argon Ion and Helium–Neon Lasers

Although use of the various emission lines from the air-cooled argon ion lasers extends the capability to image double-labeled specimens, what is actually needed is additional excitation wavelengths that do not significantly excite fluorescein (i.e., >540 nm) and more efficiently excite the rhodamines. One approach is to use instruments with multiple lasers. Typically, an inexpensive helium–neon laser with an emission line at 543 or 633 nm has been used in combination with an argon ion laser (Stelzer, 1990; Montag *et al.*, 1991; Schubert, 1991; Fricker and White, 1992; Szarowski *et al.*, 1992). The 543-nm line can efficiently excite tetramethylrhodamine (76% of its excitation maximum at 554 nm), but is still suboptimal for the longer wavelength Lissamine rhodamine and Texas Red (only 29–35% of their excitation maximums). Although most studies have used separate filter sets with these instruments, the filter configuration using 488/514-nm excitation with a 10% reflection/90% transmittance mirror (Fig. 16) can be easily modified for use with the 543-nm line. Because the 543-nm line does not significantly excite fluorescein, the red fluorescence detector can be used with an emission filter that transmits from 570 nm and above. However, with only 10% of the excitation light being reflected on the specimen, insufficient intensity may be available for the optimal excitation of the fluorophores with the lower power helium–neon lasers. Alternatively, use of a helium–neon laser with a 633-nm emission line allows the efficient excitation of a far red fluorophore such as allophycocyanin or cyanine 5.18 (60–83% of their excitation maximums).

Unfortunately, the expanded capabilities of these multiple laser instruments are offset by an increase in complexity. Beyond the requirement to keep multiple lasers in proper operation, additional optics are necessary to merge the output of the lasers into a single illumination beam. For correct operation of the confocal imaging system, the focused spot formed from the illumination sources by the objective lens must coincide both axially and transversely. If the filter configuration discussed above is used, the output of each laser must be superimposed and enter the microscope as coincidental, parallel beams to prevent misalignment of the regions scanned by the individual beams. The individual laser beams can be combined with a series of beam steering and dichroic mirrors or by passing each through the same fiber optic cable. If separate filter sets are used, this alignment is not as critical because the images will be misaligned by the switching of the dichroic mirrors.

3. Two-Color Laser Scanning Confocal Microscopy with Krypton–Argon Ion Laser

As an alternative to multiple laser instruments for the imaging of doublelabeled specimens, the use of the 488- and 568-nm lines of a single, krypton–argon ion laser is especially attractive (Brelje and Sorenson, 1992). As before, a filter configuration that allows the use of either excitation wavelength with the same dichroic mirror is required

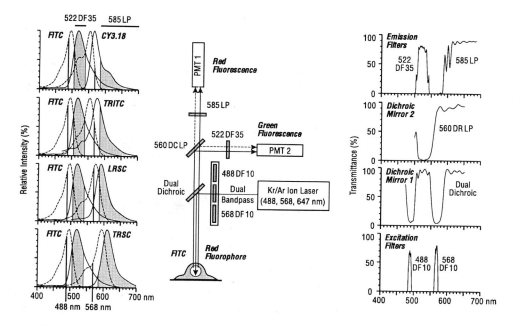

Fig. 17 Filter configuration for imaging of specimens double labeled with fluorescein and a red fluorophore, using the 488- and 568-nm lines of a krypton–argon ion laser. *Left:* The excitation (dashed) and emission (solid) spectra of various combinations of fluorescein and a red fluorophore, either cyanine 3.18 (CY3.18), tetramethylrhodamine (TRITC), Lissamine rhodamine (LRSC), or Texas Red (TRSC). The 488- and 568-nm lines are shown as vertical lines beneath the excitation spectra. The wavelengths transmitted by the band-pass emission filters for detecting green and red fluorescence are shown as shaded regions on the emission spectra. *Center:* The filter configuration for sequential or simultaneous excitation of the fluorophores by 488- and 568-nm light. Selection of the various laser wavelengths is done by positioning the appropriate excitation filter in the illumination path. The transmittance spectra for the dual-bandpass excitation filter is not shown because it is almost identical to the combination of the individual 488- and 568-nm excitation filters. *Right:* The transmittance spectra of the various excitation filters, dichroic mirrors, and emission filters are shown. See Fig. 10 for filter abbreviations.

to avoid needing to realign the microscope when changing filter sets (Fig. 17). This is possible with a dual-dichroic mirror with two reflection bands centered on the laser lines and that transmits at all other wavelengths. Although a 10% reflectance/90% transmittance mirror could be used instead, the higher efficiency reflection of each excitation wavelength onto the specimen by the dual-dichroic mirror allows the laser to be operated at lower output powers. The lower operating current should lengthen the useful life of the laser by reducing the effect of degrading mechanisms within the plasma tube. As with the previous filter configurations, a longpass dichroic mirror placed in the detection path is used to direct the green and red fluorescence into separate detectors. The excitation filters should be slightly tilted ($<2°$) to avoid interference effects caused by specular reflections and to prevent reflection of the other laser lines back into the plasma tube. If this occurs, the operating current and noise characteristics of the laser may be substantially altered.

Using this configuration, the fluorescence signals from the fluorophores can be collected by sequential or simultaneous excitation with the 488- and 568-nm lines. Although sequential excitation improves the spectral isolation of the fluorophores, simultaneous excitation is useful for the rapid surveying of specimens to select regions for further examination. Subsequently, the identified regions of the specimen can be examined with a single excitation wavelength to verify the presence of double labeling. For simultaneous imaging, it is preferable to have emission lines of similar output powers to allow relatively equal excitation of both fluorophores.

The krypton–argon ion laser has already been successfully used in several biological studies to image specimens double labeled with fluorescein and cyanine 3.18 (Elde *et al.*, 1991; Sorenson *et al.*, 1991).

4. Three-Color Laser Scanning Confocal Microscopy with Krypton–Argon Ion Laser

The filter configuration for two-color imaging with the krypton–argon ion laser can be easily modified to also allow the use of the 647-nm line for imaging (Fig. 18). This extends the usable spectral range into the far-red region of the light spectrum and permits the imaging of specimens triple labeled with fluorescein, rhodamine, and a far red fluorophore such as cyanine 5.18 or allophycocyanin (Brelje and Sorenson, 1992). The efficient excitation of each fluorophore by only one of these wavelengths and their well-separated emission peaks should allow the independent detection of fluorescence signals from each in the presence of the others. The only changes required for adapting the dual-excitation wavelength filter configuration is the replacement of the dual-dichroic mirror by a triple-dichroic mirror, and the ability to switch emission filters for the red fluorescence detector. Because the instrument we use has only two detectors (the Bio-Rad MRC-600), the appropriate excitation and emission filters must be selected to allowing imaging with either the 568- and 647-nm lines. If these emission filters are positioned between the detector aperture and the photomultiplier tube, no realignment of the detector aperture is required when switching between the emission filter.

Unfortunately, this use of two detectors restricts simultaneous imaging to only two of the three fluorophores. With the filter configuration shown in Fig. 18, only fluorescein and either rhodamine or cyanine 5.18 can be simultaneously imaged. If the second dichroic mirror is replaced by a longpass dichroic mirror centered on 640 nm, the simultaneous imaging of the green fluorescence from fluorescein or the red fluorescence from a rhodamine in one detector, and the far-red fluorescence from cyanine 5.18 in the other, is possible. Because this dichroic mirror is in the detection path, and not in the illumination path from the laser to the specimen, it can be exchanged with the previous dichroic mirror without shifting the region of the specimen scanned, but realignment of the detector apertures will be necessary.

The advantage of an instrument with three detectors is it would allow the simultaneous imaging with any combination of two or all three of the emission lines available from the krypton–argon ion laser. An additional dichroic mirror can be used to separate the red and far red fluorescence into separate detectors. This capability would be particularly

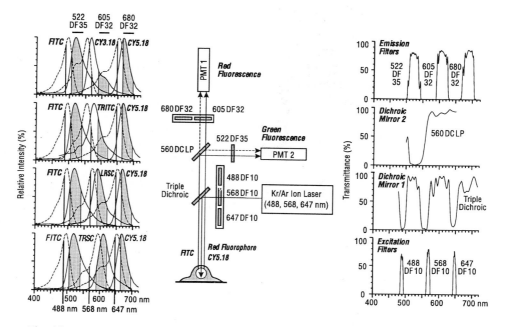

Fig. 18 Filter configuration for imaging of specimens triple labeled, using the 488-, 568-, and 647-nm lines of a krypton–argon ion laser. *Left:* The excitation (dashed) and emission (solid) spectra of various combinations of fluorescein (FITC), a red fluorophore, and cyanine 5.18. The following red fluorophores are shown: cyanine 3.18 (CY3.18), tetramethylrhodamine (TRITC), Lissamine rhodamine (LRSC), and Texas Red (TRSC). The laser lines are shown as vertical lines beneath the excitation spectra. The wavelengths transmitted by the bandpass emission filters for detecting the fluorophores are shown as shaded regions on the emission spectra. *Center:* The filter configuration for imaging of triple-labeled specimens. The triple-dichroic mirror allows the reflection of each excitation wavelength onto the specimen and the transmittance of the fluorescence from the various fluorophores with the same dichroic mirror. Because the Bio-Rad MRC-600 has only two detectors, the emission filter for one photomultiplier (PMT 1) must be switched in addition to the excitation filters for the red fluorophore and CY5.18. *Right:* The transmittance spectra of the various excitation filters, dichroic mirrors, and emission filters are shown. See Fig. 10 for filter abbreviations.

useful for the surveying of specimens for regions with a particular staining pattern or for time-resolved imaging of living specimens.

V. Practical Aspects of Multicolor Laser Scanning Confocal Microscopy

In discussing the practical application of LSCM for multicolor immunofluorescence studies, the use of the inexpensive, air-cooled argon ion and krypton–argon ion lasers as light sources will be compared. The implications of using the limited number of excitation wavelengths available from these lasers for the detection and spectral isolation of fluorophores in single- and multilabeled specimens is demonstrated. Finally, several

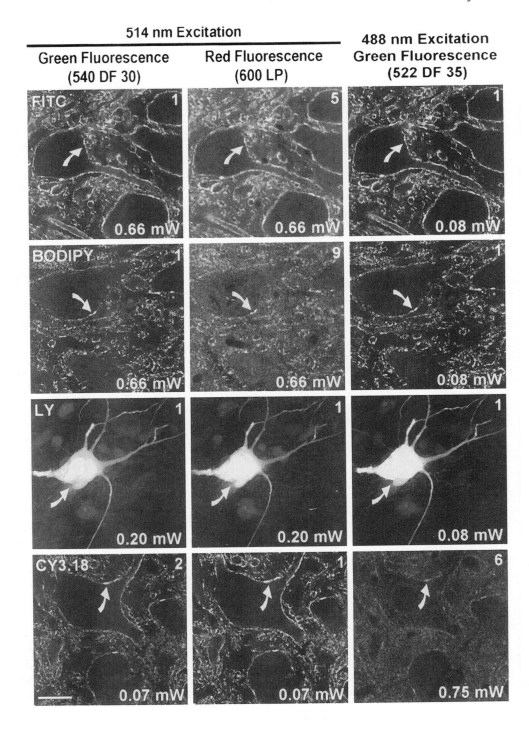

examples of the advantages of specific visualization of fluorophores with a krypton–argon ion laser are shown.

A. Multicolor Laser Scanning Confocal Microscopy with Air-Cooled Argon Ion Laser

Although a restricted range of emission lines is available from an air-cooled argon ion laser, it has been successfully used for multicolor LSCM with fluorescein and a rhodamine (Brelje *et al.,* 1989; Fehon *et al.,* 1990; Mossberg and Ericsson, 1990; Mossberg *et al.,* 1990; Schweitzer and Paddock, 1990; Stamatoglou *et al.,* 1990; Brelje and Sorenson, 1991; Vincent *et al.,* 1991; Wetmore and Elde, 1991). As will be shown below, the inefficient visualization of one or both fluorophores with this configuration limits its use to the imaging of brightly stained specimens in which the relative intensity of the staining of the fluorophores can be carefully controlled to balance their detection and the cross-talk between the detectors. However, it is difficult to specifically visualize each fluorophore. Thus it is difficult to determine whether a structure is double labeled. As a result, this approach is best used when the individual fluorophores label structures that are spatially separated (i.e., single objects are not expected to be double labeled).

1. Visualization Specificity of Fluorophores with Argon Ion Laser

The specificity of the fluorophores with respect to the microscope filters was examined by attempting to visualize each fluorophore individually with the filter configurations used for the imaging of specimens double labeled with fluorescein and a red fluorophore (Figs. 10 and 15). To provide specimens with strong fluorescent signal with a minimum of background, spinal cord sections were stained for the presynaptic marker synaptophysin with the various fluorophore-conjugated secondary antibodies. Synaptophysin is a calcium-binding glycoprotein located in the membrane of synaptic vesicles (Jahn *et al.,* 1985; Wiedenmann and Franke, 1985). Because the intracellular injection of cells is frequently used in combination with immunofluorescence, the specificity of Lucifer yellow was also examined with these filters. A Lucifer yellow-injected neuron in the sacral parasympathetic nucleus of the spinal cord was employed for that purpose. With these intensely stained specimens, the chance of observing each fluorophore with the inappropriate filters is maximized.

When using the filter configuration for simultaneous excitation by the 514-nm line of an argon ion laser (Fig. 10), the green fluorophores could be observed in both the green and red fluorescence detectors (Fig. 19). In these green fluorescence images, the

---

Fig. 19 Specificity of visualization of fluorophores, using the microscope filters for two-color imaging with the 514-nm line of an argon ion laser. Rat spinal cord sections were stained with a mouse monoclonal antibody against synaptophysin and either fluorescein (FITC)-conjugated donkey anti-mouse IgG, BODIPY-conjugated goat anti-mouse IgG, or cyanine 3.18 (CY3.18)-conjugated donkey anti-mouse IgG. Single optical sections of a region within the ventral horn are shown for each of these fluorophores. For Lucifer yellow (LY), projections of 20 images acquired at 1.0-μm intervals through an intracellularly injected neuron within the superior cervical ganglion are shown. The intensity of laser illumination used to acquire the images is given in milliwatts (mW). The relative increase in photomultiplier gain and contrast for the images of each fluorophore is indicated in the upper right of each image. Scale bar = 25 mm. Nikon × 60 NA 1.4.

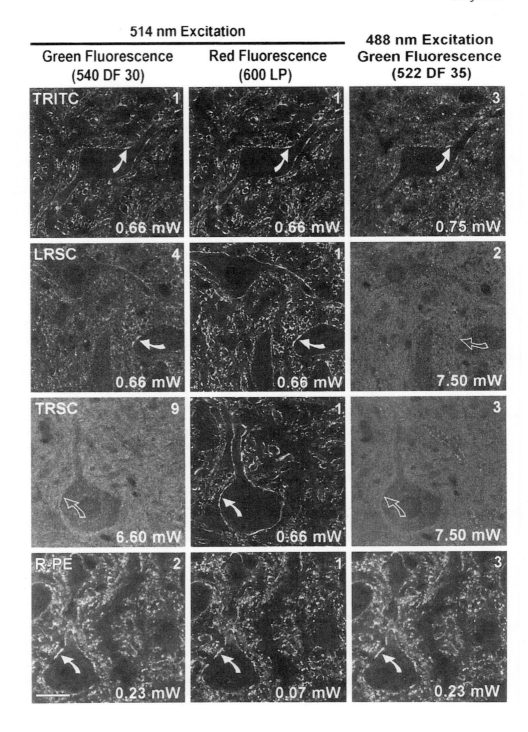

numerous varicosities containing synaptophysin-like immunoreactivity on the cell bodies and processes of the large motor neurons in the dorsal horn of the spinal cord are easily appreciated for both the fluorescein- and BODIPY-stained specimens. Although less than 5% of the fluorescein and 1% of the BODIPY emission spectra overlap with the transmittance region of the 600-nm longpass filter (Fig. 10), both fluorophores could also be observed in the red fluorescence detector with the same intensity of laser illumination. Although the fluorescence signal was weaker than that observed in the green fluorescence detector, both fluorophores could be easily imaged by increasing the gain of the photomultiplier tube used as the red fluorescence detector. Because of the narrow emission peak of BODIPY, its relative intensity observed between the red and green fluorescence detectors was half of that observed for fluorescein. In contrast, the extremely broad emission spectra of Lucifer yellow allowed its detection with almost equal efficiency in both the green and red fluorescence detectors of an intracellular-labeled neuron. This suggests that the cross-talk of fluorescence from the green fluorophore into the red fluorescence detector will be a problem for double-labeled specimens.

When using the filter configuration for sequential excitation by the 488- and 514-nm lines of an argon ion laser (Fig. 15), the more efficient excitation of the green fluorophores by the 488-nm light allowed comparable green fluorescence images to be obtained with only one-eighth of the illumination intensity required with 514-nm light (0.08 vs 0.66 mW). As expected, the use of a separate filter set for the detection of the red fluorescence with 514-nm excitation did not alter the visibility of these fluorophores (data not shown).

Similar to the lack of specificity in visualization of the green fluorophores, most of the red fluorophores could be observed in both the green and red fluorescence detectors when using the filter configuration for simultaneous excitation by the 514-nm line of an argon ion laser (Figs. 19 and 20). Because of the large overlap between the emission spectra of the shorter wavelength red fluorophores and the green fluorescence emission filter (Fig. 10), cyanine 3.18, tetramethylrhodamine, and R-phycoerythrin were easily observed with the green fluorescence detector. In addition, the R-phycoerythrin was difficult to image because of its rapid fading with repeated scanning. The longer wavelength Lissamine rhodamine was slightly less visible, and Texas Red was undetectable in the green fluorescence detector (Fig. 20). Although the increased specificity of visualization with Texas Red is desirable, it should be noted that the inefficient excitation of Texas Red by 514-nm light (<7% of its excitation maximum at 592 nm) substantially reduces the sensitivity of detection in weakly stained specimens. For more efficient detection

Fig. 20 Specificity of visualization of fluorophores, using the microscope filters for two-color imaging with the 514-nm line of an argon ion laser. Rat spinal cord sections were stained with a mouse monoclonal antibody against synaptophysin and then with tetramethylrhodamine (TRITC)-, Lissamine rhodamine (LRSC)-, Texas Red (TRSC)-, or R-phycoerythrin (R-PE)-conjugated donkey anti-mouse IgGs. Single optcal sections of a region within the ventral horn are shown for each of these fluorophores. The intensity of laser illumination used to acquire the images is given in milliwatts (mW). The relative increase in photomultiplier gain and contrast for the images of each fluorophore is indicated in the upper right of each image. Scale bar = 25 mm. Nikon × 60 NA 1.4.

of fluorescence from a red fluorophore, the shorter wavelength fluorophores would be preferred if the cross-talk into the green fluorescence detector could be reduced.

As previously suggested (see Section IV,B,1), using the filter configuration for sequential excitation by the 488- and 514-nm lines of an argon ion laser should reduce the cross-talk of the red fluorophores into the green fluorescence detector (Figs. 19 and 20). Nonetheless, the presence of significant green fluorescence from tetramethylrhodamine and R-phycoerythrin still resulted in their being observed in the green fluorescence detector. This suggests that the more efficiently excited cyanine 3.18 and Lissamine rhodamine would be preferable with this configuration. However, even though the overlap between the emission spectra of cyanine 3.18 and the green fluorescence emission filter is quite small (Fig. 10), the superior brightness of this fluorophore makes even this small amount significant for brightly stained specimens.

In summary, it was not possible to specifically visualize most of the fluoro-phores examined, using the 488-and/or 514-nm lines of an argon ion laser. This suggests that relative staining intensity of the green and red fluorophores in the double-labeled specimens must be carefully controlled to balance their detection and cross-talk between the detectors. In addition, the detection of the specific double labeling of structures will be hindered by the likely presence of artifactual double labeling resulting from structures intensely stained with a single fluorophore.

2. Observations on Two–Color Laser Scanning Confocal Microscopy with Argon Ion Laser

To demonstrate the difficulty in imaging double-labeled specimens with an argon ion laser, spinal cord sections were stained for the neurotransmitters serotonin and substance P. In the superficial dorsal horn, a smaller number of serotonin-immunoreactive nerve fibers occurs interspersed among the large number of intensely stained substance P-immunoreactive nerve fibers. Previous studies have demonstrated that these neurotransmitters exist in separate populations of fibers in the superficial dorsal horn (Wessendorf and Elde, 1985). This provides an excellent test specimen for examining, using an argon ion laser, the detection of the fluorophores in a double-labeled specimen.

As expected, separate populations of fibers were observed when stained for serotonin with a fluorescein-conjugated secondary antibody and when stained for substance P with a Lissamine rhodamine-conjugated secondary antibody (Color Plate 7). In this case, the intense staining of substance P allows its imaging with minimal cross-talk of the weaker fluorescein staining of serotonin. The few instances of apparent coexistence result from the superimposition of individual fibers separated in different optical sections during the formation of the projection image.

In contrast, a markedly different distribution of staining was observed with similar sections stained for serotonin with a Lissamine rhodamine-conjugated secondary antibody and for substance P with a fluorescein-conjugated secondary antibody (Color Plate 7). In this case, it appears that most of the observed fibers are double labeled for both neurotransmitters. The total absence of fibers stained only with fluorescein and the infrequent fibers stained only with Lissamine rhodamine suggests that cross-talk of fluorescein into the red detector occurred. This occurred because the intense staining of substance P was visualized with fluorescein, and the weaker serotonin staining was visualized with the less bright fluorophore, Lissamine rhodamine. Therefore the amount of red fluorescence from the strongly stained fluorescein fibers is the same, if not more than, the amount of red fluorescence from the Lissamine rhodamine fibers. In this case, compensation for "bleed-through" of the fluorescein signal into the red fluorescence image is not possible because the Lissamine rhodamine-stained fibers are so inefficiently imaged. This demonstrates how the inability to specifically visualize the fluorophores with an argon ion laser can alter the apparent distribution of staining. This dramatically limits the use of this approach with specimens that have relatively weak or unknown staining patterns.

3. Cross-Talk between Detectors with Double-Labeled Specimens

Because it is not possible to specifically visualize both fluorophores in double-labeled specimens with an argon ion laser, it is important that the relative staining intensity of the green and red fluorophores be carefully controlled to balance their detection and cross-talk between the detectors. To demonstrate the consequences of these adjustments, further attempts were made to specifically visualize both fluorophores when observing spinal cord sections stained for the neurotransmitters serotonin and substance P.

Because of the inefficient excitation of the red fluorophore by 514-nm light, the cross-talk of the fluorescein into the red fluorescence detector can be a problem for specimens with intense fluorescein staining. As previously shown (Color Plate 7), considerable cross-talk of the fluorescein signal into the red fluorescence detector is observed when substance P is stained with a fluorescein-conjugated secondary antibody and when serotonin is stained with a Lissamine rhodamine-conjugated secondary antibody. When these fluorophore-conjugated secondary antibodies are used at similar dilutions, the cross-talk of fluorescein into the red fluorescence detector is more than the signal from the Lissamine rhodamine (Fig. 21A). Although this cross-talk can be compensated by subtracting a fraction of the green fluorescence image from the red fluorescence image, the ability to detect serotonin fibers lightly stained with Lissamine rhodamine in the "corrected" red fluorescence image is greatly reduced (Fig. 21A). To reduce the cross-talk of fluorescein into the red fluorescence detector, the intensity of the fluorescein staining can be reduced by using higher dilutions of the fluorescein-conjugated secondary antibody (Fig. 21B and C). In these cases, the reduction in cross-talk from fluorescein is easily appreciated by the higher illumination intensity (0.66 vs 0.20 mW) necessary to obtain an image in the red fluorescence detector. Because of the improved separation of the signals from the fluorophores into the green and red fluorescence detectors, a large number of serotonin fibers in the "corrected" red fluorescence images can be observed after cross-talk compensation (Fig. 21B and C). Although this approach improved the detection of the rhodamine staining and reduced cross-talk from fluorescein into the red fluorescence image, it is probably preferable to switch the secondary antibodies so fluorescein is used to visualize the more difficult-to-detect serotonin fibers.

On the other hand, a similar cross-talk problem can occur with the red fluorophore into the green fluorescence detector when double-labeled specimens are intensely stained

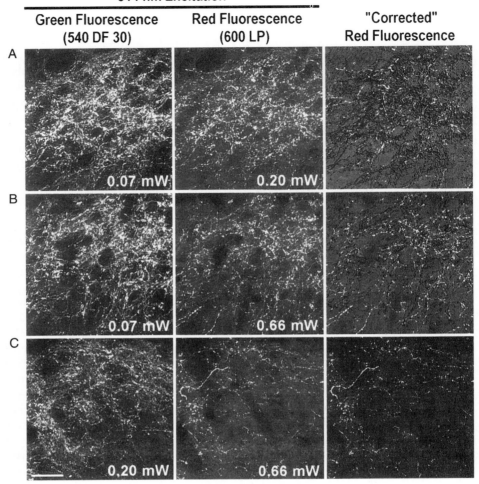

Fig. 21 Cross-talk of FITC fluorescence into the red fluorescence detector, using the 514-nm line of an argon ion laser for two-color imaging (Fig. 10). Rat spinal cord sections were stained with goat anti-serotonin and rabbit anti-substance P antisera. A 1 : 100 dilution of the Lissamine rhodamine-conjugated donkey anti-goat IgG was used with 1 : 100 (A), 1 : 500 (B), and 1 : 1000 (C) dilutions of the fluorescein-conjugated donkey anti-rabbit IgG. In this region of the spinal cord, the superficial dorsal horn, coexistence would not be expected. A corresponding decrease in the amount of fluorescein fluorescence was observed in the red fluorescence image with the increasing dilution of the fluorescein-conjugated secondary antibody. The rhodamine fluorescence was estimated by subtracting a fraction [100% for (A), 75% for (B), and 15% for (C)] of the corresponding green fluorescence image from the observed red fluorescence images. With the reduction in fluorescein staining, a corresponding increase in the number of fibers in these "corrected" red fluorescence images is observed. This reflects the difficulty in detecting the red fluorescence from Lissamine rhodamine when fluorescein contributes the majority of the signal to the red fluorescence image. Each image is a projection of 16 optical sections acquired at 1.0-μm intervals. Scale bar = 25 μm. Nikon ×60, NA 1.4.

with a red fluorophore. Although this cross-talk could also be reduced by staining substance P less intensely with the red fluorophore, this is not really practical because intense staining of the red fluorophore is required to compensate for its inefficient excitation by 514-nm light. A better approach is to use red fluorophores with emission spectra that are shifted to longer wavelengths that have less overlap with the green fluorescence emission filters (Fig. 22). When using the filter configuration for sequential excitation by the 488- and 514-nm lines of an argon ion laser (Fig. 10), the cross-talk of cyanine 3.18 and tetramethylrhodamine into the green fluorescence detector is similar in intensity to the fluorescein signal (Fig. 23). In contrast, no cross-talk of the longer wavelength

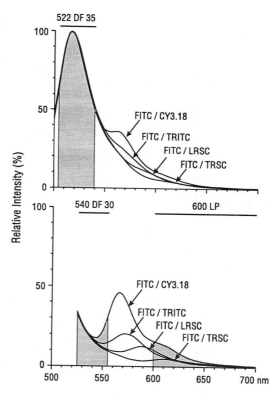

Fig. 22 Spectroscopic examination of the fluorescence observed from an equimolar solution of fluorescein (FITC)- and various red fluorophore-conjugated IgGs. (Note that the concentrations are for the fluorophores instead of the protein because of the slight differences in the fluorophore/protein ratios of the conjugates.) The following red fluorophores were used in combination with fluorescein-conjugated goat anti-mouse IgG: cyanine 3.18 (CY3.18)-, tetramethylrhodamine (TRITC)-, Lissanine rhodamine (LRSC)-, or Texas Red (TRSC)-conjugated goat anti-mouse IgGs. The emission spectra of the various combinations of fluorophores in response to excitation by 488-nm light (*top*) or 514-nm light (*bottom*) are shown. The regions of the emission spectra that are transmitted by the bandpass emission filters are also shown (shaded regions). Note the increased contribution of the shorter wavelength red fluorophores to the emission that would be observed through the green fluorescence emission filter used with 514-nm excitation (*bottom*).

514 nm Excitation

Green Fluorescence
(540 DF 30)

Red Fluorescence
(600 LP)

488 nm Excitation
Green Fluorescence
(522 DF 35)

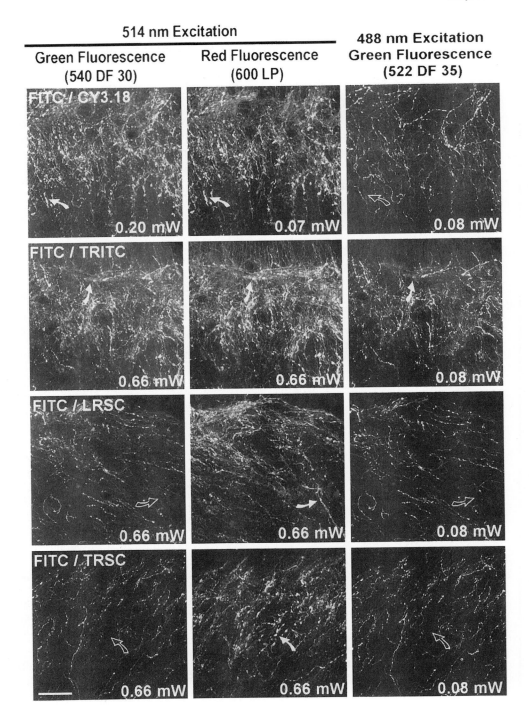

Lissamine rhodamine and Texas Red into the green fluorescence detector was observed. Because the green fluorescence emission filter is at shorter wavelengths with the filter configuration for sequential excitation by the 488- and 514-nm lines of an argon ion laser (Fig. 15), this configuration dramatically reduced the cross-talk of the red fluorophores into the green fluorescence detector (Fig. 23). In addition, it should be noted that cyanine 3.18 required a lower illumination intensity than the rhodamines (0.07 vs 0.66 mW) to obtain a red fluorescence image of the substance P staining.

As demonstrated by these specimens, two-color imaging with an argon ion laser is practical only for some combinations of fluorophores when their relative staining intensity can be carefully controlled. Usually, the less efficient imaging of the red fluorophore requires that it be used to visualize the more strongly staining antigen. However, if both substances of interest can only be weakly stained, two-color imaging with an argon ion laser of these specimens may not be possible.

4. Is Three–Color Laser Scanning Confocal Microscopy with an Argon Ion Laser Possible?

Although three-color fluorescence detection with a single excitation wavelength from an argon ion laser has been reported for flow cytometry (Festin *et al.,* 1990; Afar *et al.,* 1991; Lansdorp *et al.,* 1991), it is unclear whether this approach can be successfully applied to multicolor LSCM. Three-color flow cytometry with a single excitation wavelength is usually done with 488-nm excitation of fluorescein (peak emission wavelength of 520 nm), R-phycoerythrin (peak emission wavelength of 575 nm), and one of the phycobiliprotein tandem conjugates (peak emission wavelengths of 610–670 nm; Fig. 24). These tandem conjugates have extremely large Stokes shifts and include R-phycoerythrin/allophycocyanin (Glazer and Stryer, 1983), R-phycoerythrin/Texas Red (Festin *et al.,* 1990), R-phycoerythrin/cyanine 5.18 (Lansdorp *et al.,* 1991), or the photosynthetic pigment from dinoflagellates, peridinin chlorophyll *a* binding protein (Recktenwald, 1989; Afar *et al.,* 1991). Because the correction of the cross-talk between images becomes more difficult as the difference between the excitation and emission wavelengths increases (see Section IV,A,2), this approach should be less successful than that observed for two-color LSCM with a single excitation wavelength.

Fig. 23   Cross-talk of the red fluorophore into the green fluorescence detector, using an argon ion laser for two-color imaging. Rat spinal cord sections were stained with goat anti-serotonin and rabbit anti-substance P antisera. Fluorescein (FITC)-conjugated donkey anti-goat IgG was used in combination with either cyanine 3.18 (CY3.18)-, tetramethylrhodamine (TRITC)-, Lissamine rhodamine (LRSC)-, or Texas Red (TRSC)-conjugated donkey anti-rabbit IgGs. All secondary antibodies were used as 1 : 100 dilutions. In this region of the spinal cord, the superficial dorsal horn, coexistence would not be expected. The shorter wavelength red fluorophores, both cyanine 3.18 and tetramethyl-rhodamine, are visible in the green fluorescence images collected with 514-nm excitation (Fig. 10). However, the more optimal filter set using 488-nm excitation (Fig. 15) allows a more specific detection of the green fluorescence from fluorescein for these red fluorophores. Also, note the lower illumination intensity needed to acquire the image of the cyanine 3.18 staining compared to the rhodamines. Each image is a projection of 16 optical sections acquired at 1.0-μm intervals through the superficial dorsal horn. Scale bar $= 25$ μm. Nikon $\times 60$ NA 1.4.

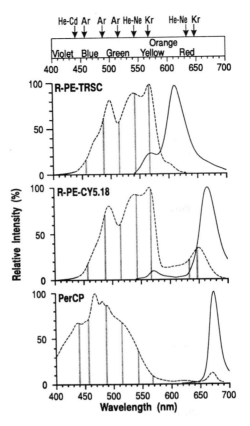

Fig. 24 Excitation (dashed) and emission (solid) spectra of large Stokes shift fluorophores derived from photosynthetic pigments. *Top:* The tandem conjugate of R-phycoerythrin and Texas Red with streptavidin (R-PE–TRSC; Gibco Life Technologies, Grand Island, NY). The excitation spectrum was scanned while measuring emission at 650 nm, and the emission spectrum was scanned while exciting at 490 nm. *Middle:* The tandem conjugate of R-phycoerythrin and cyanine 5.18 (R-PE–CY5.18) with goat anti-mouse IgG (Jackson ImmunoResearch Laboratories). The excitation spectrum was scanned while measuring emission at 710 nm, and the emission spectrum was scanned while exciting at 490 nm. *Bottom:* Peridinin chlorophyll *a*-binding protein (PerCP; provided by Diether Recktenwald of Becton Dickinson, San Jose, CA). The excitation spectrum was scanned while measuring emission at 710 nm, and the emission spectrum was scanned while exciting at 390 nm. The emission lines for argon ion (Ar), krypton ion (Kr), helium–cadmium (He–Cd), and helium–neon (He–Ne) lasers are shown as vertical lines beneath the excitation spectra.

To test this possibility, rat pancreatic islets of Langerhans were stained for the major islet hormones. This specimen was chosen because the major islet hormones (insulin, glucagon, and somatostatin) are easily stained and are synthesized within distinct cell types within the islet (β, α, and δ cells, respectively). The combination of fluorescein, cyanine 3.18, and the tandem conjugate of R-phycoerythrin/cyanine 5.18 (R-PE–CY5.18) was chosen because of the exceptional brightness of the cyanine fluorophores. After

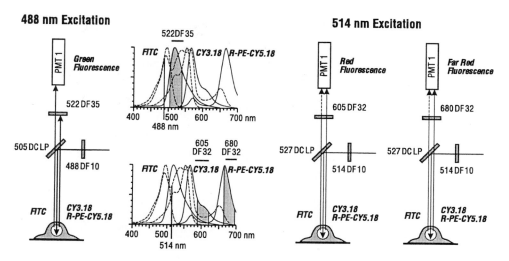

Fig. 25 Filter configurations for sequential imaging of triple-labeled specimens, using an argon ion laser. *Left:* The filter set for the detection of green fluorescence, using 488-nm excitation. *Right:* The filter sets for the detection of red and far red fluorescence, using 514-nm excitation. The excitation (dashed) and emission (solid) spectra of the fluorescein (FITC), cyanine 3.18 (CY3.18), and tandem conjugate of R-phycoerythrin/cyanine 5.18 (R-PE–CY5.18) are shown in the middle. The excitation wavelength (solid line) and regions of the emission spectra transmitted by the bandpass emission filters (shaded regions) are also shown. See Fig. 10 for filter abbreviations.

several trials examining the relative staining intensity of these fluorophores, it was determined that it was necessary in this case to reduce the intensity of the fluorescein and cyanine 3.18 staining to reduce cross-talk into the far-red fluorescence image of the R-PE–CY5.18 staining. Therefore dilutions of 1 : 500, 1 : 200, and 1 : 50 for fluorescein-, cyanine 3.18-, and R-PE–CY5.18-conjugated secondary antibodies, respectively, were used for the triple labeling of islets.

Although the use of either 488- or 514-nm excitation for imaging of all three fluorophores was possible, sequential excitation with the 488-nm line for fluorescein and the 514-nm line for cyanine 3.18 and R-PE–CY5.18 was used to increase the spectral isolation of fluorescein from the red fluorophores (Fig. 25). However, it can be seen that the greater difference between the excitation and observation wavelengths leads to a substantial decrease in the apparent "confocality" of the R-PE–CY5.18 image. In addition, the use of 488-nm excitation allowed imaging of the weak fluorescein staining required to reduce cross-talk into the images of the other two fluorophores (Fig. 26). To obtain an image of the cyanine 3.18 staining, it was still necessary to compensate for the cross-talk of fluorescein into the red fluorescence image by subtracting 40% of the green fluorescence image from the red fluorescence image. Similarly, an image of the R-PE–CY5.18 staining was obtained from the far-red fluorescence image by subtracting 20% of the fluorescein and 50% of the cyanine 3.18 images.

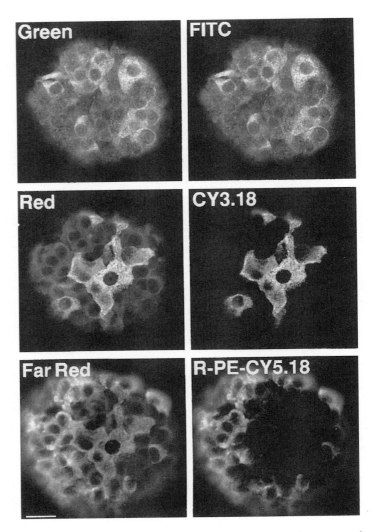

Fig. 26 Three-color imaging of a triple-labeled rat islet of Langerhans, using an argon ion laser. The major islet cell types were stained with guinea pig anti-insulin serum, rabbit anti-somatostatin serum, and a mouse anti-glucagon monoclonal antibody. The following secondary antibodies were used: fluorescein-conjugated donkey anti-guinea pig IgG, cyanine 3.18 (CY3.18)-conjugated donkey anti-rabbit IgG, and a tandem conjugate of cyanine 5.18 and R-phycoerythrin (R-PE–CY5.18) with donkey anti-mouse IgG. *Left:* Projections of six optical images acquired at 1.0-μm intervals, using the filter configurations shown in Fig. 25. Because individual islet cells produce a single hormone, the occurrence of stained cells in more than one of the images indicates the failure to spectrally isolate the fluorophores completely. *Right:* The distribution of staining after compensation for cross-talk between the observed images. FITC staining was assumed to be represented by the green fluorescence image. CY3.18 staining was estimated by subtracting 40% of the green fluorescence image from the red fluorescence image. Similarly, R-PE–CY5.18 staining was estimated by subtracting 20% of the green fluorescence image and 50% of the "CY3.18" image from the red fluorescence image. Scale bar = 25 μm. Olympus ×40 NA 0.85.

These images of a triple-labeled islet can be merged to form a color image that allows the distribution of the islet hormones to be examined. Because individual islet cells synthesize only a single hormone, the occurrence of double- and triple-labeled cells in the images as collected is artifactual (Color Plate 8A). Although compensation of the cross-talk between the images dramatically improves the separation of the fluorescence signals (Color Plate 8B), several cells still appear to be double labeled for insulin and glucagon in the "corrected" images. This is not surprising considering that the loss of resolution in the far red fluorescence image will hinder the compensation of cross-talk and determination of which cells were actually stained for glucagon. This difficulty in cross-talk compensation and the need to carefully adjust the relative staining intensity of the fluorophores demonstrates the difficulty of this approach and its susceptibility to artifactual multiple labeling.

B. Multicolor Laser Scanning Confocal Microscopy with Air–Cooled Krypton–Argon Ion Laser

The broad range of emission wavelengths available from an air-cooled krypton–argon ion laser allows the use of different excitation wavelengths and emission filters for the detection of specific fluorophores. For appropriate combinations of fluorescent dyes, this approach allows the more efficient and independent detection of more than one fluorophore in a single specimen by LSCM. In addition, the specific visualization of each fluorophore permits the investigation of specimens with staining for the coexistence substances.

1. Visualization Specificity of Fluorophores with Krypton–Argon Ion Laser

The specificity of the fluorophores with respect to the microscope filters was investigated by attempting to visualize each fluorophore with the filter configuration for the imaging of triple-labeled specimens and a krypton–argon ion laser (Fig. 18). In addition, as done for our studies on the argon ion laser (see Section V,A,1), specimens stained for synaptophysin with the various fluorophore-conjugated secondary antibodies and a Lucifer yellow-injected neuron were examined.

Unlike the argon ion laser (Fig. 19), the filter configurations for three-color imaging with a krypton–argon ion laser could be used to specifically visualize the green fluorophores, except for Lucifer yellow (Fig. 27). With the filters for 488-nm excitation, the distribution of numerous varicosities containing synaptophysin-like immunoreactivity on the cell bodies and processes of neurons is easily appreciated for both the fluorescein- and BODIPY-stained specimens. Even when the intensity of laser illumination was increased 100-fold and photon counting was used, no staining whatsoever beyond autofluorescence was observed in the red and far red fluorescence images obtained with the filters for 568- or 647-nm excitation. Unlike the 514-nm excitation used with an argon ion laser, these longer excitation wavelengths from a krypton–argon ion laser are

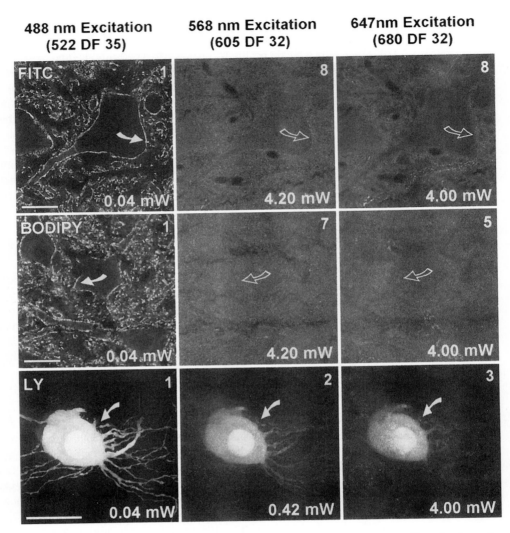

Fig. 27 Specificity of visualization of green fluorophores, using the microscope filters for three-color imaging with a krypton–argon ion laser. Rat spinal cord sections were stained with a mouse monoclonal antibody against synaptophysin and either fluorescein (FITC)-conjugated donkey anti-mouse IgG or BODIPY-conjugated goat anti-mouse IgG. Single optical sections are shown for each of these fluorophores. For Lucifer yellow (LY), projections are shown of 20 images acquired at 1.0-μm intervals through an intracellularly injected neuron within the parasympathetic nucleus. The intensity of laser illumination used to acquire the images is given in milliwatts (mW). The relative increase in photomultiplier gain and contrast for the images of each fluorophore is indicated in the upper right of each image. Scale bars = 25 mm. Nikon ×60 NA 1.4.

well past the excitation maximums of these green fluorophores. In contrast, the brightly stained cell body of a Lucifer yellow-injected neuron was still observed with the filters for 568- and 647-nm excitation (Fig. 27). Most likely this reflects the slight excitation at these wavelengths possible with Lucifer yellow because of its much broader excitation spectrum. The use of alternate intracellular markers, such as biotin derivatives and fluorophore-conjugated avidins (Horikawa and Armstrong, 1988; Kita and Armstrong, 1991), or fluorophore-conjugated dextrans (Mesce *et al.,* 1993), that can be visualized with more specific fluorophores is preferable for multicolor studies.

Although each of the red fluorophores could be imaged with the filters for 568-nm excitation, the shorter wavelength red fluorophores could not be specifically visualized by using the filter configuration for three-color imaging with a krypton–argon ion laser (Fig. 28). Similar to the argon ion laser (Figs. 19 and 20), both cyanine 3.18 and tetramethylrhodamine were visible in the green fluorescence images because of the slight overlap between their emission spectra and the green fluorescence emission filter (Fig. 18). However, it should be noted that the detection of cyanine 3.18 in the green fluorescence images has been possible only for intensely stained specimens. The longer wavelength Lissamine rhodamine and Texas Red were not observed with the filters for 488-nm excitation. In addition, red fluorescence images of cyanine 3.18 and Lissamine rhodamine could be obtained by using a lower illumination intensity (0.04 vs 0.13 mW). This probably reflects the superior brightness of cyanine 3.18 compared to the rhodamine conjugates (Wessendorf and Brelje, 1992; Mujumdar *et al.,* 1993) and the nearly maximal excitation of Lissamine rhodamine by 568-nm light. No staining beyond the normal tissue autofluorescence was observed for these red fluorophores with the filters for 647-nm excitation (Fig. 28). With the reduction in autofluorescence at longer excitation wavelengths, a 10-fold increase in laser illumination intensity was necessary to obtain comparable green and far red autofluorescence images for Lissamine rhodamine and Texas Red.

Each of the far red fluorophores examined was easily imaged with the filters for 647-nm excitation and the detection of far red fluorescence (Fig. 29). Although allophycocyanin was extremely bright when imaged with the far red filters, its broad excitation and emission spectra make it visible in both the red and far red fluorescence images, using the same illumination intensity. In contrast, a 10-fold increase in illumination intensity for 568- compared to 647-nm excitation was required to observe cyanine 5.18 in the red fluorescence image. Except for a few specimens with particularly bright staining, this has not been a problem when staining for other substances, using cyanine 5.18-conjugated antibodies. In contrast, the aluminum tetrabenztriazaporphyrin derivative could be visualized only with the far red filters, probably because of its narrow emission peak.

In summary, it appears possible to specifically visualize fluorophores with the krypton–argon ion laser if the combinations of fluorophores are carefully chosen. For two-color imaging, it is expected that FITC used in combination with Lissamine rhodamine or Texas Red should permit specific visualization of the fluorophores. Although cyanine 3.18 can be detected in the green fluorescence filters with intensely stained specimens, it should be preferable to the rhodamines when a red fluorophore must be used to visualize an

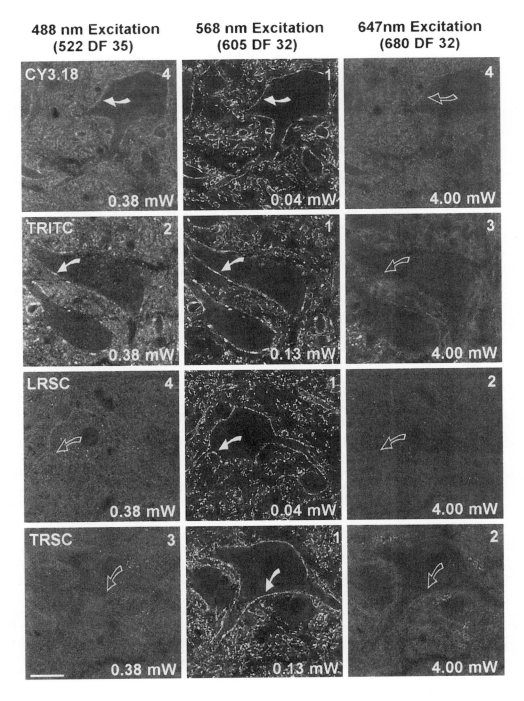

antigen that gives relatively weak staining. In this case, the superior brightness of cyanine 3.18 should improve the detection of specific staining. Similarly, it should be possible to use cyanine 5.18 or phthalocyanine derivatives as far red fluorophores for three-color imaging. However, because LSCM has sufficient sensitivity to image some of these fluorophores under suboptimal conditions, corresponding control specimens stained with a single fluorescent dye should be examined to verify the specific visualization of the fluorophores for multiple-labeled specimens. This is especially important for specimens in which the coexistence of staining will be examined.

2. Observations on Two-Color Laser Scanning Confocal Microscopy with Krypton–Argon Ion Laser

To demonstrate the superiority of specific visualization of fluorophores with the krypton–argon ion laser, rat spinal cord sections stained for serotonin and substance P were imaged with the two-color filter configuration for the krypton–argon ion laser (Fig. 17). Unlike the differences in distribution of these neuro-transmitters observed after switching the fluorophores with the argon ion laser (see Section V,A,2), both specimens appear similar except for the switching of the colors (Color Plate 7). The ability to specifically visualize each of the fluorophores removes the requirement to carefully select and carefully balance the relative staining intensity of the fluorophores used with double-labeled specimens.

It should be noted, however, that it is difficult to observe the smaller number of red fibers in the presence of the larger number of green fibers. This is not the result of actual differences in the specimens or the imaging, but results from the colors assigned to each image when printing the merged, color image (see Section VI,B).

3. Simultaneous vs Sequential Imaging of Double-Labeled Specimens

The consequences of using simultaneous vs sequential imaging of fluorophores in a double-labeled specimen can be demonstrated by using the filter configuration for two-color imaging with a krypton–argon ion laser (Fig. 17). Human islets of Langerhans were double labeled for somatostatin and glucagon, using fluorescein- and Lissamine rhodamine-conjugated secondary antibodies, respectively. By sequential excitation with 488- and 568-nm light, the distinct populations of somatostatin-containing δ cells and glucagon-containing α cells can be clearly observed (Fig. 30). The green fluorescence

---

Fig. 28 Specificity of visualization of red fluorophores, using the microscope filters for three-color imaging with a krypton–argon ion laser. Rat spinal cord sections were stained with a mouse monoclonal antibody against synaptophysin and either cyanine 3.18 (CY3.18)-, tetramethylrhodamine (TRITC)-, Lissamine rhodamine (LRSC)-, or Texas Red (TRSC)-conjugated donkey antimouse IgGs. Single optical sections are shown for each of the fluorophores. The intensity of laser illumination used to acquire the images is given in milliwatts (mW). The relative increase in photomultiplier gain and contrast for the images of each fluorophore is indicated in the upper right of each image. Scale bar = 25 μm. Nikon × 60 NA 1.4.

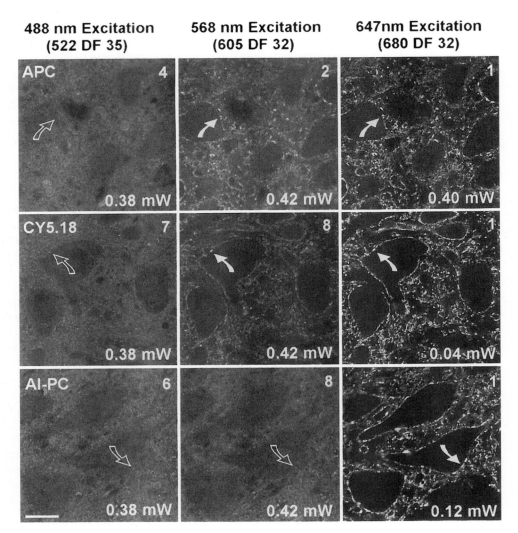

Fig. 29 Specificity of visualization of far red fluorophores, using the microscope filters for three-color imaging with a krypton–argon ion laser. Rat spinal cord sections were stained wtih a mouse monoclonal antibody against synaptophysin and either allophycocyanin (APC)-conjugated goat anti-mouse IgG (Biomeda), cyanine 5.18 (CY5.18)-conjugated donkey anti-mouse IgG (Jackson ImmunoResearch Laboratories), or aluminum tetrabenztriazaporphyrin (Al-PC)-conjugated goat anti-mouse IgG (Ultralight T680). Single optical sections are shown for each of the fluorophores. The intensity of laser illumination used to acquire the images is given in milliwatts (mW). The relative increase in photomultiplier gain and contrast for the images of each fluorophore is indicated in the upper right of each image. Scale bar = 25 μm. Nikon ×60 NA 1.4.

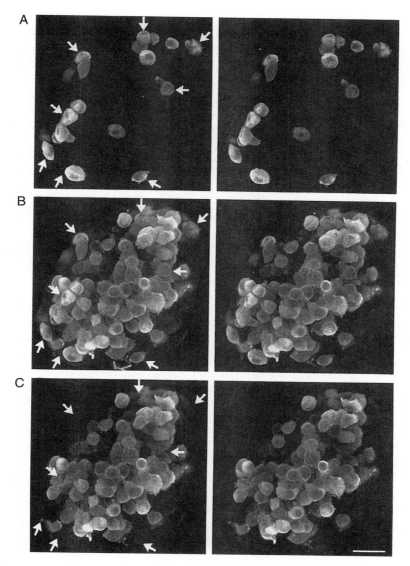

Fig. 30 Stereo images of a double-labeled human islet of Langerhans, using sequential vs simultaneous excitation with the 488- and 568-nm lines of a krypton–argon ion laser with the filter configuration shown in Fig. 17. Human islets were stained with rabbit anti-somatostatin antiserum and fluorescein-conjugated donkey anti-rabbit IgG, and a mouse anti-glucagon monoclonal antibody and Lissamine rhodamine-conjugated donkey anti-mouse IgG. (A) The somatostatin-containing δ cells are clearly visible in the green fluorescence image observed with 488-nm excitation. (B) The same cells (arrows) are also observed in the red fluorescence image observed with simultaneous excitation with the 488- and 568-nm lines. (C) The glucagon-containing α cells can, however, be specifically visualized in the red fluorescence image when observed with only 568-nm excitation. In this case, the δ cells (arrows) are not visible because of the inefficient excitation of fluorescein with 568-nm light. The stereo images were constructed from 21 optical sections acquired at 1.0-μm intervals. Scale bar $= 25 \, \mu$m. Olympus $\times$40 NA 0.85. [Human islets provided by Drs. D. Scharp and P. Lacy (Washington University, St. Louis, MO).]

image observed with 488-nm excitation shows the δ cells at the periphery of the islets stained with fluoroscein (Fig. 30A). Similarly, the red fluorescence image observed with 568-nm excitation shows the α cells (Fig. 30C). No double-labeled cells are observed when the green and red fluorescence images are compared.

In contrast, these fluorophores cannot be specifically visualized when both 488- and 568-nm light is used to simultaneously excite fluorescein and Lissamine rhodamine (Fig. 30). Because 568-nm light is at a wavelength longer than those used to obtain the green fluorescence image (505 to 539 nm), the red fluorescence simultaneously emitted by the Lissamine rhodamine is not observed in the green fluorescence image (not shown). However, the red fluorescence image contains the emissions from both Lissamine rhodamine and fluoroscein (Fig. 30B). Although less than 10% of the fluoroscein emission spectra overlaps with the red fluorescence emission filter (Fig. 17), similar fluorescence intensities for both fluorophores may result from more intense fluorescein staining or the superior brightness of fluorescein conjugates. This situation is similar to that observed with the simultaneous excitation of multiple fluorophores with an argon ion laser but with more efficient excitation of the rhodamine.

Therefore sequential excitation and imaging of each fluorophore in a multiple-labeled specimen will usually be necessary to spectrally isolate their fluorescence emissions. Unfortunately, this approach is rather time consuming because the specimen must be scanned several times with different excitation and emission wavelengths. Temporal resolution could be improved by the automatic switching of excitation filters and detectors between successive frames or even individual scan lines. Besides providing more immediate results, this approach would help minimize the problem of specimen movement changing the volume sampled for each pixel with each filter set.

4. Examples of Multicolor Laser Scanning Confocal Microscopy with Krypton–Argon Ion Laser

The usefulness of a krypton–argon ion laser for multicolor LSCM was tested by further examining lumbar spinal cord sections stained for serotonin and substance P. In the superficial dorsal horn, these neurotransmitters appear to exist in separate populations of fibers (Color Plate 9A). A few instances of apparent coexistence can be seen in the stereo images to be separate fibers superimposed on top of each other. In contrast, in the ventral horn, both neurotransmitters appear to stain the same nerve fibers and varicosities (Color Plate 9B). These processes are oriented around large cells resembling motor neurons and appear to occupy the same space when viewed in the stereo images. This suggests that the observed double labeling does not result from the apposition of different fibers of similar morphology. The difference in colors among different varicose fibers suggests that different neurons contain different molar ratios of serotonin and substance P.

The identity of the large cells surrounded by these nerve fibers in the ventral horn of the spinal cord was further investigated. Motor neurons of the lumbar spinal cord were stained by injection of the sciatic nerve with the retrograde tract-tracer hydroxystilbamidine (Fluoro-gold; Fluorochrome, Englewood, CO; Schmeud and Fallon, 1986; Wessendorf, 1991). In 300-μm vibratome sections of the spinal cord, individual hydroxystilbamidine-labeled motor neurons were identified by conventional fluorescence microscopy and intracellularly injected with Lucifer yellow. After staining for substance P, numerous nerve fibers containing substance P immunoreactivity were observed around the individual motor neurons (Color Plate 9C). These fibers appeared as a series of varicosities in close apposition to the surface of the motor neuron. From these projections, the distribution and number of varicosities containing various neurotransmitters on individual neurons can be more easily examined than by conventional fluorescence microscopy.

Control studies were undertaken to determine whether the observed double labeling was due to the neurotransmitter immunoreactivity or whether it was artifactual. The most likely source of artifactual double labeling would be the failure to specifically visualize each fluorophore or the secondary antibodies to specifically recognize only one of the primary antibodies. As previously discussed (see Section II,B), both of these artifacts can be tested by staining individual specimens with a single primary antibody and all of the fluorophore-conjugated secondary antibodies used. Spinal cord sections were singly stained with either goat anti-serotonin or rabbit anti-substance P antisera alone, followed by incubation with both the fluorescein-conjugated donkey anti-goat IgG and Lissamine rhodamine-conjugated donkey anti-rabbit IgGs. If the secondary antibodies recognized the inappropriate primary antisera or each other, it would be expected that double labeling would be observed. Similarly, double labeling would also be observed if the fluorophores could not be visualized specifically with the microscope filters. Although the latter appeared unlikely, using sequential excitation with the filter configuration for two-color imaging with a krypton–argon ion laser (see Section V,B,3), a lack of specificity will be more likely observed with brighter staining. Thus it is good practice to always test for this artifact by using the same tissue and antibodies as will be used in experimental situations.

When the sections stained for serotonin were examined, the typical pattern of serotonin immunoreactivity in the superficial dorsal horn was observed by 488-nm excitation and the green fluorescence detector (Fig. 31). Even when the illumination intensity of the 568-nm light was increased 100-fold and photon counting was used, no staining whatsoever was observed beyond normal tissue autofluorescence in the red fluorescence detector. Similarly, staining for substance P could be observed only by 568-nm excitation and the red fluorescence detector (Fig. 31). This suggests that the fluorophores can be specifically visualized and that the secondary antibodies are specific with respect to recognition of the primary antisera and have little affinity for each other. Although not shown here, these primary antibodies have already been shown to have specific immunoreactivity for the appropriate antigens (Wessendorf and Elde, 1985; Wessendorf et al., 1990a).

Similarly, the krypton–argon ion laser allows the distribution of three different neurotransmitters to be examined in triple-labeled specimens. For example, spinal cord sections were stained for serotonin, substance P, and Metenkephalin by using fluorescein-, cyanine 3.18-, and cyanine 5.18-conjugated secondary antibodies, respectively. The overall staining pattern of these neurotransmitters can be observed in a low-magnification image of a section of the lumbar spinal cord (Color Plate 10A). Although it has been infrequently mentioned, the intense and even illumination achieved with LSCM allows the acquisition of low-magnification images that would be difficult to photograph by conventional fluorescence microscopy. Because the entire section could not be viewed with

Green
Fluorescence

Red
Fluorescence

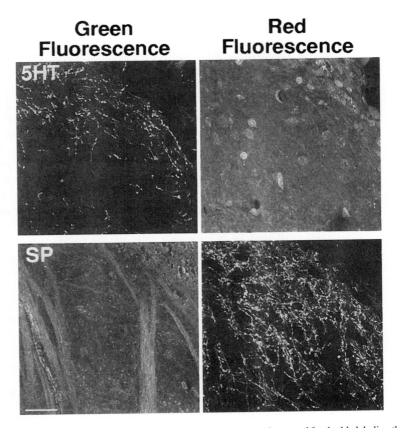

Fig. 31 Tests of the specificity of the secondary antibody and fluorophore used for double labeling the spinal cord. Rat spinal cord sections were stained with either goat anti-serotonin (5-HT) or rabbit anti-substance P (SP) antisera. All sections were then stained with both fluorescein-conjugated donkey anti-goat and Lissamine rhodamine-conjugated donkey anti-rabbit IgGs (multiple species adsorbed). The sections were examined with the filter configuration for two-color imaging with the krypton–argon ion laser (Fig. 17). If either the secondary antibodies or the fluorophores lacked specificity, double labeling would be expected. For 5-HT, only 0.05 mW of 488-nm light was required to obtain an image of the fluorescein staining in the green fluorescence detector. In contrast, even with 4.5 mW of 568-nm light and using the photon-counting mode of the Bio-Rad MRC-600, only the autofluorescence from the tissue was observed in the red fluorescence detector. Similarly, for SP only 0.05 mW of 568-nm light was required to obtain an image of the Lissamine rhodamine staining in the red fluorescence detector. In contrast, it was necessary to use 0.45 mW of 488-nm light and photon counting to obtain an image of the autofluorescence in the green fluorescence detector. This suggests that the secondary antibodies are specific in the recognition of the primary antibodies and that the fluorophores can be specifically visualized. Each image was formed from the projection of 10 optical sections acquired at 1.0-μm intervals. Scale bar = 25 μm. Nikon ×60 NA 1.4.

a ×4 objective, overlapping images of each half of the spinal cord were collected and subsequently combined into one large image for each fluorophore. Minimal photobleaching of the fluorophores was observed even with the intense illumination (∼4mW) necessary to collect these images because of the low magnification and scanning times. Although individual nerve fibers cannot be observed in low-magnification images, they can be resolved when smaller regions are examined at higher magnifications (Color Plate 10B). In this case, these neurotransmitters appear to exist primarily in separate populations of nerve fibers and varicosities in the region around the central canal of the spinal cord.

In our initial study, using LSCM to investigate the three-dimensional structure of islets of Langerhans, we observed in double-labeled specimens that the somatostatin-containing δ cells appeared adjacent to the glucagon-containing α cells (Brelje *et al.,* 1989). Although the relationship between these two cell types could be examined with the argon ion laser, it was not possible to examine simultaneously the orientation of δ cells with respect to the insulin-containing β cells. With the krypton–argon ion laser, the distribution of these islet cell types in the same islet could be easily examined by imaging islets triple labeled for insulin, somatostatin, and glucagon, using fluorescein-, cyanine 3.18-, and cyanine 5.18-conjugated secondary antibodies, respectively (Fig. 32). These images should be compared to those obtained with an argon ion laser (Fig. 26). Besides the absence of cross-talk between the fluorescence images, the more efficient imaging of the fluorophores with a krypton–argon ion laser dramatically improves the contrast and detail observed. In particular, the presence of long cytoplasmic processes on the irregularly shaped δ cells is quite noticeable (Fig. 32B). In addition, the relationship between these different islet cell types can be easily appreciated by merging the individual fluorescence images to form a color stereo image (Color Plate 11A). The somatostatin-containing δ cells are observed only in direct apposition between the glucagon-containing α cells on the exterior surface of the islet and the insulin-containing β cells that compose the majority of cells in the islet. Note that the amount of detail for each fluorophore apparent in the color merged image is less than that in the separate images (see Section VI,B,3). The advantage of three-dimensional reconstructions is particularly evident in human islets, in which the relationship between the different islet cell types is less distinguishable (Color Plate 11B).

Besides the examination of the relationship between different populations of cells and/or nerve fibers, the krypton–argon ion laser can also be used to examine the distribution of intracellular staining at higher magnifications. The utility of multicolor LSCM in approaching this type of problem can be easily appreciated from the stereo images of the cytoskeleton of individual cells from dorsal root ganglion cell cultures (Color Plate 11C). For this cell, the actin filaments were observed at the periphery of the cell near the plasma membrane and as stress fibers parallel to the surface of the culture substrate. In contrast, the microtubules were observed as a radiating network surrounding the nucleus of the cell and extending out toward the attached surfaces of the cell. In both cases, the efficient imaging of the staining allowed images with high signal-to-noise ratios to be collected.

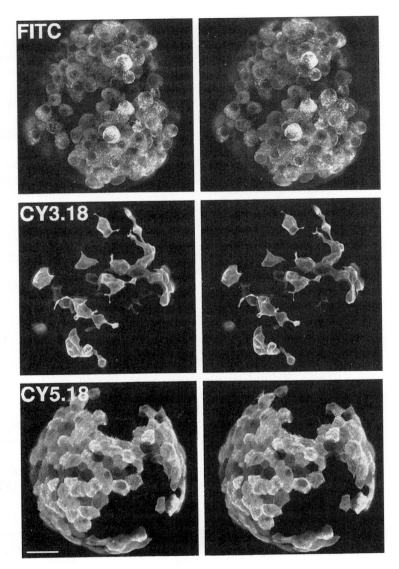

Fig. 32 Stereo images of a triple-labeled rat islet of Langerhans, using the three-color filter configuration of the krypton–argon ion laser shown in Fig. 18. Neonatal rat islets were stained with guinea pig anti-insulin serum, rabbit anti-somatostatin serum, and a mouse anti-glucagon monoclonal antibody. The staining was visualized with the following secondary antibodies: fluorescein (FITC)-conjugated donkey anti-guinea pig, cyanine 3.18 (CY3.18)-conjugated donkey anti-rabbit, and cyanine 5.18 (CY5.18)-conjugated donkey anti-mouse IgGs (multiple species adsorbed). In the individual stereo projections, note the presence of considerable detail and the absence of cells stained with the other fluorophores. The stereo projections were constructed from 25 optical sections acquired at 1.0-μm intervals across the top of the islet. Scale bar = 25 μm. Olympus $\times$40 NA 0.85.

VI. Multicolor Laser Scanning Confocal Microscopy Considerations

The general problems of imaging biological specimens by LSCM are discussed by Majlof and Forsgren ([3] in this volume). Therefore, the various artifacts incurred by the optical sectioning of thick specimens by LSCM will not be discussed in this article. Instead, the specific problems that occur when using a broad range of excitation and detection wavelengths with LSCM will be addressed.

A. Effect of Excitation/Emission Wavelength Differences

The use of different excitation/emission wavelengths with multicolor fluorescence microscopy complicates the comparison of images acquired in different spectral regions. Because many biological studies require optical resolutions approaching or beyond the diffraction limit, the wavelength dependence of resolution must be considered. The effect of chromatic and spherical aberration in the optics of the microscope are important regardless of the magnification. Although these wavelength-dependent imaging properties must be considered with conventional fluorescence microscopy, it is often much easier to observe their effects with the improved performance of confocal microscopy. Even if filter configurations are used that attempt to maintain registration, these effects will still complicate the comparison of the observed intensities of apparently corresponding pixels in images acquired with different excitation and emission wavelengths.

1. Resolution

To demonstrate the effect of different excitation/emission wavelengths on the appearance of structures near the minimal resolution possible with LSCM, intact *Giardia* cysts were double labeled with two antibodies specific for cyst wall antigens, using fluorescein- and cyanine 5.18-conjugated secondary antibodies. Previously, transmission and field emission scanning electron microscopy has shown that the cyst wall is formed by numerous 20-nm filaments in a layer with an overall thickness of 200 to 250 nm (Erlandsen *et al.*, 1990). When examined by multicolor LSCM, the green and red fluorescence images collected through the center of the same cyst appeared similar (Fig. 33). However, measuring the thickness of the cyst wall in these images showed an increase from 295 ± 50 nm ($n = 8$) in the green fluorescence image (i.e., fluorescein staining) to $388 \pm 35 (n = 8)$ in the red fluorescence image (i.e., cyanine 5.18 staining; $p < 0.05$). Although this may reflect actual differences in staining, this difference is similar to the 30% decrease in lateral resolution expected when using the longer wavelengths used to visualize the cyanine 5.18 staining (Table II). Additional evidence for this conclusion is that a similar difference in cyst wall thickness was observed when the fluorophores used to visualize the staining were switched (data not shown).

Similar to the xy images, an apparent increase in the thickness of the cyst wall also occurs in the xz images through the same cyst (Fig. 33). However, the cyst wall appears

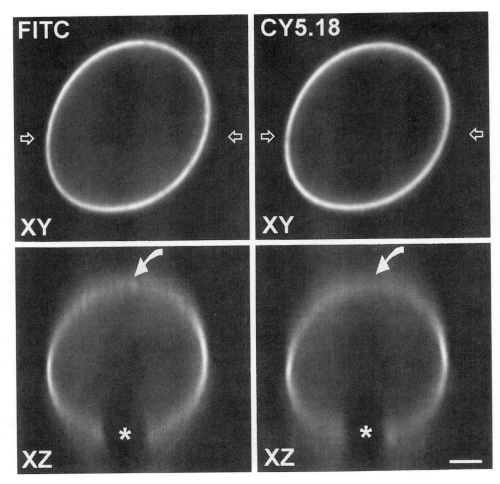

Fig. 33 The effect of using different imaging wavelengths on the observation of structures near the optical resolution of the microscope. *Giardia* cysts were stained with rabbit and mouse antibodies against cyst wall antigens followed by fluorescein (FITC)-conjugated donkey anti-mouse and cyanine 5.18 (CY5.18)-conjugated donkey anti-rabbit IgGs (multiple species adsorbed). *Top:* Single xy images acquired through the center of a double-labeled cyst. Open arrows indicate location of the xz images. *Bottom:* xz images acquired through the same cyst by collecting a single scan line at different depths along the optical (z) axis. The region across the top of the cyst with differences in intensity is indicated by solid arrows. The unstained region at the bottom of the cyst line in the xz images (asterisks) is from their attachment to the slide. Scale bar $= 2 \mu$m. Nikon $\times 60$ NA 1.4. (Specimen courtesy of S. Erlandsen, University of Minnesota.)

to be at least 700 nm thick across the top of the cyst in these xz images. This occurs because the thickness of the cyst wall is below the minimal axial resolution expected at these wavelengths (Table II). As a result, the cyst wall appears brighter when oriented parallel rather than when oriented perpendicular to the optical axis because the thickness of the cyst wall is below the axial resolution of this objective ($\sim$700 nm). This is an example of how the apparent staining intensity of structures observed with LSCM can vary depending on its orientation with respect to the optical axis (Van Der Voort and Brakenhoff, 1990).

These differences in resolution observed between images acquired at multiple wavelengths could be compensated for by adjusting the size of the detector apertures. By increasing the size of the detector apertures for the shorter wavelengths, the resolution for each imaging condition can be reduced to that observed with the longest wavelength. Although this facilitates a more direct comparison of the images acquired at different wavelengths, it is typically more desirable to adjust the detector apertures to obtain the best resolution possible with the intensity of staining for each fluorophore.

2. Spherical Aberration

Although confocal microscopy allows the collection of optical sections within thick specimens, the high-numerical aperture objectives typically used for fluorescence microscopy are corrected for the observation of specimens positioned immediately underneath a coverslip of the appropriate thickness. Spherical aberration is often introduced by focusing through a specimen mounted in a material whose refractive index varies only slightly from that of the immersion medium for an objective lens. In this case, light rays will be refracted at these boundaries and will not be properly focused on the detector aperture. This will reduce the observed intensities in the deeper optical sections and introduces discrepancies between stage and focal plane movements (see [3], this volume; see also Carlsson, 1991). Besides reducing the observed intensity of staining, the presence of extremely small amounts of spherical aberration is sufficient to produce a more substantial degradation of axial rather than lateral resolution in a confocal microscope as the focal plane is moved deeper into the specimen (Sheppard and Gu, 1991).

A dramatic example of this effect is apparent in xz images through islets of Langerhans with their peripheral α cells stained for glucagon (Fig. 34). Because the amount of spherical aberration introduced by the mismatch of refractive indexes is wavelength dependent, the attenuation of signal intensity and loss of axial resolution with deeper imaging into the islets is much more noticeable for the islet stained with fluorescein rather than cyanine 5.18. In double-labeled specimens, this effect can alter the relative intensities and position of stained structures in images collected at different excitation/emission wavelengths.

3. Chromatic Aberration

It should also be noted that even the best achromatic and apochromatic lenses have residual chromatic aberration that is more apparent when used with confocal microscopy (Wells *et al.,* 1990; Akinyemi *et al.,* 1992). Traditionally, chromatic aberration has been corrected by careful selection of objective lens elements to bring two or three selected wavelengths to the same focus. Intermediate wavelengths have slightly different focal planes depending on the lens design. Although the differences in focal plane are well within the depth of field for conventional fluorescence imaging, the lack of correction

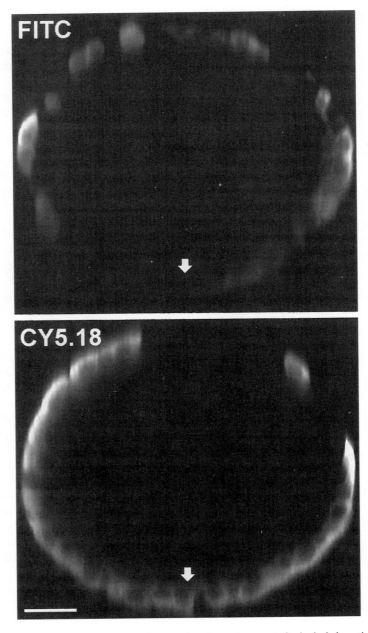

Fig. 34 The effect of using different imaging wavelengths on the extent of spherical aberration observed. Neonatal rat islets of Langerhans were stained with a mouse monoclonal antibody against glucagon and either fluorescein (FITC)- or cyanine 5.18 (CY5.18)-conjugated donkey anti-mouse IgGs. The islets were mounted in glycerol and the correction collar on the objective set for the highest contrast and detail observed midway through the fluorescein-stained islet. Each image was then corrected for refractive index differences by stretching by a factor of 1.47 in the depth dimension. The arrows indicate positively stained cells at the bottom of the islets that are barely discernible when stained with fluorescein. Scale bar = 25 μm. Olympus ×40, NA 0.85.

at these intermediate wavelengths will result in misalignment of the illumination and fluorescence paths with confocal fluorescence microscopy. Axial chromatic aberration can result in a 50% loss in the observed signal intensity and in a doubling of the practical optical section thickness at various wavelengths (Fricker and White, 1992). Although less attractive for multiline lasers, the individual lines can be separated and additional prefocusing elements used for each beam to counteract the axial chromatic aberration introduced by other optical components (Kuba et al., 1991). Similarly, lateral chromatic aberration will reduce the intensity of fluorescence observed at positions located further from the optical axis (Wells et al., 1990). Depending on the objective, the extent of this aberration can vary dramatically for the various wavelengths used for LSCM. Because these chromatic aberrations are still observable for the corrected objective lenses typically used with LSCM, the performance of individual objectives should be verified when images will be acquired at different excitation/emission wavelengths for quantitative comparisons (Wells et al., 1990).

4. Depth of Optical Sectioning

Because differences in excitation and emission wavelengths alter numerous aspects of imaging with LSCM, the apparent depth to which specimens may be optically sectioned is also affected. A common misconception is that LSCM allows the acquisition of optical sections as deep into the specimen as the working distance of the objective allows (Cheng and Summers, 1990; Visser et al., 1991).

However, the actual depth to which a specimen can be optically sectioned is dependent on absorption and scattering within the specimen being examined. Besides spherical aberration induced by mismatches between the immersion and mounting medium, structures within the specimen can produce effects similar to spherical aberration when their refractive index is different than that of the mounting medium. This will have the effect of reducing observed intensity and resolution of optical sections acquired at greater depths in the specimen. Because the extent of this induced spherical aberration is wavelength dependent, the use of longer excitation wavelengths with far red fluorophores is particularly useful for the imaging of especially thick specimens (see Section VI,A,2).

The observed intensity of stained structures in deeper optical sections can also be decreased by a reduction in the excitation (i.e., inner filter effect) or the reabsorption of fluorescence emission from fluorophores in deeper layers by overlying regions of the specimen (Van Oostveldt and Baumens, 1990; Visser et al., 1991). The simplest approach to reduce this "shadowing" of deeper structures by overlying regions is to reduce the overall intensity of staining (i.e., reduce the local absorbance value). Because autofluorescence and scattering will decrease for longer excitation wavelengths, the far red fluorophores allow the collection of images with acceptable contrast with the lower staining intensities required for the optical sectioning of especially thick specimens (Mesce et al., 1993). Although computationally intensive, a restoration filter for absorption and scattering correction in optical sections acquired with fluorescence confocal microscopy has been described (Visser et al., 1991).

B. Image Processing and Presentation

Multicolor LSCM is used to acquire monochromatic images of several fluorophores in different spectral regions. If the independent detection of signals from each fluorophore was achieved, then each image contains information on the localization of staining of different substances for the same field of view. Typically, it is advantageous to merge these monochromatic images into a single color image to facilitate the detection of double labeling or relationships between stained structures. In addition, color images are generally much more pleasant to look at than black-and-white images. Although the normal corrections required after the acquisition of digitized images are not usually required with confocal images (i.e., background subtraction, flat field correction), there are several corrections that facilitate the comparison of images acquired in different spectral regions.

1. Image Registration

Because even slight variations in the position of the dichroic mirror cause translational shifts in the resulting images, the misregistration of images acquired with multiple filter sets is a common problem. Although individual images can be translated by simple address operations when they are overlaid in different colors, the occurrence of a shift along the optical axis between the data sets must be considered for three-dimensional reconstructions. Finding the exact translation vector giving optimal registration may be further complicated by small shifts in the scanning axis with respect to the optical axis between the collection of data sets. To reduce this type of registration error, it is preferable to use filter configurations in which images of each fluorophore are acquired with the same dichroic mirror.

2. Image Contrast

The comparison of images of multiple fluorophores is frequently complicated by differences in contrast of the individual monochromatic images. Ideally, each image should properly fill the available dynamic range and be of comparable contrast before merging into a color image. The use of linear image processing transformations to more fully utilize the available dynamic range can dramatically improve the quality of each image (i.e., remapping the minimum pixel value to black and the maximum pixel value to peak white). A limitation of these transformations is that all pixels of the same intensity are similarly remapped. This is less successful if the intensity of desired details varies in different regions of an image. In this case, the use of adaptive contrast enhancement techniques that are dependent on the local characteristics of an image (e.g., subregions of $n \times n$ pixels) can improve the retention of contrast and subtle details in individual images and projections (Schomann and Jovin, 1991; Mesce *et al.*, 1993). Because these filters do not maintain relative pixel intensities, they are not suitable for quantitative studies based on the observed pixel intensities.

3. Image Display

After acquiring and processing the images, the question remains as to how to display the multiple images acquired of each fluorophore. A good solution is to merge the individual monochromatic images into a color image, so that they can be displayed simultaneously. Typically, the primary colors of the additive color system (i.e., red, green, and blue) are assigned to individual images because they produce the widest range of colors. However, it should be noted that this process quickly surpasses the capacity of the viewing medium and the observer. Often, the colors assigned to each image can have considerable effect on the appearance of the individual structures within the merged image. For example, the bright green microtubules are easily followed against the darker red background of actin filaments in the previously shown color projection of the cyto-skeleton (Color Plate 11C). In contrast, when the colors are assigned to more closely match the fluorophores (i.e., fluorescein-stained actin filaments in green and the cyanine 3.18-stained microtubules in red), it is difficult to observe the dark red microtubules against a background of bright green actin filaments. Because of these differences in discerning the relative intensities of one color against different backgrounds, it is useful to switch the colors assigned to individual images in a merged color image to prevent visualization artifacts.

Even after considering the colors assigned to each image, adjusting the brightness and contrast of each image can dramatically affect the appearance of a merged color image. Often it is difficult to obtain acceptable merged images when one of the images is colored blue because of the difficulty in detecting pure blue against a black background. This is especially true for color printers, which are usually designed to give dark primary colors. Because a lighter blue contrasts better with the black background, we typically add a fraction (0.3–0.5) of the blue image to the green image before merging the individual monochrome images into a single color image.

Even with these adjustments, a loss of detail within the individual images is usually observed in the merged color images. For example, the number of optical sections in the projections of the glucagon-containing α cells can be easily counted in the black-and-white stereo pair (Fig. 32, CY5.18). However, when colored blue in the merged three-color stereo image, it is no longer possible to count the individual optical sections even though no color reduction was done during the printing process (Color Plate 11A). Although it is obvious that lightly stained structures may not be visible in merged images, this example demonstrates that information can also be lost from brightly stained structures in merged color images. Therefore it may be preferable to show the individual monochrome images to demonstrate the actual contrast and detail present, and the merged color image to emphasize the relationships between stained structures.

The techniques for the display and analysis of the multiple data sets collected by multicolor LSCM are still being developed. Although many options currently exist for two-dimensional reconstructions, further work is required to permit the display and analysis of the multiple data sets as three-dimensional reconstructions. Hopefully, advances in the hardware and software for the manipulation of graphics will make this type of software available to more biologists in the near future. Fortunately, the development of multicolor LSCM has coincided with the availability of more capable graphics workstations and 24-bit color printers (such as the Kodak XL7700 used for the images in this article) for hard copy output.

VII. Conclusions

The use of LSCM for multicolor immunofluorescence studies is a complex task involving a thorough knowledge of immunohistochemistry, confocal microscope operation and instrumentation, and image presentation. We have presented background information that forms the basis for using these technologies for multicolor LSCM. We have also described what, in our experience, are the practical issues and limitations of doing multicolor immunofluorescence studies with the currently available confocal microscopes. From the examples shown, the advantages of using lasers with a broad selection of emission lines for the multiwavelength excitation compared to single-wavelength excitation of several fluorophores has been demonstrated. With the development of an air-cooled krypton–argon ion laser, the ability to use confocal imaging with triple-labeled specimens is now available to researchers for a cost comparable to the previous argon ion laser-equipped microscopes. Although this article is limited to the use of LSCM for multicolor immunofluorescence studies, the issues discussed in this article should enable researchers to more fully exploit the capabilities of the available instruments for other types of studies (i.e., live-cell imaging and *in situ* hybridization) that require the detection of multiple signals from a single specimen.

Acknowledgments

We would like to thank our colleague Robert Elde for continued guidance and support in the use of LSCM for multicolor immunofluorescence studies. We also wish to thank Bio-Rad Microscience (Cambridge, MA) and Ion Laser Technology (Salt Lake City, UT) for collaborating in the development of the krypton–argon ion laser suitable for LSCM; Alan Waggoner (Carnegie-Mellon University, Pittsburgh, PA) for providing the cyanine dyes and unpublished data on various fluorophores; William Stegeman (Jackson ImmunoResearch Laboratories, West Grove, PA) for performing the various conjugations with cyanine 3.18, cyanine 5.18, and the R-phycoerythrin/cyanine 5.18 tandem conjugate; Debra Schindele and George Renzoni (formerly of Ultra Diagnositics Corp., Seattle, WA) for providing samples of various Ultralight dyes; Diether Recktenwald (Becton Dickinson, San Jose, CA) for providing samples of PerCP and its conjugates; Rosaria Haugland and Sam Wells (Molecular Probes, Eugene, OR) for providing samples of various BODIPY and Cascade Blue conjugates. The many helpful discussions and critical reading of this article by William Stegeman, Cynara Ko, Alan Waggoner, and Brian Matsumoto are gratefully acknowledged. We would like to thank Paul Letourneau, Stanley Erlandsen, David Scharp, and Paul Lacy for donating various specimens. We also wish to thank Jianlin Wang, Celest Roth, Steve Schnell, and Jane Wobken for excellent technical assistance, and Jerry Sedgewick for photographic assistance.

This work was supported by PHS Grants DK-33655 from NIH (R.L.S.), DA-05466 from ADAMHA (M.W.W.), and the Department of Cell Biology and Neuroanatomy (University of Minnesota). The donation of the large number of fluorophore-conjugated secondary antibodies used in these studies by Jackson ImmunoResearch Laboratories, and the payment of the publication costs by Bio-Rad Microscience for the color plates in this article are gratefully acknowledged.

References

Afar, B., Merrill, J., and Clark, E. A. (1991). *J. Clin. Immunol.* **11**, 254–261.

Akinyemi, O., Boyde, A., Browne, M. A., Hadravsky, M., and Petran, M. (1992). *Scanning* **14**, 136–143.

Akner, G., Mossberg, K., Wikström, A. C., Sundquist, K. G., and Gustafsson, J. Å. (1991). *J. Steroid Biochem. Mol. Biol.* **39**, 419–432.

Amos, W. B. (1988). *Cell Motil. Cytoskeleton* **10**, 54–61.

Arndt-Jovin, D. J., Robert-Nicoud, M., and Jovin, T. M. (1990). *J. Microsc. (Oxford)* **157**, 61–72.

Art, J. (1990). *In* "Handbook of Biological Confocal Microscopy" (J. B. Pawley, ed.), Rev. Ed., pp. 127–139. Plenum, New York.

Aubin, J. E. (1979). *J. Histochem. Cytochem.* **27**, 36–43.

Benson, R. C., Meyer, R. A., Zaruba, M. E., and McKhann, G. M. (1979). *J. Histochem. Cytochem.* **27**, 44–48.

Blakeslee, D., and Baines, M. G. (1976). *J. Immunol. Methods* **13**, 305–320.

Bliton, C., Lechleiter, J., and Clapham, D. E. (1993). *J. Microsc.* **169**, 15–26.

Brakenhoff, G. H., Blom, P., and Barends, P. (1979). *J. Microsc. (Oxford)* **117**, 219–232.

Brakenhoff, G. H., Van Der Voort, H. T. M., Van Spronsen, E. A., and Nanninga, N. (1989a). *J. Microsc. (Oxford)* **153**, 151–159.

Brakenhoff, G. H., Van Spronsen, E. A., Van Der Voort, H. T. M., and Nanninga, N. (1989b). *Methods Cell Biol.* **30B**, 379–398.

Brakenhoff, G. H., Visscher, K., and Van Der Voort, H. T. M. (1990). *In* "Handbook of Biological Confocal Microscopy" (J. B. Pawley, ed.), 2nd Ed., pp. 87–91. Plenum, New York.

Brelje, T. C., and Sorenson, R. L. (1991). *Endocrinology* **128**, 45–57.

Brelje, T. C., and Sorenson, R. L. (1992). U.S. Pat. 5,127,730.

Brelje, T. C., Sharp, D. W., and Sorenson, R. L. (1989). *Diabetes* **38**, 808–814.

Brodin, L., Ericsson, M., Mossberg, K., Hökfelt, T., Ohta, Y., and Grillner, S. (1988). *Exp. Brain Res.* **73**, 441–446.

Buurman, E. P., Sanders, R., Draaijer, A., Gerritsen, H. C., Van Veen, J. J. F., Houpt, P. M., and Levine, Y. K. (1992). *Scanning* **14**, 155–159.

Carlsson, K. (1990). *J. Microsc. (Oxford)* **157**, 21–27.

Carlsson, K. (1991). *J. Microsc. (Oxford)* **163**, 167–178.

Carlsson, K., and Mossberg, K. (1992). *J. Microsc. (Oxford)* **167**, 23–37.

Chapple, M. R., Johnson, G. D., and Davidson, R. S. (1988). *J. Immunol. Methods* **111**, 209–217.

Chen, R. F. (1969). *Arch. Biochem. Biophys.* **133**, 263–276.

Cheng, P. C., and Summers, R. G. (1990). *In* "The Handbook of Confocal Imaging" (J. Pawley, ed.), Rev. Ed., pp. 179–195. Plenum, New York.

Cogswell, C. J., Hamilton, D. K., and Sheppard, C. J. R. (1992). *J. Microsc.* **165**, 49–60.

Coons, A. H., and Kaplan, M. H. (1950). *J. Exp. Med.* **91**, 1–13.

DeBiasio, R., Bright, G. R., Ernst, L. A., Waggoner, A. S., and Taylor, D. L. (1987). *J. Cell Biol.* **105**, 1613–1622.

Delia, D., Martinez, E., Fontanella, E., and Aiello, A. (1991). *Cytometry* **12**, 537–544.

Denk, W., Strickler, J. H., and Webb, W. W. (1990). *Science* **248**, 73–76.

Elde, R., Cao, Y., Cintra, A., Brelje, T. C., Pelto-Huikko, M., Junttila, T., Fuxe, K., Pettersson, R. F., and Hökfelt, T. (1991). *Neuron* **7**, 349–364.

Erlandsen, S. L., Bemrick, W. J., Schupp, D. E., Shields, J. M., Jarroll, E. L., Sauch, J. F., and Pawley, J. B. (1990). *J. Histochem. Cytochem.* **38**, 625–632.

Fehon, R. G., Kooh, P. J., Rebay, I., Regan, C. L., Xu, T., Mushkavitch, M. A. T., and Artavanis-Tsakonas, S. (1990). *Cell* **61**, 523–534.

Feston, R., Björkland, A., and Tötterman, T. H. (1990). *J. Immunol. Methods* **126**, 69–78.

Fox, M. H., Arndt-Jovin, D. J., Jovin, T. M., Bamann, P. H., and Robert-Nicoud, M. (1991). *J. Cell Sci.* **99**, 247–253.

Fricker, M. D., and White, N. S. (1992). *J. Microsc.* **166**, 29–42.

Fritschy, J. M., Benke, D., Mertens, S., Oertel, W. H., Bachi, T., and Möhler, H. (1992). *Proc. Natl. Acad. Sci. USA* **89**, 6726–6730.

Gailbraith, G. M., Gailbraith, R. M., and Faulk, W. P. (1978). *J. Clin. Lab. Immunol.* **1**, 163–167.

Galbraith, W., Ernst, L. A., Taylor, D. L., and Waggoner, A. S. (1989). *Soc. Photo-Opt. Instrum. Eng.* **1063**, 19–20.

Galbraith, W., Wagner, M. C. E., Chao, J., Abaza, M., Ernst, L. A., Nederlof, M. A., Hartsock, R. J., Taylor, D. L., and Waggoner, A. S. (1991). *Cytometry* **12**, 579–596.

Giloh, H., and Sedat, J. W. (1982). *Science* **217**, 1252–1255.

Glazer, A. N. (1989). *J. Biol. Chem.* **264**, 1–4.

Glazer, A. N., and Stryer, L. (1983). *Biophys. J.* **43**, 383–386.

Goller, B., and Kubbies, M. (1992). *J. Histochem. Cytochem.* **40**, 451–456.

Gratton, E., and van de Ven, M. J. (1990). *In* "Handbook of Biological Confocal Microscopy" (J. B. Pawley, ed.), 2nd Ed., pp. 53–67. Plenum, New York.

Hansen, P. A. (1967). *Acta Histochem., Suppl* **8**, 167–180.

Haugland, R. P. (1983). *In* "Excited States of Biopolymers" (R. F. Steiner, ed.), pp. 29–58. Plenum, New York.

Haugland, R. P. (1990). *In* "Optical Microscopy for Biology" (B. Herman and K. Jacobson, eds.), pp. 143–157. Wiley, New York.

Haugland, R. P. (1992). "Handbook of Fluorescent Probes and Research Chemicals." Molecular Probes, Inc., Eugene, Oregon.

Hiramoto, R., Bernecky, J., Jurand, J., and Hamlin, M. (1964). *I. Histochem. Cytochem.* **12**, 271–274.

Hoffman, R. A. (1988). *Cytometry, Suppl.* **3**, 18–22.

Horikawa, K., and Armstrong, W. E. (1988). *J. Neurosci. Methods* **25**, 1–11.

Houser, C. R., Barber, R. P., Crawford, G. D., Matthews, D. A., Phelps, P. E., Salvaterra, P. M., and Vaughn, J. E. (1984). *J. Histochem. Cytochem.* **32**, 395–402.

Jahn, R., Schiebler, W., Ouiment, C., and Greengard, P. (1985). *Proc. Natl. Acad. Sci. USA* **82**, 4137–4141.

Johnson, G. D., and deC. Nogueira Araujo, G. M. (1981). *J. Immunol. Methods* **43**, 349–360.

Johnson, G. D., Davidson, R. S., McNamee, K. C., Russel, G., Goodwin, D., and Holborow, E. J. (1982). *J. Immunol. Methods* **55**, 231–242.

Khalfan, H., Abuknesha, R., Rand-Weaver, M., Price, R. G., and Robinson, D. (1986). *Histochem. J.* **18**, 497–499.

Kita, H., and Armstrong, W. (1991). *J. Neurosci. Methods* **37**, 141–150.

Kronick, M. N. (1986). *J. Immunol. Methods* **92**, 1–13.

Kuba, K., Hua, S. Y., and Nohmi, M. (1991). *Neurosci. Res.* **10**, 245–256.

Lang, M., Stober, F., and Lichtenthaler, H. K. (1991). *Radiat. Environ. Biophys.* **30**, 333–347.

Lanier, L. L., and Recktenwald, D. J. (1991). *Methods: Companion to Methods in Enzymol.* **2**, 192–199.

Lansdorp, P. M., Smith, C., Safford, M., Terstappen, L., and Thomas, T. E. (1991). *Cytometry* **12**, 723–730.

Larsson, L.-I. (1983). *In* "Handbook of Chemical Neuroanatomy" (A. Björklund and T. Hökfelt, eds.), Vol. 1, pp. 147–209. Elsevier, Amsterdam.

Lechleiter, J., and Clapham, D. E. (1992). *Cell* **69**, 283–294.

Loken, M. R., Parks, D. R., and Herzenberg, L. A. (1977). *J. Histochem. Cytochem.* **25**, 899–907.

Lynch, R. M., Fogarty, K. E., and Fay, F. S. (1991). *J. Cell Biol.* **112**, 385–395.

McKay, I. C., Forman, D., and White, R. G. (1981). *Immunology* **43**, 591–602.

Mathies, R. A., and Stryer, L. (1986). *In* "Applications of Fluorescence in the Biomedical Systems" (D. L. Taylor, A. S. Waggoner, F. Lanni, R. F. Murphy, and R. R. Birge, eds.), pp. 129–140. Alan R. Liss, New York.

Mesce, K. A., Klukas, K. A., and Brelje, T. C. (1993). *Cell Tissue Res.* **271**, 381–397.

Montag, M., Kukulies, J., Jorgens, R., Gundlach, H., Trendelenburg, M. F., and Spring, H. (1991). *J. Microsc. (Oxford)* **163**, 201–210.

Morgan, C. G., Mitchell, A. C., and Murray, J. G. (1992). *J. Microsc. (Oxford)* **165**, 49–60.

Mossberg, K., and Ericsson, M. (1990). *J. Microsc. (Oxford)* **158**, 215–224.

Mossberg, K., Arvidsson, U., and Ulfhake, B. (1990). *J. Histochem. Cytochem.* **38**, 179–190.

Mujumdar, R. B., Ernst, L. A., Mujumdar, S. R., Lewis, C. J., and Waggoner, A. S. (1993). *Bioconjugate* (in press).

Oi, V. T., Glazer, A. N., and Stryer, L. (1982). *J. Cell Biol.* **93**, 981–986.

Paddock, S. W. (1991). *Proc. Soc. Exp. Biol. Med.* **198,** 772–780.

Pagano, R. E., Martin, O. C., Kang, H. C., and Haugland, R. P. (1991). *J. Cell Biol.* **113,** 11267–1279.

Pawley, J. B. (1990). *In* "Handbook of Biological Confocal Microscopy" (J. B. Pawley, ed.), 2nd Ed., pp. 15–26. Plenum, New York.

Pearse, A. G. E. (1980). *In* "Histochemistry: Theoretical and Applied," 4th Ed. Vol. 1, pp. 159–252. Churchill Livingstone, London.

Recktenwald, D. J. (1989). U.S. Pat. 4,876,190.

Renzoni, G. E., Schindele, D. C., Theodore, L. J., Leznoff, C. C., Fearon, K. L., and Pepich, B. V. (1992). U.S. Pat. 5,135,717.

Robitaille, R., Adler, E. M., and Charlton, M. P. (1990). Strategic location of calcium channels at transmitter release sites of frog neuromuscular synapses. *Neuron* **5,** 773–779.

Schindele, D. C., and Renzoni, G. E. (1990). *J. Clin. Immunoassay* **13,** 182–186.

Schmeud, L. C., and Fallon, J. H. (1986). *Brain Res.* **377,** 147–154.

Schormann, T., and Jovin, T. M. (1991). *J. Microsc. (Oxford)* **166,** 155–168.

Schotten, D. M. (1989). *J. Cell Sci.* **94,** 175–206.

Schubert, W. (1991). *Eur. J. Cell Biol.* **55,** 272–285.

Schweitzer, E. S., and Paddock, S. (1990). *J. Cell Sci.* **96,** 375–381.

Shapiro, H. M. (1988). "Practical Flow Cytometry," 2nd Ed. Alan R. Liss, New York.

Sheppard, C. J. R., and Gu, M. (1991). *Appl. Opt.* **30,** 3563–3568.

Sorenson, R. L., Carry, D., and Brelje, T. C. (1991). *Diabetes* **40,** 1365–1374.

Southwick, P. L., Ernst, L. A., Tauriello, E. W., Parker, S. R., Mujumdar, R. B., Mujumdar, S. R., Clever, H. A., and Waggoner, A. S. (1990). *Cytometry* **11,** 418–430.

Spack, E. G., Packard, B., Wier, M. C., and Edidin, M. (1986). *Anal. Biochem.* **158,** 233–237.

Staines, W. A., Meister, B., Melander, T., Nagy, J. I., and Hökfelt, T. (1988). *J. Histochem. Cytochem.* **36,** 145–151.

Stamatoglou, S. C., Sullivan, K. H., Johansson, S., Bayley, P. M., Burdett, I. J., and Hughes, R. C. (1990). *J. Cell Sci.* **97,** 595–606.

Stelzer, E. H. K. (1990). *In* "Handbook of Biological Confocal Microscopy" (J. B. Pawley, ed.), 2nd Ed., pp. 93–103. Plenun, New York.

Sternberger, L. A. (1986). "Immunocytochemistry," 3rd Ed. Wiley, New York.

Stewart, W. (1978). *Cell* **14,** 741–759.

Szarowski, D. H., Smith, K. L., Herchenroder, A., Matuszek, G., Swann, J. W., and Turner, J. N. (1992). *Scanning* **14,** 104–111.

Titus, J. A., Haugland, R., Sharrow, S. O., and Segal, D. M. (1982). *J. Immunol. Methods* **50,** 193–204.

Tsien, R. Y., and Waggoner, A. (1990). *In* "Handbook of Biological Confocal Microscopy" (J. Pawley, ed.), 2nd Ed., pp. 169–178. Plenum, New York.

Ulfhake, B., Carlsson, K., Mossberg, K., Arvidsson, U., and Helm, P. J. (1991). *J. Neurosci.* **40,** 39–48.

Van Dekken, H., Van Rotterdam, A., Jonker, R., Van Der Voort, H. T. M., Brakenhoff, G. J., and Bauman, J. G. J. (1990). *J. Microsc. (Oxford)* **158,** 207–214.

Van Der Voort, H. T. M., and Brakenhoff, G. J. (1990). *J. Microsc. (Oxford)* **158,** 43–54.

Van Oostveldt, P., and Baumens, S. (1990). *J. Microsc. (Oxford)* **158,** 121–132.

Vassy, J., Rigaut, J. P., Hill, A. M., and Foucrier, J. (1990). *J. Microsc. (Oxford)* **157,** 91–104.

Vincent, S. L., Sorensen, I., and Benes, F. M. (1991). *BioTechniques* **11,** 628–634.

Visser, T. D., Groen, F. C. A., and Brakenhoff, G. J. (1991). *J. Microsc. (Oxford)* **163,** 189–200.

Waggoner, A., DeBiasio, R., Conrad, P., Bright, G. R., Ernst, L., Ryan, K., Nederlof, M., and Taylor, D. (1989). *Methods Cell Biol.* **30B,** 449–478.

Wang, X., and Kurtz, I. (1990). *Am. J. Physiol.* **259,** C365–C373.

Weidenmann, B., and Franke, W. W. (1985). *Cell* **41,** 1017–1028.

Wells, K. S., Sandison, D. R., Strickler, J., and Webb, W. W. (1990). *In* "Handbook of Biological Confocal Microscopy" (J. B. Pawley, ed.), 2nd Ed., pp. 27–39. Plenum, New York.

Wells, S., and Johnson, I. (1993). *In* "Three-Dimensional Confocal Microscopy" (J. K. Stevens, L. R. Mills, and J. E. Trogadis, eds.), Academic Press, San Diego. In Press.

Wessendorf, M. W. (1990). *In* "Handbook of Chemical Neuroanatomy" (A. Björklund, T. Hökfelt, F. G. Wouterlood, and A. N. van den Pol, eds.), Vol. 8, pp. 1–45. Elsevier, Amsterdam.

Wessendorf, M. W. (1991). *Brain Res.* **553,** 135–148.

Wessendorf, M. W., and Brelje, T. C. (1992). *Histochemistry* **98,** 81–85.

Wessendorf, M. W., and Elde, R. P. (1985). *J. Neurosci.* **7,** 2352–2363.

Wessendorf, M. W., Appel, N. M., Molitor, T. W., and Elde, R. P. (1990a). *J. Histochem. Cytochem.* **38,** 1859–1877.

Wessendorf, M. W., Tallaksen-Greene, S. J., and Wohlhueter, R. M. (1990b). *J. Histochem. Cytochem.* **38,** 87–94.

Wessendorf, M. W., Tallaksen-Greene, S. J., and Wohlhueter, R. M. (1990c). *J. Histochem. Cytochem.* **38,** 741.

Wetmore, C., and Elde, R. (1991). *J. Comp. Neurol.* **305,** 148–163.

Whitaker, J. E., Haugland, R. P., Moore, P. L., Hewitt, P. C., Reese, M., and Haugland, R. P. (1991a). *Anal. Biochem.* **198,** 119–130.

Whitaker, J. E., Haugland, R. P., and Prendergast, F. G. (1991b). *Anal. Biochem.* **194,** 330–344.

White, J. C., and Stryer, L. (1987). *Anal. Biochem.* **161,** 442–452.

White, J. G., Amos, W. B., and Fordham, M. (1987). *J. Cell Biol.* **105,** 41–48.

Wilson, T. (1989). *J. Microsc. (Oxford)* **154,** 143–156.

Wilson, T., and Sheppard, C. (1984). "Theory and Practice of Scanning Optical Microscopy." Academic Press, San Diego.

Wories, H. J., Koek, J. H., Lodder, G., Lugtenburg, J., Fokkens, R., Driessen, O., and Mohn, G. R. (1985). *Recl. Trav. Chim. Pays-Bas* **104,** 288–291.

Yu, H., Ernst, L., Wagner, M., and Waggoner, A. (1992). *Nucleic Acids Res.* **20,** 83–88.

CHAPTER 6

Practical Aspects of Objective Lens Selection for Confocal and Multiphoton Digital Imaging Techniques

Gerald S. Benham

BioScience Confocal Systems
Nikon Instruments Inc.
Melville, New York, 11747

METHODS IN CELL BIOLOGY, VOL. 70
Copyright 2002, Elsevier Science (USA). All rights reserved.
0091-679X/02 $35.00

This chapter is dedicated to the memory of the late Larry Stottlemyer.

Larry added showmanship to his many humorous and entertaining stories.

Larry always seemed to move forward with a positive persuasion.

Larry showed me how to be a microscope salesperson.

Larry tenured with Nikon, Olympus, and Zeiss.

Larry had a clever ability to mix wit with IQ.

Larry was a respected professional.

Larry helped me to find my best.

Larry was my friend.

I shall miss him.

═══════════ # I. Introduction

The well-established technology of confocal microscopy (Minsky, 1957, 1988) is the topic of this second edition and numerous other books and publications providing credibility that confocal microscopy has become a standard research tool in cell biology and neuroscience (Stelzer, 1993). See Paddock (1999) for a detailed review of the improvements to confocal imaging systems during the past decade. Today, few biological research centers are without at least one confocal instrument from one manufacturer or another (Fine, 2000). Confocal microscopy used in fluorescence applications is now widely accepted. In addition, the advancement of two-photon (Denk *et al.,* 1990) and multiphoton technologies is reviewed in Chapter 3 of this edition and by Denk (2000) and Lemasters (2000). As our experience in confocal and now multiphoton digital imaging increases through the coming years, our views on the correct objective lens selection for various applications should be ever-changing.

Numerical aperture, dry versus immersion, and low versus high magnification have been the traditional criteria in previous objective lens selection situations. However, new technologies, continuous optical improvements offered by the major microscope manufacturers, and new research application requirements dictate that additional microscope objective lens characteristics are equally important considerations in microscopy and digital imaging today. With all of the importance now being attributed to the objective lens in confocal and multiphoton imaging, it is surprising to observe how rarely the literature refers to the actual objective(s) used. It is opportune at this time to briefly review the types of confocal scanner systems commercially available.

A. "On-Axis" versus "Off-Axis" Confocal Scanning

An early example of an "on-axis" confocal specimen scanner was described by Sheppard (1977). This technology, known as the scanning optical microscope, was the first to employ a laser as an illumination source. With this scanner, the beam remained stationary on the optical axis (nonscanned); the microscope stage or the objective lens was scanned. A single pinhole was placed before the detector to effect confocality. A more detailed discussion of this technology was reported by Sheppard and Wilson (1984) and Wilson and Sheppard (1984). This technology is not commercially available in the biological marketplace today.

Today, the most numerous and widely accepted commercial biological confocal systems are the laser-based, "off-axis" beam scanners. These were first described at several European laboratories by Brakenhoff *et al.* (1979), Wijnaedts van Resandt *et al.* (1985), and Carlsson *et al.* (1985). This type of confocal scanner employs a computer-controlled deflector such as a pair of galvanometer mirrors or a cardanic single-mirror to scan the illuminating laser beam through the objective lens and across a stationary object on the microscope stage. This design was first commecialized by Sarastro (later, Molecular Dynamics, now, Amersham), Bio-Rad Laboratories, and Meridian Instruments. This group is considered to represent the "slow scanners" where image acquisition time is approximately 2 s per frame of 1024×1024 pixels.

In a effort to increase confocal acquisition frame-rate, various scanning mechanisms have been employed for scanning the beam at video or near-video rates, such as a bilateral (double-sided galvanometer) mirror (Brakenhoff *et al.,* 1979), a rotating polygon mirror (White *et al.,* 1987), an acousto-optical deflector (AOD) (Draaijier and Houpt, 1988), and a galvanometer mirror coupled with a resonant galvanometer (Tsien and Harootunian, 1990). Today commercial successes in the biological market place with "off-axis" confocal beam scanners are being experienced by Bio-Rad, Leica, Nikon, Olympus, and Zeiss.

According to Cogswell *et al.* (1990), both designs may appear similar in their imaging capabilities; however, there are no similarities with respect to what constitutes the optimum objective for obtaining maximum resolution in both lateral and axial directions when using either an "on-axis" or "off-axis" confocal.

Wright *et al.* (1993) described a third type of confocal microscope which utilizes a spinning Nipkow disk with a stationary light source and stage. A white-light, spinning Nipkow disk, confocal scanner worthy of commercialization was first demonstrated by Egger and Petran (1967) and Petran *et al.* (1968). Boyde (1985, 1986) later provided biological applications data for the commercialization of this technology. This design, however, seems to have found its niche within the semiconductor industry where commercial successes are being experienced by Nikon Inc., Carl Zeiss, Inc., and Zygo Corporation. Variations of this design were commercially marketed for biological applications by Tracor Northern (now Noran Instruments), Technical Instrument Company (now Zygo Corporation), Bio-Rad Laboratories, and Nikon Inc. Today only the Yukogawa and Carl Zeiss, Inc., confocal spinning disk systems, utilizing laser illumination, are commercially available. They have had limited successes in the biological marketplace.

B. Digital Imaging Considerations

Optical and quantitative microscopy utilizing confocal, multiphoton, differential interference contrast (DIC), and epifluorescence imaging techniques are dependent on the ability of high-magnification and high-numerical-aperture immersion objectives to provide a final high-resolution, digital image.

Majlof and Forsgren (1993) explained that there were many factors specifically affecting confocal imaging such as the refractive index of the medium, objective lens selection, depth discrimination, objective lens type, objective working distance and magnification, and detector characteristics. In addition, immersion liquids, the cover glass, and the mounting medium were also an important part of the "optical train" affecting the final image (Inoué and Oldenbourg, 1995; Keller, 1995). Keller (1995) reconfirmed by Piston (1998) further elaborated the objective lens to be the most important component of the optical microscope.

C. Importance of Selecting a Suitable Objective Lens

The importance in selecting the correct objective lens for use in confocal imaging techniques is well documented (Taylor and Salmon, 1989; Cogswell *et al.,* 1990; Majlof and Forsgren, 1993; Keller, 1995; Piston, 1998; Dunn and Wang, 2000). The classic criterion for lens selection has primarily been the objective numerical aperture (NA); however, considering NA alone does not account for the two most important problems in multiple color confocal digital imaging: spherical and chromatic aberrations. The author previously believed that only two objectives were needed to obtain quality confocal images: a low-magnification, high-NA dry objective, i.e., Plan Apochromat 10×/0.45 or 20×/0.75, and a high-magnification, high-NA Plan Apochromat 60×/1.4 oil immersion objective. This is not the case today with increased demands from digital imaging in numerous specific applications where there are a variety of objective lenses available for selection from the major microscope manufacturers. The need to image the specimen not only at the surface next to the cover glass, but deep down to 200 μm or more, can be greatly assisted by the selection and use of the best objective lens such as a highly corrected water immersion objective (Brenner, 1999). The proper choice of an objective lens can have a profound influence on the outcome of a confocal microscopy experiment. Before reviewing the various objective types and their aberrations, a quick discussion of both optical and digital resolutions is appropriate.

II. Confocal Microscopy Resolution

A. Optical Resolution in Wide-Field Microscopy

1. Lateral (*X/Y* axis) Resolution

Resolution has been widely accepted as the minimum separation necessary between two features in a specimen for each to be seen as two separate objects in the image (Inoué and Spring, 1997a; Piston, 1998; Larson, 1999a). In a diffraction-limited environment with a specimen positioned directly beneath the cover-glass, image resolution is achieved in accordance with

$$r = 1.22\lambda/\text{NA (objective)} + \text{NA (condenser)}, \tag{1}$$

complying with Rayleigh's criterion for transmission imaging. According to Abbe's diffraction theory of image formation (Abbe, 1884; Keller, 1995; Lanni and Keller, 2000), a wide-field microscope image is formed at the image plane from the Airy disk projections of all sources of light passing through the specimen. In Eq. (1), r equals the radius of the Airy disk (the radius of the first minimum) according to Rayleigh's criterion (Girard and Clapham, 1994), where the light may be reflected off of the specimen or originate within it as in the case of fluorescence.

Epifluorescence is the most important imaging technique associated with biological confocal microscopy. Specimen-emitted fluorescence originates from a collection of point light sources. The Airy disk image from a point light source can be significantly larger in diameter than the resolution limit of the diffraction-limited objective lens and may resemble something like a bright target on a black background.

The minimum separation necessary between two objects for them to be seen as two separate objects in the image is half the diameter of the Airy disk. The separate identity of two objects is just barely discernible. The theoretical resolution for epifluorescence can be calculated employing Rayleigh's criterion whereby the objective acts as both objective and condenser, slightly changing Eq. (1) to Eq. (2) where r equals resolution in microns (the radius of the Airy disk):

$$r = 1.22\lambda/2 \times \text{NA (Objective)}. \tag{2}$$

λ is the wavelength of the emission photons, and NA is the numerical aperture of the objective lens. Using Eq. (2) for an objective lens with a numerical aperture of 1.40, the maximum resolution of an ideal diffraction-limited optical system imaging with 605-nm emission light is 264 nm or 0.26 μm. Consequently, imaging at 457 nm in reflection with the same 1.4-NA objective would produce a resolution equivalent to 0.20 μm. When a distance greater than the Rayleigh criterion separates subresolution objects, contrast increases, as the space between them becomes dark. When the objects are brought closer together, contrast begins to decrease as the Airy disks generated by each of the objects overlap. As the objects become closer than the Rayleigh criterion, the space between them is sufficiently filled with light, and they may appear as a single elongated object (Larson, 1999a).

2. Axial (Z–Axis) Resolution

Objects imaged with a microscope also occupy space along the axial or Z-axis. That is, they have dimension when a stack of images is collected. Even subresolution objects will occupy space in a number of images within the stack or Z-series if the step size is fine enough. The parameter measured in fluorescence microscopy is section thickness or depth, which is the same as measuring the point-spread function (PSF) of a subresolution object in three dimensions. If a subresolution object such as a 0.1-μm fluorescent bead is imaged sufficiently at high magnification, with small enough pixels in the detector, an image resembling an Airy disk appears. An X/Z projection of a slice through the center of the Z-series acquired using a wide-field microscope with suitable detector through the same 0.1-μm bead in 0.1-μm steps a few microns above and below the bead produces an image resembling an hourglass. The X/Z projection of a subresolution spherical object is elongated in Z and is always many times higher than it is wide at its waist, even in an ideal optical system. The image of the bead appears diffuse and spread out in X and Y in sections above and below it. Aberrations and other optical imperfections may elongate the point spread even more, especially in Z. Photons originating from secondary objects above and below the plane of the primary object will contribute to background in that part of the image if they are close enough laterally. Since there is no way to exclude them, all photons collected by the objective lens not emitted by the object of interest contribute to background. For practical purposes, the sectioning depth in the Z-axis in wide-field fluorescence microscopes can be no better than five times the Airy disk diameter. And there is still no guarantee that the object of interest will be imaged with sufficient contrast to be seen over background (Larson, 1999a).

B. Optical Resolution in Confocal Microscopes

It is widely accepted that confocal microscopes improve lateral resolution by excluding light from outside the zero order (central spot in the Airy disk) with an appropriately sized pinhole (Sheppard and Wilson, 1984; Wilson and Sheppard, 1984; Wilson, 1990; Boxall *et al.*, 1993; Kitagawa, 1993; Wright *et al.*, 1993; Piston, 1998; Periasamy *et al.*, 1999). The theoretical improvement in lateral resolution can be as great as 40% if the pinhole size selected is exactly optimal and optical aberrations or distortions do not degrade the image. In practice, the improvement in resolution is most often less. By excluding light from beyond the zero order, smaller subresolution objects can be detected with a confocal than with a wide-field microscope. The most dramatic improvement in resolution, however, is in the images from section planes above and below the object. Unlike the hourglass-shaped cross section in Z created by a wide-field microscope, the pinhole of the confocal microscope creates an elliptical cross section in Z.

A confocal microscope improves axial or Z resolution the same way it improves lateral resolution. Photons originating outside the ideal focal plane are blocked by the pinhole. When a high-numerical-aperture objective lens is used, objects appear three to four times thicker in Z than their X and Y dimensions when imaged by a confocal microscope with the correct pinhole aperture selected (Larson, 1999a). Confocal axial resolution was reviewed extensively by Inoué (1995), and it is generally accepted that axial resolution is about half that of the lateral resolution with all parameters remaining the same (Girard and Clapham, 1994). This represents a significant improvement in optical sectioning capability compared with wide-field fluorescence microscopes. It is all the more significant because by excluding photons from outside the ideal focal plane, the signal-to-noise ratio is also improved dramatically. Photons originating from nearby points in the sample do not degrade contrast the way they do in wide-field fluorescence microscopy.

Carlsson and Liljeborg (1989) discussed the design of their confocal system which used two pinhole diameters, i.e., 25 and 50 μm, which were interchangeably placed in front of a photomultiplier tube (PMT). They commonly used the 50-μm pinhole which was somewhat smaller in diameter than the Airy disk of the $40\times/1.0$ and the $63\times/1.4$ objectives frequently used. The 50-μm pinhole corresponded to about half the size of the Airy disk of the $100\times/1.3$ objective used. Carlsson and Liljeborg (1989) observed no improvement in the Z-axis resolution when using the 25-μm pinhole. Wilson and Carlini (1987) explained that a reduction in Z-axis resolution would be expected when the pinhole aperture became larger than approximately two-thirds of the Airy disk diameter. Refer to Larson (2000) for a complete guide to pinhole size selection in relation to selected Nikon CFI60 objectives suitable for use with the Nikon PCM 2000 confocal imaging system with similar pinhole diameters.

Many factors influence resolution in epifluorescence confocal systems (Oldenbourg *et al.*, 1993). Three of the most important parameters generally accepted for achieving the best resolution are: (1) the highest objective NA possible, (2) the smallest pinhole aperture size so it adequately fills the objective back aperture, and (3) the shortest emission wavelength. In applications using brightfield confocal reflection imaging techniques, better resolution would be achieved from the shortest laser excitation wavelength used.

C. Electronic Scan Zoom and Digital Resolution

Up to this point, we have dealt only with resolution as it relates to the image formed at the primary image plane. In microscopy, we never directly use the image formed at the primary image plane. Instead, eyepieces are used to relay it to our retinas. We also project it on film, or we place a detector at the image plane and, with a monitor, view the image the detector produces. The concept of digital resolution applies to the image produced by the detector. Although this discussion applies specifically to the Nikon PCM-2000 confocal imaging system, many of the concepts presented here can be generalized to other manufacturers' digital imaging technologies.

The PCM-2000 produces a high-resolution 1024×1024 pixel image. An image may be captured from a field measuring approximately 200 μm by 200 μm when using a $60\times$ objective lens with the electronic (optical) scan zoom set to default at $1\times$. This translates into a pixel size of 0.2 μm. Most paired-galvanometer mirror scanning confocal systems have a zoom capability to $10\times$ in increments of 0.1 such as in the PCM 2000. This seems like a perfect combination for high-resolution imaging; however, as Larson (1999a) pointed out, image quality is less about the bright parts of the image than it is about the spaces between them. In digital as in optical resolution, our ability to differentiate between features in the specimen depends on our ability to separate them in space. This means that if nearby objects fluoresce at the same intensity, there must be some space between the two objects visualized in the image at a lower gray level than the objects of interest for them to be distinguishable as two separate objects. Unfortunately, with a pixel size of only 0.2 μm, two subresolution structures separated by 0.2 μm will be almost indistinguishable in the final image. They will look like a single ovoid structure and will not be resolved (Larson, 1999a).

The pixel size must be reduced sufficiently so that at least one pixel at a measurably lower intensity separates the objects of interest. For practical purposes, a pixel 2.3 times smaller than the optical resolution limit of the system is needed to digitally resolve that separation. This is known as the Nyquist criterion. In the case of the PCM-2000 with a Plan Apochromat $60\times/1.4$ oil immersion, the Nyquist criterion can be accomplished if the electronic zoom is set to $2.3\times$. The resulting pixel size, 0.087 (0.09) μm, meets the requirement (Larson, 1999a). Refer to the bulletin prepared by the Nikon BioScience Department (1999) for the minimum magnification recommendations for digital imaging.

At 0.2-μm pixel size, it is impossible to digitally resolve a 0.2-μm separation between two objects, no matter how the image is digitized. At 0.1-μm pixel size, the separation may or not be resolved, depending upon where the objects lie in the digitized field. With the Nyquist criterion satisfied by a 0.09-μm pixel size, the separation is resolved no matter where the objects lie in the field. Increased digital resolution does not come without cost. The electronic scan zoom produces an increase in digital resolution by scanning a reduced field in the same 2 s it takes to acquire an image at a zoom of $1\times$. The image acquired from this field is spread over the same 1024×1024 pixels. In the case where a $60\times$ magnification objective lens is used on a PCM-2000 at $2.3\times$ electronic scan zoom, the laser illuminates a field 90 μm on a side. All points in that reduced field are illuminated nearly five times longer than at $1\times$ electronic scan zoom, resulting in much greater localized photodamage.

D. Point–Spread Function

The point-spread function (PSF) is a fundamental measurement characterizing any digital imaging system whether it be wide-field deconvolution, confocal, or multiphoton. The resolution of a refractive system is set by its PSF whereby the central zone of a PSF is the Airy disk, and the distance to the first minimum is the index of refraction for the object (Fine, 2000). The measured PSF can be used to improve the estimate of an object's structure provided by the observed image (Shaw and Rawlins, 1991). The PSF must first be calculated from theory or directly determined to reverse distorting effects from the microscope (Keating and Cork, 1994). The PSF volume measurement not only serves as a benchmark for evaluating the optical properties of an optical imaging system by establishing the theoretical resolution; it also provides input data for computational inverse algorithms (Diaspro *et al.,* 1999a).

Ideal objectives for PSF examinations should be unaffected by spherical and chromatic aberrations as much as possible. Typically used objectives are high-NA immersion lenses with long working distances such as a CFI*60* Plan Apochromat 60×/1.4 oil immersion or a CFI*60* Plan Apochromat 100×/1.4 oil immersion with respective working distances of 0.21 mm and 0.13 mm. Depending upon the specimen medium, a CFI*60* Plan Apochromat 60×/1.20 WI (water immersion) with a working distance of 0.22 mm may prove to be more useful with beads mounted in water. Point-spread functions are usually measured by imaging a subresolution fluorescence bead (coated with fluorescein or other fluorochromes) through a Z-series at approximately 0.10- or 0.2 μm steps beginning from just below the bead and sectioning all the way through to just above the bead. The data set can then be analyzed with three-dimensional software, and an X-Z profile can be generated displaying the PSF. The optical transfer function (OTF) of an objective lens can later be derived from the PSF calculations (Shaw and Rawlins, 1991). Shaw and Rawlins (1991) also found it to be more beneficial to take numerous scans with restricted laser power over fewer scans with increased laser power. They presumed that lower laser power was more effective because of nonlinearity of fluorescence emission (fluorochrome saturation effects).

In a comparison study of deconvolution, two-photon, and confocal imaging techniques, Periasamy *et al.* (1999) used a Nikon CFI Plan Fluor 20×/0.75 multiimmersion (water and oil) objective lens for image acquisition. Periasamy *et al.* (1999) stated that it was important to adjust the illumination by adjusting the laser power level depending upon the specimen, preparation, and the depth of penetration.

When working with a deconvolution imaging system, a point-spread function should be calculated for each objective versus working distance and specimen medium for acurate results. When measuring PSF in a confocal microscope, in addition to the above criteria, varying pinhole size should be taken into consideration. The role of the pinhole size becomes important with low fluorescence light levels (Wilson, 1990).

E. Signal–to–Noise Ratio

One of the most important considerations in confocal microscopy is the emission signal intensity since very few photons reach the detector. Confocal digital imaging depends upon a pinhole of varying diameter or a selection of fixed pinhole diameters to restrict out-of-focus information or signal intensity. Therefore pinhole diameter size, optical aberrations, and scattering can all reduce the intensity signal (Fine, 2000). In addition, random photon flux fluctuations can overpower the signal, reducing image contrast and resolution.

Two types of noise have been described. The first is "shot noise" which results from statistical variations in the number of detected photons (Sheppard *et al.*, 1995). The second source of noise is "background noise" emanating from stray light in the measured signal. A source of background noise can be autofluorescence from the cements and coating of the optical components of the objective lens selected. This may be more prominent in older generations of objective lenses. Additional background noise can also be generated from uneven laser light sources.

It becomes imperative to increase the signal and reduce the noise, thereby placing a lot of emphasis on the signal-to-noise (S/N) ratio. Fine (2000) pointed out that intrinsic photon noise increases with the square root of the mean light intensity. The S/N ratio can be improved by increasing the signal light intensity. This may be accomplished by averaging multiple frames, optimizing the optical microscope components, i.e., objective selection, varying scan speeds which determine the pixel dwell time, increasing the diameter of the pinhole, using higher dye concentrations, and increasing laser illumination. Increasing the laser intensity should always be done with discretion as increased photobleaching and live-cell phototoxicity may occur.

Proper selection of excitation light intensity, PMT detector sensitivity, pinhole size, and the concentration of the probe within the specimen gave a confocal signal-to-noise (S/N) ratio superior to that of wide-field microscopy (Periasamy *et al.*, 1999).

III. Infinity versus Finite Lens Systems

In a finite optical system, after light from an object passes through the objective, it is directed toward the primary image plane located at the eyepiece focus plane and converges there. In an infinity optical system, however, light becomes a flux of parallel relays after passing through the objective and does not converge until after passing through a tube lens. This does not mean, however, that there is an infinite space available inside the optical system (Nikon Corporation, 1997; Nihoshi, 2000). However, the four major microscope manufacturers correct for infinity in somewhat different ways.

All infinity systems use an objective and a tube lens. The infinity space and focal lengths are different, i.e., Nikon (200 mm), Leica (200 mm), Olympus (180 mm), and Zeiss (160 mm). The correction for chromatic aberration in the Nikon CFI*60* objectives is done completely in the objective, and the tube lens is neutral. Both Leica and Zeiss remove varying percentages of residual chromatic aberration not compensated for by their objectives through their tube lens design.

The major microscope manufacturers continually strive to improve optical and objective lens performance with optimum flexibility. Detailed reviews of microscope objective lenses from various manufacturers can be found in the literature from Keller (1995), Inoué and Spring (1997b), Piston (1998), and Steyger (1999). Heretofore, there has not been

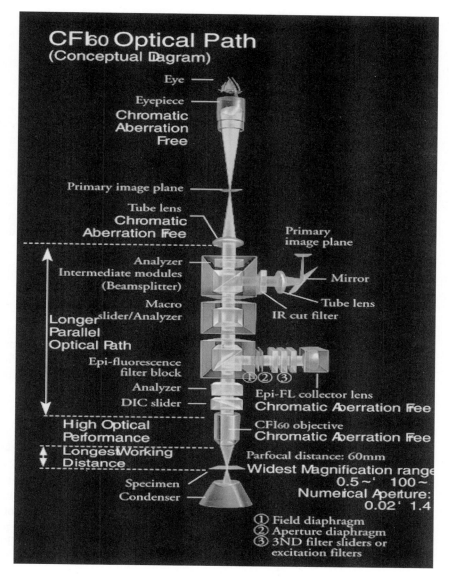

Fig. 1 Nikon CFI optical path (conceptual diagram). Courtesy of Nihoshi (2000). (See Color Plate.)

any detailed information published discussing the Nikon line of objective lenses. It is therefore appropriate to review in some detail the new infinity line of objective lenses offered by Nikon. Nikon Corporation has achieved optimum flexibility and performance by creating the CFI*60* Eclipse Series of objectives.

Nikon had two distinct goals with the creation of the CFI optical system (Fig. 1; see color plate) for advanced biological research microscopes which are used in both confocal and multiphoton imaging applications. The first goal was to improve optical performance, and the second to boost the overall flexibility as a system when various microscope attachments were added to the system (Schwartz, 2000). By using a tube lens focal length of 200 mm and objectives having a parfocal distance of 60 mm with a larger thread diameter of 25 mm, Nikon succeeded in realizing both higher NA and longer working distances over the entire magnification range. A detailed description of Nikon's new CFI60 infinity optical system was presented by Nikon Corporation (1998).

Both axial and lateral chromatic aberrations have been corrected independently in the CFI60 objective and the tube lens. The 200-mm lens creates a smaller angle between light rays passing through the center and those off axis. This minimizes shifts in light rays on the image plane between the center of the field of view and its periphery, dramatically reducing blurring during DIC and epifluorescence wide-field microscopy. This technology can only help to improve confocal imaging. The CFI optical design has made possible the creation of 0.5× and 1× ultralow magnification objectives providing a range from 0.5× to 100× magnifications with numerical apertures from 0.02 to 1.4.

IV. Objective Lens Nomenclature and Features

A. Types of Objectives

There have been so many improvements and corrections in objective lens manufacturing in recent years that some of the original standard definitions of various objective types do not necessarily hold true now. The classical definition of the objective correction is given plus the current improvements from Nikon Corporation for each objective type (Nikon Corporation, 1999).

1. Achromat

Early achromat objectives consisted of two or more lens elements usually made of crown glass (converging thin edge) and flint glass (diverging thick edge). Crown optical glass has a low dispersion and usually a low refractive index. Flint glass is a heavy, brilliant optical glass with high dispersion and usually a high refractive index. In the classic achromatic objective, chromatic aberration has been corrected for two colors (red and blue), and spherical aberration has been corrected for one color (green). The correction for chromatic aberration has been dramatically improved in the CFI60 Achromat Series of objectives and are at the same correction levels as the Plan Achromat Series. The CFI60 Achromat Series has also been corrected for spherical aberration (two colors) and coma, and the image flatness is excellent for standard wide-field eyepieces with a 22-mm field of view.

2. Plan Achromat

In the classic design, the plan achromat objective would have the same chromatic and spherical aberration corrections as the classic achromat objective. In addition it would be corrected for curvature of field, an aberration where the surface in focus appears as a curved surface and not a plane. The edges of the field of view would appear blurred when the central portion of the field is in focus. The term "plan" has been applied to signify that the objective was further corrected so that the field of view was flat from edge to edge.

The CFI*60* Plan Achromat Series provides image flatness over the entire 25-mm field of view with a 10× ultrawide-field eyepiece. These objectives are ideal for brightfield observations and wide-field photomicrography.

3. Plan Fluor

Fluorite objectives were manufactured with a percentage of crystal, fluorite, or a synthetic optical component combined with optical glass to further improve upon chromatic (red, blue, and green) and spherical (two colors) aberrations. These objectives were found to provide a higher percent of transmission in the ultraviolet (UV) range and be optimal for fluorescence examinations. Different manufacturers apply various descriptive names to these objectives such as Fluor, Plan Fluor, Plan Fluotar, Neofluar, or Plan Neofluar. Again the term "plan" shows the added correction for curvature of field. Fluorite objectives were sometimes called semi-apochromats (Ploem and Tanke, 1987).

The CFI*60* Plan Fluor Series features an extra high transmission rate especially in the UV down to 340 nm and a flatness of field comparable to that of the CFI*60* Plan Achromat Series. With these improvements to the classic fluorite objective, the CFI*60* Plan Fluor Series are outstanding for brightfield, fluorescence, and DIC observations with ultrawide-field eyepieces (25-mm field of view).

The CFI*60* Plan Fluor 20× MI (multiimmersion) objective has a high numerical aperture (NA 0.75) and superior contrast. The use of immersion media helps eliminate surface reflections and provides an image with higher contrast. This objective has a correction collar which allows it to be used with oil, glycerin, and water (with or without a cover glass). It is therefore also ideal for brightfield fluorescence, DIC, confocal, and multiphoton microscopy.

4. Plan Aprochromat

In the classic description, apochromat has always been associated with an objective of highest quality. These objectives are chromatically corrected for three wavelengths (red, blue, and green), and spherical aberration has also been corrected for three colors. The term "plan" indicates that field curvature has also been corrected. These objectives traditionally have the most complex construction with increased lens elements and the highest NA demanding the highest price within a manufacturer's line of objectives.

The CFI*60* Plan Apochromat Series feature objectives with longer working distances and high numerical apertures and have been designed to correct for all optical aberrations through the visible spectrum from violet to red from center to edges across the entire 25-mm field of view. These objectives have superior image flatness, color reproduction, and resolving power at the theoretical limits of today's optical technology. They are suitable for use with ultrawide-field eyepieces. Thus with high NA and longer working distance, these objectives are ideal for brightfield, fluorescence, DIC, confocal, and multiphoton microscopy.

5. Super Fluor

The CFI*60* S (Super) Fluor series is a new line of objectives offered by Nikon manufactured with quartz and special glass components. The CFI*60* S Fluor Series ensure the highest transmission rate of UV wavelengths down to 340 nm. These objectives have an improved signal-to-noise ratio for the shorter wavelengths and high numerical apertures producing sharper and brighter fluorescence images. Because of glass configurations and the manufacturing process, chromatic aberration is highly corrected with this series of objectives. They are designed for maximum transmission and are flat in the center. These objectives are extremely efficient through a wide range of wavelengths making them ideal for fluorescence, UV confocal imaging, and multi-photon.

6. Water Dipping

Most manufacturers today offer a series of objectives for water immersion or water dipping. Water dipping objectives do not require that the specimen be observed with a cover glass, thus making them highly desirable for the study of live cells and physiology. The CFI*60* Fluor Water Dipping Series have been designed for physiological applications without cover glasses. They have a narrow top with a 2.0-mm working distance (WD) to enable microelectrodes to be applied to the specimen at a 45° angle. The objective nose cone is constructed of ceramic and glass to isolate them thermally, chemically, and electronically. Such fluor objectives have a very high transmission rate in the UV and IR wavelengths and exhibit a high signal-to-noise ratio. They are ideally designed for live specimen experiments in brightfield, fluorescence, confocal, and multiphoton microscopy.

B. Magnification and Color Code

Each objective has clear markings identifying the magnification of the lens. In addition, Nikon displays a color-coded ring on the objective barrel identifying the magnification as follows:

Magnification	1×	2×	4×	10×	20×	40×	50×	60×	100×
Color code	Black	Gray	Red	Yellow	Green	Light blue	Light blue	Cobalt blue	White

C. Numerical Aperture

The numerical aperture (NA) is the most important factor in defining the performance characteristics of an objective lens as shown in Eq. (3),

$$NA = n \sin O \tag{3}$$

where n is the refractive index (RI) of the medium between the specimen and the objective at d-line (587 nm). O equals half the cone angle of light emitted from the condenser and accepted by the objective and is said to be the half-angle of the incident rays to the top lens of the objective. For a dry objective lens, $n = 1.000$ (air). For a typical oil immersion objective lens, $n = 1.515$ (oil). For a water immersion objective lens, $n = 1.360$ (water) at 38°C. The higher the NA becomes, the higher the resolving power of the objective, when the resolving power is defined as the power to recognize two points. The higher the NA, the brighter the image acquired. In turn, the higher the NA, the shallower the depth of focus on a detector.

D. Working Distance

Working distance (WD) defines the maximum distance between the last lens of the objective and the surface of a 0.17-mm cover glass. The WD defines the amount of travel distance in millimeters over which the objective may gather information (photons) beneath the cover glass. For example, an objective lens with a WD of 0.22 mm (220 μm) would travel 220 μm downward beneath the cover glass into the specimen of interest before the objective lens top element would come to rest on the upper surface of the coverslip. Piston (1998) discussed the importance of magnification, numerical aperture, and working distance for digital optical microscopy.

E. Correction Collar

Dry objectives with high numerical aperture are susceptible to spherical and other aberrations which can impair resolution and contrast when used with a cover glass whose thickness differs from the specified value. A 1.5 cover glass (0.17 mm thick) should be used as standard; however, not all 1.5 cover glasses are exactly 0.17 mm, and many specimens have media between them and the cover glass. The correction collar is used to adjust for these subtle differences to ensure the optimum objective performance (Nikon Corporation, 1998). The rotation of the correction collar changes the position of some of the internal lens components, compensating for spherical aberrations introduced by inconsistencies in cover-glass thickness and mounting-medium thickness.

F. Cover Glass and Cover-Glass Correction

For optimum performance, the thickness of the cover glass should be 0.17 mm (170 μm). Even if the correct nominal thickness of cover glass is selected, there can be large variations between cover-glass thickness from the same box. In practice, the actual thickness of cover glasses can vary by more than 40 μm (Dunn and Wang, 2000). For example, at NA = 0.95, a 0.01-mm difference in the thickness reduces image formation by 45% from the ideal image.

G. Retraction Stopper

Some objectives for oil immersion have a retraction stopper. In order to prevent clean slides from being accidentally smeared with immersion oil, the retraction assembly can be engaged by pushing in the front element support housing and twisting it to the right. This will lock the objective front element in the up position so it will not leave immersion oil on a clean slide as the nosepiece is rotated. Twisting to the left will release the retracted objective for use (Nikon Corporation, 1998).

H. Application Markings

Most objectives display an application code. Nikon uses this marking to differentiate DIC and phase contrast objectives. An objective designated DIC has been selected to be used for differential interference contrast. Phase contrast objective lenses are marked as follows: (1) DM, dark contrast middle type, (2) DL, dark contrast light type, and (3) DLL, lower contrast type.

Most microscope manufacturers offer a wide variety of objectives. To assist the user, Nikon and the other microscope manufacturers clearly mark each objective with detailed information on the objective barrel (Fig. 2; see color plate) assisting in its proper use.

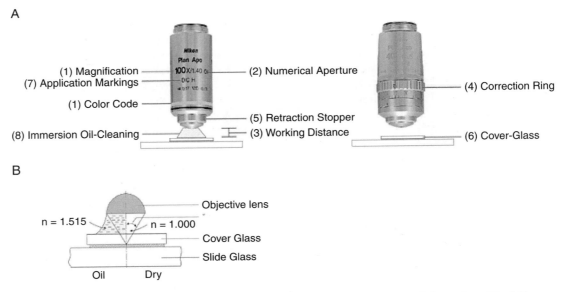

Fig. 2 (*Top*) Picture of two objective lenses showing all markings. (*Bottom*) Comparison of dry (air) versus immersion objective in relation to specimen slide and cover glass. Courtesy of Nihoshi (2000). (See Color Plate.)

V. Objective Lens Aberrations

It is extremely important to use an objective lens under the precise optical conditions for which it was designed (Brakenhoff *et al.,* 1990). Most importantly, the use of the correct imaging refractive index medium can be critical. In some biological applications, the precise refractive index may be uncertain and may vary along the optical path (Visser *et al.,* 1990).

All lenses suffer from aberrations, which if not corrected to some degree would prevent imaging an exact replica of the object. There are six basic aberrations effecting the microscope objective lens. Spherical and chromatic aberrations affect the entire field, while coma, astigmatism, field curvature, and distortion are more prominent in off-axis image points. A good description of the latter four aberrations was presented by Bradbury (1984), Wilson (1990), and Keller (1995). Spherical and chromatic aberrations play a more important role in confocal and multiphoton digital imaging. Refer to Hell and Stelzer (1995) for a detailed review of the aberrations found in confocal fluorescence microscopy. In order for a microscope to function properly, however, all of these aberrations must be corrected as much as possible. There is really no cure for problems with chromatic correction other than by selecting the appropriately corrected lens. However, in the case of spherical aberration introduced by the specimen thickness and mounting medium, this can be corrected to some extent by using correction collars, multiimmersion lenses, and water immersion lenses.

A. Spherical Aberration

Since the surfaces of lens components, which make up an objective, are spherical, light rays passing through marginal zones will be refracted more than those passing through the middle zones, and in turn, will be refracted more than those passing through the axial zones (Bradbury, 1984). The result is that a point source will be imaged from each zone at an increased distance from the lens causing a blurring of the image from that point source of light. Spherical aberration results where the axial and peripheral rays are focused differently, whereby light is focused to a broad region rather than to a single point (Dunn and Maxfield, 1998). Besides the components inherent in the objective itself, spherical aberration can be generated by changing an objective's designated tube length within the microscope (Keller, 1995).

High-NA dry objectives are optimized for observations immediately below a 0.17-mm cover glass, and if the specimen is mounted deep beneath the cover glass, spherical aberration will become prominent (Matsumoto and Kramer, 1994). Hale and Matsumoto (1993) also reported that factors such as immersion medium, cover-glass variations, the specimen mounting medium, and the tissue itself may have profound influences on the amount of spherical aberration present in an optical system. Cogswell and Sheppard (1990) found that improvements to spherical aberration could be realized by inserting minimal correcting lenses in the optical beam path.

However, the principal source of spherical aberration in high-resolution studies is the difference between the refractive indices of the immersion and the mounting media, i.e., oil immersion objective and an aqueous solution of living cells (Hell *et al.,* 1991). When this mismatching of oil and water indices occurs, the full NA of an oil immersion objective is not realized; i.e., 1.4 NA is reduced to 1.33 NA (Brenner, 1999). Since water has a significantly lower refractive index than most immersion oils, spherical aberration has limited the depth of confocal examinations of live tissues until the introduction of highly corrected, high-NA water immersion objectives.

With such water immersion objectives enabling refractive index matching in aqueous samples, sampling depth has become independent of spherical aberration. Working distances of such water immersion objectives allow sampling to depths greater than 200 μm. Dunn and Wang (2000) provided data comparing sampling depth using oil and water immersion objectives while examining endosomes in aqueous solution.

Since spherical aberration is a major optical problem in confocal digital imaging of living cells with oil immersion objectives, and spherical aberration increases proportionately with depth into the aqueous medium and cellular material, it is highly recommended to switch from using an oil immersion objective to a water immersion lens when imaging specimens suspended in an aqueous solution greater than 15 μm beneath the cover glass. Cannell and Soeller (2000) reported strong signal decay due to refractive index mismatch when using an oil immersion objective while collecting calcium spark data when imaging at depths greater than 30 μm in an aqueous solution. The effects of spherical aberration can be observed as (1) a loss of intensity and contrast, (2) an inability to collect and resolve small spatial frequencies, and (3) reduced accuracy of reproduction in an optical system, all in direct proportion to the distance from the cover glass.

B. Chromatic Aberration

An ideal lens would focus all colors to a single point. However, the RI of glass varies with the wavelength of light. The RI for blue is greater than that for green, and green has a greater RI than red. This is called dispersion. Light originating from an axial point will not come to a common focal plane. The blue rays will become focused closest to the lens, and the longer wavelengths will become focused farther from the lens. This phenomenon, axial chromatic aberration, can be noticed as fringes of varying colors as one focuses through the image when viewing through the eyepieces.

In multicolor confocal microscopy, axial chromatic aberration results from different excitation wavelengths being focused to different points in the sample, whereby different emission wavelengths are gathered from different points in the sample. This aberration is generally found throughout the field of view. This can have a profound effect on confocal microscopy by limiting the instrument's ability to demonstrate colocalization in the Z-axis. The correction is usually accomplished by combining a few to several lens elements of varying design during the manufacture of the objective.

Even after correction, some objectives may still show some degree of aberration; this is called lateral chromatic aberration or chromatic difference of magnification. This becomes noticeable from off-axis points and may be present as a result of the path length in one glass surface to another (Bradbury, 1984). Chromatic difference of magnification must be corrected in the lens system to produce an equal and opposite chromatic difference of magnification in the final image formed. This aberration can be minimized by collecting data from the center of the microscope's field, which is commonly done in "off-axis" beam laser scanning confocal microscopy. Care should be taken when panning or scanning toward the periphery.

VI. Dry (Nonimmersion) and Immersion Objective Lenses

A. Dry Objectives (NA < 1.00)

There are a wide variety of dry objective lenses (objectives with a NA of less than 1.00) available to choose from. These objectives are designed to be used with air between them and the cover glass. The highest achievable NA for a dry objective is 0.95, and high NA objectives may have shorter working distances and be more sensitive to spherical aberration. A Plan Apochromat 40×/0.95 objective may have a correction collar for cover glass thickness inconsistencies enabling the lens to work around the effects of spherical aberration. Generally a low NA dry objective is selected for lower power observation to locate a specific region of interest. However, working distance, desired magnification, NA and resolving power, and the nature of the specimen (such as live cell versus fixed tissue) may dictate which objective is actually the best choice.

B. Oil Immersion Objectives

Higher numerical aperture immersion objectives collect more light than dry objectives. This is due to the reduction of light scattering brought about by the air-to-cover-glass interface when an immersion medium is used. Many objectives have been designed to be used with a medium other than air between the objective front element and the cover glass. Typical immersion media are oil, glycerin, and water. By using immersion media, the NA of the objective can be increased thus improving the maximum resolving power of the objective. It becomes very important to match the refractive index (RI) of the immersion medium with that of the mounting medium beneath the cover glass to prevent the image distortions brought about by spherical aberration due to variance in cover glass thicknesses. Most immersion objectives are designed to be used with an immersion oil with a RI and dispersion of that approximating the cover glass (RI = 1.515).

C. Cleaning Objective Lenses

It might be appropriate to review the proper technique for cleaning objective lenses. Herman (1998) recommended the following: (1) work in a clean environment, (2) keep the confocal-microscope system covered when not in use, (3) avoid optical component contact with dust, moisture, salt water, and corrosive solvents at all times, and (4) do not smoke. An objective lens may be inspected with the use of an inverted ocular (eyepiece); this technique provides a clear view of the objective top lens and any droplets, lint, or dust particles upon the lens surface.

Care should be taken not to touch the top lens of a dry objective at any time. Dust particles in the air or on the top lens surface can cause scratches when rubbed even with lens paper. It is always recommended to use a solvent for cleaning. Herman (1998) listed such cleaning agents as an acetone–ether–ethanol mixture, pure alcohol, Sparkle, distilled water, or a mixture of cloroform–alcohol–water. Extreme caution should be taken when using highly flammable and potentially carcinogenic solvents such as benzene, toluene, or xylene.

After using immersion oil, blot the lens dry with lens tissue. Then slightly moisten a piece of lens tissue with a selected cleaning fluid (each manufacturer will recommend a solvent of choice) that will dissolve or disperse what resides on the top lens of the objective. Clean off all traces of the oil from the immersion objective. It is important not to use a solvent that will dissolve the cement maintaining the lens internal elements making up the objective lens. Please remember that some solvents listed above may be considered to be carcinogenic. Cleaning is also essential for water immersion objectives as well. After use, blot the water off the top lens with lens tissue (Nikon Corporation, 1998).

D. Working with Spherical Aberration

The major problem when using a high-resolution, highly corrected, oil immersion objective is that the deeper into the specimen and aqueous medium one images, the more spherical aberration occurs since oil immersion objectives are optimized for focusing directly beneath the cover glass. Focusing deep may greatly affect the epifluorescence intensity. Often the cellular details or events one wishes to record are more than 20 μm deep. This is especially true for many confocal and multiphoton applications. Therefore, spherical aberration occurs with oil immersion objectives when focusing deeper along the z-axis into living cells (Brenner, 1999).

Inoué (1989) discussed the effects of spherical aberration on images of cilia from sea urchin embryos, whereby structures could easily be misinterpreted. Light loss due to spherical aberration can be quite substantial. This inability to get sufficient light to the detector can become a major limitation in confocal and multiphoton imaging of live cells.

There are two types of objectives manufactured for working with specimens in a water environment. The first is a water immersion objective where water is used as the immersion medium instead of oil. A drop or two of water is placed between the cover glass and the objective. These objective lenses are highly effective for specimens which have a refractive index similar to that of water, such as live cells or embryos. The second type of water objective is the dipping objective lens. The dipping objectives are tapered, corrosion-free lenses with very long working distances and are dipped directly into the water medium.

1. Water Immersion Objectives

Highly corrected water immersion objectives are now often required to perform combinations of techniques. The Nikon CFI*60* Plan Apochromat 60×/1.2 WI objective with 220-μm working distance is a good example of a highly corrected water immersion objective. The objective has a correction collar that accommodates for cover glasses from 0.15 mm to 0.18 mm in thickness. In addition to eliminating spherical aberration problems, the lens is chromatically corrected and has high transmission in the near ultraviolet through the red spectrum, making it useful for many confocal, multiphoton, and wide-field microscopy applications. Therefore, consider the possibility of using a highly corrected water immersion objective in order to acquire high-resolution images from specimens several to many microns thick or in an aqueous medium.

2. Dipping Objectives for Live Cell Physiology and Biophysics

There are also other considerations since the demands of research get more complex with every passing year. In order to work with cells *in vivo* or *in vitro*, the maintenance of normal physiological conditions is mandatory. The cells or tissue must be held in a chamber or vessel surrounded by physiological solution. Sometimes this requires physiological dipping objectives with long working distances.

It should also be mentioned that mechanical requirements, such as chambers, micromanipulation, and physiological requirements of the specimen, must also be considered. Physiology water dipping objectives should be considered for some experiments which require the specimen to be maintained in aqueous solution. The ability of a narrow-nose dipping objective to get into small chambers and petri dishes can be as important a factor as its ability to produce high-resolution images (Brenner, 1999).

In this case the objective is dipped directly into the physiological solution without a cover glass, and numerical aperture is sacrificed in favor of working distance. Nikon's CFI60 Water Fluor 60×/1.0 with a 2-mm working distance is a pretty good compromise for confocal, multiphoton, and epifluorescence microscopy.

3. Using the Objective Correction Collar

Another way to reduce the effects of spherical aberration when using a high-magnification, high-NA dry lens is to use a lens with a correction collar. A correction collar can compensate for cover-glass thickness variances and for spherical aberration as one focuses deeper into a specimen. The proper method for using a correction collar is as follows:

(a) Bring the specimen into focus with the correction collar set at 0.17, since the thickness of the standard cover glass averages 0.17 mm (170 μm).

(b) Turn the correction collar to a point near its limit.

(c) Refocus on the specimen.

(d) Pick out a group (of at least two) closely spaced fine structures.

(e) Rotate the correction collar by a small amount and refocus the lens.

(f) Observe the structures selected and look for an improvement in your ability to see them.

(g) Do not be alarmed if you cannot see a change.

(h) Repeat steps (e) and (f) until the structures appear sharp and at high contrast.

(i) Continue repeating steps (e) and (f) until you see the first signs of image degradation.

(j) Turn the correction collar in the reverse direction in increments of 1/2 to 1/3 until optimum resolving power and contrast of the structures are achieved.

E. Multiimmersion Objectives

One should also be aware of the existence of multiimmersion objectives. These high-NA lenses operate with water (CG and NCG), glycerin, and oil. Applications are wide ranging from neurobiology and embryology to developmental biology and microcirculation. Using the glycerin immersion instead of the oil immersion mode can eliminate the problem caused by differences in refractive indices between mounting and immersion media (Hale and Matsumoto, 1993). The same high-numerical objective lens would also provide a water immersion mode which would eliminate refractive indices differences when examining living specimens in aqueous solutions. The magnification and NA of multiimmersion objectives range widely among the microscope manufacturers. Nikon's multiimmersion offering for confocal microscopy is a Plan Fluor 20×/0.75 MI (W,G,O).

F. Practical Applications

In some circumstances it might be easier to examine one's specimens using the same immersion medium for three different objective lens magnifications. From a convenience point of view, the medium is not changed, whereby a confocal examination could be conducted using a 20× oil, a 40× oil, and a 60× oil immersion objective in sequence. It is sometimes more efficient to use multiple objective lenses with increasing magnifications than to rely on the confocal electronic scan zoom. By doing so, one can usually benefit by increased NA with each objective as the magnification increases. In addition, when there are time restraints, one never has to worry about dipping a dry objective into oil while changing from one objective to another. The same can be applied to water immersion objective lenses. However, when using all oil immersion objectives, spherical aberration must always be considered to be a potential problem when examining deep into the specimen, especially in an aqueous medium.

VII. Epi (Reflected Light) Objectives

The removal of out-of-focus flare caused by fluorescence background and the ability to obtain optical sections through thick specimens are the main reasons confocal imaging systems are used in biological applications. Improvements in resolution become secondary in importance. However, improvements in resolution capabilities are of ultimate importance in the use of confocal imaging systems for measuring in semiconductor and other industrial applications (Larson, 1998).

In industrial applications, objective lenses are known as reflected-light, brightfield, brightfield/darkfield (BD), brightfield/darkfield/DIC (BDDIC), epi-plan reflected, or industrial-type objectives and are supplied by the major microscope manufacturers. "Epi-" objectives are similar to their biological counterparts, except they are corrected for use without a cover glass, may have special antireflection coatings, and in some cases have longer working distances.

Nikon introduced an infinity line of objective lenses for the industrial line of microscopes in 1993. Nikon later introduced the Eclipse Series of microscopes with the CFI optical system in 1996 directed predominantly at the life science market place. Harrison (1999) reviewed the infinity line of industrial objectives and explained depth of field and field of view with Eqs. (4) and (5),

$$\text{Depth of field in microns} = \frac{n \times \lambda}{2 \times \text{NA}^2} + \frac{n \times 1000}{7 \times \text{NA} \times M} \tag{4}$$

$$\text{Field of view in mm} = \text{Eyepiece field} \#/m \tag{5}$$

where

$\text{NA} = $ Numerical aperture

$n = 1$ (refractive index for air/dry objectives)

$\lambda = $ Wavelength of light (0.55 μm is standard)

$M = $ Total magnification (eyepiece and objective)

$m = $ Objective magnification.

During the year 2000, Nikon introduced the new "LU" CFI*60* line of industrial "Epi-" objectives throughout the industrial microscope line for reflected light applications (Chambers, 2000). According to Chambers (2000), "L" is for luminous, as the new objectives have increased brightness; "U" is for universal, as all objectives in the line are strain-free and are appropriate for DIC; "CF" is for chromatic-aberration free, as all the objectives have been corrected for lateral and axial chromatic aberration; "I" is for infinity, providing more flexibility; and "60" is for the 60-mm parfocal length. Now there is an industrial line of Nikon objectives possessing all the advantages of the biological line. High-contrast images are formed by controlling flare (non-image-forming light), vastly improving the signal-to-noise ratio S/N. These objectives have longer working distances, and contamination is reduced due to a new antistatic construction which reduces the attraction of dust and other contaminates (important in clean-room environments). Lastly, Chambers (2000) stated that the LU line of objectives are environmentally friendly since the lenses incorporate "Eco-Glass" which is made without toxic materials such as arsenic and lead.

Some of these objectives may be useful for biological confocal applications, since the new LU Epi Plan line of objectives can be used on transmitted-light Eclipse microscopes, without an adapter for biological non-cover-glass applications where longer working distances may be desirable. Refer to Table I for specifications of the new LU CFI*60* Epi-objectives.

Table I
Nikon LU and L CFI60 Epi and BD Reflected–Light Objectives

Objective[a]	Magnification	Numerical aperture	Working distance (mm)	Recommended for
CFI LU Plan Epi	5×	0.15	23.50	Confocal
CFI LU Plan Epi	10×	0.30	17.30	
CFI LU Plan Epi	20×	0.45	4.50	
CFI LU Plan Epi	50×	0.80	1.00	Confocal
CFI LU Plan Epi	100×	0.90	1.00	
CFI LU Plan Epi ELWD	20×	0.40	13.00	
CFI LU Plan Epi ELWD	50×	0.55	10.10	Confocal
CFI LU Plan Epi ELWD	100×	0.80	3.50	
CFI L Plan Epi ESLWD	20×	0.35	24.00	Confocal
CFI L Plan Epi SLWD	50×	0.45	17.00	
CFI L Plan Epi SLWD	100×	0.70	6.50	
CFI LU Plan Apochromat Epi	100×	0.95	0.40	Confocal
CFI LU Plan Apochromat Epi	150×	0.95	0.30	Confocal
CFI L Plan Apochromat Epi	150× WI	1.25	0.25	Confocal
CFI LU Plan BD	5×	0.15	18.00	
CFI LU Plan BD	10×	0.30	15.00	
CFI LU Plan BD	20×	0.45	4.50	
CFI LU Plan BD	50×	0.80	1.00	
CFI LU Plan BD	100×	0.90	1.00	
CFI LU Plan BD ELWD	20×	0.40	13.00	
CFI LU Plan BD ELWD	50×	0.55	9.80	
CFI LU Plan BD ELWD	100×	0.80	3.50	
CFI LU Plan Apochromat BD	100×	0.90	0.51	
CFI LU Plan Apochromat BD	150×	0.90	0.40	

[a] BD, brightfield/darkfield; ELWD, extra long working distance.
From Chambers (2000).

VIII. Live Cell and Fixed Tissue Confocal Applications and Objective Lens Selection

It has been more than 15 years since the first successful commercial confocal system was marketed for biological applications (Amos *et al.,* 1987; White *et al.,* 1987), and there have been volumes of publications on the topic. Earlier reviews were reported by Dixon and Benham (1987), Smallcombe and Benham (1993), and Schwartz *et al.* (1996). It is not the purpose of this chapter to review them all, but rather to cite selected examples which explain the objectives that were chosen for particular research studies.

Hale and Matsumoto (1993) reviewed the problems associated with confocal imaging of thick intact tissues and the complex parameters which affect the signal intensity detection in a laser scanning confocal imaging system (LSCM) and the fluorescence intensity of the selected dye or dyes in a multiple lable preparation. Included variables which influence the performance of a LSCM for a given fluorochrome or multiple fluorochromes are the binding efficiency of the probe(s), the abundance of the molecule(s), the effect of the tissue itself in relation to preparation, depth of probe penetration within the tissue, the fluorochrome(s) selected, probability of absorption, quantum yield, absorption/emission maxima, emission lifetime(s), laser emission wavelength(s), wavelength selectivity of the emission filter(s), wavelength selectivity of the detector PMT(s), and the factors affecting photostability of the dyes themselves. Considering all of these variables, the objective lens becomes the most important microscope component in excitation and transmittance of fluorescence signals to the LSCM detector. The most appropriate objective lenses for confocal and multiphoton microscopy and digital imaging applications reported from selected references are shown in Table II. Refer to Terasaki and Dailey (1995) for an additional representative list of objectives used during confocal microscopy in the study of living cells.

Besides the prerequisite of a fixed stage microscope, patch-clamping experiments require a water immersion objective with a long working distance and a high NA (Dodt *et al.*, 2000). They listed two water immersion objectives from each of the four major microscope manufacturers ideal for patch-clamping studies; some objectives were described as having high transmission in both the UV and IR wavelengths.

It is also difficult to categorize research reports as either live cell work or fixed tissue today since most research may require both types of examinations. For the purposes of reporting objective lens selection, the two categories are combined as live cell and fixed tissue applications in the following.

A. UV Imaging

Lechleiter and Clapham (1992) observed spiral waves of intracellular calcium in *Xenopus* oocytes using a D Plan Apochromat UV 10×/0.40 excited by simultaneous confocal UV and visible light excitation and detection with a LSCM modified by Bliton *et al.* (1993). Bliton and co-workers generated a model proposing intracellular calcium signaling where wave propagation were controlled by IP3 mediated calcium release by injecting second messengers into the oocytes and releasing caged compounds by UV laser scanning (Bliton *et al.*, 1993). Their modifications converting a visible light confocal into a simultaneous UV and visible confocal system were extensively reviewed by Girard and Clapham (1994).

Diliberto *et al.* (1994) reviewed the major variables when selecting confocal visible and UV calcium imaging methodology and recommended high-NA 40× and 60× objectives for achieving vertical spatial resolution from about 0.5 to 0.7 μm.

While determining the role of calcium transients in *Drosophila* photoreceptor cell function, Ranganathan *et al.* (1994) used a Plan Fluor 40×/1.3 or a Plan Apochromat 60×/1.4 oil immersion objective with their real-time UV confocal imaging system. Ranganathan and co-workers found that calcium levels rose earlier and higher in the rhabdomeres than in the rest of the cell.

Tsien and Bacskai (1995) mostly used a Plan Fluor 40×/1.0 oil immersion or a Fluor 40×/1.0 Wl objective with their prototype design of a video-rate, UV-visible CLSM.

Table II
Confocal and Multiphoton Applications and Objective Lens Selection

Applications	Objective lenses of choice[a]	References
Point-spread function	Plan Apochromat 60×/1.4 oil Plan Apochromat 60×/1.2 WI	Dunn and Wang (2000)
Fluorescence beads	Plan Apochromat 63×/1.4 Plan Apochromat 60×/1.4 oil Plan Apochromat 40×/1.0 oil	Cogswell (1995) Author's personal experience Majlof and Forsgren (1993)
High-resolution GFP	Plan Apochromat 10×, Plan Apochromat 20× Plan Fluor 20× MI, Plan Fluor 40× oil Plan Apochromat 20×, Plan Fluor 40× oil High NA 60× WI, High NA 60×, and 100× oil	Chalfie and Kain (1998) Paddock (1999) Miesenbock (2000) Kohler et al. (1997)
Biochemistry	100×/1.3 oil	Glazer et al. (1990)
Pharmacology and calcium	High NA 10× and 20× High NA 40×, 60×, and 100× oil High NA 20× MI, 40× WI, 60× WI	Author's personal experience Diliberto et al. (1994) Backsai et al. (1995)
Early mammalian development	Plan Apochromat 10×, Plan Apochromat 20× Plan Fluor 20× MI, Plan Fluor 40× oil, Fluor 100×/1.3 oil	Author's personal experience Girard and Clapham (1994)
Early Drosophila development	Plan Apochromat 10×, Plan Apochromat 20× Plan Fluor 20× MI, Plan Fluor 40× oil	Paddock (1991) Author's personal experience
CNS in-vivo development	High NA 10× and 20×, Plan Apo 40× 1.3 oil 20×/0.8 glycerine, 25×/0.65 oil	Cline et al. (2000) Petran et al. (1990)
DNA and genone research	Neofluar 40×/0.75, Plan Neofluar 63×/1.2 oil	Arndt-Jovin et al. (1990) Paddock et al. (1991)
Embryology	Plan Apochromat 10×/0.45 Fluor 40×/1.3 oil Plan Apochromat 60×/1.4 oil Plan Neofluor 25×/0.80, Fluor 40×/1.3 oil	Wright et al. (1993) Wright et al. (1993) Summers et al. (1993)
Patch-clamping	Fluor 40×/1.0 Dipping, Plan Apochromat 60×/1.2 WI	Dodt et al. (2000)
Cell motility	Fluor 40×/1.0 WI, Plan Apochromat 10×, Plan Apochromat 20× High NA 20× MI, 40× WI, 60× WI	Bacskai et al. (1995)
Cell biology and cytoskeleton	Plan Apochromat 63×/1.4 oil Plan Apochromat 60×/1.4 oil Apochromat 40×/1.3 oil, Plan Neofluor 40×/0.9 MI	Carlsson and Liljeborg (1989) Vassy et al. (1990) Hale and Matsumoto (1993)
Cell signaling and UV calcium	DPlan Apochromat 10× UV/0.40 Plan Fluor 40×/1.3, Plan Apochromat 60×/1.4 oil	Lechleiter and Clapham (1992) Ranganathan et al. (1994)
Neurophysiology	Plan Apochromat 40×/1.0 oil High NA Objectives	Carlsson and Liljeborg (1989) Stelzer and Wijanaendts-van-Resandt (1990)
Bone cell biology	10×/0.50 glycerine, 20×/0.80 glycerine, 40×/1.3 glycerine, 100×/1.3 glycerine	Boyde et al. (1990b) Boyde et al. (1990b)
Botanical and plant biology imaging	Plan Apochromat 40×/1.0 oil, Plan Apochromat 100×/1.3 oil Fluor 10×/0.50, Plan Apochromat 60×/1.4 oil	Carlsson and Liljeborg (1989) Brakenhoff et al. (1990) Fricker and White (1992) Shaw et al. (1992) Paddock (1991)
Reflection imaging	High NA, Long WD, water immersion High Dry NA = 0.85 (NCG) Plan Fluor 40×/1.3, Plan Apochromat 60×/1.4 Plan Apochromat 100×/1.4	Cogswell and Sheppard (1990) Larson (1999b)

Table II (*continued*)

Bacteria	High NA 60× oil and/or WI, High NA 100× oil and/or WI	Author's personal experience
Biomaterials	High NA 10×, 20×, 40×, LWD 20×, 20× WI	Author's personal experience
	20× "Epi-" Brightfield or NCG	Author's personal experience
	40× "Epi-" Brightfield or NCG	
Ocular imaging	50×/1.0 WI	Masters (1990)
	High dry NA = 0.85 (NCG)	Masters and Paddock (1990)
		Cogswell and Sheppard (1990)
Medical and surgical research	High NA LWD Objectives	Author's personal experience
Dental research	High NA 10×, 20×, 40×, LWD 20×, 20× WI	Boyde *et al.* (1990a)
	Plan Apo 60×/1.4 oil, 50×/1.0 WI,	Boyde *et al.* (1990b)
	Water dipping objectives	Watson (1990)
	Fluor 100×/1.3 oil	Petran *et al.* (1990)
Microbiology—yeast	High NA Dry 40×, High NA 60× oil and/or WI	Brakenhoff *et al.* (1989)
	High NA 100× oil and/or WI	Author's personal experience
Microbiology—biofilms	20× and/or 40× Epi-Brightfield or NCG	Author's personal experience
	ULWD 40×	Tan (2000)
	Plan Apochromat 40×/0.95	Bruhn (2000)
	Plan Apochromat 100×/1.4 oil	
	20×/0.6, 40×/0.55, 40×/0.75	Lawrence *et al.* (2000)
	63×/0.9 water dipping, 63×/1.2 WI	

[a] Objective notation: Magnification/NA

These objectives were selected because of their high NA and brightness along with high UV transmission, and the cost-to-performance relationship. Their confocal design which was commercialized as the Nikon RCM 8000 was designed to demonstrate high-speed imaging of diffusible ion indicators without pushing the limits of spatial resolution (Tsien and Bacskai, 1995). Today, the new CFI*60* S Fluor objectives are ideal for UV excitation of 340 nm and above or whenever there is a need for high-transmission, brighter fluorescence imaging.

Bliton and Lechleiter (1995) discussed objectives with regard to UV transmission, antireflection coatings, chromatic errors, immersion medium errors, and aberrations at UV wavelengths. They also reported the transmission percentages of a few UV-capable objectives manufactured by Nikon, Olympus, and Zeiss.

B. Visible Calcium Ion Imaging

Using the visible, long-wavelength, calcium indicator Fluo-3, Hernandez-Cruz *et al.* (1990) monitored over time both sustained and rapidly dissipating calcium gradients in sympathetic neurons isolated from the bullfrog. Both full frame and line-scan imaging techniques were employed to collect data. The confocal system included an inverted microscope, and imaging was conducted with a water immersion Apochromat 40×/0.75 objective (Hernandez-Cruz *et al.,* 1990).

In order to determine whether anoxia increased calcium concentrations in adult and neonatal neurons, Friedman and Haddad (1993) monitored 28-day-old rat CA1 hippocampal neurons using Fluo-3. Four important conclusions were made from confocal investigations using an inverted microscope with a Plan Neofluar 25×/0.85 oil immersion objective. In addition, propidium iodide was used to stain the nuclear region following anoxia, and scanning transmission images were also acquired concurrently to observe changes in morphology at the same time points as with calcium (Friedman and Haddad, 1993).

Using a Fluor 40×/1.0 WI objective, Bacskai *et al.* (1995) introduced the calcium-sensitive dyes calcium green or Fluo-3 by iontophoresis into the somal region of a single cell 150 μm into an intact vertebrate spinal cord preparation. Their findings suggested periodic calcium fluctuation involvement in motoneuron axons during action potential generations.

Cannell and Soeller (2000) observed calcium sparks using a water immersion objective with a long working distance and high NA. They explained the line scan technique whereby a single line was scanned through the cell repetitively over time. Each line was then displayed beneath the previous 2-ms line in a vertical montage, and the time course of the spark was observed and plotted.

C. Cytoskeleton

Imunofluorescence confocal microscopy for the study of cytoskeleton can be a powerful tool for the observation of fine details and three-dimensional reconstruction provided that individual microtubules are not too densely packed (Hale and Matsumoto, 1993). They concluded that most problems resulted either from the poor optical quality of the tissue (light refraction, reflection, and scattering between the plane of focus and the top lens of the objective) or from the varying characteristic within an objective lens to correct for spherical aberration. The farther down within the tissue images are acquired, the more likely resolution will decline and spherical aberration correction will diminish. Since axial resolution is less than lateral resolution by approximately a factor of 2 of the Rayleigh criterion, the ability to resolve fine details will be dependent upon an objective with a high NA, adequate magnification, long working distance, and proper correction for spherical aberration, even though filaments of subresolution size may be detected.

D. Immunogold and Silver Grains

In a transmitted brightfield or darkfield conventional microscopy comparison with reflected, brightfield confocal microscopy, Paddock *et al.* (1991) found that reflected light confocal microscopy significantly improved the ability to visualize silver grains in autoradiograms of specimens prepared by *in situ* hybridization. Their data was collected from samples of HIV-infected human peripheral blood cells, tissue sections of human placenta, and human skin routinely using three objective lenses: a Plan Apochromat 10×/0.45, a Plan Apochromat 40×/1.2, and a Plan Apochromat 60×/1.4 (Paddock *et al.*, 1991).

There are occasions when a fluorescent dye and immunogold may be imaged simultaneously using a confocal microscope when an appropriate combination of filters and mirrors are employed. Berthoud and Powley (1993) injected the fluorescent carbocyanine dye DiI into the dorsal motor nucleus of vagal efferent preganglionic fibers, and also labeled the ganglion *in toto* with Fluorogold providing a counterstain with an emission wavelength well separated from that of DiI. The objective typically used for this study was a 40×with electronic scan zoom magnifications of up to 6×. They offered conclusions regarding the labeled vagal structures as terminals and whether they were of efferent or of afferent origin (Berthoud and Powley, 1993).

E. Bone and Dental Cell Biology

Boyde *et al.* (1990a) demonstrated a new and easier method for measuring the functional capacity of isolated osteoclasts by determining the volume of bone or dentine resorption. Their method of examination was described in detail using a tandem scanning confocal microscope. Boyde *et al.* (1990b) found that they could use a 60×/1.4 oil or a 100×/1.4 oil objective on relatively level resorption pits. Even a 50×/1.0 WI or 100×/1.2 WI objective could be used on a "not-too-steep" sided resorption pit. They also concluded that high-NA dry objectives could be used. Today's generation of high-NA, long-working-distance, dry objectives might even prove to be easier for use and provide equally or more acceptable data.

F. Genetics

Albertson *et al.* (1991) found that confocal microscopy could be used routinely for gene mapping as long as the chromosome bands could be visualized. Using propidium iodide and FITC-labeled antibodies, they successfully mapped single-copy probes to human chromosome bands. Results reported were from confocal data acquired using an upright microscope and a Plan Apochromat 60×/1.4 oil immersion objective (Albertson *et al.,* 1991).

In a comparison of wide-field and confocal microscopy to study subcellular localization of creatine kinase in cultured astrocytes, Manos and Edmond (1992) found the enzyme, important in energy homeostasis, to be distributed uniformly throughout the cytoplasm, and especially intense in the nucleus. For wide-field microscopy, fluorescence was routinely observed with either a Fluor 20×/0.75 or a Fluor 40×/0.85 objective. For confocal imaging, a Plan Apochromat 60×/1.4 oil immersion objective was exclusively used. The comparisons were performed using an upright microscope and the fluorescent dyes, FITC and DiI (Manos and Edmond, 1992).

G. Brain Tissue, Neuroscience, and Retinal Imaging

Freire and Boyde (1990) reported using a tandem scanning, reflected-light, real-time, direct-view confocal microscope (Egger and Petran, 1967) for the examination of Golgi-impregnated material. With such a spinning-disk type confocal, incoming light passed through the objective before scanning the field of view. Freire and Boyde (1990) found that standard high-NA, immersion objectives in the reflection mode were suitable for observing Golgi-impregnated neurons. Stereo pairs were produced with the addition of appropriate software.

Using the carbocyanine dye DiI, O'Rourke and Fraser (1990) were able to model in three dimensions over time retinal ganglion-cell terminal arbors within the tecta of live, intact *Xenopus* larvae. Visualizing the same arbor over a period of several days, they were able to reveal the growth and ramification of optic nerve fiber within the brain. A Fluor 40×/1.3 oil immersion objective was used to observe the arbors, while a Plan Apochromat 60×/1.4 oil immersion lens, was used to observe fine structures such as growth cones (O'Rourke and Fraser, 1990).

O'Rourke *et al.* (1992) used time-lapse confocal microscopy imaging every 1 to 2 min to observe dynamic behaviors of migrating cells during early development of the mammalian cerebral cortex. Migrating cells in the intermediate zone were viewed with the use of the following objectives: a Fluor 20×, a Plan Neofluar 25×, and a UV D-Apochromat 40× (O'Rourke *et al.,* 1992).

Mason *et al.* (1992) used confocal microscopy to map the distribution of appositions between intracellularly labeled neurons in rat rostral ventromedial medulla and immunoreactive (ENK-IR) profiles at 1- to 2-μm increments in the Z-axis. Experiments were conducted using FITC and Texas Red fluorochromes and were examined through a Plan Neofluar 63×/1.3 oil immersion objective (Mason *et al.,* 1992).

Hale and Matsumoto (1993) discussed cytoskeletal organization in the intact retina. By examing the intact fluorescently labeled retinal cells, they were able to observe the orientation and distribution of retinal cells, compare neighboring cell features, and trace cell-to-cell contacts and synaptic connections. While using Cy-3-conjugated secondary antibodies for high signal and deep retinal penetration, the objectives of choice by Hale and Matsumoto (1993) were an Achromat 63×/1.3 oil immersion and an UV Apochromat 40×/1.3 oil immersion. The increased transmission gain of the achromatic lens made up for the correction sacrificed by not using a Plan Apochromat with a 1.4 NA. Since they used the electronic scan zoom feature (scanning in the center of the lens), they did not require "plan" lenses.

H. Ocular Imaging

Masters (1990) reviewed how confocal microscopy could be utilized in the examination of the unstained, living, *in vivo* eye. He discussed the development of optical sectioning microscopes in ophthalmology, ophthalmic confocal applications, and the real-time systems available during that period. The cornea and ocular lens have minimum contrast and high transparency, whereas the retina provides little transparency displaying high contrast (Masters, 1990). Thus this region of the eye lends itself to reflection confocal imaging.

Masters and Paddock (1990) and Masters (1991) chose a 50×/1.0 Wl objective. They were particularly able to observe the regions between Descemet's membrane and the epithelial basal lamina without the objective actually touching the cornea. Observations were made with a non-real-time CLSM in the reflected-light and fluorescence modes.

I. Microbiology

The field of microbiology encompasses a vast number of disciplines including bacteriology, mycology, virology, nematology, parasitology, epidemiology, plant and animal pathology, and algology. Confocal microscopy has become a standard imaging technique and most recently the use of two-photon imaging has been explored. The following sections provide objective lens selection information for the confocal study of yeast and for the expanding field of biofilms research.

1. Yeast

The relatively small size (5–8 μm), the presence of a cell wall making microinjection of fluorescent markers difficult, and the relatively low contrast in culture present shortcomings when using yeast as model organisms (Kahana and Silver, 1998). Kahana and Silver found that a high level of magnification was necessary when working in fluorescence microscopy. Confocal microscopy of yeast was best conducted with a Plan Apochromat 60×/1.40 oil immersion or a Plan Apochromat 60×/1.2 WI objective. Also, it may be necessary to use a Plan Apochromat 100×/1.40 oil immersion objective for ease in locating yeast while in the wide-field mode.

2. Biofilms

Very few laboratories specialize in this area of microbiology; however, the field of biofilm research is receiving a lot of attention of late. Biofilms are sessile communities of microbial cells that develop on surfaces in virtually all aquatic environments. Biofilm microorganisms are quite resistant, and bacterial species are having an increasing impact on industrial, water, and medical ecosystems. Recent advances in biofilm science and engineering include the description of complex structure and transport in biofilms, biofilm phenotyping, resistant mechanisms to antimicrobial agents, and the role of cell-to-cell communication in biofilm development (Lange, 2000). Research requires involvement of various multidisciplinary fields such as microbiology, all areas of engineering, chemistry, mathematics and computer sciences, physics, and plant and soil sciences. New techniques for the study and microscopical examination of such biofilm formation and bacterial growth are evolving to include confocal microscopy. Since 1991, there have been an increasing number of publications using confocal imaging for the study of biofilms where the CLSM has become the standard tool for the investigation of interfacial microbial communities (Neu and Lawrence, 2000).

Lawrence *et al.* (1998) and Lawrence and Neu (1999) provided excellent reviews of biofilms research using the CLSM. They concluded that the CLSM permitted optical sectioning of intact, fully hydrated biofilm material providing enhanced resolution, clarity, spatial distribution of biofilm properties, and a digital database for image processing and analysis. Second only to having a perfectly aligned CLSM and wide-field microscope, Lawrence *et al.* (1998) emphasized that the proper choice of objectives was most critical. They felt that confocality was lost with the use of an objective with a NA of less than 0.50, whereas an objective with a NA of 1.4 provided excellent confocality.

A

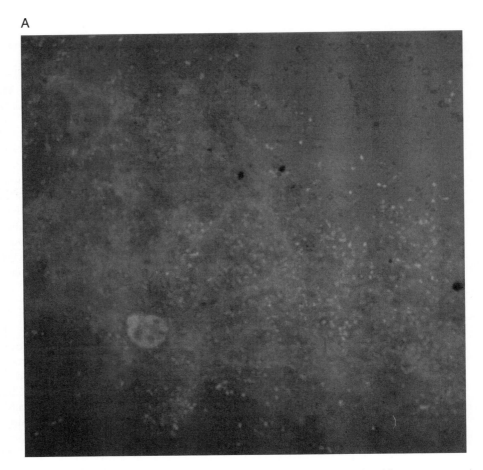

Fig. 3 (A) Sulfate-reducing bacterium, *Pseudomonas aeruginosa,* stained with acridine orange grown in 0.1×TSB on a zirconium coupon. Imaged with a Plan Apochromat 100×/1.40 oil immersion objective. (B) Sulfate-reducing bacterium, *Pseudomonas aeruginosa,* stained with acridine orange grown in water with mineral salts on aluminum coupon. Imaged with a Plan Apochromat 100×/1.40 oil immersion objective. Courtesy of Bruhn (2000). (See Color Plate.)

Thomas *et al.* (2000) characterized organism size and morphotype of 10 periodontal biofilm microorganisms with SYTO9 and propidium iodide fluorescence dyes. Resolving biofilm microorganisms of this type almost always warrants the use of 60× and 100× high-NA objectives. On the other hand, Graham *et al.* (2000), designed a flow cell (closed dome-shaped system to mimic conditions of the human eye) to study bacterial adherence on whole contact lenses. Preliminary bacterial examination of the lens apex was conducted with a 10× objective with wide-field microscopy and a CLSM. In order to penetrate the chamber portion of the flow cell using a CLSM, an

B

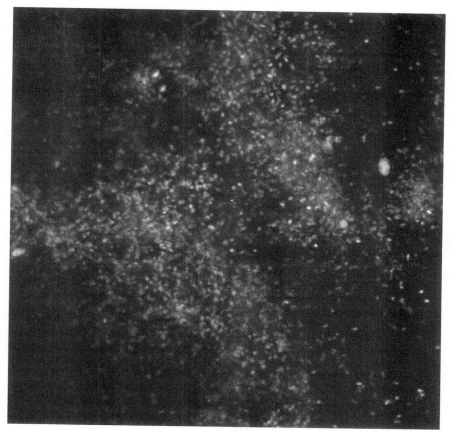

Fig. 3 *(continued).*

ultralong-working-distance (ULWD) Nikon 40× with a 10.3-mm WD was required (Tan, 2000). The images obtained with the ULWD objective were quite impressive.

While developing and implementing methods to quantify morphological features from biofilm images, Yang *et al.* (2000) examined biofilms with wide-field microscopy and a CLSM. Important features that their algorithms concentrated on were porosity, fractal dimension, diffusional length, angular second moment, inverse difference moment, and textural entropy. Such reflected-light CLSM biofilm examinations required the use of a 4× low magnification objective routinely, and occasional use of 10× and 20× lenses.

In a study to learn more of the role of biofilms in waste water treatment, i.e., their architecture, exopolymer chemistry, and populations, Lawrence *et al.* (2000) used a variety of objectives with a CLSM, such as 20×/0.6, 40×/0.55, 40×/0.75, 63×/0.7, 63×/0.9 water dipping, and 63×/1.2 WI.

While studying metal surfaces in relationship to the sulfate-reducing bacterium *Pseudomonas aeruginosa* with the fluorochrome acridine orange, Bruhn (2000) used a 40×/0.95 and a Plan Apo 100×/1.4 immersion oil objective. With the use of the electronic scan zoom function, she was able to examine the adhesive biofilm–metal surface interface. Aluminum and zirconium were examined and are shown in Figs. 3A and 3B (Bruhn, 2000) (See Color Plate).

Decho and Kawaguchi (1999) developed a methodology to examine *in-situ* natural microbial communities with a CLSM and two-photon laser scanning microscope. They used a 40× dry lens and a 100× oil immersion objective. For CLSM examinations of subgingival biofilms, most recently, Neu and Lawrence (2000) have employed two-photon technology to examine thick biofilms with nucleic acid specific fluorochromes such as acridine orange, DAPI, and SYTO using both upright and inverted microscopes. Their objective lenses of choice were a 63×/0.9 water immersible (dipping) and a 63×/1.2 WI.

IX. Possible Solutions for Potential Problematic Applications

Today microscope manufacturers correct for chromatic aberrations in different ways. Therefore multicolor (multiple-label) experiments, i.e., colocalization of two different probes, should be only conducted with optical systems that have been designed to work together (Piston, 1998). The concept of swapping objectives and other optics from various manufacturers to solve problematic applications is no longer recommended since the major microscope manufacturers each have different tube lengths and have different corrections for their respective tube lenses.

A. Multiple-Label Confocal Fluorescence Imaging

Confocal manufacturers have moved away from providing the multiline krypton–argon "mixed-gas" (Brelje *et al.,* 1993) single laser as the only source of illumination for multiple-label confocal fluorescence imaging (Smallcombe and Benham, 1993). Today, users have the three-laser option where an argon ion laser (488 nm), a green helium–neon laser (543 nm), and a red helium–neon (633 nm) or red diode laser (635 nm) are used in unison to acquire simultaneous three-color images via three separate PMT detectors (Paddock, 1999).

Nikon's new CFI*60* Super Fluor objectives (S Fluor Series manufactured with quartz and special glass components) have high numerical apertures and superior signal-to-noise ratios. These objectives are extremely low in autofluorescence, optimizing S/N. A design with fewer lens elements and air/glass interfaces optimizes luminous flux. The special optical elements and antireflection coatings used in the CFI*60* S Fluor series objectives have been designed to permit the objectives to be used through an extended range of fluorescence excitations from the UV to the IR. Since these lenses are not corrected for curvature of field and not designated with a "plan" marking, another lens might be more suitable when flatness of the entire field is considered to be critical.

1. Image Brightness

Immunofluorescence imaging with the confocal microscope where photons are restricted by the pinhole places additional demands on the objective. Detection of submicron details are dependent upon an objective's ability to collect and transmit small amounts of light. The need for high transmission will sometimes outweigh the requirement for aberration correction (Hale and Matsumoto, 1993; Keller, 1995). The Nikon CFI60 S Fluor series objectives are chromatically corrected so their performance is nearly that of apochromats. This optical correction makes them excellent for use with multiprobe fluorescent specimens. The higher numerical apertures contribute to the image brightness as well as the enhanced resolution and corrections of these objectives. Image brightness is determined by

$$\text{Image brightness} = (\text{NA})^4 / \text{Magnification}^2. \tag{6}$$

It is important to select an objective lens of appropriate magnification with the highest possible NA for that magnification (working distance may also be a consideration). Also select the best objective correction for maximum wavelength transmission at the chosen magnification. Herman (1998) reminded us that specimen intensity decreases as the objective lens magnification increases, where intensity (I) is proportional to the inverse square of the magnification:

$$I = 1/\text{Mag}^2. \tag{7}$$

Hale and Matsumoto (1993) reviewed their experiences of objective lens selection while imaging subcellular details in thick tissue sections. They found that a compromise between high-numerical-aperture and moderate aberration-corrected immersion lenses of lower magnification worked well for their examinations. Therefore it is almost never necessary to use an oil immersion 100× objective during confocal examinations when there is a built-in electronic scan zoom feature. More light will be collected with an oil immersion 60× lens, and more again with an oil immersion 40× objective. The NA of the three different magnifications (40×, 60×, 100×) offered from most microscope manufacturers will differ by only about 0.1 in oil immersion objectives, and in most cases the aberration corrections and working distances will remain nearly the same for all three magnification choices (Nikon Corporation, 1998).

B. X-Z and 3-D Imaging

Full three-dimensional images can be collected by scanning X-Y planes as a function of Z (Stelzer and Wijanaendts-van-Resandt, 1990). It is possible to scan X-Z images directly. This technique could be compared to optically slicing through the three-dimensional aspect of the specimen. It is simply scanning a single line in the X direction continuously at different Z depths (Paddock, 1999). Stelzer and Wijanaendts-van-Resandt (1990) recommended using high-NA objectives for X-Z scanning. They further stated that the minimum step size along the optical axis should be greater than the lateral resolution. Oversampling should be avoided or kept to a minimum to prevent phototoxicity.

Carrington *et al.* (1990) showed that depth-dependent distortion could be corrected and accurate Z-axis measurements of vertical resolution could be made using a water immersion objective lens. They found that the point-spread function for the WI objective was symmetrical above and below the point of focus, indicating little or no detectable spherical aberration. Tests also indicated that the axial resolution of the objective matched theoretical calculations for a WI, NA 1.20 objective. Carrington *et al.* (1990) also noted significant improvements in the ability to use image deconvolution on three-dimensional specimens over traditional high-NA oil immersion objectives. In addition, the water immersion objective proved to be brighter than oil immersion lenses of equal or slightly higher NA when imaging more than 20 μm into aqueous media and tissue, because spherical aberration was virtually nonexistent with the use of a Plan Apochromat WI objective lens.

It is possible to compensate mathematically for distortions with deconvolution software; however, deconvolution does nothing to remove distortion and is easily corrupted by scattered light. These methods require accurate measurement of the point-spread function. If resolution varies with depth, then measurements must be taken at different depths, which complicates the process.

Since the CLSM collects data as a series of optical sections (Z-series), the data can be combined as a projection. The projection can then be displayed as a montage or a three-dimensional reconstruction (Paddock, 1999).

C. Backscattered Light Reflection

Cogswell (1995) reviewed the backscattering light reflection technique and applied it to the imaging of immunogold labels. He also compared the effects of three different excitation wavelengths using red, green, and blue lasers. Cogswell (1995) found that the green laser provided brighter images of African blood lily (*Haemanthus katherinae*) endosperm microtubule bundles. He used a Plan Aporchromat 63×/1.4 oil immersion objectives to detect immunogold particles.

Most galvanometer-mirror scanning confocal systems offer a number of interesting ways to gather useful information about cells and tissues, which go beyond fluorescence confocal microscopy. Bodies are called refractile when their refractive index is different from that of the surrounding medium. Many organelles inside the cell, such as vacuoles, secretory granules, and the endoplasmic reticulum, are refractile. Refractile bodies have the property of reflecting back some portion of the light which falls on them, similar in principle to when one views a reflection in a windowpane.

Backscattered light (BSL) reflection relies on the confocal detection of laser light scattered back from refractile bodies in the cell. When working with live fluorescently stained cells, it becomes critical to extract the most information possible from as many photons striking the specimen since stained specimens are far less tolerant of laser illumination than unstained ones (Pawley *et al.,* 1993). They further stated that if a surface illuminated with incident light was optically smooth, the scattering was coherent and became reflection.

This light is collected through the pinhole and is assembled into a BSL image in the same way as fluorescence (Larson, 1999b). Pawley *et al.* (1993) noted that the emission filter must be removed when imaging with BSL, and that precautions must be taken to prevent specular reflections from appearing as concentric bright rings obscuring the BSL signal. To correct for such reflection artifacts, Pawley *et al.* (1993) inserted a half-wave plate after the laser to rotate the laser polarization and a rotatable quarter-wave plate between the ocular and the objective so that the illumination would be reflected down the optical axis.

D. Interference Reflection Contrast

Interference reflection contrast imaging can be useful for the study of contact foci between the cell and its substrate. This technique is accomplished in a similiar fashion to BSL except only the 488-nm laser line should be used for illumination. As one slowly focuses from a plane in the center of the cell toward the cover glass, typical interference reflection patterns associated with a cell's contact patch will appear. As one focuses through the contact patch onto the surface of the cover glass, the gain should be adjusted downward so as not to oversaturate the detector PMT, allowing the image to white out.

To enhance the contrast of the interference reflection image, it often becomes necessary to adjust the CLSM's gain and black level so that the gray-level histogram will be confined to the midrange. Larson (1999b) found that the CFI*60* Plan Fluor 40×/1.3 oil immersion, the CFI*60* Plan Apochromat 60×/1.4 oil immersion, and the CFI*60* Plan Apochromat 100×/1.4 oil immersion objectives provided excellent interference reflection images. According to Larson (1999b), there is no real advantage to an iris objective lens because the CLSM's pinhole is selectable. In practical use, an increase in the electronic scan zoom magnification may be required when using a 40× or 60× magnification objective (Larson, 1999b).

E. Scanning Transmission Phase Contrast

The scanning illuminating laser beam of a confocal instrument does not stop when it reaches the specimen focal plane, but rather passes on completely through the specimen proceeding along the transmitted optical path all the way to the wide-field transmission illuminating bulb. A receptor may be placed at an ideal location in the optical path between the bottom of the condenser and the transmission lamp housing to capture the laser illumination after it has passed through the condenser.

Two of the most important changes which occur as light passes through a transparent specimen are changes in amplitude and phase. When light passes through a medium of different refractive indices, the amplitude of the wave is unchanged; however, the velocity and wavelength change. The amplitude of undiffracted light must be shifted in phase to that of the diffracted light. For a detailed explanation of phase contrast, please refer to Inoué and Spring (1997a) and Larson (1999b).

Thus, a nonconfocal scanning transmission phase-contrast image can be generated by scanning the illuminating laser beam through a phase-contrast objective lens possessing a phase ring in the back focal plane and through a phase-contrast condenser with a matching phase annulus. The scanning light after passing through the phase condenser is intercepted by an appropriately placed transmission accessory receptor. The signal is then transmitted through a fiber optic and sent to a detector which in most cases is a PMT or photodiode.

A separate detection channel for scanning transmission detection is highly recommended. Confocal instruments with one, two, three, or more detection channels may be used for nonconfocal phase-contrast imaging, where one of the detector channels is used for the scanning transmission image. Scanning transmission phase-contrast and confocal fluorescence images can be acquired simultaneously or sequentially and merged following their acquisition. If the scanning transmission image is collected sequentially, it is preferable to collect the fluorescence image first especially if the fluorescence is emitting at low light levels, since this laser scanning technique causes photobleaching and fading.

Most confocal manufacturer's scanning transmission accessories have ND filters on the collectors to reduce intensity and PMT or detector over-saturation. See Tables III–VI for lists of objectives available from microscope manufacturers for use with scanning transmission phase contrast imaging accessories.

F. Scanning Transmission Differential Interference Contrast

With this second type of transmission imaging, the same accessories are required for nonconfocal scanning transmission differential interference contrast as for widefield DIC microscopy. Sometimes when the polarizer and analyzer are in the extinction position, the resultant image may look more like a polarized light image than a DIC when viewed on the monitor. It is sometimes beneficial to back the nosepiece prism out slightly in order to obtain a normal appearing DIC image (Larson, 1999b). Typically, the gain, black level, and detector contrast channel assigned to scanning DIC transmission imaging of most confocal systems can be adjusted to emphasize highlights or dark regions in the image. For a detailed discussion of the scanning transmission DIC technique, refer to Larson (1999b).

Since confocal imaging may photobleach the specimen while in the phase contrast scanning transmission mode, and the DIC analyzer and polarizer attenuate scanning laser illumination, it is recommended that the fluorescence images be acquired first, followed by the scanning transmission DIC image. Using an appropriate software program provided with most confocal systems, all the sequentially acquired images may be merged into one image where the fluorescence image(s) can be superimposed upon the transmission DIC image in perfect registration with one another.

G. Ultrawide Low-Magnification Confocal Imaging

Some confocal microscopists may opt to use very low magnification objective lenses such as the CFI60 Plan Achromat UW 2×/0.06, the CFI60 Plan Achromat UW 1×/0.04, or the CFI60 Macro Plan Apochromat UW 0.5×/0.02. Farkas (2000) has obtained

Table III
Nikon Objective Lenses for Confocal and Multiphoton Digital Imaging

Objective	Magnification	Numerical aperture	Coverglass (mm)	Working distance (mm)	Transmission % at 350 nm[a]	Transmission % at 580 nm[a]	Transmission % at 900 nm[a]	Contrast methods	Recommended for
CFI Plan Apo Macro Ultra Wide	0.5x	0.02	0.17	7.00	E			BF	Confocal
CFI Plan Apochromat	2x	0.10	0.17	8.50	E			BF	Confocal
CFI Plan Apochromat	4x	0.20	0.17	15.70	E			BF	Confocal
CFI Plan Apochromat	10x	0.45	0.17	4.00	E	A		BF	Confocal
CFI Plan Apochromat	20x	0.75	0.17	1.00	E	A	D	BF	Confocal
CFI Plan Apochromat	40x Corr. Coll.	0.95	0.11–0.23	0.14	E	A	B	BF	Confocal
CFI Plan Apochromat	40x Oil	1.00	0.17	0.16	E	A	B	BF/PH/DIC	Confocal
CFI Plan Apochromat WI	60x WI Corr. Coll.	0.95	0.17	0.15	E	A	C	BF/PH/DIC	Confocal
CFI Plan Apochromat	60x	1.20	0.15–0.18	0.22	E	A	B	BF/DIC/IR	Confocal and two-photon
CFI Plan Apochromat	60x A Oil	1.40	0.17	0.21	E	A	C	BF/PH/DIC/IR	Confocal and two-photon
CFI Plan Apochromat	100x Oil	1.40	0.17	0.13	E	A	C	BF/PH/DIC	Confocal
CFI Plan Apochromat	100x Oil (NCG)	1.40		0.13	E	A	C	BF	Confocal
CFI Plan Apochromat IR	100x Oil IR	1.40	0.17	0.13	E	A	B	BF/DIC/IR	Confocal and two-photon
CFI Plan Fluor	10x	0.30	0.17	16.00	B	A	B	BF/PH/DIC/UV/IR	Confocal
CFI Plan Fluor	20x	0.50	0.17	3.00	B	A	A	BF/PH/DIC/UV/IR	Confocal and two-photon
CFI Plan Fluor	20x MI (O,G,W)	0.75	0.17	0.35 (Oil) 0.34 (Glycerine) 0.33(Water)	B	A	B	BF/DIC	Confocal and two-photon
CFI Plan Fluor	40x	0.70	0.17	0.72	A	A	A	BF/PH/DIC/UV/IR	Confocal and two-photon
CFI Plan Fluor	40x Oil	1.30	0.17	0.20	C	A	B	BF/DIC/IR	Confocal and two-photon
CFI Plan Fluor	60x Corr. Coll.	0.85	0.17	0.30	C	A	B	BF/DIC/IR	Confocal and two-photon
CFI Plan Fluor	60x Oil	1.25	0.17	0.20	C	A	B	BF/DIC/IR	Confocal and two-photon
CFI Plan Fluor	100x Oil	1.30	0.17	0.20	B	A	A	BF/PH/DIC/UV/IR	Confocal and two-photon
CFI ELWD Plan Fluor	20x Corr. Coll.	0.45	0–2.00	8.10–7.00	A			BF/PH/DIC	Confocal
CFI ELWD Plan Fluor	40x Corr. Coll.	0.60	0–2.00	3.70–2.70				BF/PH/DIC	Confocal
CFI ELWD Plan Fluor	60x Corr. Coll.	0.70	0.50–1.50	2.10–1.50				BF/PH/DIC	Confocal
CFI Super (S) Fluor	10x	0.50	0.17	1.20	B	A	A	BF/UV/IR	Confocal and two-photon
CFI Super (S) Fluor	20x	0.75	0.17	1.00	B	A	A	BF/UV/IR	Confocal and two-photon
CFI Super (S) Fluor	40x Corr. Coll.	0.90	0.17	0.30	B	A	B	BF/UV/IR	Confocal and two-photon
CFI Super (S) Fluor	40x Oil	1.30	0.17	0.22	B	A	A	BF/UV/IR	Confocal and two-photon
CFI Super (S) Fluor	100x Oil	1.30	0.17	0.20	A	A	C	BF/UV/IR	Confocal and two-photon
CFI Fluor Water Dipping	10x WD	0.30		2.00	B	A	B	BF/DIC/UV/IR	Confocal and two-photon
CFI Fluor Water Dipping	40x WD	0.80		2.00	B	A	A	BF/DIC/UV/IR	Confocal and two-photon
CFI Fluor Water Dipping	60x WD	1.00		2.00	B	A	A	BF/DIC/UV/IR	Confocal and two-photon
CFI Plan Achromat UW	1x	0.04	0.17	3.20	E		B	BF	Confocal
CFI Plan Achromat UW	2x	0.06	0.17	7.50	E		B	BF	Confocal
CFI Plan Achromat	40x (NCG)	0.65		0.48	E	A	B	BF/IR	Confocal and two-photon
CFI Plan Achromat	50x Oil	0.90	0.17	0.40	E	A	B	BF/IR	Confocal and two-photon
CFI Plan Achromat	100x (NCG)	0.90		0.26	E	A	A	BF/IR	Confocal and two-photon

[a] 10 nm.

The transmission data are average, typical values. They can vary slightly from lens to lens. For transmission percentages, A = >71%, B = 51–70%, C = 31–50%, D = 16–30%, E = <15%. BF, brightfield; PH, phase contrast; DIC, differential interference contrast; IR, infrared; UV, ultraviolet; NCG, no cover glass; MI, multiimmersion. From Okugawa (2000).

Table IV
Zeiss Objective Lenses for Confocal and Multiphoton Digital Imaging

Objective	Magnification	Numerical aperture	Cover glass (mm)	Working distance (mm)	Transmission % at 350 nm[a]	Transmission % at 550 nm[a]	Transmission % at 900 nm[a]	Contrast methods	Recommended for
C-Apo Water	10×	0.45	0.17	1.82	B	A	A	BF/UV/DIC	Confocal
C-Apo Water	40×	1.20	0.14–0.18	0.23	C	A	B	BF/DIC	Confocal
C-Apo Water	63×	1.20	0.14–0.18	0.24	B	A	B	BF/UV/DIC	Confocal
Fluar	10×	0.50	0.17	2.00	B	A	A	BF/D/PH/DIC/UV/IR	Confocal and two-photon
Fluar	20×	0.75	0.17	0.66	A	A	A	BF/D/PH/DIC/UV/IR	Confocal and two-photon
Fluar	40× Oil	1.30	0.17	0.14	A	A	A	BF/D/PH/DIC/UV/IR	Confocal and two-photon
Fluar	100× Oil	1.30	0.17	0.17	A	A	B	BF/D/PH/DIC/UV/IR	Confocal and two-photon
Plan-Neofluar	20×	0.50	0.17	2.00	B	A	A	BF/D/PH/DIC/UV/IR	Confocal and two-photon
Plan-Neofluar	40×	0.75	0.17	0.50	B	A	B	BF/D/PH/DIC/UV/IR	Confocal and two-photon
Plan-Neofluar	40× Oil	1.30	0.17	0.20	B	A	B	BF/D/PH/DIC/UV/IR	Confocal and two-photon
Plan-Neofluar	63× Corr. Coll.	0.95	0.10–0.20	0.12	B	A	B	BF/UV/IR	Confocal and two-photon
Plan-Neofluar	63× Oil	1.25	0.17	0.10	D	B	B	BF/PH/DIC/IR	Confocal and two-photon
Plan-Neofluar	100× Oil	1.30	0.17	0.20	C	A	B	BF/PH/DIC/IR	Confocal and two-photon
Plan-Apochromat	10×	0.45	0.17	2.80	D	A	A	BF/D/PH/DIC/IR	Confocal and two-photon
Plan-Apochromat	20×	0.75	0.17	0.61	E	A	B	BF/D/PH/DIC/IR	Confocal and two-photon
Plan-Apochromat	40× Corr. Coll.	0.95	0.10–0.20	0.16	E	A	B	BF/PH/IR	Confocal and two-photon
Plan-Apochromat	40× Oil	1.00	0.17	0.31	E	A	B	BF/PH/IR	Confocal and two-photon
Plan-Apochromat	63× Oil	1.40	0.17	0.10	E	A	C	BF/PH/DIC/IR	Confocal and two-photon
Plan-Apochromat	100× Oil	1.40	0.17	0.09	E	A	C	BF/PH/DIC/IR	Confocal and two-photon
Achroplan	10× Water	0.30	0.17	3.10	C	A	A	BF/DIC/IR	Confocal and two-photon
Achroplan	20× Water	0.50	0.17	2.00	C	A	A	BF/DIC/IR	Confocal and two-photon
Achroplan	40× Water	0.80	0.17	3.30	B	A	A	BF/PH/DIC/UV/IR	Confocal and two-photon
Achroplan	40× Water IR	0.80	0.17	3.30	E	A	A to 1200 nm	BF/DIC/IR	Two-photon
Achroplan	63× Water	0.95	0.17	2.00	C	A	A	BF/PH/DIC/IR	Confocal and two-photon
Achroplan	63× Water IR	0.90	0.17	2.00	E	A	A to 1200 nm	BF/DIC/IR	Two-photon
Achroplan	100× Water	1.00	0.17	1.00	B	A	B	BF/PH/DIC/UV/IR	Confocal and two-photon

[a] 10 nm.

The magnifications are based on $f = 160$ mm for the tube lens. The transmission data are average, typical values. They can vary slightly from lens to lens. Transmission percentages: A = >71%, B = 51–70%, C = 31–50%, D = 16–30%, E = <15%. BF, brightfield; D, darkfield; PH, phase contrast; DIC, differential interference contrast; UV, ultraviolet; IR, infrared.
From Keller (2000).

Table V

Olympus Objective Lenses for Confocal and Multi-Photon Digital Imaging

Objective	Magnification	Numerical aperture	Cover glass (mm)	Working distance (mm)	Transmission % at 350 nm[a]	Transmission % at 550 nm[a]	Transmission % at 900 nm[a]	Contrast methods	Recommended for
UPlanApo 10×	10×	0.40	0.17	3.10	B	A	C	BF/DF/DIC/UV	Confocal
UPlanApo 20×	20×	0.70	0.17	0.65	B	A	C	BF/DF/DIC/UV	Confocal and two-photon
UPlanApo 40× Oil Iris	40×	0.50–1.00	0.17	0.13	C	A	C	BF/DF/DIC	Confocal and two-photon
UPlanApo 60× W/IR	60×	1.20	0.13–0.21	0.31–0.25	C	A	B	BF/DIC/IR	Confocal and two-photon
UPlanApo 100× Oil Iris	100×	0.5–1.35	0.17	0.10	C	A	C	BF/DIC	Confocal and two-photon
PlanApo 60× Oil	60×	1.40	0.17	0.15	—	A	C	BF/DIC	Confocal and two-photon
PlanApo 40× WLSM	40×	0.90	0.17	0.16	—	A	C	BF/DIC	Confocal (FRET)
PlanApo 60× WLSM	60×	1.00	0.17	0.15	—	A	C	BF	Confocal (FRET)
Apo 20× WLSMUV	20×	0.40	0.17	0.14	B	A	B	BF/UV	UV Confocal
Apo 40× WLSMUV	40×	0.90	0.17	0.12	B	A	C	BF/UV	UV Confocal
LUMPlanFI 40× W/IR2	40×	0.80	0 (Water Dipping)	3.30	B	A	B	BF/DIC/UV/IR2	Confocal and two-photon
LUMPlanFI 60× W/IR2	60×	0.90	0 (Water Dipping)	2.00	C	A	B	BF/DIC/IR	Confocal and two-photon
LUMPlanFI 100× W	100×	1.00	0 (Water Dipping)	1.50	B	A	B	BF/DIC/UV/IR	Confocal and two-photon

[a] 10 nm.

The transmission data are average, typical values. They can vary slightly from lens to lens. Transmission percentages: A = >71%, B = 51–70%, C = 31–50%, D = 16–30%, E = <15%. BF, brightfield; DIC, differential interference contrast; UV, ultraviolet; IR, infrared.

From Matsuba (2000).

Table VI

Leica Objective Lenses for Confocal and Multi-Photon Digital Imaging

Objective	Magnification	Numerical aperture	Cover glass[a]	Working distance[a]	Transmission % at 350 nm[a]	Transmission % at 550 nm[a]	Transmission % at 900 nm[a]	Recommended for
HC Plan Apochromat	10× Imm. CS	0.40	0.17	NA	C	A	A	Confocal/two-photon
HC Plan Apochromat	20× Imm. CS Corr. Coll.	0.70	0.17	NA	C	A	A	Confocal/two-photon
HCX Plan Apochromat U-V-I	40×	0.75	0.17	NA	B	A	B	Confocal/two-photon
HCX Plan Apochromat	40× Oil CS	0.75–1.25	0.17	NA	D	A	B	Confocal/two-photon
HCX Plan Apochromat	63× Oil CS	0.60–1.32	0.17	NA	D	A	B	Confocal/two-photon
HCX Plan Apochromat	63× Water CS Corr. Coll.	1.20	0.17	NA	D	A	B	Confocal/two-photon
HCX Plan Apochromat	100× Oil CS	0.70–1.40	0.17	NA	D	A	B	Confocal/two-photon
HCX Apochromat L U-V-I	10× Water	0.30	0.17	NA	A	A	A	Confocal/two-photon
HCX Apochromat L U-V-I	20× Water	0.50	0.17	NA	B	A	A	Confocal/two-photon
HCX Apochromat L U-V-I	40× Water	0.80	0.17	NA	B	A	A	Confocal/two-photon
HCX Apochromat L U-V-I	63× Water	0.90	0.17	NA	B	A	A	Confocal/two-photon

[a] Measurements are in millimeters. NA, Not available from manufacturer. The transmission data are average, typical values. They can vary slightly from lens to lens. Transmission percentages: A = >71%, B = 51–70%, C = 31–50%, D = 16–30%, E = <15%.
From Hinsch (2000).

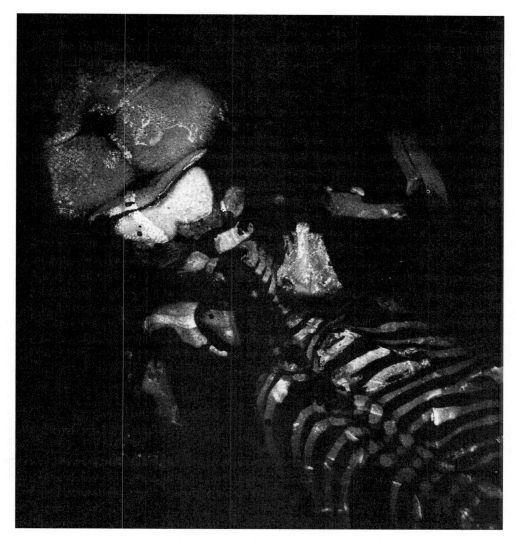

Fig. 4   Mouse embryo imaged using a CFI*60* Macro Plan Apochromat 0.5× objective with an integrated Nikon E800M (Marco) microscope and a Nikon PCM 2000 confocal imaging system. Courtesy of Farkus (2000).

incredible low-power confocal and digital images (Fig. 4) using the Macro Plan Apochromat ultrawide 0.5× objective with a Nikon Eclipse E800M (Macro) microscope and Nikon PCM 2000; similar images may be seen on The University of Pittsburgh, Imaging Center Web page (Farkas, 2000).

X. Optimum Objective Lenses for Multiphoton Applications

A. Introduction to Multiphoton

Confocal (single-photon) microscopy excites fluorophores throughout the Z-axis, however, detects only from a small fraction of the entire fluorescence generated. In confocal microscopy, optical sectioning is based on the rejection of out-of-focus information by the confocal pinhole whereby only fluorescence light is accepted from the focal plane (Denk, 2000). In multiphoton microscopy, all the advantages of confocal microscopy (linear absorption) such as optical sectioning capabilities and rejection of out-of-focus fluorescence are maintained. However, only the fluorophores from the excited plane of focus are collected. For detailed reviews of two-photon and multiphoton (three-photon) imaging, refer to Denk *et al.* (1995), Diaspro *et al.* (1999b), Denk (2000), Potter (2000), Wise (2000), Xu (2000), Lemasters (2000), and Chapter 3 of this edition. Even though multiphoton and nonlinear optical excitation have been shown to be new and powerful tools for the investigation of dynamic cell processes, the technology requires further instrumentation development before it finds widespread acceptance (Wise, 2000).

B. Advantages and Disadvantages of Multiphoton Microscopy

Denk (2000) listed three pathways to establishing a multiphoton excitation microscope and the advantages and disadvantages of each. Briefly, one can: (1) build an instrument from scratch, (2) purchase a complete commercially available system, or (3) convert an existing laser scanning microscope. Denk (2000) elaborates on the value of option 3; and references several successes. Examples of converted systems have been described in detail by Diaspro *et al.* (1999b), Potter (2000), and Mohler and Squirrell (2000).

Lemasters (2000) summarized the advantages and disadvantages of multiphoton microscopy. The advantages are: (1) thin optical sectioning capability without the pinhole requirement, (2) photobleaching (photodamage) confined to the plane of focus, (3) extended viability in thick preparations of living cells and tissues, (4) deeper penetration than single-photon into thick specimens since excitation wavelengths are red to infrared, (5) less effect of scattering and chromatic aberration since the emission does not pass through a pinhole, and (6) ability to perform bleaching and uncaging experiments in small volumes.

The disadvantages of multiphoton excitation are: (1) multiphoton excitation spectra are difficult to predict and measure, (2) sample heating can result from water absorption at multiphoton wavelengths, (3) damage to protein residues in cellular monolayers is greater in three-photon than single-photon confocal microscopy, and (4) femtosecond pulsed lasers are more expensive and difficult to maintain (Lemasters, 2000).

C. Suitable Objectives for Multiphoton Imaging

Diaspro *et al.* (1999b) modified a Nikon PCM 2000 confocal system into a dual-mode imaging system with both confocal and two-photon excitation (TPE) capabilities.

Diaspro (2000) prefered high-numerical-aperture objectives for TPE and confocal imaging. The objectives of choice were a Nikon CFI*60* Plan Fluor 100×/1.3 oil immersion and a new CFI*60* Plan Apochromat 100×/1.4 IR oil immrsion. They found the new 100× IR oil immersion objective to be optimum for TPE data acquisition. Potter (2000) reported using a Fluor 40×/0.75 dry objective with a Plan Neofluar 63×/1.2 WI to examine hippocampal brain slices. Diaspro (2000) also found the CFI*60* Plan Fluor 40×/1.3 oil immersion lens to be very useful for both TPE and wide-field microscopy. Diaspro (2000) examined large samples, i.e., oocytes, where the use of dry objectives at magnifications of 4× and 10× with high numerical apertures was found to be necessary. Another issue of concern is the possible refraction index mismatching with the sample when using immersion objectives. Last, it should be remembered that large numerical apertures are important in TPE imaging, because the fluorescence intensity varies with the fourth power of the NA of the objective.

According to Diaspro (2000), in TPE imaging, one must know the relationship between the amount of glass and the effects on pulse width, because it is not trivial to measure. For example, Diaspo (2000) is working on two measurement devices: a pulse-width sensor and a power meter in the focal region. These measurements monitor exactly the loss of power within an objective for excitation and for emission (power levels at the entrance are an order of magnitude different). The PSF measurement should always be performed.

XI. Transmission Data for Confocal and Multiphoton Objectives

Specialized applications such as IR DIC (740–950 nm), laser trapping (700–1056 nm), multiphoton excitation (700–1100 nm), fluorochromes in the lower UV to near IR (Fura-2 to Cy-7), and requirements for extended-range chromatic corrections for FISH have led manufacturers to dramatically extend the capabilities and transmission ranges of their optical systems. Therefore it sometimes becomes necessary to check on the specific spectral characteristics of an objective because of the technique involved. Improved transmissions have been affected by improved antireflection lens coatings and new computer-aided lens component design.

A very important issue for contemporary researchers is knowing the wavelength transmission values of the specific objective lenses they use for the UV (350 nm), visible (550 nm), and/or IR (900 nm) ranges. The objective lens manufacturers have been shy in publicizing the complete spectral characterization of the objectives as a function of wavelength and input power. Until now, publication of such detailed data has remained a problem since most microscope manufacturers are competitive with one another and have continued to be reluctant to publish such detailed proprietary data.

Manufacturers' objective spectral characterizations may be subject to misinterpretation since the data obtained from different manufacturers may have been tested by different techniques. Even though the method of objective transmission testing by Zeiss was reported by Keller (1995), it still cannot be ascertained that all the manufacturers test their objective lenses using exactly the same method. According to Bliton and Lechleiter

(1995), their objective transmission test criteria and specifications were not standardized between the microscope manufacturers. Some end users may perform transmission tests with the objective still mounted on the microscope where prisms and other glass components may affect the final output reading. In addition, some transmission data may be normalized providing increased transmission values when compared to others. Last, some objective data may be theoretical only (not tested). With all this potential for inequality, it seemed only fair, knowing all the variables between manufacturers, that a range of lens transmission values be published (A = >71%, B = 51–70%, C = 31–50%, D = 16–30%, E = <15%). This way the transmission data should not be misinterpreted and judgements should not be made as to one manufacturer having better objectives than another. The data should provide, however, valuable information as to which objective lens from each manufacturer is better for specific transmission values at 350 nm, 550 nm, and 900 nm. An objective with a transmission value of A or B (>50%) should be considered to have good to excellent transmission capabilities at the wavelengths specified.

Objectives for multiphoton must have high transmission capabilities in the IR range. More complete lists of suitable objective lenses for confocal and TPE or multiphoton use offered by microscope manufacturers are shown in Tables III (Nikon), IV (Zeiss), V (Olympus), and VI (Leica).

XII. Conclusions

In confocal microscopy, optimizing digital image quality requires that the imaging specialist maintain a critical balance among intensity, field size, magnification, and resolution. As for the importance of the selection of an objective lens, one should take into consideration its magnification and objective color code, numerical aperture (NA), working distance (WD), the presence and appreciation of a correction collar, the presence of a retraction stopper, and any additional application markings displayed on the objective lens. An appreciation of cover glass presence and thickness is equally important. Basically, any objective could be used for confocal and multiphoton digital imaging; however, high NA and maximum aberration correction would dictate a choice from among Plan Fluor and Plan Apochromat objectives at magnifications greater than 10×. Last, the excitation and emission wavelengths should be considered, along with the pinhole diameter in relation to the objective pupil in the emission imaging path as it becomes critical in minimizing chromatic aberration. A confocal instrument can exceed the resolution of the wide-field microscope by up to 40% (Wilson and Sheppard, 1984; Scott, 1992). Objective lens flatness of field and freedom from distortions are extremely important in reconstructing accurate three-dimensional models from confocal data sets. According to Brakenhoff et al. (1989), the three-dimensional shape of the of illumination and detection sensitivity distributions was influenced by the optical properties of the objectives used and their inherent aberrations. According to Scott (1992), achieving the highest level of resolution is dependent upon using objectives which do not cause distortion, loss of light, or flare or possess any other aberrations.

Cogswell *et al.* (1990) concluded that "on-axis" specimen scanning did not require plan objectives; however, image quality was impaired if spherical aberration became prominent. Also, low-power imaging could be obtained with high-NA, low-magnification objectives, and there would to be no falloff in brightness to the field edge. With "off-axis" beam scanning, Cogswell *et al.* (1990) noted that plan objectives were a requirement to correct for curvature of field, and that beam scanning was more susceptible to lens aberrations. They also found that it was best to use low-magnification, low-NA objectives for low-power confocal imaging to prevent loss in lateral and axial resolution. This may not necessarily be true with contemporary beam scanning confocal systems using the latest generations of microscope objectives.

Larson (2000) recommended the following Nikon CFI*60* objectives as most optimum for the Nikon PCM 2000 confocal microscope based on Airy disk measurements in relation to pinhole size at emission wavelengths of 515 nm, 595 nm, and 675 nm. These were the Plan Apochromat 20×/0.75, Plan Fluor 20×/0.75 MI, Plan Fluor 40×/0.60 LWD, Plan Fluor 40×/1.30 oil immersion, Plan Apochromat 60×/0.95, Fluor 60×/1.00 W, Plan Fluor 100×/1.30 oil immersion, and the Plan Apochromat 100×/1.40 oil immersion objectives.

From a practical-experience point of view, my personal preference is similar to that of Larson (2000) with only a few substitutions. Therefore, in an ideal situation, I would choose the following objectives lenses for my routine confocal microscopy examinations using a six-position nosepiece:

- CFI Plan Apochromat 20×/0.75 (ideal with DIC)
- CFI Plan Fluor 20×/0.75 MI (multiimmersion—oil, glycerine, water)
- CFI Fluor 40×/0.80 W Dipping Objective (ideal with DIC)
- CFI S Fluor 40×/1.30 oil immersion (ideal with DIC)
- CFI Plan Apochromat 60×/1.20 WI with cover glass correction ring
- CFI Plan Apochromat 60×/1.40 oil immersion (ideal with DIC)

This selection of objective lenses is ideal for the core facility imaging center where one is able to image nearly every application without objective replacement. It also is a marvelous selection for performing a confocal demonstration to a new prospect when there is uncertainty of the imaging challenges about to arise. Some of the listed objectives may also be used in conjunction with laser scanning transmission, differential interference contrast (DIC). Some applications may require a high-numerical-aperture, long-working-distance, low-power objective such as a Plan Aporchromat 10× for panning and finding important regions of interest. Also, there may be occasions when a Plan Apochromat 100×/1.40 oil immersion objective may be required to view very low magnification specimens such as bacteria through the wide-field microscope prior to confocal examination. Scott (1992) summed it up: objective lenses should be chosen for confocal microscopy that have the least chromatic aberration, the highest NA, and the longest working distance appropriate for the application to be studied.

Highly corrected water immersion objectives (WI) permit researchers to perform confocal, DIC, and fluorescence microscopy techniques deep within aqueous environments and living tissues obtaining images and data beyond that of oil immersion objectives. The elimination of spherical aberration can allow confocal microscopists to penetrate deeper into tissues, obtain increased apparent X-Y resolution, and create X-Z scanned images while maintaining better morphological accuracy (Brenner, 1999).

It is extremely important to appreciate that high-NA oil immersion lenses were designed to be used to a depth of no greater than 15 to 20 μm below the coverslip (Hale and Matsumoto, 1993). It might be helpful if there were more magnification variations of MI objectives, especially those working in glycerol and water environments. Inoué (1990) first suggested and Hale and Matsumoto (1993) repeated that high-NA oil immersion objectives with computer-controlled, motorized correction collars would permit automated adjustments for spherical aberration as the depth of the focal plane changed throughout the WD of the particular lens.

Dunn and Wang (2000) pointed out how chromatic and spherical aberrations play a critical role in the optical performance of the fluorescence confocal microscope. They went on to report that some lenses might be better than others because of objective design parameters dictated by the objective's intended application. From preliminary experience, Dunn and Wang (2000) found chromatic aberration in the new generation Nikon CFI60 optics to be much improved from the previous finite generation.

The research community has a responsibility to continue to communicate their requirements to the major manufacturers, and the microscope manufacturers have, in turn, the responsibility of listening and responding with positive action. The future advancement of confocal and multiphoton microscopy will be dependent upon more and better-designed immersion objectives, especially water lenses. Immediate objective requirements are: (1) lower magnifications with higher NAs, (2) increased range of multiimmersion objectives, and (3) an extended range of high-performance IR and UV objectives.

Acknowledgments

I thank Jeff Larson (Product Manager, BioScience Confocal Systems, Nikon Instruments Inc., Melville, NY) for reviewing the manuscript and offering valuable comments and contributions for preparing this manuscript. I also thank Toshiaki Nohishi (Optical Design Department, Nikon Corporation, Ofuna, JAPAN) for contributing valuable information on Nikon CFI60 objective lenses and for providing important illustrations on CFI60 optics and objectives. The critical manuscript review and technical assistance of Mel Brenner (Technical Manager, BioScience Department, Nikon Instruments Inc., Melville, NY) has been essential to this work. I am grateful to Stanley Schwartz (Manager, BioScience Department, Nikon Instruments Inc., Melville, NY) and to Ed Lieser and Kate Hendricks (Regional Managers, Nikon Instruments Inc.) for their reviews and especially for valuable discussions on CFI infinity optics specifically and technical suggestions for several other sections of this manuscript. I also thank my colleagues Yoshi Ishikawa, General Manager; Koji Hashimoto, Planning Manager; and Lee Shuett, Executive Vice President (Nikon Instruments Inc., Melville, NY) for their continued support during the writing of this manuscript. Funding for the color illustrations in this manuscript was supported in part by the BioScience Department (Nikon Instruments Inc., Melville, NY). I thank Alberto Diaspro (Department of Physics, University of Genoa, Genoa, Italy), Daniel Farkas (Bio-Imaging Laboratory, University of Pittsburgh), and D. F. Bruhn (Idaho National Engineering and Environmental Laboratory, Idaho Falls, Idaho) for making available unpublished material and for e-mail discussions during the preparation of this manuscript. Special appreciation is extended to Hisashi Okugawa (Nikon), Ernst Keller (Zeiss), Kenji Matsuba (Olympus), and Jan Hinsch (Leica) for convincing their respective factory managements to release objective transmission data for publication. The many helpful comments, suggestions, and critical reading of this manuscript by Brian Matsumoto are greatly appreciated. I also thank my wife, Bernadette, for her patience during the writing of the manuscript and for her critical final review.

References

Abbe, E. (1884). Note on the proper definition of the amplifying power of a lens or a lens-system. *J. Royal Microsc. Soc.* **4**, 348–351.

Albertson, D. G., Sherrington, P., and Vaudin, M. (1991). Mapping nonisotopically labeled DNA probes to human chromosome bands by confocal microscopy. *Genomics* **10**, 143–150.

Amos, W. B., White, J. G., and Fordam, M. (1987). Use of confocal imaging in the study of biological structures. *Appl. Opt.* **26**, 3239–3243.

Arndt-Jovin, D. J., Robert-Nicoud, M., and Jovin, T. M. (1990). Probing DNA structure and function with a multi-wavelength fluorescence confocal laser microscope. *J. Microsc.* **157**, 61–72.

Bacskai, B. J., Wallen, P., Lev-Ram, V., Grillner, S., and Tsien, R. Y. (1995). Activity-related calcium dynamics in lamprey motoneurons as revealed by video-rate confocal microscopy. *Neuron* **14**, 19–28.

Berthoud, H.-R., and Powley, T. L. (1993). Characterization of vagal innervation to the rat celiac, suprarenal and mesenteric ganglia. *J. Autonom. Nervous Syst.* **42**, 153–170.

Bliton, A. C., and Lechleiter, J. D. (1995). Optical considerations at ultraviolet wavelengths in confocal microscopy. *In* "Handbook of Biological Confocal Microscopy" (J. B. Pawley, Ed.), 2nd ed., pp. 431–444. Plenum Press, New York and London.

Bliton, C., Lechleiter, J., and Clapham, D. (1993). Optical modifications enabling simultaneous confocal imaging with dyes excited by ultraviolet- and visible-wavelength light. *J. Microsc.* **169**, 15–26.

Boxall, E. S., White, N. S., and Benham, G. S. (1993). The processing of three-dimensional confocal data sets. *In* "Multidimensional Microscopy" (P. C. Cheng, T. H. Lin, W. L. Wu and J. L. Wu, Eds.), pp. 251–266. Springer-Verlag, New York.

Boyde, A. (1985). Stereoscopic images in confocal (tandem scanning) microscopy. *Science* **230**, 1270–1272.

Boyde, A. (1986). Applications of tandem scanning reflected light microscopy and three-dimensional imaging. *Ann. N.Y. Acad. Sci.* **483**, 428–439.

Boyde, A., Dillon, C. E., and Jones, S. J. (1990a). Measurement of osteoclastic resorption pits with a tandem scanning microscope. *J. Microsc.* **158**, 261–265.

Boyde, A., Jones, S. J., Taylor, M. L., Wolfe, L. A., and Watson, T. F. (1990b). Fluorescence in the tandem scanning microscope. *J. Microsc.* **157**, 39–49.

Bradbury, S. (1984). "An Introduction to the Optical Microscope." Royal Microscopical Society Handbook 01. Oxford University Press, Oxford.

Brakenhoff, G. J., Blom, P., and Barends, P. (1979). Confocal scanning light microscopy with high aperture immersion lenses. *J. Microsc.* **117**, 219–232.

Brakenhoff, G. J., van der Voort, H. T. M., and Oud, J. L. (1990). Three-dimensional image representation in confocal microscopy. *In* "Confocal Microscopy" (T. Wilson, Ed.), pp. 185–197. Academic Press, London.

Brakenhoff, G. J., van der Voort, H. T. M., van Spronsen, E. A., and Nanninga, N. (1989). Three-dimensional imaging in fluorescence by confocal scanning microscopy. *J. Microsc.* **153**, 151–159.

Brelje, T. C., Wessendorf, M. W., and Sorenson, R. L. (1993). Multicolor laser scanning confocal immunofluorescence microscopy: practical application and limitations. *In* "Cell Biological Applications of Confocal Microscopy" (B. Matsumoto, Ed.), pp. 97–181. Academic Press, San Diego.

Brenner, M. (1999). Nikon water immersion objectives. Invited Presentation. M.S.A., Annual Meeting. Portland, Oregon.

Bruhn, D. (2000). Personal communication. Imaging Laboratory, Idaho National Engineering Laboratory, Idaho Falls, ID.

Cannell, M. B., and Soeller, C. (2000). Imaging calcium sparks in excitable cells. *In* "Imaging Neurons— A Laboratory Manual" (R. Yuste, F. Lanni, and A. Konnerth, Eds.), pp. 16.1–10. Cold Spring Harbor Laboratory Press, Cold Spring Harbor, NY.

Carlsson, K., and Liljeborg, A. (1989). A confocal laser microscope scanner for digital recording of optical serial sections. *J. Microsc.* **153,** 171–180.

Carlsson, K., Danielsson, P., Lenz, R., Lijeborg, A., Majlof, L., and Aslund, N. (1985). Three-dimensional microscopy using a confocal laser scanning microscope. *Opt. Lett.* **10,** 53–55.

Carrington, W., Fogarty, K., Lifschitz, L., and Fay, F. (1990). 3D imaging on confocal and wide-field microscopes. *In* "Handbook of Biological Confocal Microscopy" (J. B. Pawley, Ed.), revised ed., pp. 151–161. Plenum Press, New York and London.

Chalfie, M., and Kain, S. (1998). "Green Fluorescent Protein—Properties, Applications, and Protocols." Wiley-Liss, New York.

Chambers, W. (2000). LU series CFI60 objectives for reflected light. Dealer Technical Bulletin # DB076–2000A. Nikon Science and Technologies Group. Melville. 4 pp.

Cline, H. T., Edwards, J. E., Rajan, I., Wu, G. I., and Zou, D. J. (2000). In vivo imaging of CNS neuron development. *In* "Imaging Neurons—A Laboratory Manual" (R. Yuste, F. Lanni, and A. Konnerth, Eds.), pp. 13. 1–12. Cold Spring Harbor Laboratory Press, Cold Spring Harbor, NY.

Cogswell, C. J. (1995). Imaging immunogold labels with confocal microscopy. *In* "Handbook of Biological Confocal Microscopy" (J. B. Pawley, Ed.), 2nd ed., pp. 507–513. Plenum Press, New York and London.

Cogswell, C. J., and Sheppard, C. J. R. (1990). Confocal brightfield imaging techniques using an on-axis scanning optical microscope. *In* "Confocal Microscopy" (T. Wilson, Ed.), pp. 213–243. Academic Press, London.

Cogswell, C. J., Sheppard, C. J. R., Moss, M. C., and Howard, C. V. (1990). A method for evaluating microscope objectives to optimize performance of confocal systems. *J. Microsc.* **158,** 177–185.

Decho, A. W., and Kawaguchi, T. (1999). Confocal imaging of in-situ natural microbial communities and their extracellular polymeric secretion using nanoplast resin. *Bio Technics* **27,** 1246–1252.

Denk, W. (2000). Principles of multiphoton-excitation fluorescence microscopy. *In* "Imaging Neurons—A Laboratory Manual" (R. Yuste, F. Lanni, and A. Konnerth, Eds.), pp. 17. 1–8. Cold Spring Harbor Laboratory Press, Cold Spring Harbor, NY.

Denk, W., Piston, D. W., and Webb, W. W. (1995). Two-photon molecular excitation in laser-scanning microscopy. *In* "Handbook of Biological Confocal Microscopy" (J. B. Pawley, Ed.), revised ed., pp. 445–458. Plenum Press, New York and London.

Denk, W., Strickler, J. H., and Webb, W. W. (1990). Two-photon laser scanning fluorescence microscopy. *Science* **248,** 73–76.

Diaspro, A. (2000). Personal communication. Department of Physics. University of Genoa, Genoa, Italy.

Diaspro, A., Annunziata, S., Raimondo, M., and Robello, M. (1999a). Three-dimensional optical behaviour of a confocal microscope with single illumination and detection pinhole through imaging of subresolution beads. *Microsc. Res. Technique* **45,** 130–131.

Diaspro, A., Corosu, M., Ramoino, P., and Robello, M. (1999b). Adapting a compact confocal microscope system to a two-photon excitation fluorescence imaging architecture. *Microsc. Res. Technique* **47,** 196–205.

Diliberto, P. A., Wang, X. F., and Herman, B. (1994). Confocal imaging of Ca^{2+} in cells. *In* "A Practical Guide to the Study of Calcium in Living Cells" (R. Nuccitelli, Ed.), pp. 243–262. Academic Press, San Diego.

Dixon, A. J., and Benham, G. S. (1987). Applications of the confocal scanning fluorescence microscope—in biomedical research. *Am. Biotechnol. Lab.* **5,** 20–25.

Dodt, H.-U., D'Arcangelo, G., and Zieglgansberger, W. (2000). Infrared videomicroscopy. *In* "Imaging Neurons—A Laboratory Manual" (R. Yuste, F. Lanni, and A. Konnerth, Eds.), pp. 7. 1–8. Cold Spring Harbor Laboratory Press, Cold Spring Harbor, NY.

Draajier, A., and Houpt, P. M. (1988). A standard video-rate confocal laser-scanning reflected and fluorescence microscope. *Scanning* **10,** 139–145.

Dunn, K., and Maxfield, F. R. (1998). Ratio imaging instrumentation. *In* "Video Microscopy" (G. Sluder and D. E. Wolf, Eds.), pp. 217–236. Academic Press, San Diego.

Dunn, K. D., and Wang, E. (2000). Optical aberrations and objective choice in multicolor confocal microscopy. *BioTechniques* **28,** 542–550.

Egger, M. D., and Petran, M. (1967). New reflected-light microscope for viewing unstained brain and ganglion cells. *Science* **157,** 305–307.

Farkas, D. (2000). Personal communication. University of Pittsburgh, Pittsburgh, PA.

Fine, A. (2000). Confocal microscopy; principles and practice. *In* "Imaging Neurons—A Laboratory Manual" (R. Yuste, F. Lanni, and A. Konnerth, Eds.), pp. 11.1–11. Cold Spring Harbor Laboratory Press, Cold Spring Harbor, NY.

Freire, M., and Boyde, A. (1990). Study of golgi-impregnated material using the confocal tandem scanning reflected light microscope. *J. Microsc.* **190,** 285–290.

Fricker, M. D., and White, N. S. (1992). Wavelength considerations in confocal microscopy of botanical specimens. *J. Microsc.* **166,** 29–42.

Friedman, J. E., and Haddad, G. G. (1993). Major differences in Ca^{++} response to anoxia between neonatal and adult rat CA1 neurons: role of Ca^{++} and Na^{+}. *J. Neuosci.* **13,** 63–72.

Girard, S., and Clapham, D. E. (1994). Simultaneous near ultraviolet and visible excitation confocal microscopy of calcium transients in *Xenopus* oocytes. *In* "A Practical Guide to the Study of Calcium in Living Cells" (R. Nuccitelli, Ed.), pp. 263–284. Academic Press, San Diego.

Glazer, A. N., Peck, K., and Mathies, R. A. (1990). A stable double-stranded DNA–ethidium homodimer complex: application to picogram fluorescence detection of DNA in agarose gels. *Proc. Natl. Acad. Sci. USA* **87,** 3851–4855.

Graham, M., Palmer, R. J., Carney, F. P., Tan, I., Winterton, L., Andino, R., and Littlefield, S. (2000). The influence of tear composition and flow rate on bacterial adherence to soft contact lenses. ASM Conference on Biofilms 2000. Big Sky, MT. Poster # 42.

Hale, I. L., and Matsumoto, B. (1993). Resolution of subcellular detail in thick tissue sections: immunohistochemical preparation and fluorescence confocal microscopy. *In* "Cell Biological Applications of Confocal Microscopy" (B. Matsumoto, Ed.), pp. 289–324. Academic Press, San Diego.

Harrison, M. (1999). Infinity Objective Specifications. Technical Bulletin # TB008–1999A. Nikon Science and Technologies Group. Melville. 3 pp.

Hell, S. W., and Stelzer, E. H. K. (1995). Lens aberration in confocal fluorescence microscopy. *In* "Handbook of Biological Confocal Microscopy" (J. B. Pawley, Ed.), 2nd ed., pp. 347–354. Plenum Press, New York and London.

Hell, S., Reiner, G., Cremer, C., and Stelzer, E. H. K. (1991). Aberrations in confocal fluorescence microscopy induced by mismatches in refractive index. *J. Microsc.* **169,** 391–405.

Herman, B. (1998). "Fluorescence Microscopy," 2nd ed. Springer-Verlag, New York.

Hernadez-Cruz, A., Sala, F., and Adams, P. (1990). Subcellular calcium transients visualized by confocal microscopy in a voltage-clamped vertebrate neuron. *Science* **247,** 858–862.

Hinsch, J. (2000). Personal communication. Leica Microsystems Inc., Allendale, NJ.

Inoué, S. (1989). Imaging of unresolved objects, superresolution, and precision of distance measurement, with video microscopy. *In* "Fluorescence Microscopy of Living Cells in Culture, Part B: Quantitative Fluorescence Microscopy—Imaging and Spectroscopy" (D. L. Taylor and Y.-L. Wang, Eds.), pp. 85–112. Academic Press, San Diego.

Inoué, S. (1990). Foundations of confocal scanned imaging in light microscopy. *In* "Handbook of Biological Confocal Microscopy" (J. B. Pawley, Ed.), revised ed., pp. 1–14. Plenum Press, New York and London.

Inoué, S. (1995). Foundations of confocal scanned imaging in light microscopy. *In* "Handbook of Biological Confocal Microscopy" (J. B. Pawley, Ed.), 2nd ed., pp. 1–17. Plenum Press, New York and London.

Inoué, S., and Oldenborg, R. (1995). Microscopes. *In* "Handbook of Optics" (M. Bass, Ed.), pp. 17.1–52. McGraw-Hill, New York.

Inoué, S., and Spring, K. (1997a). Microscope image formation. *In* "Video Microscopy: The Fundamentals," 2nd ed., pp. 13–117. Plenum Press, New York.

Inoué, S., and Spring, K. (1997b). Practical aspects of microscopy. *In* "Video Microscopy: The Fundamentals." 2nd ed., pp. 119–162. Plenum Press, New York.

Kahana, J. A., and Silver, P. A. (1998). The uses of green fluorescent protein in yeasts. *In* "Green Fluorescent Protein—Properties, Applications, and Protocols" (M. Chalfie and S. Kain, Eds.), pp. 139–152. Wiley-Liss, New York.

Keating, T. J., and Cork, R. J. (1994). Improved spatial resolution in ratio images using computational confocal techniques. *In* "A Practical Guide to the Study of Calcium in Living Cells" (R. Nuccitelli, Ed.), pp. 221–241. Academic Press, San Diego.

Keller, H. E. (1995). Objective lenses for confocal microscopy. *In* "Handbook of Biological Confocal Microscopy" (J. B. Pawley, Ed,), 2nd ed., pp. 111–126. Plenum Press, New York and London.

Keller, H. E. (2000). Personal communication. Carl Zeiss Inc., Thornwood, NY.

Kitagawa, H. (1993). Theory and principal technologies of the laser scanning confocal microscope. *In* "Multidimentional Microscopy" (P. C. Cheng, T. H. Lin, W. L. Wu and J. L. Wu, eds.), pp. 53–71. Springer-Verlag, New York.

Kohler, R. H., Cao, J., Zipfel, W. R., Webb, W. W., and Hanson, M. R. (1997). Exchange of protein molecules through connections between higher plant plastids. *Science* **276,** 2039–2042.

Lange, K. (2000). Brochure. Center for Biofilm Engineering, Montana State University, Bozeman, MT.

Lanni, F., and Keller, H. E. (2000). Microscopy and microscope optical systems. *In* "Imaging Neurons—A Laboratory Manual" (R. Yuste, F. Lanni, and A. Konnerth, Eds.), pp. 1.1–72. Cold Spring Harbor Laboratory Press, Cold Spring Harbor, NY.

Larson, J. (1998). Use of immersion type objective lenses on Optiphot-200C. Field Notes. Nikon Science and Technologies Group. Melville, New York. FN 2/98, pp. 5–6.

Larson, J. (1999a). Understanding optical and digital resolution. Technical Bulletin. Nikon Science and Technologies Group. Melville. 6 pp.

Larson, J. (1999b). Using PCM 2000 for non-fluorescent imaging techniques. Technical Bulletin. Nikon Science and Technologies Group. Melville. 5 pp.

Larson, J. (2000). Confocal pinhole selection for PCM 2000. Technical Bulletin. Nikon Science and Technologies Group. Melville. 3 pp.

Lawrence, J. R., and Neu, T. R. (1999). Confocal laser scanning microscopy for analysis of microbial biofilms. *In* "Methods in Enzymology" (R. J. Doyle, Ed.), pp. 131–144. Academic Press, New York.

Lawrence, J. R., Wolfaardt, G. M., and Neu, T. R. (1998). The study of biofilms using confocal laser scanning microscopy. *In* "Digital Image Analysis of Microbes: Imaging, Morphometry, Fluorometry and Motility Techniques and Applications" (M. H. F. Wilkinson and F. Schut, Eds.), pp. 431–465. John Wiley & Sons, New York.

Lawrence, J. R., Manz, W., Swerhone, G. D. W., Winkler, M., and Neu, T. R. (2000). Analyses of methanol fed biofilms cultured in rotating annular bioreactors. ASM Conference on Biofilms 2000. Big Sky, MT. Poster # 119.

Lechleiter, J. D., and Clapham, D. E. (1992). Molecular mechanisms of intracellular calcium excitability in *X. laevis* oocytes. *Cell* **69,** 283–294.

Lemasters, J. J. (2000). Two-photon excitation. *In* "Light Microscopy for the Biomedical Sciences" (J. J. Lemasters, K. Jacobson, and E. D. Salmon, Eds.), pp. L15–L22. Carolina Workshops, University of North Carolina, Chapel Hill.

Majlof, L., and Forsgren, P.-O. (1993). Confocal microscopy: important considerations for accurate imaging. *In* "Cell Biological Applications of Confocal Microscopy" (B. Matsumoto, Ed.), pp. 79–95. Academic Press, San Diego.

Manos, P., and Edmond, J. (1992). Immunofluorescent analysis of creatine kinase in cultured astrocytes by conventional and confocal microscopy: a nuclear localization. *J. Compar. Neurol.* **326,** 272–282.

Mason, P., Back, S. A., and Fields, H. L. (1992). A confocal laser microscopic study of enkephalin-immunoreactive appositions onto physiologically identified neurons in the rostral ventromedial medulla. *J. Neurosci.* **12,** 4023–4036.

Masters, B. R. (1990). Confocal microscopy of ocular tissue. *In* "Confocal Microscopy" (T. Wilson, Ed.), pp. 305–324. Academic Press, London.

Masters, B. R. (1991). Confocal microscopy of the in-situ crystalline lens. *J. Microsc.* **165,** 159–167.

Masters, B. R., and Paddock, S. (1990). In vitro confocal imaging of the rabbit cornea. *J. Microsc.* **158,** 267–274.

Matsuba, K. (2000). Personal communication. Imaging Systems. Olympus America Inc., Melville.

Matsumoto, B., and Kramer, T. (1994). Theory and applications of confocal microscopy. *Cell Vision* **1,** 190–198.

Miesenbock, G. (2000). Imaging exocytosis with pH-sensitive green fluorescent proteins. *In* "Imaging Neurons—A Laboratory Manual" (R. Yuste, F. Lanni, and A. Konnerth, Eds.), pp. 59. 1–12. Cold Spring Harbor Laboratory Press, Cold Spring Harbor, NY.

Minsky, M. (1957). Microscopy apparatus. U.S. Patent No. 3,013,467.

Minsky, M. (1988). Memoir on inventing the confocal scanning microscope. *Scaning* **10,** 128–138.

Mohler, W. A, and Squirrell, J. M. (2000). Multiphoton imaging of embryonic development. *In* "Imaging Neurons—A Laboratory Manual" (R. Yuste, F. Lanni, and A. Konnerth, Eds.), pp. 21. 1–11. Cold Spring Harbor Laboratory Press, Cold Spring Harbor, NY.

Neu, T. R., and Lawrence, J. R. (2000). 2-photon laser scanning microscopy (2-plsm) of biofilm systems. ASM Conference on Biofilms 2000. Big Sky, MT. Poster # 118.

Nihoshi, T. (2000). Personal communication. Optical Design Department. Nikon Corporation. Ofuna, Japan.

Nikon BioScience Department. (1999). Digital imaging: determining minimum magnification for video and digital imaging. "Eclipse Tech Notes EN010." Nikon Science and Technologies Group. Melville, New York.

Nikon Corporation (1997). "Why Nikon has chosen the CFI*60* 200/60/25 specification for the biomedical microscope." Printed in Japan (9704–20)T.

Nikon Corporation (1998). "CFI*60* Optics." Printed in Japan (9903–05)T.

Nikon Corporation (1999). "Microscope components for transmitted light applications." Printed in Japan (9905–03)T.

Okugawa, H. (2000). Personal communication. Optical Design Department. Nikon Corporation. Ofuna, Japan.

Oldenbourg, R., Terada, H., Tiberio, R., and Inoue, S. (1993). Image sharpness and contrast transfer in coherent confocal microscopy. *J. Microsc.* **172,** 31–39.

O'Rourke, N. A., and Fraser, S. E. (1990). Dynamic changes in optic fiber terminal arbors lead to retinotopic map formations: an *in vivo* confocal microscopic study. *Neuron* **5,** 159–171.

O'Rourke, N. A., Dailey, M. E., Smith, S. J., and McConnell, S. K. (1992). Diverse migratory pathways in the developing cerebral cortex. *Science* **258,** 299–301.

Paddock, S. W. (1991). The laser-scanning confocal microscope in biomedical research. Minireview. *Soc. Exp. Biol. Med.,* 772–779.

Paddock, S. W. (1999). Confocal laser scanning microscopy. *BioTechniques* **27,** 992–1002.

Paddock, S., Mahoney, S., Minshall, M., Smith, L., Duvic, M., and Lewis, D. (1991). Improved detection of *in situ* hybridization by laser scanning confocal microscopy. *BioTechniques* **11,** 486–493.

Pawley, J. B., Amos, W. B., Dixon, A., and Brelje, T. C. (1993). Simultaneous, non-interfering collection of optimal fluorescent and backscattered light signals on the MRC 500/600 Proc. 51st Annual Meeting of the Microscopical Society of America, pp. 156–157.

Periasamy, A., Skoglund, P., Noakes, C., and Keller, R. (1999). An evaluation of two-photon excitation versus confocal and digital deconvolution fluorescence microscopy imaging in *Xenopus* morphogenesis. *Microsc. Res. Technique* **47,** 172–181.

Petran, M., Hadravsky, M., Egger, M. D., and Galambos, R. (1968). Tandem scanning reflected light microscope. *J. Opt. Soc. Am.* **58,** 661–664.

Petran, M., Boyde, A., and Hadravsky, M. (1990). Direct view confocal microscopy. *In* "Confocal Microscopy" (T. Wilson, Ed.), pp. 245–283. Academic Press, London.

Piston, D. W. (1998). Choosing objective lenses: the importance of numerical aperture and magnification in digital optical microscopy. *Biol. Bull.* **195,** 1–4.

Ploem, J. S., and Tanke, H. J. (1987). "Introduction to Fluorescence Microscopy." Royal Microscopical Society Handbook 10. Oxford University Press, Oxford.

Potter, S. M. (2000). Two-photon microscopy for 4D imaging of living neurons. *In* "Imaging Neurons—A Laboratory Manual" (R. Yuste, F. Lanni, and A. Konnerth, Eds.), pp. 20. 1–16. Cold Spring Harbor Laboratory Press, Cold Spring Harbor, NY.

Ranganathan, R., Bacskai, B. J., Tsien, R., and Zuker, C. S. (1994). Cytosolic calcium transients: spatial localization and role in *Drosophila* photoreceptor cell function. *Neuron* **13,** 837–848.

Schwartz, S. (2000). Personal communication. BioScience Department, Nikon Instruments Inc., Melville.

Schwartz, S., Brenner, M., and Benham, G. S. (1996). Confocal imaging for the individual researcher: acquiring the tools in a competitive funding environment. *Am. Lab.* **28,** 52–58.

Scott, M. L. (1992). Choosing optics for confocal microscopy. *Am. Lab.* **24,** 32–37.

Shaw, P. J., and Rawlins, D. J. (1991). The point-spread function of a confocal microscope: its measurement and use in deconvolution of 3-D data. *J. Microsc.* **163,** 151–165.

Shaw, P., Highett, M., and Rawlins, D. (1992). Confocal microscopy and image processing in the study of plant nuclear structure. *J. Microsc.* **166,** 87–97.

Sheppard, C. J. R. (1977). The scanning optical microscope. *IEEE J. Quantum Electron* **QE-13** 100.

Sheppard, C. J. R., and Wilson, T. (1984). Image formation in confocal scanning microscopes. *Optik* **55,** 331–342.

Sheppard, C. J. R., Xiaosong, G., Gu, M., and Roy, M. (1995). Signal-to-noise in confocal microscopes. *In* "Handbook of Biological Confocal Microscopy" (J. B. Pawley, Ed.), revised ed., pp. 363–371. Plenum Press, New York and London.

Smallcombe, A., and Benham, G. S. (1993). Effective laser lines for biological confocal microscopy. *In* "Multidimensional Microscopy" (P. C. Cheng, T. H. Lin, W. L. Wu and J. L. Wu, eds.), pp. 267–290. Springer-Verlag, New York.

Stelzer, E. H. K. (1993). Designing a confocal fluorescence microscope. *In* "Multidimensional Microscopy" (P. C. Cheng, T. H. Lin, W. L. Wu and J. L. Wu, eds.), pp. 33–51. Springer-Verlag, New York.

Stelzer, E. H. K., and Wijanaendts-van-Resandt, R. W. (1990). Optical cell splicing with the confocal fluorescence microscope: microtomoscopy. *In* "Confocal Microscopy" (T. Wilson, Ed.), pp. 199–212. Academic Press, London.

Steyger, P. (1999). Assessing confocal microscopy systems for purchase. *Methods* **18,** 435–446.

Summers, R. G., Stricker, S. A., and Cameron, R. A. (1993). Applications of confocal microscopy to studies of sea urchin embryogenesis. *In* "Cell Biological Applications of Confocal Microscopy" (B. Matsumoto, Ed.), pp. 265–287. Academic Press, San Diego.

Tan, I. (2000). Personal communication. CIBA Vision Corporation. Duluth, GA.

Taylor, D. L., and Salmon, E. D. (1989). *In* "Fluorescence Microscopy of Living Cells in Culture, Part A: Fluorescent Analogs, Labeling Cells, and Basic Microscopy" (Y.-L. Wang and D. L. Taylor, Eds.), pp. 207–237. Academic Press, San Diego.

Terasaki, M., and Dailey, M. E. (1995). Confocal microscopy of living cells. *In* "Handbook of Biological Confocal Microscopy" (J. B. Pawley, Ed.), 2nd ed., pp. 327–346. Plenum Press, New York and London.

Thomas, J., Karakiozis, J., Capehart, K., Crout, R., Sincock, S., and Robinson, P. (2000). Classifying periodontal subgingival biofilms directly with vital probes: flow cytometry and confocal laser scanning microscopy. ASM Conference on Biofilms 2000. Big Sky, MT. Poster # 40.

Tsien, R. Y., and Bacskai, B. J. (1995). Video-rate confocal microscopy. *In* "Handbook of Biological Confocal Microscopy" (J. B Pawley, Ed.), 2nd ed., pp. 459–478. Plenum Press, New York and London.

Tsien, R. Y., and Harootunian, A. T. (1990). Practical design criteria for a dynamic ratio imaging system. *Cell Calcium* **11,** 93–109.

Vassy, J., Rigaut, J. P., Hill, A.-M., and Foucrier, J. (1990). Analysis by confocal scanning laser microscopy imaging of the spatial distribution of intermediate filaments in foetal and adult rat liver cells. *J. Microsc.* **157,** 91–104.

Visser, T. D., Oud J. L., and Brakenhoff, G. J. (1990). Refractive index and axial distance measurements in 3-D microscopy. *Optik* **90,** 17–18.

Watson, T. F. (1990). The application of real-time confocal microscopy to the study of high-speed dental-burr-tooth-cutting interactions. *J. Microsc.* **157,** 51–60.

White, J. G., Amos, W. B., and Fordham, M. (1987). An evaluation of confocal versus conventional imaging of biological structures by fluorescence light microscopy. *J. Cell Biol.* **105,** 41–48.

Wijnaendts van Resandt, R. W., Marsman, H. J. B., Kaplan, R., Davoust, J., Stelzer, E. H. K, and Strickler, R. (1985). Optical fluorescence microscopy in three-dimensions: microtomoscpy. *J. Microsc.* **138,** 29–34.

Wilson, T. (1990). Optical aspects of confocal microscopy. *In* "Confocal Microscopy" (T. Wilson, Ed.), pp. 93–141. Academic Press, London.

Wilson, T., and Carlini, A. R. (1987). Size of the detector in confocal imaging systems. *Opt. Lett.* **12,** 227–229.

Wilson, T., and Sheppard, C. (1984). "Theory and Practice of Scanning Optical Microscopy." Academic Press, London.

Wise, F. W. (2000). Lasers for multiphoton microscopy. *In* "Imaging Neurons—A Laboratory Manual" (R. Yuste, F. Lanni, and A. Konnerth, Eds.), pp. 18. 1–9. Cold Spring Harbor Laboratory Press, Cold Spring Harbor, NY.

Wright, S. J., Centonze, V. E., Stricker, S. A., DeVries, P. J., Paddock, S. W., and Schatten, G. (1993). Introduction to confocal microscopy and three-dimensional reconstruction. *In* "Cell Biological Applications of Confocal Microscopy" (B. Matsumoto, Ed.), pp. 1–45. Academic Press, San Diego.

Xu, C. (2000). Two-photon cross sections of indicators. *In* "Imaging Neurons—A Laboratory Manual" (R. Yuste, F. Lanni, and A. Konnerth, Eds.), pp. 19. 1–9. Cold Spring Harbor Laboratory Press, Cold Spring Harbor, NY.

Yang, X., Beyenal, H., Harkin, G., and Lewandowski, Z. (2000). Quantifying biofilm structure using image analysis. *J. Microbiol. Methods* **39,** 109–119.

CHAPTER 7

Resolution of Subcellular Detail in Thick Tissue Sections: Immunohistochemical Preparation and Fluorescence Confocal Microscopy

Irene L. Hale and Brian Matsumoto

Department of Biological Sciences and Neuroscience Research Institute
University of California, Santa Barbara
Santa Barbara, California 93106

METHODS IN CELL BIOLOGY, VOL. 70

I. Introduction: Development of Fluorescence Microscopy Techniques for Resolution of Subcellular Detail

A. Study of Cytoskeletal Organization in Cultured Cells

A question such as "how is the cytoskeleton organized?" can be answered adequately only by using high-resolution microscopy techniques that also provide three-dimensional information. Transmission electron microscopy (TEM) satisfies the first criterion, but is not convenient for three-dimensional analysis. Conversely, brightfield light microscopy, because of its shallow depth of field, can provide three-dimensional information conveniently by "optical sectioning" of thick specimens, but makes detection and resolution of small structures, such as individual cytoskeletal filaments, difficult. The theoretical limit of resolution of the light microscope is approximately 200 nm; however, much smaller structures can be visualized by enhancing their optical signal (Inoué and Spring, 1997). For example, the sensitivity and high contrast of fluorescence microscopy, coupled with the specificity of antibody labeling, have permitted the detection of individual cytoskeletal filaments even though their diameter is smaller than the limit of resolution. Osborn *et al.* (1978) compared immunofluorescence and TEM images of the same cultured cell and demonstrated that, under optimal conditions, single microtubules can be visualized by immunofluorescence microscopy. When measured by TEM, the diameter of these microtubules, which were labeled by indirect immunofluorescence (microtubule plus two IgG layers), was approximately 55 nm, much less than the limit of resolution of two point sources (R) by Rayleigh's criterion:

$$R = 1.22\lambda/2\text{NA} \tag{1}$$

This value, which applies to lateral (xy) resolution, is approximately 230 nm for the wavelength of light ($\lambda = 520$ nm, i.e., fluorescein-conjugated antibody) and the objective lens numerical aperture (NA 1.4; 2NA applies to epiillumination) used in this study.

In this example, microtubules will have to be farther than 230 nm apart in the horizontal plane to be resolved. Axial (z) resolving power is lower than lateral resolving power. In the vertical dimension, two point sources must be two times farther than the Rayleigh limit apart, 460 nm in this example, to be resolved (see Inoué and Spring, 1997, for a discussion of resolution). Hence, immunofluorescence microscopy is a powerful tool for studying the cytoskeleton, enabling the cell biologist to observe both fine details and the three-dimensional structure of cytoskeletal arrays, as long as individual filaments are not too densely packed.

Historically, this level of detection and resolution has been attainable only with cultured cells. The main reason for this restriction is that monolayers of cells are an almost ideal optical pathway for fluorescence microscopy, whereas thicker tissue sections are not. For instance, in a monolayer, generally there are few fluorescent structures above or below the plane of focus, so that in-focus structures are seen against a dark background and the potential for high contrast with the fluorescent signal is realized. Another reason higher resolution images have been obtained with cultured cells is that the specimen preparation methods for monolayers are less damaging than many of those for intact tissues. Finally, the deleterious effects of spherical aberration are minimized by the thinness of the preparation (see Chapter 4 of this volume).

B. Problems Encountered with Intact Tissues

The thickness of intact tissues presents several problems. The first two are related to the poor optical quality of tissue. It is not homogeneous with respect to optical properties such as refractive index and absorption, so light can be refracted, reflected, and scattered by structures between the plane of focus and the objective lens. One result is that light is attenuated as it passes through the tissue. Compounding this problem, when tissue is labeled with fluorescent probes, fluorochromes overlying the object of interest absorb some of the illuminating light. In addition, light from outside the plane of focus creates a high background signal in conventional fluorescence microscopy. Light spreads with distance along the vertical (optical) axis as a solid cone from fluorochromes in the focal plane, as well as from those outside it, which also are excited by the illuminating light. Autofluorescence can be another source of background. The resultant decrease in intensity of both exciting and emitted light and the increase in intensity of background light with thickness reduce sensitivity and contrast (Cheng and Summers, 1990; Inoué, 1990). This reduction in signal-to-noise ratio hinders detection of small objects and prevents discrimination of neighboring structures that, theoretically, should be resolved by a high numerical aperture lens. In short, the in-focus structures merge with the background.

The second problem is that objective lenses correct for spherical aberration only under specified conditions (Keller, 1990). These conditions include the refractive indices and dispersions of the components of the optical path between the object of interest and the objective, and the distance through these media that light traverses to reach the objective. Each component also must be optically homogeneous. Components may include immersion medium, cover glass, specimen mounting medium, and the tissue itself. For instance, the specified refractive index for oil immersion lenses is 1.51, which is the refractive index of most immersion oils, cover glasses, and mounting media such as Permount. Focusing down into a thick, unembedded specimen introduces an increasingly thick layer of tissue, which, overall, has a refractive index closer, depending on the mounting medium, to that of water (1.33) or glycerol (1.47) and more importantly is not uniform in its optical properties. As the optical path is altered, correction for spherical aberration is lost. Consequently, resolution declines with depth (Keller, 1990; Chapter 4 of this volume). The loss of correction also results in a decline in image intensity, in addition to the factors discussed above. A third problem is that lenses with a high numerical aperture for high-resolution work have short working distances, about 100–150 μm. A fourth problem is that antibodies and many other probes cannot penetrate far into tissues.

The problems associated with thickness necessitate mechanical sectioning of tissues for both antibody labeling and imaging purposes. Although sectioning solves these problems, it can introduce others. First, to prepare tissues for sectioning, they often are embedded in paraffin or resin. Structural alteration, most notably shrinkage, occurs with embedding procedures. When volume is decreased, structures are more tightly packed, reducing the ability to resolve them. Infiltration of paraffin and polymerization of some resins require temperatures hot enough to induce autofluorescence. Furthermore, embedding procedures may block antibody access to binding sites, alter antigen structure, or extract antigens, in turn decreasing fluorescence signal intensity. Therefore, pre- or nonembedding labeling methods often are preferable. Second, without some technique for removing the out-of-focus fluorescence signal, sections often need to be 5 μm thick or less for good contrast. Because some cells are larger and/or extend processes greater distances than this, serial sectioning may be required to view whole cells. Mechanical sectioning introduces artifacts and makes it difficult to gather three-dimensional information.

C. Solutions to Problems with Intact Tissues

One of the goals of our research in retinal cell biology has been to observe cytoskeletal organization in the intact retina, rather than in enzymatically or mechanically dissociated cells (Nagle *et al.,* 1986; Vaughan and Fisher, 1987; Sale *et al.,* 1988), which previously had been the best preparation for resolving fine detail and at the same time obtaining three-dimensional information. Although individual, fluorescently labeled retinal cells can be imaged with good contrast, important information about the intact tissue is lost. In the intact retina it is possible to (1) see the orientation and distribution of cells within the retina, (2) compare cytoskeletal and other features among neighboring cells within a population, which is especially important for assessing responses to drug treatments and other manipulations, and (3) trace cell–cell contacts and synaptic connections. Another advantage is that cells remain whole, whereas during dissociation they can lose processes or, in the case of photoreceptors, break apart just distal to the nucleus. Additionally, the plasma membranes of the outer segment, the photosensitive compartment, and the inner segment, the metabolic compartment, remain as distinct domains. In rod photoreceptors these membranes can fuse during dissociation (Spencer *et al.,* 1988). We have been able to realize our goal by taking advantage of techniques that solve some of the problems outlined in Section I.B. Gentle procedures for specimen preparation and techniques for removing the out-of-focus fluorescence signal have been equally important for maximizing detection and effective resolution.

We initially developed a method combining preembedding labeling of thick tissue sections with resin embedment and thin sectioning. Thick sectioning permits antibody labeling throughout the tissue, while mechanical thin sectioning physically removes much of the out-of-focus fluorescence. We sectioned retinal tissue for fluorescent probe labeling with a Vibratome (model 1000; Technical Products International, Polysciences, Warrington, PA), as an alternative to frozen sectioning. [See Priestley (1984) and Larsson (1988) for reviews of Vibratome methods.] The Vibratome cuts thick sections ($>20\,\mu$m) of tissues with a transversely vibrating blade, so that soft tissues sustain minimal compression and can be supported with a noninfiltrating matrix, such as agarose. Vibratome sectioning produces minimal ultrastructural damage. It has been used successfully for TEM examination of immunolabeling, for example, with antibodies to neuropeptides and neurotransmitters and their synthesizing enzymes (reviewed in Priestley and Cuello, 1983), a Golgi apparatus enzyme (Novikoff *et al.*, 1983), and microtubules and microtubule-associated proteins in neurons (Matus *et al.*, 1981). It is gentle enough for preparing retinal tissue for culture (Mack and Fernald, 1991). We embedded labeled Vibratome sections in JB-4 (Polysciences), a hydrophilic resin, because it does not require tissue dehydration with organic solvents. JB-4 embedding seems to cause less extraction in general and, essential for our work, does not extract phalloidin, which is removed by alcohols. In addition, this resin exhibits low autofluorescence and can be polymerized on ice, to avoid tissue autofluorescence induced by heating. Sections 3–5 μm thick can be cut, providing adequate contrast for epifluorescence microscopy. Contrast can be enhanced further by using an objective lens with a diaphragm, which can be partially closed to reduce flare, and by using video image enhancement techniques.

Matsumoto *et al.* (1987) first employed these procedures to label actin with fluorescent phalloidin, a probe specific for filamentous actin, and to localize filaments in pseudopodia of the retinal pigment epithelium during phagocytosis of rod outer segment membranes. This combination of techniques has provided sufficient structural preservation and contrast to reveal domains of the actin (phalloidin labeling) and tubulin (antibody labeling) cytoskeletons in frog photoreceptors that are difficult to observe in other intact retinal preparations. Matsumoto (unpublished observations) (Fig. 1A) was able to detect the concentration of actin filaments in the ciliary stalk and resolve it from the nearby actin-filled, calycal processes. Vaughan *et al.* (1989) used this method to view the microtubule array in the ellipsoid region (shown in Figs. 2A and 2B), which is composed of a number of individual microtubules or small bundles that radiate from one of the basal bodies at the base of the ciliary stalk.

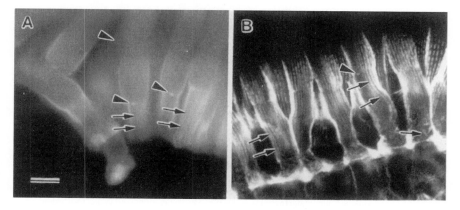

Fig. 1 Comparison of epifluorescence and laser scanning confocal microscopy (LSCM) for viewing the actin cytoskeleton in photoreceptors. In both (A) and (B), 100-μm-thick Vibratome sections of frog retina fixed in 0.1% glutaraldehyde–4% formaldehyde were labeled with rhodamine-conjugated phalloidin. Vibratome sections were cut parallel to the long axis of the photoreceptors. (A) An unenhanced epifluorescence image of a thin section (5 μm) cut from a Vibratome section embedded in JB-4 resin. (B) An LSCM image (single optical section, approximately 0.75 μm thick) of an unembedded thick section. The greater contrast with LSCM makes it easier to resolve actin cables and smaller bundles (arrows) in the photoreceptor inner segment. The actin network in the ciliary stalk (arrowheads) can be detected with both techniques. Scale bar = 10 μm.

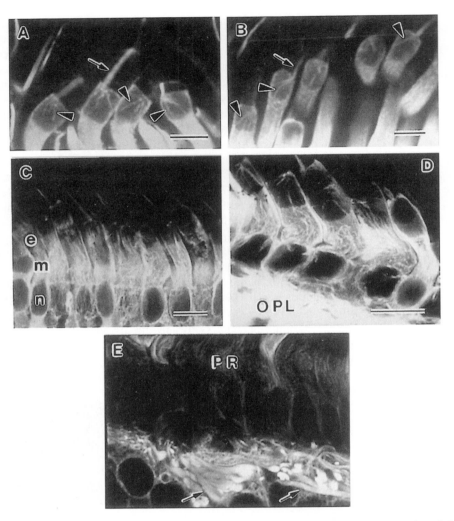

Fig. 2 Antitubulin labeling of photoreceptors and neuronal processes. (A) and (B) show the microtubule array in the ellipsoid regions of longitudinally sectioned frog rod photoreceptors, which consists of a number of what are probably individual microtubules radiating from one of the basal bodies beneath the ciliary stalk. LSCM provides sufficient contrast to visualize these microtubules in thick tissue sections. Arrows indicate the ciliary microtubules in the photoreceptor outer segment. These images also illustrate that detergent treatment does not damage these microtubules. This tissue was fixed in 4% formaldehyde alone. The length and number of microtubules (arrowheads) in the ellipsoid region appear to be the same in images of Vibratome sections labeled without [e.g., (A)] or with [e.g., (B)] detergent (1.0% Triton X-100) present throughout the antibody incubations (30 h) and rinses. Scale bar = 10 μm. (C) and (D) demonstrate the ability of the LSCM to resolve the closely spaced microtubules in the myoid region (m) of frog photoreceptors. For orientation, the ellipsoid region (e) and nucleus (n) are also marked. Two optical sections (about 0.75 μm thick) were projected one on top of the other to generate these images. Extending the depth of field in this way allows structures such as microtubules to be followed for greater distances, as they pass in and out of two or more thin optical

We have been able to substitute optical sectioning with the laser scanning confocal microscope (LSCM) (MRC-500; Bio-Rad, Cambridge, MA) for mechanical sectioning to generate the thin sections necessary for high contrast and resolution. In point-scanning confocal microscopes such as this one, the illuminating light, a narrow laser beam, is scanned over the specimen and a pinhole aperture prevents light arising from outside the plane of focus from reaching the detector, a photomultiplier tube (reviewed in Brakenhoff *et al.,* 1989; White *et al.,* 1990; Inoué, 1990; Chapter 1 of this volume) (see Fig. 4 in Chapter 1 for a light ray diagram). Although out-of-focus light cannot be eliminated completely, in practice the fluorescence collected can be confined to an optical section as thin as 0.5–0.7 μm (objective lens NA 1.4) (Chapters 1 and 10 of this volume). This thickness is close to the axial resolution limit, for example, 0.46 μm for fluorescein emission (see Section I.A). Laser scanning confocal microscopy allows the shallow depth of field of the light microscope to be used effectively. The increased contrast acheived with the LSCM can be seen in Fig. 1; actin bundles in the photoreceptor inner segment that are indistinct in a 5-μm-thick JB-4 section without image enhancement (A) are evident in an optical section 4 μm below the surface of a 100-μm-thick Vibratome section (B). There are two other benefits of confocal microscopy to our work. First, resin or paraffin embedding is no longer necessary, and consequently structural alteration and autofluorescence caused by these procedures is eliminated (see Section II.B). Second, three-dimensional information can be gathered with ease. The combination of immunofluorescence and LSCM makes the goal of viewing cytoskeletal arrays and other subcellular detail in three dimensions in intact tissues not only theoretically possible for the cell biologist, but also feasible.

To take advantage of this potential, we have attempted to work out specimen preparation parameters, drawing on the preembedding Vibratome method, and confocal imaging parameters for resolution of subcellular detail in thick tissue sections. Of course, there is no single best tissue preparation and labeling method applicable to all probes or every antigen–antibody combination. In this article we present a convenient nonembedding method that requires no special equipment other than a Vibratome. This method is suitable for antigens that are not sensitive to aldehyde fixation and are not extracted by liquid-phase fixation or subsequent detergent permeabilization. With it we have successfully labeled a variety of antigens, including cytoskeletal, integral membrane, and soluble proteins, using antibodies as well as phalloidin and lectins. There are trade-offs

---

sections. (C) An image collected 15 μm below the surface of a thick Vibratome section. Image quality is still good at this depth, compared to (D), which was collected at the surface. Scale bar in (C) = 10 μm. (B), (D), and (E) demonstrate the extremely wide intensity range possible with fluorescently labeled tissue. In (B), labeling intensity in the myoid is sufficiently great that pixel intensities in this region are all near maximum when parameters are set to image preferentially the weak ellipsoid microtubule signal. In (D), the outer plexiform layer (OPL) is so bright that neuronal processes are obscured. (E) Beam intensity and gain had to be substantially reduced, to the point that the signal in the photoreceptors (PR) is almost undetectable, in order for individual horizontal cell processes (arrows) and other finer processes to be seen in the OPL. The power of LSCM to resolve fine detail in thick tissue sections is displayed again by this single optical section (about 0.75 μm). (D) and (E) are neighboring fields at the same magnification. Scale bar in (D) = 10 μm.

between many of the parameters involved, so the challenge has been to achieve a collective compromise between them. These parameters and their impact on each other are discussed in Sections II and III. Detailed procedures are provided in Section V.

II. Immunohistochemical Preparation of Thick Tissue Sections

A. Fixation

Immunocytochemists always have been faced with a trade-off between the quality of tissue preservation and the retention of antigenicity. Preembedding techniques alleviate this problem, because they require only a relatively light fixation prior to antibody labeling, and because the dehydration and embedding procedures themselves can reduce antigenicity or extract antigens (Larsson, 1988). Preembedding immunocytochemistry introduces an additional trade-off, however, one between preservation and depth of penetration of antibodies and other probes into the tissue (Larsson, 1988; Priestley *et al.*, 1992). Both the quality of tissue preservation and the intensity of the fluorescent signal, partly determined by retention of antigenicity or of the antigen in the tissue, limit attainable resolution in laser scanning confocal microscopy (LSCM), while the depth of probe penetration limits the three-dimensional information available. A knowledge of the properties of various fixative agents, with these trade-offs in mind, is helpful when choosing a fixative, but ultimately the choice must be made empirically for each antigen–antibody combination and each type of tissue. For several proteins we have found that an acceptable balance between these demands can be achieved with light aldehyde fixation. It may be desirable to include agents such as taxol (Chapter 10 of this volume) or ethylene glycol–bis(β-aminoethyl ether)–N,N,N',N'-tetraacetic acid (EGTA) (Osborn and Weber, 1982) to preserve microtubules optimally in some tissues. For complete discussions of other fixatives and conditions for particular immunocytochemical applications, the reader is referred to Osborn and Weber (1982) for microtubules, Priestley (1984) for neurotransmitters, enzyme markers, and neuropeptides, and Larsson (1988) and Priestley *et al.* (1992) for a general review.

Formaldehyde, alone or in combination with other agents, is a commonly used fixative for preembedding immunocytochemistry, because it is relatively mild and partially reversible (reviewed in Sternberger, 1979; Larsson, 1988). It forms stable methylene cross-bridges (reviewed in Puchtler and Meloan, 1985), but cross-links less extensively than glutaraldehyde. Thus it retains antigenicity well for many antigens and does not impede antibody penetration greatly. At the same time, formaldehyde is effective enough to preserve ultrastructure well, as observed when combined with ultracryotomy (Larsson, 1988). Its effectiveness is due partly to its rapid penetration. Cells are killed quickly enough to prevent autolytic damage (Pease, 1962). We have successfully used buffered 4% formaldehyde alone, pH 7.3 (granular paraformaldehyde; Electron Microscopy Sciences, Ft. Washington, PA) to fix retinal tissue for phalloidin staining (Molecular Probes, Eugene, OR) of filamentous actin (not shown) and for immunolabeling of cytoskeletal filaments [actin, not shown; tubulin, Fig. 2; glial fibrillary acidic

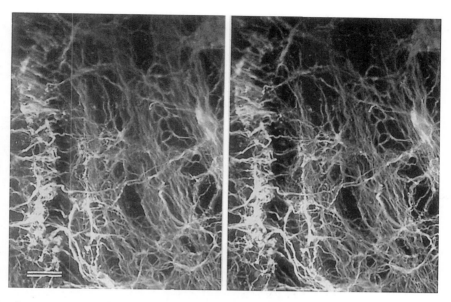

Fig. 3   Stereo pair of astrocytes labeled with an antibody to GFAP, a glial cell-specific intermediate filament protein. The tissue was fixed in 4% formaldehyde and permeabilized with 0.1% Triton X-100. LSCM images were collected at 0.5-μm increments through 12 μm near the vitreal surface of a whole mount of cat retina. The optical sections are parallel to the vitreal surface, giving an *en face* view. To create the stereo pair, images were projected with a pixel shift of -0.5 for the left and $+0.5$ for the right. This stereo pair shows the power of using LSCM to optically section unembedded tissue for studying the distribution and orientation of cell processes in three dimensions. The fine processes radiating from several astrocytes are distinct, and their three-dimensional organization is easily traced. Several processes extend to contact a blood vessel on the left. Scale bar $= 20$ μm.

protein (GFAP), Fig. 3; vimentin, not shown], integral membrane proteins (opsin, Fig. 4A–4C; high molecular weight rim protein, not shown), and soluble proteins [cellular retinaldehyde-binding protein (CRALBP), Fig. 5; actin monomers, not shown]. (See Table I for a complete description of these primary antibodies.) In our studies this fixative provided the highest anti-tubulin labeling intensity, when compared to glutaraldehyde (0.1–1.0%), glutaraldehyde–formaldehyde combinations, or cold ($-10°$C) methanol. Labeling with actin, opsin, and rim protein antibodies was higher with formaldehyde alone than with glutaraldehyde–formaldehyde. Formaldehyde fixation retained CRALBP in the Müller cell, a type of glial cell, and the retinal pigment epithelium (Fig. 5A), and soluble actin in the photoreceptor. In addition, it preserved cytoskeletal filaments and the Golgi apparatus in the rod photoreceptor well. In fact, although overall cellular morphology was better preserved with glutaraldehyde–formaldehyde than with formaldehyde alone, the appearance of the microtubule array in the ellipsoid region and of the Golgi apparatus was indistinguishable between the two fixatives (comparisons not shown; Figs. 2A and 4C, respectively, are examples of these structures). Microtubules were not disorganized, fragmented, or lost, and the Golgi apparatus was not collapsed

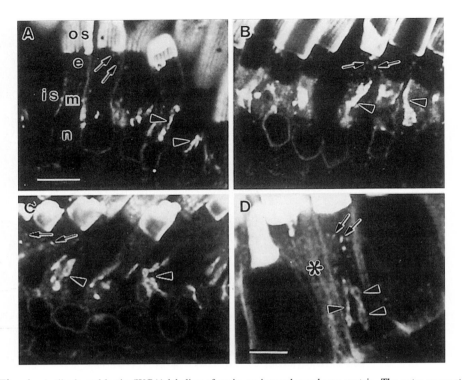

Fig. 4 Antibody and lectin (WGA) labeling of opsin, an integral membrane protein. The outer segment (os), inner segment (is), ellipsoid (e) and myoid (m) regions, and nucleus (n) are marked in (A) for orientation. Vibratome sections were cut parallel to the long axis of the photoreceptors. (A–C) Effects of detergents on anti-opsin (15–18) labeling intensity in formaldehyde-fixed frog rod photoreceptors. Antibody solutions contained either detergent (A), 0.1% saponin (B), or 0.1% Triton X-100 (C). To allow comparisons of labeling intensity, the aperture size, neural density filter, gain, and black level were held constant and set such that the maximum intensity value for the brightest specimens was just below the upper limit of the scale. All images then were photographed and printed at the same exposure. Optical section thickness was about 1.0 μm. Labeling of the Golgi apparatus (arrowheads) and the area surrounding it, which probably is the endoplasmic reticulum (see Section II.C.1), was most intense with saponin (B). Labeling of vesicles in the ellipsoid region (small arrows) that may be post-Golgi transport vesicles was more intense with both detergents, as shown in (B) and (C). Scale bar = 10 μm. (D) Wheat germ agglutinin labeling of rod photoreceptor inner segment membranes is similar to that seen with anti-opsin, except for the absence of endoplasmic reticulum staining. Additionally, the extracellular matrix surrounding the photoreceptors is labeled as well, as seen between the photoreceptors and along the surface of the rod marked with an asterisk. This image is a projection of four optical sections so that the equivalent of an approximately 3-μm section through the cell is obtained. The advantage of such projections is that a greater portion of large organelles can be seen at the same time without the use of three-dimensional display techniques. Arrowheads indicate three segments of the Golgi that are at different depths. Scale bar = 5 μm.

or vesiculated. Formaldehyde alone is an excellent fixative for many antigen–antibody combinations and is a good preliminary choice for testing this Vibratome technique.

We obtained the best preservation if tissues were fixed for about 8 h or longer, and if they were stored in formaldehyde rather than buffer. Retinal tissues stored in buffer for longer than 1 week prior to labeling became soft and showed decreased fluorescence image sharpness. Formaldehyde fixation has been reported to be partially reversible by washing. Because formaldehyde forms stable cross-bridges, however, much of the fixative is not washed out by buffer. Perhaps storage in formaldehyde stabilizes cellular components that are not cross-linked but are more loosely bound by the fixative.

Because glutaraldehyde (a dialdehyde) is a more effective cross-linker than formaldehyde alone, it achieves superior preservation of cell shape (Sabatini *et al.*, 1963) and certain structures, for example, microtubules (Eckert and Snyder, 1978; Osborn and Weber, 1982), and better retention of soluble molecules, for example, neuropeptides (Larsson, 1988) and dyes. The cross-linking activity of glutaraldehyde, however, can reduce enzyme activity (Sabatini *et al.*, 1963), antigenicity and antibody penetration (Priestley, 1984; Larsson, 1988; Chapter 10, this volume). We have observed reduced labeling with both anti-tubulin and anti-opsin, especially with glutaraldehyde concentrations of 0.5% or greater. For these reasons, many researchers use low concentrations (<0.5%) of this fixative in combination with formaldehyde. In our studies, a brief (<45 min) fixation in 0.1% glutaraldehyde (8% EM grade; Polysciences)–4% formaldehyde, pH 7.3, noticeably improved overall preservation, as judged by photoreceptor cell shape, photoreceptor outer segment membrane organization, and extraction visible by differential interference contrast. With formaldehyde fixation, photoreceptors sometimes appear stretched lengthwise, and the membranous disks that make up the photosensitive outer segment become less tightly stacked (photoreceptor domains are marked in Fig. 4A). Even this light glutaraldehyde fixation, however, reduced antibody labeling intensity as much as threefold and the depth of tubulin antibody labeling, partly determined by the depth of antibody penetration, by 25–30% (see Section II.C.1). For some antibodies, for example, anti-tubulin, a decrease in labeling intensity is tolerable, because good image contrast can still be obtained (Figs. 2D and 2E). To allow comparisons of labeling intensity, the aperture size, illumination intensity, gain, and black level were held constant for all specimens and set such that the maximum intensity value for the brightest specimens overall or for the brightest structure(s) of interest did not exceed the upper limit of the intensity scale. The average pixel intensity over several areas containing the structure(s) of interest was then calculated for comparison.

Another problem with glutaraldehyde is that it produces both nonspecific labeling and/or autofluorescence, which can greatly reduce contrast. High background fluorescence in the plane of focus negates the gain in contrast achieved by the removal of out-of-focus fluorescence with LSCM. We have found that, fortunately, the level of background fluorescence is relatively low with light glutaraldehyde fixation. It should be emphasized that this is true only if a dilute (10% or less) stock solution sealed under nitrogen is used, the stock solution is stored frozen after opening, and the glutaradehyde is added to the fixative solution the same day it is used. Background fluorescence can be minimized further by (1) fixation on ice (not advisable for microtubules, which can be

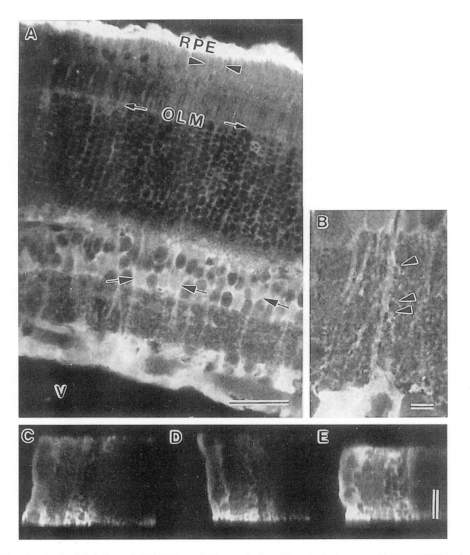

Fig. 5 Antibody labeling of CRALBP, a soluble protein found in retinal pigment epithelium (RPE) and Müller glial cells. (A) Cat retinal tissue was fixed in 4% formaldehyde and permeabilized with 0.1% Triton X-100. The Vibratome section was cut parallel to the long axis of the photoreceptors. The position and orientation of these cell types within the cat retina, as well as their shape, are revealed in great detail, as CRALBP appears to be distributed throughout the entire cytoplasm. The RPE apposes the photoreceptors, and fine villous processes (arrowheads) can be seen projecting among the photoreceptor outer segments. Müller cells extend from just above the photoreceptor nuclei, where they form a barrier termed the outer limiting membrane (OLM, small arrows), to the vitreous (V), where they expand to form large end feet in the ganglion cell layer. Labeled Müller cell nuclei (large arrows) are present in the inner nuclear layer. Scale bar = 20 μm. (B) Müller cells also extend fine processes laterally. Several processes (arrowheads) can be traced as they branch from a cell in this higher magnification image. For orientation, part of the end-foot region is visible in the lower left.

Table I
Primary Antibodies

Anti-actin: Recognizes nonmuscle actin. Polyclonal to chicken gizzard, dilution = 1/100 (Biomedical Technologies, Inc., Stoughton, MA).

Anti-cellular retinaldehyde-binding protein (CRALBP): Specific for the vitamin A carrier protein present in retinal pigment epithelium and Müller glial cells. Polyclonal to bovine, dilution = 1/400 (purified IgG) (Bunt-Milam and Saari, 1983).

Anti-glial fibrillary acidic protein (GFAP): Specific for an intermediate filament protein found in glial cells. Polyclonal to bovine, dilution = 1/400 (Dako, Carpinteria, CA).

Anti-high molecular weight rim protein: Recognizes an integral membrane protein restricted to the rims of the membranous disks that make up the photoreceptor outer segment. Polyclonal to frog, dilution = 1/200 (serum) (Papermaster et al., 1978).

Anti-opsin (15-18): Binds the photopigment of rod and the single-cone photoreceptors. Monoclonal to turtle opsin (surface loop between helices IV and V, lumenal/extracellular membrane face), dilution = 1/50 (hybridoma culture fluid) (Gaur et al., 1988).

Anti-opsin (rho 4D2): Binds the photopigment of rod photoreceptors. Monoclonal to bovine rhodopsin (NH_2 terminus, lumenal/extracellular membrane face), dilution = 1/50 (hybridoma culture medium) (Hicks and Molday, 1986).

Anti-β-tubulin: Recognizes the β subunit of tubulin. Monoclonal antibody, dilution = 1/1000 (ascites fluid) (Chu and Klymkowsky, 1989).

Anti-vimentin: Specific for an intermediate filament protein found in glial cells. Monoclonal to human, dilution = 1/400 (Dako, Carpinteria, CA).

---

cold labile), (2) secondary fixation with paraformaldehyde alone, (3) thorough rinsing (Vaughn et al., 1981), and most effectively, by (4) reduction with sodium borohydride (Fisher Scientific, Pittsburg, PA) (Weber et al., 1978; Bacallao et al., 1990). Thus, if a light fixation with glutaraldehyde does not block access to or alter the antigenic determinant in question, preservation can be enhanced without too great of a decrease in signal-to-noise ratio.

Cold ($-10°C$) methanol, a precipitating fixative, preserves cytoskeletal components well, and because it extracts lipids it enhances antibody penetration (Osborn and Weber, 1982; Larsson, 1988). We have not adopted methanol fixation for retinal tissue for two reasons. First, retinas fixed in cold methanol alone are soft, making it difficult to handle and section them. Second, we obtained lower labeling intensity with this tissue, using anti-tubulin, when compared to retinal tissue fixed in formaldehyde alone. Because methanol is a precipitating fixative, the decreased labeling intensity may reflect blocking of antibody access to the antigenic determinant. We have not tried methanol as a secondary fixative, following formaldehyde.

Scale bar = 5 μm. (C–E) Comparison of anti-CRALBP penetration into formaldehyde-fixed cat retinal tissue with no detergent (C), 0.1% saponin (D), or 1.0% Triton X-100 (E). Images are vertical scans through Vibratome sections approximately 100 μm thick, collected at constant intensity settings, as described for Figs. 4A–4C. These images show the increase in immunolabeling depth obtained with Triton X-100, but not with saponin. They also show that labeling intensity is not changed by detergent treatment. [Note that only one surface of the sections in (C) and (D) is well stained, probably as a result of inadequate stirring during incubation in antibody solutions.] Scale bar (vertical) = 40 μm.

B. Vibratome Sectioning

Most tissues cannot be stained or viewed whole, and therefore must be sectioned. Thick sections for immunofluorescence LSCM can be obtained by sectioning resin- or paraffin-embedded tissue, by cryosectioning frozen tissue (the technique most widely used), or by sectioning unembedded tissue with the Vibratome. If resin is used for postembedding labeling, it is necessary to etch away the plastic matrix so that sections can be stained throughout their depth. Hydrophobic resins, such as Spurr's or Epon, can be removed with sodium ethoxide treatment; however, this can reduce the tissue's antigenicity. To our knowledge, hydrophilic resins used commonly for postembedding immunocytochemistry, including LR White and Lowicryl, cannot be etched. We have adopted a nonembedding method for three reasons: superior structural preservation, superior contrast, and convenience. Structural alteration with embedding techniques is caused by extraction and shrinkage. Shrinkage obscures fine detail by collapsing structural components so that they become packed too close together and can no longer be resolved. We have measured an approximately 30% decrease in photoreceptor inner segment diameter (photoreceptor domains are marked in Fig. 4), which is about a 50% decrease in cross-sectional area, with paraffin and JB-4 embedment, compared to formaldehyde fixation with no embedment. Measurements were taken near the center of about 15 cells from 3 sections per treatment in 2 experiments. The ability to resolve structures is affected also by contrast, which depends on high labeling intensity and low background. We compared opsin antibody labeling [antibody (15–18); see Table I] of photoreceptor inner segments in formaldehyde-fixed retinas without (Vibratome sections) or with embedment. Labeling was more intense if the tissue was not embedded than after it was embedded in epoxy resin (etched sections). Labeling in unembedded and paraffin-embedded (dewaxed sections) (Paraplast X-tra; Ted Pella, Tustin, CA) retinas was comparable if Vibratome sections were not detergent permeabilized. When detergent was added to the antibody incubation solution, however, labeling intensity increased, for instance, about twofold for opsin immunolabeling with saponin treatment (see Section II.C.1; Figs. 4A and 4C). With tubulin, GFAP, and CRALBP antibodies, labeling intensity was reduced slightly after paraffin embedment. A much more serious problem with these embedding procedures is that they require heating to 55–60°C, which can induce autofluorescence. We have observed increases in background fluorescence great enough to begin obscuring small structures and fine detail. The importance of minimizing shrinkage and autofluorescence cannot be overemphasized.

Both freezing and Vibratome techniques can preserve cellular structure and antigenicity well. In the case of antigens that are extracted by liquid-phase fixation or are sensitive to certain fixatives, freeze substitution or freeze-drying are the techniques of choice (see Larsson, 1988, for a discussion of these techniques). For other antigens, freezing and Vibratome techniques are convenient and probably equally effective, as long as ice crystal formation during freezing is prevented. We chose Vibratome sectioning because Vibratomes are relatively inexpensive and easy to use. The other sections in this article are applicable to either technique.

The Vibratome technique is convenient; it is straightforward and consists of only a few short steps. After fixation, tissue pieces first are immersed in an agarose or gelatin solution, for example, a 5% solution of low gelling temperature (37°C) agarose (type XI; Sigma, St. Louis, MO), which does not actually infiltrate the tissue, but simply gels around it to provide support. We use low gelling temperature agarose to prevent heat damage and a high level of tissue autofluorescence. Then the agarose is cut into blocks, each containing a tissue piece, and glued onto supports. The block and the Vibratome blade are submerged in a liquid bath in the boat of the Vibratome during sectioning. No further processing is necessary before sections are labeled with fluorescent probes. Sections between 25 and a few hundred microns can be cut; however, practical considerations limit section thickness. We have found that retinal sections less than 40 μm thick can be too fragile to handle. Sections thicker than 40–50 μm may not be useful if individual cells or populations of cells that occupy the entire thickness must be viewed, because antibody penetration is limited and because image intensity and resolution decline with depth. In this case, because high-resolution work is restricted to the upper 15–20 μm (see Section III.A), sections could be viewed from both surfaces.

C. Labeling with Fluorochrome–Conjugated Probes

1. Penetration of Antibodies

To meet the goal of obtaining three-dimensional information, it must be possible for probes to penetrate some distance into thick sections. Limited penetration of antibodies and other probes into Vibratome sections is possible without permeabilization treatments; however, this may be insufficient for three-dimensional analysis. Penetration of 15 μm, that is, 1–1.5 hepatocytes, into sections of liver tissue has been reported by Novikoff *et al.* (1983) for unconjugated antibody followed by horseradish peroxidase-conjugated protein A. We have observed penetration of approximately 7 to 22 μm into retinal sections for unconjugated followed by fluorochrome-conjugated antibodies, depending on the antibody and tissue region (see, e.g., Figs. 5C and 6A). Note that these values actually are measurements of the depth of antibody labeling. Antibodies penetrate farther than this, but the level of binding depends on the concentration at a given depth in relation to antibody affinity and antigen concentration (Priestley *et al.*, 1992). Cell membranes can be permeabilized by several methods, including freeze–thawing and extracting lipids with organic solvents (dehydration–rehydration) or detergents (surfactants). (See Priestley, 1984, and Larsson, 1988, for reviews of permeabilization methods.) To enhance penetration and thoroughly rinse out unbound probe, we add detergent to all incubation and rinse buffers. We have not tried permeabilizing with detergent before or at the same time as fixation, although extraction of unpolymerized cytoskeletal subunits increases contrast and such procedures may further improve penetration (procedures reviewed in Osborn and Weber, 1982; Larsson, 1988). We have found that whereas probes as large as fluorochrome-conjugated avidin (m_r 66,000] penetrate at least 50 μm into tissue permeabilized with 0.1% Triton X-100 (Sigma) (biotechnology reagent, electrophoresis grade; Fisher), penetration of larger molecules, for example, antibodies (IgG; $m_r \sim 150,000$), is more restricted.

We compared depth of antibody labeling in Vibratome-sectioned retinal tissue without detergent and with L-α-(palmitoyl)lysophosphatidylcholine[(palmitoyl)lysolecithin], a

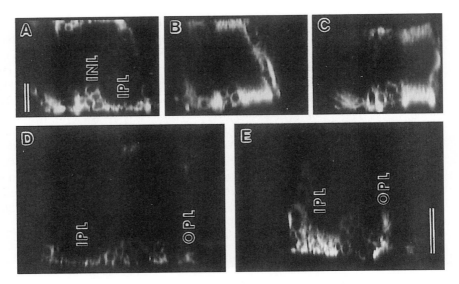

Fig. 6 Increase in tubulin antibody penetration and labeling intensity with Triton X-100 treatment. Images are vertical (xz) scans through thick (75–100 μm) Vibratome sections of frog retina. (A–C) The increase in depth of immunolabeling in the inner plexiform layer (IPL) when sections are permeabilized with Triton X-100. The following treatments are shown: (A) no detergent; (B) 1–2 h pretreatment with 0.1% Triton X-100; and (C) 1.0% Triton X-100 present during antibody incubations (30 h) and rinses. (A) also shows the variation in depth among different regions of the frog retina, for example, between the inner nuclear layer (INL) and the inner plexiform layer (IPL), which is composed principally of neuronal processes. Scale bar (vertical) = 25 μm. (D and E) Comparison of antitubulin labeling intensity without detergent (D) or with 1.0% Triton X-100 (E). Intensity settings and photographic conditions were kept constant, as described for Figs. 4A–4C. Labeling in the plexiform layers (IPL and OPL) was about twice as intense when Triton X-100 (1.0%) was present during antibody incubations. [Note that for (A), the gain was increased to match the intensity of (C).] Scale bar = 40 μm.

zwitterionic detergent (Sigma); Triton X-100, a nonionic detergent; and saponin, a sapogenin glycoside (Sigma). Zwitterionic and nonionic detergents stabilize protein conformation when used to solubilize membrane proteins. Lysolecithin, a natural component of cell membranes, has been used for live cell work (Barak *et al.,* 1980). Saponin, which may selectively extract cholesterol, has been reported to be less extractive or damaging to membranes than Triton X-100 by the criterion of ultrastructural appearance (Ohtsuki *et al.,* 1978; Willingham and Yamada, 1979). We tested three concentrations of Triton X-100: 0.1, 0.2, and 1.0%. We used saponin and lysolecithin at 0.1%. The results of this comparison are compiled in Table II; percentage increases in immunolabeling depth, compared to no detergent treatment, are averages of values from several Vibratome sections in two experiments. Depth was measured from vertical (xz) scans through the sections, such as those in Figs. 6A–6C. The depth of intense labeling was measured, although weak labeling extends farther. Sections were incubated in primary antibodies for 18–20 h and in secondary antibodies for 12–14 h. Depth was decreased with shorter incubation times, whereas it was not substantially increased with longer incubation times.

Table II

Increase in Depth of Immunolabeling with Detergent Permeabilization

Detergent	Percentage increase[a]			
	Antitubulin[b]		Anti-CRALBP[c] Form.[d]	Anti-opsin Form.[d]
	Form.[d]	Glut.[d]		
Triton X-100 (0.1%, 1- to 2-h pretreatment)	80	40	NM[e]	NM
Triton X-100 (0.1%)	100	55	40	125
Triton X-100 (0.2%)	170	60	NM	NM
Triton X-100 (1.0%)	200	140	45	NM
Saponin (0.1%)	75	15	0	80
Lysophosphatidylcholine (0.1%)	75	20	NM	0

[a] Increase over no detergent treatment.

[b] Depth in inner plexiform layer (mainly neuronal processes; see Fig. 2B).

[c] Depth in inner plexiform layer (Müller glial cells; see Fig. 6A).

[d] Fixation: Form, 4% formaldehyde; Glut., 0.1% glutaraldehyde–4% formaldehyde.

[e] NM, Not measured.

Detergent was present during all incubations and rinses, unless otherwise stated. All incubations were performed at 4°C. For other details, see Section V.C. Sections of frog retina fixed in 4% formaldehyde or 0.1% glutaraldehyde–4% formaldehyde were used for anti-tubulin and sections of formaldehyde-fixed frog retina were used for anti-opsin, whereas sections of formaldehyde-fixed cat retina were used for anti-CRALBP.

Detergent permeabilization enhanced penetration of anti-β-tubulin, -CRALBP, and -opsin antibodies, with two exceptions. No increase was apparent with saponin treatment for anti-CRALBP or with lysolecithin for anti-opsin. Overall, saponin and lysolecithin were less effective than Triton X-100, presumably because they removed fewer membrane components. For the antibody to tubulin, labeling depth in the inner plexiform layer (Figs. 6A–6C), which is composed mainly of neuronal processes, increased with Triton X-100 concentration. In formaldehyde-fixed tissue, the depth with 0.1% Triton X-100 was twice that with no detergent, and with a concentration of 1.0% was three times greater. A 1- to 2-h pretreatment with 0.1% Triton X-100 was surprisingly effective; immunolabeling depth was increased almost as much as when detergent was present throughout antibody incubations. Brief detergent pretreatment may be useful if lengthy incubation in Triton X-100 extracts membrane components of interest. For antitubulin, depth of labeling in the inner plexiform layer in formaldehyde-fixed tissue ranged from approximately 7 μm without detergent to about 21 μm with 1.0% Triton X-100. Penetration is more restricted with fixation in 0.1% glutaraldehyde–4% formaldehyde. Without detergent, the depth of labeling was only slightly reduced; however, with 1.0% Triton X-100 it was limited to about 16 μm, a 25% reduction.

There were also differences in immunolabeling depth among tissue regions and among antibodies. For example, labeling with the antibody to tubulin extended deeper into the inner nuclear layer, which is composed mainly of cell bodies, than the inner plexiform layer, which is composed mainly of neurites; in formaldehyde-fixed tissue without detergent, labeling depths were approximately 22 μm (about three cell bodies) and 7 μm, respectively (Fig. 6A). In the inner nuclear layer, labeling depth was increased only slightly with detergent treatment. Glutaraldehyde–formaldehyde fixation reduced the depth of anti-tubulin labeling in the inner nuclear layer by approximately 30% for all treatments. Detergent treatment does not overcome the barrier to penetration presented by increased cross-linking with glutaraldehyde. Anti-CRALBP labeling extended deepest overall, even without detergent. The depth in the inner retina, where the Müller glial cells are located (shown in Fig. 5A), was about 20 μm with no detergent and about 29 μm with 1.0% Triton X-100 (Figs. 5C and 5E). In the retinal pigment epithelium (identified in Fig. 5A) it was lower, about 15 μm, and was not enhanced by detergent treatment. Anti-opsin labeling extended approximately 7 μm (about one cell diameter) into the photoreceptor layer with no detergent and 16 μm (about 2.5 cell diameters) with 0.1% Triton X-100. These variations indicate that factors other than detergent type and concentration, possibly including the type of antigen/organelle being labeled, structural features of the cell/tissue, and the degree of fixative cross-linking, as well as the concentration of the antigen and affinity of the antibody, affect antibody penetration.

In addition to the depth of antibody labeling, the intensity of labeling was also increased by detergent treatment for anti-tubulin and anti-opsin. The method of estimating intensity is described in Section II.A. Labeling of microtubules was increased uniformly in the plexiform and photoreceptor layers by the three detergents tested, up to about twofold with Triton X-100 (all concentrations), as shown in Figs. 6D and 6E. The effect of detergents on the intensity of labeling of the integral membrane protein, opsin, in rod photoreceptors was more complex.

The results for antibody (15–18) (see Table I) are as follows. Lysolecithin (0.1%) enhanced only Golgi apparatus labeling, when compared to no detergent (not shown). Saponin and Triton X-100 (both at 0.1%) altered opsin labeling intensity differentially for different membranous organelles. These two detergents increased labeling intensity for some organelles, while leaving others unchanged or reduced. Both enhanced labeling of the ellipsoid region vesicles and Golgi apparatus, saponin more so than Triton X-100 (Figs. 4A–4C). Saponin treatment also dramatically enhanced labeling of what likely is the endoplasmic reticulum (Fig. 4B). This fairly uniform immunolabeling throughout the myoid region appeared identical to that seen with the lectin, concanavalin A, in this compartment (data not shown). This lectin recognizes α-D-mannose and α-D-glucose, which are components of the immature N-linked oligosaccharides added to secretory and membrane glycoproteins in the endoplasmic reticulum. Concanavalin A has been demonstrated to be a marker for this organelle in several cell types with high secretory activity (reviewed in Lis and Sharon, 1986). Opsin contains two N-linked oligosaccharides and is a major biosynthetic product of the photoreceptor. Hence, concanavalin A would be expected to label the endoplasmic reticulum of this cell (reviewed in Bok, 1985). Labeling of the endoplasmic reticulum with antibody (15–18) was difficult to detect with the other treatments (e.g., Figs. 4A and 4C). Thus, whereas saponin enhanced immunolabeling of endoplasmic reticulum membranes, Triton X-100 and lysolecithin did not. Detergent treatment previously has been shown to substantially improve immunolabeling of opsin (Hicks and Barnstable, 1987) and other membrane proteins (Pickel,

1981). Perhaps these detergents do so for antibody (15–18) by altering the membrane environment in such a way that antibody access or binding efficiency is improved, but only for particular, and different, organelles. On the other hand, treatment with saponin and Triton X-100 reduced inner segment plasma membrane labeling. Although no single treatment, including incubation in antibodies without detergent, was optimal, saponin provided the most comprehensive localization information for this monoclonal antibody to opsin.

Adding to the complexity, different results were obtained with another monoclonal antibody to opsin, rho 4D2 (see Table I) (data not shown). Faint endoplasmic reticulum labeling with this antibody was apparent without detergent treatment and was increased to a lesser extent than that of other organelles by both Triton X-100 and saponin. In this case, Triton X-100 enhanced labeling of vesicles and Golgi slightly more than saponin. Variable effects of these detergents have been reported previously by Goldenthal *et al.* (1985), who observed reduced labeling of integral membrane proteins with Triton X-100 versus saponin treatment for some antibodies, but not others. Therefore it is worth the effort to test antibodies to membrane proteins without and with several detergents to obtain accurate assessments of protein presence and content in various membranous organelles. The choice of detergent for simultaneous labeling of a membrane protein and, for example, cytoskeletal elements should be based on this type of information.

We did not observe by fluorescence microscopy any damaging effects of the detergent treatments described here. At 0.1%, none of the detergents altered Golgi apparatus morphology in any obvious way, nor was there definitive evidence that they extracted opsin from photoreceptor membranes. Using the microtubules in the ellipsoid region of rods as test structures, we could not detect fragmentation or loss of microtubules in formaldehyde-fixed tissue, even at the highest Triton X-100 concentration (1.0%) (Figs. 2A and 2B). We had worried that membrane permeabilization might result in loss of soluble proteins from cells in formaldehyde-fixed tissue, but labeling intensity of CRALBP, a 33-kDa soluble protein, was not reduced by the detergent treatments we tested, including 1.0% Triton X-100 (compare Fig. 5C with Figs. 5D and 5E). Apparently, there is no trade-off between penetration and retention of structural integrity and integral membrane or soluble proteins at these concentrations. The depth of immunolabeling can vary substantially depending on how long sections are incubated in antibody and how well the incubation solutions are stirred (see Section V.C), as well as with the factors discussed above. In addition, although sections are incubated free floating in antibody solutions, sometimes only one surface of the section is well stained (compare, e.g., Figs. 5C and 5D with Fig. 5E). For these reasons, we recommend performing an initial vertical (*xz*) scan to quickly assess penetration depth, and to determine whether the well-stained surface is facing the coverslip.

2. Fluorochromes

Signal intensity in LSCM varies not only with the binding efficiency of the probe, the abundance of the molecule being labeled, and any effect of tissue processing on these factors, but also with the fluorochrome used. There are a variety of fluorescent labels that can be conjugated to probes. The fluorescence intensity of a dye is determined by the product of the extinction coefficient, that is, probability of absorption, and the quantum yield, that is, probability of fluorescent emission, at a given wavelength. In addition, other properties such as absorption/emission maximum, emission lifetime, and photo-stability, together with the emission wavelengths of the laser, the wavelength selectivity of barrier filters, the wavelength sensitivity of the detector, and the environment of the fluorochrome, all determine the performance in LSCM of a given fluorochrome. The environment can be critical. Factors including the proximity to each other of dye molecules conjugated to a probe can have a dramatic effect on quantum yield. Environmental factors such as pH affect the photostability of some dyes. (See Tsien and Waggoner, 1990, and Chapter 5 of this volume for reviews of fluorochromes and their properties.) For multiple-label imaging, the signals from two or more fluorochromes must be balanced to minimize crosstalk through barrier filters. (See Chapter 5 of this volume for a discussion of the parameters important in multiple-label confocal imaging.) Finally, contrast depends on the level of background fluorescence, and the molecules responsible for autofluorescence have characteristic absorption and emission spectra that also must be considered. This section touches only briefly on this complex topic and is intended to give an idea of the interplay among some of these elements.

Such interplay means that, in the end, the choice of fluorochrome(s) must be made empirically for each system. For example, features of the confocal system we use and the cell type we study present conflicting requirements. The Bio-Rad MRC-500 light source is an argon ion laser and the detector is a photomultiplier tube. Hence, fluorescein (FITC; absorption maximum 490 nm, emission maximum 520 nm) should be a better choice than rhodamine (TRITC; absorption maximum 554 nm; emission maximum 573 nm), because its absorption maximum is better matched to the emission wavelengths (488 and 514 nm) of the argon ion laser and its emission maximum is better matched to the spectral sensitivity of the photomultiplier tube, which is two to three times more sensitive to green light (see Fig. 12 of Chapter 4, this volume). Moreover, fluorescein has a higher quantum yield. In spite of this, we often have found it preferable to use rhodamine for immunolabeling of structures in the ellipsoid region of photoreceptors (photoreceptor domains are marked in Fig. 2), because there is strong autofluorescence from the mitochondria, which are concentrated in this region of the cell, that overlaps with and thus masks fluorescein emission. Although contrast is increased with rhodamine, we still have some difficulty detecting small structures, such as microtubules and transport vesicles, because of the low emission and detection of this dye in the MRC-500 confocal system. One solution to these problems is to use lasers that excite rhodamine more efficiently than the 514-nm line of the argon ion laser. The krypton–argon ion laser used with the MRC-600 and the green helium–neon laser have emission lines at 568 nm and 543 nm, respectively. A different type of solution is a sulfoindocyanine dye, CY3.18 (CY3), developed in Waggoner's laboratory (Southwick *et al.,* 1990; Yu *et al.,* 1992). It has absorption/emission maxima similar to that of rhodamine, but a significantly greater extinction coefficient. More importantly, it was designed to interact less when several dye molecules are bound to a single antibody, as dye interactions usually reduce quantum yield substantially (Tsien and Waggoner, 1990). The CY3-conjugated secondary antibodies (available from Jackson ImmunoResearch, West Grove, PA) provide a signal intensity comparable to that of fluorescein in a region of the spectrum where autofluorescence intensity is low. These properties make CY3 especially useful for labeling molecules that compose small structures, as above, or are in low abundance, and for imaging at deeper optical planes within thick tissue sections (see Section III.A). However, the greater extinction coefficient of CY3 potentially limits its usefulness for imaging deep within tissue, because of self-shadowing. In addition, CY3 is more photostable than fluorescein. Protein conjugates of another cyanine dye, CY2, with absorption/emission maxima similar to that of fluorescein (Amersham, Piscataway, NJ; Jackson ImmunoResearch, West Grove, PA) and of the Alexa 488 and 546 (emission maxima) dyes (Molecular Probes, Eugene, OR) also exhibit robust fluorescence emission and photostability (Jackson ImmunoResearch, http://www.jacksonimmuno.com/2001site/home/catalog/f-cy3-5.htm; Panchuk-Voloshina *et al.*, 1999).

Photostability is important, because multiple scans of each optical plane must be collected for averaging or integration, and multiple planes need to be imaged to gather three-dimensional information. In the latter case, as fluorochromes outside the plane of focus also receive some illumination, photobleaching can be a serious problem when many optical sections are collected. While rhodamine, CY2, CY3, and the Alexa dyes are pH insensitive, for fluorescein the pH of the mounting medium greatly affects photostability. This dye is much more stable at or above pH 8 than it is at pH 7 (see Chapter 5). We rinse fluorescein-labeled sections in pH 8 phosphate-buffered saline before mounting. Mounting in a solution containing an antioxidant, such as *n*-propyl gallate (Giloh and Sedat, 1982) or *p*-phenylenediamine (Johnson *et al.*, 1982), protects against bleaching and photodamage to the specimen. We have found that for many dyes bleaching is noticeably decreased in a hydrophobic environment, for example, resin (embedded tissue) or glycerol. Therefore, we use a 5% solution of *n*-propyl gallate in 100% glycerol, which is diluted slightly by the buffer in the tissue sections. These steps dramatically reduce photobleaching with our specimens. Because dyes are effected by the mounting environment, care must be taken when choosing a mounting medium for a specific dye. For example, some Alexa dyes perform best in an aqueous medium, whereas others perform best in mountants containing glycerol and an antifade reagent(s) (Panchuk-Voloshina *et al.*, 1999). As another example, the anti-fade reagent *p*-phenylenediamine cleaves the cyanine dyes; CY2 especially is sensitive (Jackson ImmunoResearch, http://www.jacksonimmuno.com/2001site/home/techq_3.htm). The fluorescent signal becomes weak and diffuse with storage. The alternative reagents, *n*-propyl gallate and DABCO, are recommended.

III. Confocal Imaging Parameters for Maximum Resolution

A. Microscope

We have worked out these parameters using a Bio-Rad MRC-500 laser scanning confocal microscope with an argon ion laser light source and a photomultiplier tube detector. The same imaging parameters apply to other point-scanning systems. The most critical variable for successful confocal imaging is alignment of the emitted beam with the pinhole aperture, which is achieved by adjusting the positions of the laser scanning mirrors. Even slight misalignment substantially reduces image intensity and contrast. For this reason, final alignment should be performed at the objective lens magnification and electronic zoom setting that will be used, because fine adjustments often are necessary after initial alignment with low-power lenses.

Three other imaging parameters are especially important to attaining the goal of maximizing contrast with LSCM, thereby maximizing detection and attainable resolution. The first is the size of the pinhole aperture, which determines the extent to which out-of-focus light is excluded. Aperture size is variable in this system. At the minimum aperture size, the thickness of the optical section is minimized, approaching the axial resolution limit of the objective lens (see Section I.C). As the confocal aperture is opened, increasing levels of out-of focus light obscure fine detail (see Section I.B). Tables III and V in Chapter 10 provide information to determine the range of optical section thicknesses possible for several objective-lens numerical apertures. The second imaging parameter is incident light intensity, which can be reduced by neutral-density filters. Because the signal from fluorochromes saturates at a lower illumination intensity than autofluorescence, and scattered light signals do not saturate, keeping the illumination below the level of fluorochrome saturation maximizes contrast (Tsien and Waggoner, 1990). A compromise usually must be reached when setting these two parameters, such that neither is optimum by itself, but the combination produces the sharpest images. In general, we have obtained the best results by opening the aperture enough to allow the use of a neutral density filter that transmits 3% or less of the light. The third parameter is gain. We have found it preferable to set the gain somewhat below maximum to reduce electronic noise.

When these three parameters are adjusted for maximum contrast, the LSCM has minimum sensitivity. Remember that sensitivity is as important as contrast for detecting small structures. This trade-off creates the need for high labeling intensity and objective lenses with high light gathering capability and transmission (see Section III.B). In our experience, CY3-conjugated secondary antibodies give the required high signal. For tubulin primary and CY3-conjugated secondary antibodies, we have been able to use an NA 1.4 objective lens with a 1% transmission filter and with the aperture at minimum size. This was not possible with rhodamine or fluorescein. For this reason, CY3 noticeably improves our ability to resolve detail in the photoreceptor myoid region. Figure 2D is an image obtained with these parameters. A benefit of using a 3% or less transmission filter and a medium gain setting is that fewer scans per image are required when background signals (autofluorescence and scattering) and electronic noise are low, reducing photobleaching and saving time.

A fourth parameter affecting xy resolution is the image scale, that is, pixel size, which is determined by the magnification of the objective lens and the electronic zoom setting. According to the Nyquist criterion, the zoom setting should be chosen to give a pixel size less than or equal to one-half the theoretical resolution of the objective (Inoué and Spring, 1997). If the pixel size is too large, the maximum resolution is not achieved (see Chapter 4); if it is too small, unnecessary photobleaching results.

The problem of exchanging sensitivity for contrast with LSCM is aggravated by a decline in image intensity with increasing depth into the specimen (see Section I.B). The decline in intensity with depth can be compensated for by increasing the gain or switching to a higher transmission neutral-density filter at deeper optical planes. Increasing the number of scans accumulated for each image is then desirable, in order to compensate for the resultant increase in noise. Increasing the gain is helpful only up to the point at which electronic noise interferes with image contrast. Similarly, increasing illumination intensity is beneficial only until autofluorescence and scattering begin to obscure the primary signal, or until photobleaching becomes severe. For these reasons, Art (1990) suggests direct photon counting as a better solution for weak signals. A photon-counting circuit greatly reduces noise, so that weak signals are easier to detect above background. This option is available on the Bio-Rad MRC-600.

When collecting serial optical sections, termed Z series, expanding the intensity scale after each image is collected can help to maintain constant image intensity. This can be done with a pixel remap function. Pixel remapping is not effective if the intensity range across the scan area is broad; as long as the brightest areas remain at maximum value, the scale will not be adjusted. Note that remapped images no longer provide accurate quantitative intensity information. If sections are to be collected through a depth of more than 5–10 μm adjacent series of several microns each may have to be collected, so that the gain can be adjusted. The effect of photobleaching in out-of-focus planes while other planes are being imaged can be minimized by collecting Z series beginning with the deepest optical section, as it has the lowest image intensity to start with.

An additional consideration is that fluorescence intensity usually is not even across the tissue section, because labeled structures are not uniformly distributed within the tissue or individual cells. The intrascene dynamic range of the photomultiplier tube can be too narrow to accommodate the wide range of fluorescence intensities in a given field of view. For this reason, we often find it necessary to set the illumination intensity, gain, and black level to bring out detail in a particular area of interest, even though other regions of the cell or tissue are obscured in that image. An example of this can be seen for anti-tubulin labeling in Fig. 2; compare the photoreceptor myoid region in Figs. 2B and 2D and the plexiform layer in Figs. 2D and 2E. When individual microtubules in the myoid region can be distinguished (Fig. 2D), microtubules in the ellipsoid region are difficult to detect. Likewise, when the beam intensity and gain are reduced so that neuronal processes in the plexiform layer can be distinguished (Fig. 2E), the weaker signal in the photoreceptors is almost below the threshold of detection. Conversely, when the beam intensity and gain are increased to lower the threshold so that labeling in the photoreceptors is easily detected, bright areas, such as the plexiform layers, become completely white, as in Fig. 2D. Remapping pixel intensities by modifying the look-up table (LUT) after image collection is a good solution if the intensity range is not too broad. A nonlinear intensity scale can be constructed to expand that portion of the range corresponding to the area of interest while other portions are contracted. In the tubulin labeling example, we expand the low to midportion of the range, corresponding to pixel intensities for the ellipsoid microtubules in Fig. 2C or the photoreceptors in Fig. 2D, and contract the upper end. In this way, the lower intensity values are increased, while the higher values are kept below maximum. In other words, we remap the linear scale to a plateauing curve (see also Chapter 10). If autofluorescence is present, a sigmoidal curve can be useful, to avoid raising the lowest values, which correspond to background fluorescence. This type of adjustment is more powerful if performed during collection. It is important, however, to compare images collected with and without adjustment to guard against information loss. The Bio-Rad MRC-600 has a nonlinear collection feature (Chapter 10).

We have encountered the following two difficulties with interpretation of confocal images of small structures. They arise from the greater thickness of optical sections (>0.5 μm), compared to the size of the object. For example, thin structures, such as microtubules (antibody decorated, 55 nm), oriented perpendicular to the vertical axis are fainter than those that are parallel to it and thus span the depth of the optical section (van der Voort and Brakenhoff, 1990). Figure 1B illustrates a similar point for actin bundles. The actin cables that run under the photoreceptor inner segment and extend into the calycal processes appear fainter where the surface of the cell is perpendicular to the vertical axis than at the sides, where several cables are in line along the vertical axis and their signals are additive. The same is true of planar structures, such as the plasma membrane. The second difficulty also is the result of light from the object being collected in adjacent optical sections. In Z series, the size of small structures can appear substantially greater in the vertical dimension. We advise cutting both longitudinal and cross sections of uniformly oriented cells, such as photoreceptors and epithelia, with the Vibratome to detect labeling of thin structures more easily and to verify the shape of small structures. The concentration of actin filaments in the photoreceptor ciliary stalk serves as an example. This actin network appears as a small fluorescent spot (Fig. 1) in single optical sections of either longitudinal- or cross-sectioned retinas. Yet it often can be followed through two or even three adjacent optical sections, roughly 0.75 μm thick, collected at 0.75-μm intervals.

Although image enhancement and the manipulation, display, and analysis of three-dimensional images will not be discussed further here, what is done with images after they are collected can be as important to obtaining accurate information as how they are collected. For discussions of these topics, the reader is referred to Inoué and Spring (1997) and Chen et al. (1990) (see also Chapters 1 and 10 of this volume).

B. Lenses

Objective lens parameters and the ways in which they affect imaging in confocal microscopy are discussed in detail by Keller (1990) and Majlof and Forsgren (Chapter 4 of this volume). In this section, we will touch only briefly on three issues that have particular application to high-resolution imaging in thick tissue sections. These issues are (1) the special importance of objective lens numerical aperture to resolution in LSCM, (2) the trade-off that can be made between correction and high lens transmission, and (3) ways to counteract the loss of correction for spherical aberration with depth into thick tissue sections. In microscopy in general, the highest resolution is possible by using lenses with the best correction for aberrations and the greatest numerical aperture [Eq. (1)]. In LSCM, optical section thinness depends similarly on objective lens numerical aperture and correction. Objective lenses also must be aligned with and matched to the other optical components of the microscope. In fluorescence microscopy, detection of small structures depends on how much light the objective can gather and transmit. These factors are even more critical in LSCM, where light is restricted by the pinhole aperture. The need for adequate correction for aberrations now is counterbalanced by the need for high transmission (Keller, 1990). The fewer lens elements a lens has, the more light it will transmit, but the less well corrected it will be. Lenses with the greatest numerical aperture also provide the highest light-gathering capability. Image brightness in epifluorescence microscopy varies as

$$\text{Image brightness} = NA^4/\text{magnification}^2. \tag{2}$$

Thus magnification becomes an additional consideration. Lower magnification objectives are desirable.

We have been able to reach a good compromise by using high numerical aperture, moderately to well corrected immersion lenses of the lowest magnification required. Immersion lenses, with their high numerical aperture, gather more light than dry lenses and give a sharper image by reducing light scatter induced by an air–cover glass interface. It is worth sacrificing some correction to gain transmission with objectives over ×40. For example, although it is only partially corrected for spherical aberration at red wavelengths, we prefer a ×63/NA 1.3 Achromat (Fluoreszenz, oil; Leitz Wetzlar) over a ×63/NA 1.4 Plan-Apochromat (oil; Nikon) for rhodamine- and CY3-conjugated probes for imaging deeper than a few microns from the surface. Its greater transmission allows image collection deeper within the tissue section before image intensity decreases substantially (see Section I.B). In our experience, the loss in contrast from opening the pinhole aperture appeared to diminish image sharpness more than the effect of spherical aberration. Because the scan area can be restricted to the center of the field by means of an electronic zoom feature, use of Plan, or flat-field, objectives is not always critical. The electronic zoom feature also limits the need for a ×100 objective lens, which would give a dimmer image than lower magnification lenses of the same numerical aperture. For our studies, a ×40 NA 1.3 Apochromat (UVFl 40; Olympus) oil immersion lens provided the best images overall, because it combines a high numerical aperture, moderate magnification, and high transmission.

The shallow working distance (approximately 100–150 μm) of high-numerical-aperture objectives is a limitation for work with thick tissue. More importantly, with oil immersion lenses the ability to resolve fine detail is lost well before the lens actually contacts the cover glass. In our experience, image quality deteriorates substantially below about 15 μm from the surface. We have implemented two partial solutions to counteract the decline in image intensity and resolution with depth caused by spherical aberration induced by the specimen (see Section I.B). The first is to mount the specimen directly on the cover glass. This minimizes the space between the section and the cover glass to make the most of the limited depth into glycerol/tissue in which correction for spherical aberration is maintained. In addition, while most immersion lenses are designed for use with No. 1½ cover glasses (0.17 mm thick), thinner cover glasses [No. 1 (0.12–0.13 mm) or No. 0 (0.10 mm)] can be used to increase working distance and image intensity at greater depths. These gains often well outweigh the cost of loss of ideal correction for aberrations. The second is to use a glycerol immersion lens ($\times$ 40/0.9 NA, Plan-Neofluar, oil/glycerol/water; Zeiss). With a glycerol immersion lens, the refractive index of a specimen mounted in glycerol is closer to that of the specified immersion medium. We found that image intensity was maintained at deeper focal planes with this glycerol lens than with a $\times$ 40/NA 1.4 oil immersion lens (see above); however, the image was noticeably less sharp. A better solution, then, would be a higher numerical aperture glycerol or water lens. Indeed, the $\times$60/NA 1.25 water immersion lenses (Olympus or Nikon) maintain both image intensity and sharpness at depths of at least 25 μm in aqueous mounts. This still is not a perfect solution, because biological material is optically inhomogeneous and therefore requires differing corrections for spherical aberration as one focuses into the tissue. Objectives with a correction collar can be adjusted, to a limited degree, to obtain sharp images at various depths within the specimen. Ideally, alterations in spherical aberration correction would be automatically and continuously applied as one focuses through the specimen. Unfortunately, such lenses are not currently available.

IV. Applications

A. Retinal Cell Biology

We initially applied the Vibratome technique and LSCM to the study of cytoskeletal organization in cells in the vertebrate retina. This technique has considerably improved our ability to detect individual cytoskeletal filaments or small bundles, as well as fine cellular processes filled with filaments, in sections of intact retina at some depth below the surface of thick sections or whole mounts of the retina. For example, we have been able to obtain images of retinal astrocytes in the cat retina labeled with an antibody to the intermediate filament protein, GFAP, that show distinctly many fine processes radiating from the cells. Figure 3 is a stereo pair collected between 1 and 13 μm below the surface of a Vibratome section. This stereo pair demonstrates the power of collecting a series of high-contrast optical thin sections and creating a three-dimensional image from them. It is easy to trace even the most delicate GFAP-filled processes of these astrocytes in three dimensions. This figure also demonstrates the potential for gaining information about both identity, because GFAP is present specifically in glial cells, and morphology.

Previously, with JB-4 sections and video microscopy, it was not possible to distinguish the filamentous nature of microtubules in the photoreceptor myoid region. Tubulin labeling appeared nearly uniform. With the nonembedding procedure and LSCM, microtubule bundles in the myoid clearly can be resolved in images collected within the upper 15 μm of thick sections (Figs. 2C and 2D). This has enabled us to assess the effect of microtubule poisons on microtubules in this region, where the endoplasmic reticulum and Golgi apparatus are located, giving us further insight into our earlier studies on the role of microtubules in opsin transport (Vaughan *et al.,* 1989). A drug-sensitive population of microtubules is disassembled by colchicine, leaving uniform monomer labeling, while a relatively insensitive population remains near the nucleus (I. L. Hale and B. Matsumoto, unpublished observations). In addition, with this method we were able to resolve fine neuronal processes in the plexiform layers of the retina, as shown in Fig. 2E for processes in the outer plexiform layer labeled with antitubulin.

This method also has proved to be well suited for visualizing the distribution of integral membrane proteins and soluble proteins. The high contrast has allowed us to obtain images in which opsin-immunoreactive vesicles in the ellipsoid region of frog photoreceptors (Figs. 4B and 4C), which may be transport vesicles (Papermaster *et al.,* 1985), and the plasma membrane (Fig. 4A) are well defined (see also Matsumoto and Hale, 1993). In addition to antibodies, lectins, which specifically recognize carbohydrate residues, can be used with Vibratome sections. Wheat germ agglutinin (WGA) is specific for the terminal N-acetyl-β-D-glucosamine residues and N-acetyl-neuraminic acid residues found on mature N-linked oligosaccharides. Therefore, WGA labels the Golgi apparatus, transport vesicles, and plasma membrane (Lis and Sharon, 1986). In photoreceptors, this lectin labels the outer segment, along with these organelles (Fig. 4D), because the mature carbohydrate groups of opsin (and possibly other outer segment glycoconjugates) contain a terminal N-acetyl-β-D-glucosamine residue (Wood and Napier-Marshall, 1985). The intense opsin immunolabeling in the myoid region thought to be Golgi labeling, for example, in Fig. 4C, at least partly colocalizes with WGA Golgi apparatus labeling, such as that seen in Fig. 4D. Optical sectioning has revealed important information about the morphology of the Golgi apparatus and the number and distribution of the putative transport vesicles.

Images of anti-CRALBP labeling demonstrate the usefulness of this method not only for soluble antigens, but for cell-specific ones as well. CRALBP is a cytoplasmic vitamin A carrier. The locations of the CRALBP-positive cell types, retinal pigment epithelium, and Müller glial cells within the retina and their shapes can be seen in Fig. 5A. CRALBP is present in and allows tracing of fine lateral processes as they branch from the Müller cells (Fig. 5B).

B. Drawbacks

The thick Vibratome section method has some drawbacks, as does any technique. These relate to the tissue itself, to the confocal microscope system, or to both. A relatively minor drawback is the impermanence of the tissue sections. They can be kept longer if stored in formaldehyde to prevent tissue degradation, but tissue autofluorescence gradually increases over time. We have found that image quality is substantially reduced after 2 weeks at 4°C. Sections equilibrated with 50% glycerol/50% PBS can be stored at −20°C for 1 month. Another drawback is that when contrast is maximized in LSCM sensitivity is low, because the narrow pinhole aperture greatly restricts the amount of light reaching the detector. The lack of sensitivity poses a problem with weak fluorescent signals. A photon-counting circuit reduces noise, and therefore can improve detection of weak signals. The problem of the loss of spherical aberration correction with depth into the specimen currently restricts high-resolution imaging to approximately the upper 15–20 μm of thick sections. In addition, the limited penetration of antibodies into thick tissue sections also can restrict image collection to the upper 15 μm. For viewing large cells, sections twice the limiting thickness can be cut and the sections viewed from both surfaces. Antibody F(ab')$_2$ fragments have the potential to solve the problem of antibody penetration. Changes in objective lens design to allow correction to be maintained through a range of depths could allow advantage to be taken of increases in antibody penetration. Perhaps the biggest drawback we have experienced arises from the limited intrascene dynamic range of the photomultiplier tube compared to the wide intensity range often encountered with intact tissue and compartmentalized cells. Solutions include nonlinear remapping of pixel intensities after image collection and nonlinear collection. A more dramatic improvement in this area is achieved by the Nipkow disk confocal design, because the image can be viewed directly and the human eye has superior intrascene dynamic range. Possible solutions to some of these problems are discussed in more detail in the next section.

C. Future Prospects

The large size of antibodies limits their penetration into Vibratome sections. Use of the smaller F(ab')$_2$ fragments (m_r 100,000) produced by enzymatic digestion should allow increased penetration. The extent of avidin penetration supports this possibility. This 66-kDa protein can penetrate at least 50 μm into Vibratome sections permeabilized with 0.1% Triton X-100. Of course, for indirect immunolabeling it will be necessary to use F(ab')$_2$ fragments of both primary and secondary antibodies.

Changes in lens design that will allow high-numerical-aperture lenses to be used at depths below 15–20 μm from the surface of thick specimens may be forthcoming. High-numerical-aperture glycerol immersion objectives would reduce the need to use oil immersion lenses, and thereby eliminate the problem caused by the difference in refractive index between the mounting medium and the cover glass/immersion medium currently encountered with oil immersion lenses. High-numerical-aperture water immersion lenses are providing the same solution for aqueous mounting media. As suggested by Inoué (1990), high-numerical-aperture oil immersion lenses with a motorized correction collar would allow automated correction as the depth of the focal plane is changed.

The video-rate scanning confocal microscopes, such as the Yokogawa Nipkow disk system, promise to facilitate confocal imaging of thick tissue sections (see Chapter 2). The fluorescence image can be viewed directly by eye with these microscopes, and recorded by using either a 35-mm camera with high-speed film or a sensitive video camera. Direct viewing provides three advantages. First, the time required to view specimens by confocal microscopy is dramatically reduced. The entire section can be scanned and three-dimensional information can be gathered quickly, just as it can in conventional epifluorescence microscopy, simply by observing through the eyepieces while moving the stage and focusing up and down through the tissue. The researcher does not have to wait for several scans to be averaged before a sharp image can be obtained for examination or documentation. Second, the human eye is still the best detector in terms of intrascene dynamic range and, although not as good as the human eye, photographic film is better than video cameras or photomultiplier tubes. Thus detail can be seen in bright and dim areas of the field of view at the same time. Third, for double-label work, simultaneous viewing of, for example, green and red fluorescence is possible without splitting the emitted light for detection by separate detectors. Maintaining the full intensity aids double-label work with weak signals. Another benefit is that with the eye or a sensitive color video camera, the wavelength (i.e., color) information can be utilized more effectively to ascertain true colocalization; for instance, structures stained with both fluorescein and rhodamine appear yellow. With separate monochromatic detectors, the inefficiency of barrier filters can make it difficult to distinguish double fluorescence emission from bleed-through of a single strong emission.

V. Appendix: Procedures for Immunohistochemical Preparation

A. Fixing Tissue

We do not perform an initial fixation by vascular perfusion, but this step is desirable for tissues that cannot be rapidly dissected. Cacodylate is preferable to a phosphate buffer for fixation, because its greater buffering capacity can maintain a stable pH. We do not use cacodylate buffer with sodium borohydride, because the combination gives off an unpleasant odor. (See Chapter 10 for phosphate-buffered saline formulation.) Glutaraldehyde fixative should be made with an 8% stock solution (EM grade; Polysciences). We have found that solutions of higher concentration can cause intense autofluorescence, possibly because of lower purity. After formaldehyde fixation, tissues should be stored in fixative for the best preservation, as formaldehyde fixation/stabilization partially reverses in buffer (see Section II.A). Step-by-step instructions are given in Table III.

Table III
Procedure for Fixing Tissue

Formaldehyde
1. Fix small pieces of tissue (1 mm thick—frog retinal thickness is about 300 μm) by immersion in 4% formaldehyde (granular paraformaldehyde; Electron Microscopy Sciences) in 0.1 M sodium cacodylate, pH 7.3, at 4°C for at least 8 h. Store tissue in this solution at 4°C.
2. Rinse tissue in two changes, 10 min each, of 0.1 M sodium cacodylate. Transfer to phosphate-buffered saline (PBS), pH 7.3, for 10 min. Rinsing will continue during sectioning. Tissue can be rinsed overnight.

Glutaraldehyde–Formaldehyde
1. Fix in 0.1% glutaraldehyde–4% formaldehyde in 0.1 M sodium cacodylate, pH 7.3, on ice for 30–35 min before transferring them to 4% formaldehyde at 4°C for an additional 4 h or more.
2. Rinse tissue in two changes, 10 min each, of 0.1 M sodium cacodylate. Transfer to PBS, pH 7.3, for 10 min.
3. Incubate tissues in three changes, 10 min each, of 0.1% sodium borohydride (Fisher Scientific), a reducing agent, in PBS, pH 8.0, at room temperature (Bacallao *et al.,* 1990). (Add the sodium borohydride to PBS, pH 8.0, just before use.) *Note:* The tissue will float, because bubbles collect on its surface.
4. Rinse tissues in three changes, 10 min each, of PBS, pH 7.3. If tissue is still floating, hold at 4°C until it sinks. Tissue can be rinsed overnight.

B. Agarose Embedding and Vibratome Sectioning

The agarose solution should be cooled to below 40°C, just above the gelling temperature (37°C), before use. An elevated agarose temperature is a major contributor to tissue autofluorescence. Small weighing boats can be used as molds. Be careful not to let tissue embedded in agarose dry out; always keep in a humid chamber. We use a Technical Products International Vibratome 1000 (Polysciences). We have found that low speed, intermediate amplitude settings, and a blade angle of 20–25° perform best for the retina with the sclera removed; other tissues may require different settings. See Section II.B for a discussion of section thickness. Step-by-step instructions are given in Table IV.

C. Labeling with Fluorochrome-Conjugated Probes

Sections are incubated free floating. Solutions need to be stirred to achieve even penetration of probes into tissue sections, as well as adequate rinsing. It is also desirable to use small volumes in order to conserve probes. We have found that sections can be immersed and adequately, yet gently, stirred on a rotator (50 rpm) in 400 μl of solution, using disposable microbeakers (10-ml size; Fisher Scientific), or in 200 μl or less, using 1.5-ml microcentrifuge tubes or small glass vials lying on their sides. We have found that penetration is better when microbeakers are used, even if microcentrifuge tubes are filled with 400 μl. All incubations and rinses should be performed at 4°C. In addition, sections should be kept in the dark during incubations in fluorescent probes and subsequent storage. We recommend centrifuging diluted antibody and lectin solutions at 14,000 rpm in

Table IV
Procedure for Agarose Embedding and Vibratome Sectioning

1. Transfer the tissue to a mold filled with buffer. Remove as much liquid as possible with a Pasteur pipette without letting the tissue become dry. Cover the tissue with a 5% solution of low gelling temperature agarose (type XI; Sigma) in PBS, pH 7.3, cooled to just below 40°C. Position tissue as desired. The agarose will gel quickly.
2. Once the agarose gels, transfer to 4°C until firm.
3. Cut the agarose with a scalpel or razor blade into a small cube around each piece of tissue. Orientation of the tissue in the agarose block determines the sectioning plane, because the orientation of the block and the blade in the Vibratome is fixed.
4. Blot the surface of the agarose to be glued and glue the cubes onto supports with cyanoacrylate (Krazy Glue).
5. Fill the boat of the Vibratome with chilled PBS.
6. Break a double-edge razor blade in half and clean with ethanol to remove the protective oil coating. Mount the blade and specimen support.
7. Set speed and amplitude, and section at desired thickness.
8. Collect sections with a small spatula or camel's hair brush. Store sections in PBS at 4°C. As in Section V.A. for storage of formaldehyde-fixed tissue longer than a few days, transfer to formaldehyde fixative solution.

Table V
Procedure for Labeling with Fluorochrome-Conjugated Probes

Antibodies[a]

1. Incubate in normal goat serum (Sigma), diluted 1/50, for 1–2 h, to block nonspecific staining seen with some antibodies.
2. Rinse twice, 10 min each.
3. Incubate in primary antibody solution for 18–24 h (overnight). Dilutions must be determined empirically for each antibody; concentrations previously used for sections of embedded or frozen tissue are useful as a starting point.
4. Rinse three times, 15 min each.
5. Incubate in secondary antibody solution for 12–18 h. We have not seen nonspecific labeling with these long incubation times. In our experience, fluorescein (FITC)- and rhodamine (TRITC)-conjugated, affinity-purified goat anti-mouse or anti-rabbit antibodies (Cappel Research Products, Organon Teknika Corporation, Durham, NC) work well at a concentration of 50 μg/ml. Because the signal with CY3-conjugated secondary antibodies (Jackson ImmunoResearch) is more intense, we suggest a concentration of 7.5 μg/ml or less.
6. Rinse for at least 2–3 h in several changes of buffer containing detergent before viewing; unbound fluorescent antibody produces an undesirable level of background fluorescence in thick sections.
7. Store in plain PBS after the final rinse to minimize the length of time sections are exposed to detergent.
8. Store in 4% formaldehyde if sections of formaldehyde-fixed tissue will be kept for more than a few days.

Phalloidin

1. Incubate sections in fluorochrome-conjugated phalloidin (Molecular Probes) at a concentration of 10 units/ml in PBS, pH 7.3, containing 0.1% Triton X-100 for 2 h.
2. Rinse for at least 1 h with several changes of buffer before viewing.
3. Store sections as described for indirect immunofluorescence.

Lectins

1. Incubate sections in fluorochrome-conjugated lectin in standard antibody buffer with 0.1% Triton X-100 and any additional cations recommended by the supplier for 12–18 h. We have used fluorescent WGA and concanavalin A (Vector Laboratories, Burlingame, CA) at 5 and 10 μg/ml, respectively.
2. Rinse for at least 2–3 h in several changes of buffer.
3. Store sections as described for indirect immunofluorescence.

[a] A standard incubation and rinse buffer consisting of 0.5% bovine serum albumin (BSA), detergent of choice (see Section II.C.1), and 0.01% sodium azide (preservative) in PBS, pH 7.3, can be used throughout.

a Beckman (Fullerton, CA) microfuge for 5 min to remove large aggregates. The purity and age of the detergent stock used can be critical to good penetration and labeling intensity. Detergents that have been purified for biological applications are recommended. We have found that Triton X-100 loses effectiveness and may become somewhat autofluorescent with time, and suggest storing at 4°C and replacing after several months. Step-by-step instructions are given in Table V.

D. Mounting Sections

We mount sections no more than a few hours prior to viewing, as background fluorescence can increase after a prolonged period in mountant. Large cover glasses should not be used, because they flex when immersion objectives are used, interfering with Z-series collection. Thick tissue slices do not appear to be deformed by mounting under

Table VI
Procedure for Mounting Sections

1. Transfer one or two sections with a spatula to a drop of PBS on an 18×18 mm No. $1\frac{1}{2}$ cover glass (0.17 mm thick). Thinner coverslips [No. 1 (0.12–0.13 mm) or No. 0 (0.10 mm)] can be used to increase working distance and image intensity at greater depths, but at the cost of some loss of correction for aberrations. Remove buffer by blotting with a Kimwipe.
2. Add a small drop of 100% glycerol containing 5% *n*-propyl gallate to the section(s).
3. Invert the cover glass onto a glass slide slowly to avoid forming bubbles.
4. Affix the cover glass to the slide with nail polish; otherwise it may move when immersion lenses are used. It should be noted that nail polish is highly autofluorescent. *Caution:* Allow the nail polish to dry completely before viewing to avoid getting it on the objective lens.

a cover glass, probably because the surrounding agarose prevents compression. When oil immersion lenses are used, sections should be mounted directly on the cover glass (see Section III.B). Step-by-step instructions are given in Table VI.

Acknowledgments

We thank Mr. Robert Gill for expert technical assistance, Dr. Steven Fisher and Mr. Kevin Long for critical reading of the manuscript and valuable suggestions, Dr. William Stegeman at Jackson ImmunoResearch for generously providing CY3- and CY2-conjugated secondary antibodies before they were commercially available, Dr. Jack Saari for the gift of the CRALBP antibody, Drs. Vijay Sarthy and Robert Molday for the gift of the rhodopsin antibodies, Dr. Michael Klymkowsky for the gift of the β-tubulin antibody, and Meridian Instruments, Inc. and Bio-Rad Microscience Division for demonstrations of their slit scanning confocal microscopes. This research was supported by National Institutes of Health Grant EY-07191 to B. Matsumoto.

References

Art, J. (1990). Photon detectors for confocal microscopy. *In* "Handbook of Biological Confocal Microscopy" (J. B. Pawley, Ed.), revised ed., pp. 127–139. Plenum, New York.

Bacallao, R., Bomsel, M., Stelzer, E. H. K., and De Mey, J. (1990). Guiding principles of specimen preservation for confocal fluorescence microscopy. *In* "Handbook of Biological Confocal Microscopy" (J. B. Pawley, Ed.), revised ed., pp. 197–205. Plenum, New York.

Barak, L. S., Yocum, R. R., Nothnagel, E. A., and Webb, W. W. (1980). Fluorescence staining of the actin cytoskeleton in living cells with 7-nitrobenz-2-oxa-1,3-diazole-phallacidin. *Proc. Natl. Acad. Sci. USA* **77,** 980–984.

Bok, D. (1985). Retinal photoreceptor–pigment epithelium interactions. Friedenwald lecture. *Invest. Ophthalmol. Vis. Sci.* **26,** 1659–1694.

Brakenhoff, G. J., van Spronsen, E. A., van der Voort, H. T., and Nanninga, N. (1989). Three-dimensional confocal fluorescence microscopy. *In* "Methods in Cell Biology" (D. L. Taylor and Y.-L. Wang, Ed.), Vol. 30, pp. 379–398. Academic Press, San Diego.

Bunt-Milam, A. H., and Saari, J. C. (1983). Immunocytochemical localization of two retinoid-binding proteins in vertebrate retina. *J. Cell Biol.* **97,** 703–712.

Chen, H., Sedat, J. W., and Agard, D. A. (1990). Manipulation, display, and analysis of three-dimensional biological images. *In* "Handbook of Biological Confocal Microscopy" (J. B. Pawley, Ed.), revised ed., pp. 141–150. Plenum, New York.

Cheng, P. C., and Summers, R. G. (1990). Image contrast in confocal microscopy. *In* "Handbook of Biological Confocal Microscopy" (J. B. Pawley, Ed.), revised ed., pp. 179–195. Plenum, New York.

Chu, D. T., and Klymkowsky, M. W. (1989). The appearance of acetylated alpha-tubulin during early development and cellular differentiation in *Xenopus. Dev. Biol.* **136,** 104–117.

Eckert, B. S., and Snyder, J. A. (1978). Combined immunofluorescence and high-voltage electron microscopy of cultured mammalian cells, using an antibody that binds to glutaraldehyde-treated tubulin. *Proc. Natl. Acad. Sci. USA* **75,** 334–338.

Gaur, V. P., Adamus, G., Arendt, A., Eldred, W., Possin, D. E., McDowell, J. H., Hargrave, P. A., and Sarthy, P. V. (1988). A monoclonal antibody that binds to photoreceptors in the turtle retina. *Vision Res.* **28,** 765–776.

Giloh, H., and Sedat, J. W. (1982). Fluorescence microscopy: reduced photobleaching of rhodamine and fluorescein protein conjugates by *n*-propyl gallate. *Science* **217,** 1252–1255.

Goldenthal, K. L., Hedman, K., Chen, J. W., August, J. T., and Willingham, M. C. (1985). Postfixation detergent treatment for immunofluorescence suppresses localization of some integral membrane proteins. *J. Histochem. Cytochem.* **33,** 813–820.

Hicks, D., and Barnstable, C. J. (1987). Different rhodopsin monoclonal antibodies reveal different binding patterns on developing and adult rat retina. *J. Histochem. Cytochem.* **35,** 1317–1328.

Hicks, D., and Molday, R. S. (1986). Differential immunogold-dextran labeling of bovine and frog rod and cone cells using monoclonal antibodies against bovine rhodopsin. *Exp. Eye Res.* **42,** 55–71.

Inoué, S. (1990). Foundations of confocal scanned imaging in light microscopy. *In* "Handbook of Biological Confocal Microscopy" (J. B. Pawley, Ed.), revised ed., pp. 1–14. Plenum, New York.

Inoué, S., and Spring, K. R. (1997). "Video Microscopy: The Fundamentals," 2nd ed. Plenum, New York.

Johnson, G. D., Davidson, R. S., McNamee, K. C., Russell, G., Goodwin, D., and Holborow, E. J. (1982). Fading of immunofluorescence during microscopy: a study of the phenomenon and its remedy. *J. Immunol. Methods* **55,** 231–242.

Keller, H. E. (1990). Objective lenses for confocal microscopy. *In* "Handbook of Biological Confocal Microscopy" (J. B. Pawley, Ed.), revised ed., pp. 77–86. Plenum, New York.

Larsson, L.-I. (1988). "Immunocytochemistry: Theory and Practice." CRC Press, Boca Raton, FL.

Lis, H., and Sharon, N. (1986). *In* "The Lectins: Properties, Functions, and Applications in Biology and Medicine" (I. E. Liener, N. Sharon, I. J. Goldstein, Eds.), pp. 293–370. Academic Press, New York.

Mack, A. F., and Fernald, R. D. (1991). Thin slices of teleost retina continue to grow in culture. *J. Neurosci. Methods* **36,** 195–202.

Matsumoto, B., and Hale, I. L. (1993). Preparation of retinas for studying photoreceptors with confocal microscopy. *In* "Methods in Neuroscience" (P. C. Hargrave, Ed.), Vol. 15, pp. 54–71. Academic Press, San Diego.

Matsumoto, B., Defoe, D. M., and Besharse, J. C. (1987). Membrane turnover in rod photoreceptors: ensheathment and phagocytosis of outer segment distal tips by pseudopodia of the retinal pigment epithelium. *Proc. R. Soc. Lond. B Biol. Sci.* **230,** 339–354.

Matus, A., Bernhardt, R., and Hugh-Jones, T. (1981). High molecular weight microtubule-associated proteins are preferentially associated with dendritic microtubules in brain. *Proc. Natl. Acad. Sci. USA* **78,** 3010–3014.

Nagle, B. W., Okamoto, C., Taggart, B., and Burnside, B. (1986). The teleost cone cytoskeleton. Localization of actin, microtubules, and intermediate filaments. *Invest. Ophthalmol. Vis. Sci.* **27,** 689–701.

Novikoff, P. M., Tulsiani, D. R., Touster, O., Yam, A., and Novikoff, A. B. (1983). Immunocytochemical localization of alpha-D-mannosidase II in the Golgi apparatus of rat liver. *Proc. Natl. Acad. Sci. USA* **80,** 4364–4368.

Ohtsuki, I., Manzi, R. M., Palade, G. E., and Jamieson, J. D. (1978). Entry of macromolecular tracers into cells fixed with low concentrations of aldehydes. *Biol. Cell* **31,** 119–126.

O'Rourke, N. A., Dailey, M. E., Smith, S. J., and McConnell, S. K. (1992). Diverse migratory pathways in the developing cerebral cortex. *Science* **258,** 299–302.

Osborn, M., and Weber, K. (1982). Immunofluorescence and immunocytochemical procedures with affinity purified antibodies: tubulin-containing structures. *In* "Methods in Cell Biology" (L. Wilson, Ed.), Vol. 24, pp. 97–132. Academic press, San Diego.

Osborn, M., Webster, R. E., and Weber, K. (1978). Individual microtubules viewed by immunofluorescence and electron microscopy in the same PtK2 cell. *J. Cell Biol.* **77,** R27–34.

Panchuk-Voloshina, N., Haugland, R. P., Bishop-Stewart, J., Bhalgat, M. K., Millard, P. J., Mao, F., and Leung, W. Y. (1999). Alexa dyes, a series of new fluorescent dyes that yield exceptionally bright, photostable conjugates. *J. Histochem. Cytochem.* **47,** 1179–1188.

Papermaster, D. S., Schneider, B. G., Zorn, M. A., and Kraehenbuhl, J. P. (1978). Immunocytochemical localization of a large intrinsic membrane protein to the incisures and margins of frog rod outer segment disks. *J. Cell Biol.* **78,** 415–425.

Papermaster, D. S., Schneider, B. G., and Besharse, J. C. (1985). Vesicular transport of newly synthesized opsin from the Golgi apparatus toward the rod outer segment. Ultrastructural immunocytochemical and autoradiographic evidence in *Xenopus* retinas. *Invest. Ophthalmol. Vis. Sci.* **26,** 1386–1404.

Pease, D. C. (1962). Buffered formaldehyde as a killing agent and primary fixative for electron microscopy. *Anat. Rec.* **142,** 342.

Pickel, V. M. (1981). Immunocytochemical methods. *In* "Neuroanatomical Tract Tracing Methods" (L. Heimer and M. J. Robards, Eds.), pp. 483–509. Plenum, New York.

Priestley, J. V. (1984). Pre-embedding ultrastructural immunocytochemistry: immunoenzyme techniques. *In* "Immunolabeling for Electron Microscopy" (J. M. Polak and I. M. Varndell, Eds.), pp. 37–52. Elsevier, New York.

Priestley, J. V., and Cuello, A. C. (1983). Electron microscopic immunocytochemistry for CNS transmitters and transmitter markers. *In* "Immunohistochemistry" (A. C. Cuello, Ed.), pp. 273–321. Wiley, Chichester.

Priestley, J. V., Alvarez, F. J., and Averill, S. (1992). Pre-embedding electron microscopic immunocytochemistry. *In* "Electron Microscopic Immunocytochemistry: Principles and Practice" (J. M. Polak and J. V. Priestley, Eds.), pp. 89–121. Oxford Univ. Press, Oxford.

Puchtler, H., and Meloan, S. N. (1985). On the chemistry of formaldehyde fixation and its effects on immuno-histochemical reactions. *Histochemistry* **82,** 201–204.

Sabatini, D. D., Bensch, K., and Barnett, R. J. (1963). Cytochemistry and electron microscopy. The preservation of cellular ultrastructure and enzymatic activity by aldehyde fixation. *J. Cell Biol.* **17,** 19–58.

Sale, W. S., Besharse, J. C., and Piperno, G. (1988). Distribution of acetylated alpha-tubulin in retina and in vitro-assembled microtubules. *Cell Motil. Cytoskeleton* **9,** 243–253.

Southwick, P. L., Ernst, L. A., Tauriello, E. W., Parker, S. R., Mujumdar, R. B., Mujumdar, S. R., Clever, H. A., and Waggoner, A. S. (1990). Cyanine dye labeling reagents—carboxymethylindocyanine succinimidyl esters. *Cytometry* **11,** 418–430.

Spencer, M., Detwiler, P. B., and Bunt-Milam, A. H. (1988). Distribution of membrane proteins in mechanically dissociated retinal rods. *Invest. Ophthalmol. Vis. Sci.* **29,** 1012–1020.

Sternberger, L. A. (1979). "Immunocytochemistry," 2nd ed. Wiley, New York.

Tsien, R. Y., and Waggoner, A. (1990). Fluorophores for confocal microscopy: Photophysics and photochem-istry. *In* "Handbook of Biological Confocal Microscopy" (J. B. Pawley, Ed.), revised ed., pp. 169–177. Plenum, New York.

van der Voort, H. T. M., and Brakenhoff, G. J. (1990). 3-D image formation in high-aperture fluorescence confocal microscopy: A numerical analysis. *J. Microsc. (Oxford)* **158,** 43–54.

Vaughan, D. K., and Fisher, S. K. (1987). The distribution of F-actin in cells isolated from vertebrate retinas. *Exp. Eye Res.* **44,** 393–406.

Vaughan, D. K., Fisher, S. K., Bernstein, S. A., Hale, I. L., Linberg, K. A., and Matsumoto, B. (1989). Evidence that microtubules do not mediate opsin vesicle transport in photoreceptors. *J. Cell Biol.* **109,** 3053–3062.

Vaughn, J. E., Barber, R. P., Ribak, C. E., and Houser, C. R. (1981). Methods for the immunocytochemical localisation of proteins and peptides involved in neurotransmission. *In* "Current Trends in Morphological Techniques" (J. E. Johnson, Ed.), Vol. 3, pp. 33–70. CRC Press, Boca Raton, FL.

Weber, K., Rathke, P. C., and Osborn, M. (1978). Cytoplasmic microtubular images in glutaraldehyde-fixed tissue culture cells by electron microscopy and by immunofluorescence microscopy. *Proc. Natl. Acad. Sci. USA* **75,** 1820–1824.

White, J. G., Amos, W. B., Durbin, R., and Fordham, M. (1990). Development of a confocal imaging system for biological epifluorescence applications. *In* "Optical Microscopy for Biology" (B. Herman and K. Jacobson, Eds.), pp. 1–18. Alan R. Liss, New York.

Willingham, M. C., and Yamada, S. S. (1979). Development of a new primary fixative for electron microscopic immunocytochemical localization of intracellular antigens in cultured cells. *J. Histochem. Cytochem.* **27,** 947–960.

Wood, J. G., and Napier-Marshall, L. (1985). Cytochemical analysis of oligosaccharide processing in frog photoreceptors. *Histochem. J.* **17,** 585–594.

Yu, H., Ernst, L., Wagner, M., and Waggoner, A. (1992). Sensitive detection of RNAs in single cells by flow cytometry. *Nucleic Acids Res.* **20,** 83–88.

CHAPTER 8

Confocal Fluorescence Microscopy for *Tetrahymena thermophila*

E. S. Cole,[1] K. R. Stuart,[1] T. C. Marsh,[2] K. Aufderheide,[3] and W. Ringlien[4]

[1] Biology Department
St. Olaf College
Northfield, Minnesota 55057

[2] Department of Pharmacology
University of Minnesota
Minneapolis, Minnesota 55455

[3] Department of Biology
Texas A & M University
College Station, Texas 77843

[4] Instrument Maker
Carleton College
Northfield, Minnesota 55057

I. Introduction

Confocal fluorescence microscopy has proven to be an invaluable tool for analysis of numerous biological problems being investigated in the unicellular, ciliated protist *Tetrahymena thermophila*. Researchers studying the dynamics and localization of cytoplasmic and cytoskeletal proteins (Lee *et al.*, 1999; Fujiu and Numata, 1999, 2000; Gaertig, 2000), the synthesis, transport, and exocytosis of dense-core secretory granules (Chilcoat *et al.*, 1996), and nuclear and chromosomal behaviors during meiosis and conjugation (Kiersnowska *et al.*, 2000; Stuart and Cole, 1999), as well as intracellular localization of specific nucleic acid sequences (Ward *et al.*, 1997; Marsh *et al.*, 2001) have all gained useful insights through the application of computer-facilitated, confocal fluorescence microscopy. This chapter has been written to provide methodology for visualizing three types of intracellular targets: (1) proteins labeled via indirect immunofluorescence methodology, (2) nucleic acids labeled via *in situ* hybridization of biotinylated oligonucleotide probes, and (3) fluorescently labeled targets in living cells (vital dyes for nuclei, mitochondria, and GFP-tagged proteins). The first two targets involve methods of fixing and permeabilizing the cells, and consequently provide a static view of intracellular localizations. The third topic involves visualization of fluorescent targets in live cells. Since living *Tetrahymena* cells are extremely mobile, a nondestructive method for restraining them is necessary. For this, we provide rough sketches and detailed descriptions of two mechanical devices that immobilize living cells under the microscope. We hope that with these details (and a skilled mechanical engineer), researchers will be able to utilize this technology for improved visualization of fluorescently tagged intracellular markers in living systems.

II. Indirect Immunofluorescence and Confocal Microscopy

In the course of our work with *Tetrahymena thermophila* we have discovered that *Tetrahymena* do not tolerate many of the fixation protocols published for viewing cytoskeletal structures in other eukaryotic cells. We have learned that for optimal preservation of microtubule-based structures, a paraformaldehyde fixative is best. This procedure works well for most of the antisera we have tested.

For confocal microscopy cells should be attached to poly-L-lysine-coated coverslips. The reason for this is that the focus motor of the confocal microscope applies sufficient suction to the coverslip (via contact between the objective lens and the immersion oil) to move suspended cells in and out of focus as different optical sections are imaged. With cells fixed to the coverslip, this problem can be reduced or eliminated.

The fixation method using paraformaldehyde is modified from that of Dr. Jacek Gaertig (see Gaertig and Fleury, 1992). For alternative fixations that may favor particular antisera, see Stuart and Cole (1999). Sample images appear in Figs. 1–3 (see color plates).

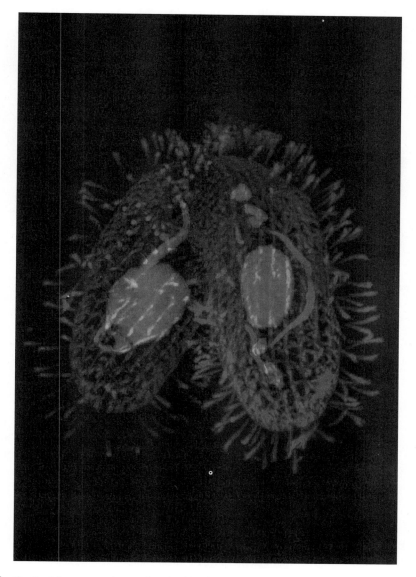

Fig. 1 Confocal fluorescence image of mating *Tetrahymena* cells. Microtubules (green) were labeled with a beta-tubulin monoclonal antiserum (12G10 kindly provided by Dr. Joseph Frankel, University of Iowa) and visualized with Cy-5 conjugated goat anti-mouse secondary antiserum (Amersham). Nuclei (red) were visualized with Sytox nuclear stain (Molecular Probes, Inc.). Cells were scanned under a Bio-Rad MRC1000/1024 laser confocal microscope operated by Dr. Mark Sanders, Program Director of the Imaging Center, Department of Genetics and Cell Biology, University of Minnesota. (See Color Plate.)

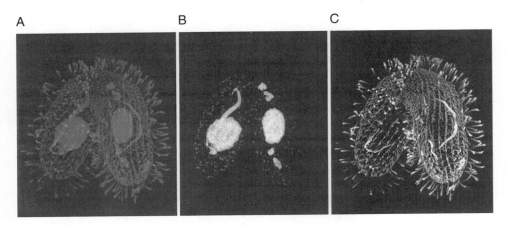

Fig. 2 The same mating pair shown in Fig. 1, with the various fluorescent channels separated. (A) The entire image, (B) the nuclei, and (C) the microtubules. (See Color Plate.)

A. Reagents

PHEM buffer pH 6.9 (for 250 ml):

60 mM Pipes (4.54 g)
25 mM Hepes (1.5 g)
10 mM EGTA (950 mg)
2 mM MgCl$_2$ (100 mg)
(dH$_2$O to 250 ml)

Practical notes: The Pipes will not dissolve until the pH is raised to at least 6. Add Pipes to 225 ml of dH$_2$O. Raise the pH to 6.9 (with 5 M NaOH), add the remaining reagents, and add dH$_2$O to bring to 250 ml. Raise to a final pH of 6.9 (with 5 M NaOH).

Phosphate-buffered saline (PBS), pH 6.9:

9 g NaCl
1.66 g Na$_2$HPO$_4$·7H$_2$O
820 mg NaH$_2$PO$_4$·H$_2$O
Bring to 1 L with dH2O
Adjust the final pH to 6.9 (with 5 M NaOH)

Fixative:

3% Paraformaldehyde, 0.2–0.5% Triton X-100, in PHEM buffer
40 ml PHEM buffer pH 6.9 (see above)
120 μl Triton X-100
10 ml 16% Paraformaldehyde solution (Electron Microscopy Sciences #15710)

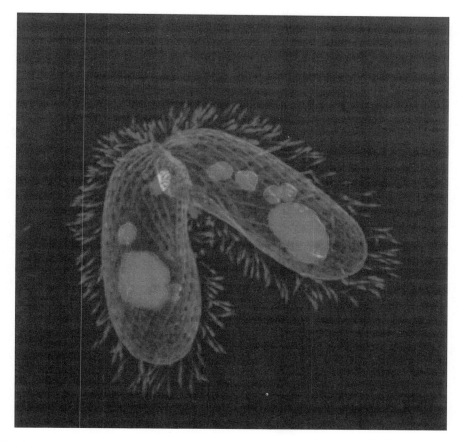

Fig. 3 Mating *Tetrahymena* viewed under laser confocal microscopy. Microtubules (blue) were labeled with 12G10 beta tubulin antiserum (courtesy of Dr. Joseph Frankel) and a Cy-5 conjugated secondary antiserum. The subcortical layer (pink) was labeled with a polyclonal antiserum directed against TCBP-25 (courtesy of Dr. Osamu Numata) and a Texas Red conjugated secondary antiserum. Nuclei (green) have been labeled with Sytox. (See Color Plate.)

Practical notes: Use a wide-tip 200-μl micropipette tip or a 1000 μl micropipette tip for measuring the Triton X-100 (it is viscous). Add Triton X-100 to PHEM buffer and shake or stir well for several minutes before adding paraformaldehyde. Aliquot and freeze (15-ml disposable centrifuge tubes work well, 4 to 6 ml in each). Thaw before needed.

1. Reagents to Be Made the Day of Use

Antiquenching mounting medium (Sanders, 1995)

DABCO: 250 mg DABCO (Sigma D-2522) in 1 ml dH2O; add to 9 ml glycerol. This can be easily mixed (by inversion) in a 15 ml centrifuge tube.

1% BSA–PBS (for each sample; increase as needed): 30 mg BSA (bovine serum albumin: Sigma A-7906) in 3 mls PBS (phosphate-buffered saline). Shake to dissolve.

0.1% BSA–PBS (for each sample; increase as needed): 30 mg BSA in 30 ml PBS. Shake or stir to dissolve.

Polylysine-coated coverslips

1. Fill a coverslip staining jar (small Coplin jar) with 9 ml poly-L-lysine solution (0.1% w/v in H_2O, Sigma P8920). Alternatively, put 20 ml poly-L-lysine in a 50-ml centrifuge tube. The polylysine can be stored in the coverslip staining jar or centrifuge tube for at least several weeks.
2. Soak the coverslip in polylysine for 5′ (up to 7 coverslips in the staining jar: 4 straight across, and 3 on the diagonals; one at a time is best in the centrifuge tube).
3. Set on edge, on absorbent paper to dry. Use within a few hours.

SYTOX (Molecular Probes): We have experimented with the optimal concentration of SYTOX and found that the minimum effective concentration is 10 nM. We have come to use 25 nM routinely, and have noticed that background cytoplasmic staining appears at 50 nM. Sytox comes in a 5 mM concentrate. We dilute 1/1000 with dH_2O to make a stock solution.

B. Preparation of Samples

1. Protocol Outline

      1. Pellet cells in 15-ml centrifuge tube.
      2. Wash cells with PHEM buffer.
      3. Fix cells.
      4. Pellet cells and remove fixative.
      5, 6. Two rinses with 0.1% BSA–PBS.
      7. Add 1 ml 0.1% BSA–PBS and transfer to microfuge tube.
      8. Pellet cells.
      9. Add primary antiserum.
     10. Two rinses with 0.1% BSA–PBS.
     11. Add secondary antiserum.
     12. Pellet and remove antiserum.
     13. Counterstain.
     14. Two rinses with 0.1% BSA–PBS.
     15. Resuspend in 300–1200 μl 0.1% BSA–PBS and allow to settle on coverslips.
     16. Meanwhile, prepare slides.
     17. Wick away buffer and place onto prepared slides.
     18. Blot edges and seal with Permount.

2. Protocol in Detail

1. Pipet 2–5 mls of cells (at 200,000 cells per ml) into a 15 ml centrifuge tube. Centrifuge (IEC clinical: #4 for 2 min), decant supernatant, and resuspend the pellet (by finger flicking) in the liquid that remains.

2. Add 5 ml PHEM buffer to wash cells. Centrifuge, decant supernatant, and resuspend in the liquid that remains.

3. Working in a fume hood, add 1–2 ml of paraformaldehyde fixative to cells, and let sit at room temperature for 30 min. Increase the amount of fixative if the sample is large—the ratio of fixative to pelleted sample should be about 4 to 1 or greater. Cells may be fixed for as long as 60 min, beyond that, however, the detergent begins to cause too much cell degradation.

4. Centrifuge, decant supernatant, and loosen pellet by finger flicking. Dispose of supernatant (containing paraformaldehyde) appropriately.

5. Add 5 ml 0.1% BSA–PBS for 5 min to rinse cells,. Centrifuge, decant supernatant, and resuspend.

6. Repeat step 5, but do not resuspend the pellet by flicking.

7. Add 1 ml 0.1% BSA–PBS, mix with the pipetor, and transfer solution to a microfuge (Eppendorf) tube.

8. Centrifuge (IEC clinical #4 for 2 min or equivalent) and aspirate supernatant. Resuspend the pellet in the liquid that remains by flicking.

9. For primary staining, add 200 μl (or up to 1 ml) diluted primary antiserum (in 1% BSA–PBS) to the pellet. Time will vary depending on the antiserum, and incubation at 31°C or 37°C may be beneficial. (30 to 60 min at 31°C is typical for the antisera we use.) Two different primary antisera can be combined in this step.

10. After incubation, centrifuge, remove the supernatant by aspiration, and resuspend the pellet. Rinse the sample (centrifuge, aspirate, resuspend) twice with 1 ml 0.1% BSA–PBS each time.

11. For secondary staining, add 200 μl to 1 ml diluted secondary antiserum (in 1% BSA–PBS) to the pellet. As with primary staining, time will vary depending on the antiserum, and incubation at a higher temperature may be beneficial.

12. After incubation, centrifuge, remove the supernatant by aspiration and resuspend the pellet.

13. If counterstaining with SYTOX nuclear stain (SYTOX, Molecular Probes S-7020), add 1 ml 0.1% BSA–PBS to pellet along with 5 μl of stock SYTOX for 5 sec.

14. After 5 min, rinse the sample (centrifuge, aspirate, resuspend) twice with 1 ml 0.1% BSA–PBS each time. If not counterstaining, simply rinse twice.

15. Add 300 μl per intended coverslip (typically 1200 μl per sample for 4 coverslips). Place 300 μl on each poly-L-lysine-coated coverslip and allow to settle (in a moist chamber if needed) for 20 min.

16. During the 20 min, prepare slides. To prevent crushing of the cells, put four tiny drops of Permount on the slide, approximating where the corners of the coverslip will fall. Put one or two drops of antiquenching mounting medium on the slide.

17. After 20 min, wick away the liquid from the coverslip with pieces of torn filter paper. Blotting from two sides simultaneously will allow a maximum number of cells to adhere to the coverslip. Quickly (before it begins to dry) place the coverslip onto the prepared slide.

18. Blot excess moisture from the edges of the coverslip with pieces of torn filter paper, and seal the edges of the coverslip with Permount.

Practical notes: The number of washes given here is minimal, as is the time for staining. The number of washes could be increased to reduce nonspecific staining (as many as six between and after staining). The length of time for staining could be increased (from 4 h to overnight) to aid in binding, or so that more dilute antisera can be used.

A moist chamber can be made by placing a circle of filter paper in the lid of a Petri dish and moistening it with H2O. Parafilm-M or filter paper in the bottom of the dish will prevent the coverslips from sticking to the bottom.

Antisera generated against different epitopes can require totally different (even antithetical) fixation procedures (e.g., paraformaldehyde versus ethanol). It is worth noting that the fixative one uses when screening for effective antisera can define and limit conditions for later analysis.

III. Fluorescence *in Situ* Hybridization Detection of *Tetrahymena thermophila* rDNA for Confocal Microscopy

The primary goal of this *in situ* hybridization method is to preserve the structural integrity of the cell so that the subcellular localization of a sequence can be determined. *In situ* techniques often involve drying cells onto a microscope slide prior to hybridization, which can distort and rupture cells and the structures within them. Furthermore, because the cell and its contents flatten out, possible substructure information is lost. The alternative method described here was adapted from a protocol published by Ward *et al.* (1997). A sample appears in Fig. 4.

A. Sample Preparation

All of the steps are performed in a single 1.5-ml microfuge tube. Supernatants are removed using 1-ml disposable plastic pipette tips instead of decanting, thereby minimizing sample loss. The small-scale prep described here should produce enough cells for two slides.

1. Transfer 1 ml of cells (density of 1.0×10^5 cells/ml) into a 1.5-ml microfuge tube. Concentrate the cells by gentle centrifugation for 2.5 min using a benchtop centrifuge (IEC clinical) with a swinging bucket rotor. A 15-ml centrifuge tube may be used as an adapter to hold the microfuge tube. Use tape to narrow the opening of the larger tube until the smaller tube fits snugly at the top.

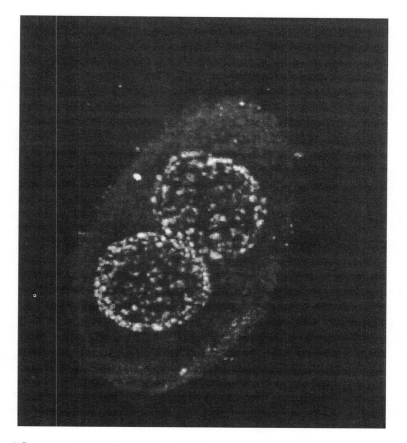

Fig. 4 A fluorescence *in situ* hybridization confocal image of rDNA localization in the *Tetrahymena thermophila* developmental mutant, rad51−/−. Hybridization of the biotinylated rDNA probe was detected using FITC–avidin stain. A 5-μm cross-sectional image was obtained by confocal microscopy, which shows peripheral localization of abundant rDNA (nucleoli) in the macronuclei of a rad51−/− cell arrested at this late developmental stage.

2. Wash the cells once in 10 mM Tris-HCl pH 7.5; resuspend the loose pellet in remaining Tris buffer.

3. Fix cells using 0.5 ml Schaudin's reagent (2 parts saturated aqueous HgCl$_2$ to 1 part absolute ethanol, 1% acetic acid) for 10 min at room temperature. Pellet cells using low-speed centrifugation for 1 min. The lowest speed setting should be sufficient to concentrate the cells at the bottom of the tube. Carefully remove as much supernatant as possible and dispose of it in an appropriate container.

4. Wash cells 1× in 70% ethanol followed by 2× in 100 mM triethanolamine pH 8.0. Acetylate cells using 0.2 ml 0.1% acetic anhydride in 100 mM triethanolamine pH 8.0 for 10 min at room temperature.

5. Wash cells 1× in 2× SSC (0.3 M NaCl, 0.03 M sodium citrate). Resuspend cells in 0.1 ml 70% formamide, 2× SSC and heat to 70°C for 1 min, then rapidly cool in ice bath for 10 min. Dilute cell suspension to 0.5 ml with ice-cold H_2O and immediately harvest cells by centrifugation. At this point, the sample is a nearly transparent pellet.

6. Mix cells in 50-μl hybridization solution (30% formamide, 4× SSC, 0.1% BSA, 10% dextran sulfate, 5% SDS, 10 mM EDTA pH 8.0) and incubate overnight at 37°C with ~50 ng biotinylated rDNA probe (see below).

7. Wash cells by adding 0.5 ml 40% formamide, 2× SSC at 37°C for 30 min followed by a 10 min wash in 2× SSC then 1× SSC at room temperature.

8. Incubate cells in 100 μl detection solution [2 μg FITC or Texas Red conjugated Avidin (Vector Laboratories) in 100 mM sodium carbonate pH 8.2, 0.5 M NaCl] for 30 min at room temperature.

9. Centrifuge and wash cells sequentially in 4× SSC (0.6 M NaCl, 0.06 M sodium citrate); 4× SSC, 0.2% Triton X-100; 4× SSC for 10 min each at room temperature.

10. Remove as much supernatant as possible from the final wash and add one drop of VectaShield w/DAPI (Vector Laboratories) to the cell pellet. Other antiquench mounting media and counterstain such as SYTOX may also be used. Thoroughly mix the antiquench with the cells using a 200-μl pipette tip.

11. Prepare microscope slides. Trace an outline of a 22 × 22 mm microscope coverslip onto a piece of paper. Position a glass slide midway over the outline of the coverslip and place a small dot of Permount or nail polish (Avon) just inside each corner of the outline. The dots will provide adequate space between the slide and coverslip to prevent crushing the cells.

12. Pipette 20 μl of cells in mounting medium to a prepared slide and add coverslip. Blot excess mounting medium with filter paper and seal the edges with Permount or nail polish. Slides may be stored at 4°C for several days.

B. rDNA Probe Synthesis

Biotinylated rDNA probe was synthesized using the Nick translation protocol described in Sambrook, Fritsch, and Maniatis (1989). It is recommended that a series of test reactions be performed to determine optimal DNase I concentration.

C. Nick Translation Reaction

Mix ingredients in the following order and keep on ice:

2.5 μl 10× Nick translation buffer
2.0 μl (pD5H8 1 mg/ml)
1.0 μl 2.5 mM dNTP-dTTP
1.0 μl 1 mM Biotin-16-dUTP (Roche)
15 μl H_2O

2.5 μl 1/20-1/30 dilution of RQ1 DNase (Promega)

1.0 μl DNA polymerase I (New England Biolabs)

1. Incubate reaction for 60 min at 16°C.

2. Stop the reaction by adding 1 μl 0.5 mM EDTA pH 8.0 and precipitate by adding 2 μl 10 mg/ml herring sperm DNA and 1 volume of 2-propanol.

3. Wash 1× with 70% ethanol, briefly air dry pellet, then dissolve in 20 μl H_2O.

IV. Live Fluorescence Microscopy

In order to view fluorescent targets in living cells, cells must be immobilized. Various cell compression devices have been invented and reinvented over the past century. We describe the assembly and operation of two such devices as well as some simple suggestions for fluorescently labeling live cells. The first device, the Aufderheide Rotocompressor, has the advantages of a low profile and the ability to perform oil immersion with both the objective lens and the substage condenser lens for Kohler illumination and optimal resolution. This latter feature is valuable for applications with various kinds of brightfield, darkfield, and interference microscopy. The second device, the Ringlien Differential Screw Thread Compressor, has the advantage of quick assembly and disassembly and is free of lubricants, and the two coverslips are brought together with a direct (nonrotary) action. Finally we offer some suggestions for *in vivo* labeling with fluorescent probes.

A. The Aufderheide Rotocompressor Device

1. Background

Light microscopic investigations of living microorganisms, especially ciliate protozoa, have required the application of some means of slowing down or immobilizing the cells to permit high-resolution, high-magnification imaging. Many chemical, physical, and mechanical techniques have been developed which provide varying degrees of success at immobilization (reviews: Aufderheide, 1992; Foissner, 1991; Hanson, 1974; Repak, 1992; Wichterman, 1953).

The "ideal" immobilizing technique would require that the cell be physiologically unaffected by the immobilization, undistorted, viable, reversibly immobilized, available for retrieval for culture or other handling, and with optimal optical characteristics. No technique meets all these criteria successfully, but some approaches are much better in performance than others. In general, chemical techniques involve paralysis or narcotization of the cell. Many chemical treatments are poorly reversible without extensive washing of the cell and the physiology/viability of the cell is sometimes severely affected. Physical techniques typically involve placing the cell in a viscous medium or attaching the cell to a prepared surface. Retrieval of the cell from the viscous material is sometimes difficult, not all such materials are nontoxic, and the optical characteristics are distorted by some materials (Spoon *et al.*, 1977). Mechanical immobilization techniques involve trapping the cell between closely apposed surfaces or in a complex fibrous environment. Mechanical trapping can be hard to control to avoid crushing the cell, and it can cause considerable distortion to the cell while it is trapped. Nevertheless, reversal of the mechanical trapping and recovery of the cell is more easily accomplished than with other approaches, and the techniques can also provide a significant improvement of the optical path through the cell. Carefully controlled mechanical trapping techniques seem to be the most useful for many microscopic applications and species of organisms. Many classical and modern studies have made extensive use of a mechanical trapping device for qualitative and quantitative observations and manipulations of the cells (e.g., Aufderheide *et al.*, 1993, 1999; Fenchel, 1967; Grimes, 1976; Hanson and Ungerleider, 1973; Janetopoulos *et al.*, 1999; Uhlig, 1972).

The simplest form of mechanical immobilization involves drying a wet mount slide until the coverslip traps the cells against the slide surface. The observer usually has a few minutes to study the cell until continued drying crushes it. Control of the degree of distortion of the cell is not easily accomplished, nor is recovery of the cell. Early refinements of this approach involved applying a controllable spring device to the coverslip, allowing the user some degree of control over the amount of pressure applied to immobilize the cell (e.g., Allman, 1855). A modern variation on the coverslip technique involves using small squares of Handiwrap plastic instead of a glass coverslip (Spoon, 1976). These approaches, although simple, are not satisfactory for critical research applications.

Modern devices for mechanical immobilization are essentially a coverslip attached to a micrometer device, allowing critical and delicate control over the distance between the coverslip and the slide. These devices are known as microcompressors or rotocompressors. The specimen is trapped between the coverslip and the slide. Ideally, the specimen is held with a minimum of distortion. A few design versions have been reported in the past 50 years. All are marvels of precision machining and are tributes to the skills of the mechanics who can make them.

The "classical" rotocompressor was developed by Dr. Asa Schaeffer, apparently in the 1930s (Wichterman, 1953; Hanson, 1974), and was available commercially for several years. It is now rare, and people who have them generally hoard them. Although it had several design limitations, it was nevertheless a rugged and useful instrument and it has served as the model for many subsequent design refinements.

There are three modern designs of rotocompressors: the model of Spoon (1978), the model of Uhlig and Heimberg (1981), and the model of Aufderheide (1986, and described here). Only the Uhlig and Heimberg model is available commercially. Spoon's model is difficult to construct because it contains 22 separate precision parts, but it does use commonly available glass slides and coverslips. Uhlig and Heimberg's model is large and it uses specialized coverslips. It is unique in that it is designed to work on either an upright or an inverted microscope. Both the Spoon and the Uhlig and Heimberg models require that the coverslips be cemented to the metal surfaces. If the coverslip is broken, replacement will require careful cleaning of the metal surfaces. Both of those units depend upon their glass surfaces being strictly parallel, so the installation of a fresh coverslip requires precision placement.

Aufderheide's model is a highly modified Schaeffer unit and it requires specialized slides and coverslips that are commercially available but somewhat difficult to find. Unlike the other two models, its coverslip is held in place mechanically, so that cement use is avoided. The coverslip is also distorted slightly in its center when it is properly mounted, so that the problem of maintaining strictly parallel glass surfaces is also avoided. Both the Aufderheide and the Uhlig and Heimberg units permit one to set the optics for Köhler illumination, even with double oil immersion; this setting is not possible with either the Spoon unit or the original Schaeffer unit. The glass surfaces of the Aufderheide unit are also siliconized with a commercial siliconization preparation. This significantly reduces the degree of adhesion/shear between the specimen and the glass, and thus the probability of damage to the specimen.

In all models, the specimen is placed in a very small droplet in the center of the viewing area of the unit. The top is applied and the unit is screwed down until the specimen is immobilized. Further adjustment of the distance between coverslip and slide can be made at any time while the specimen is being viewed. For instance, the distance between the slide and the coverslip needed for secure, long-term immobilization of *Paramecium* is about 10 μm. The glass and metal surfaces of all versions of the rotocompressor must be kept scrupulously clean and properly lubricated.

The successful use of the rotocompressor depends upon the organism being studied and the experience and skills of the investigator using the instrument. Considering the success of many studies using the instrument, the investments in cost of construction and time in learning how to use the instrument properly are undoubtedly well repaid.

2. Parts Description

The rotocompressor is constructed from six parts, five of which are precision machined from brass or stainless steel (see Fig. 5). Proper care must be given to all these pieces, as you would give to any precision scientific instrument.

a. Baseplate

The baseplate is a commercial 2-inch $\times$ 3-inch $\times$ 1-mm glass microscope slide (Corning #2947) with a small platform constructed in its center. The platform is formed by cementing two 12-mm diameter coverslips (one of No. 1 thickness, and the other of No. 2 thickness), one on top of the other, in the center of the slide. The platform is thus approximately 0.3 mm thick. The top surface of the platform is siliconized.

b. Outer Ring

The outer ring is machined metal and is glued to the baseplate. The cement used is Permount dissolved in xylene. (An equivalent such slide cement will probably do—but it must be easily removable.) Its threaded inner face accepts the inner ring.

Sectional View AA

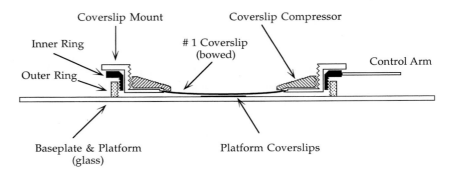

Plan View

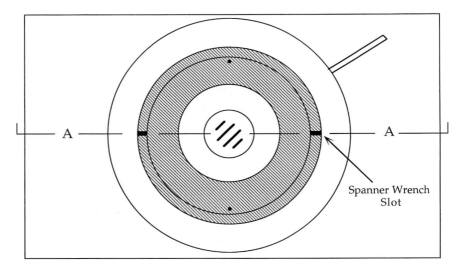

Fig. 5   Schematic cross-section and plan views of the Aufderheide rotocompressor.

c. Inner Ring

The inner ring is threaded metal and is screwed into the outer ring. Its smooth inner face accepts the coverslip mount. A stainless steel pin (control arm) is inserted into its outer face and will facilitate control of the compression of the unit.

d. Coverslip Mount

The coverslip mount is machined metal. Its smooth outer face is slipped into the inner ring. The threaded inner face of the coverslip mount accepts the coverslip compressor, and the machined step in the middle of the mount creates a niche for the 25-mm glass coverslip.

e. Coverslip Compressor

The coverslip compressor is machined and threaded metal and is screwed into the coverslip mount. Its lower face is specially machined to bow the coverslip downward in the coverslip mount. Its upper face has grooves to accept the spanner wrench.

f. Spanner Wrench

The spanner wrench assists in proper adjustment of the coverslip compressor in the coverslip mount.

g. Coverslips

The coverslips used are round, 25-mm diameter, #1 thickness. They are siliconized before use. Although commercially available, this size is getting harder to find; they can currently be obtained from Carolina Biological Supply Co. You may wish to get a "lifetime supply" while they are available.

3. Assembly, Adjustment, and Repair

As described in the Parts section, the outer ring is cemented to the baseplate. It should be concentric with the platform on the baseplate. The inner ring is threaded into the outer ring. Do not screw down the inner ring too far, or it will break the cement holding the outer ring to the baseplate. The baseplate and the inner and outer rings together form what will be called the baseplate assembly.

A coverslip should be centered in its machined niche in the coverslip mount, and the coverslip compressor screwed into the mount. Use the spanner wrench for this adjustment. The coverslip compressor should be tightened sufficiently to distort the coverslip downward a fraction of a millimeter. *This is intentional and critical to the function of the unit.* The coverslip mount, the coverslip compressor, and a coverslip together form the coverslip assembly. The coverslip assembly may then be slipped into the inner ring as far as it will go. The unit is thus completely assembled and ready for use.

With the coverslip assembly firmly seated in the inner ring, slowly rotate the threaded inner ring clockwise (this will bring the coverslip toward the platform) until you can just see newton rings between the coverslip and the platform of the baseplate. Make a mark on the baseplate indicating the position of the pin where this level is reached. This mark indicates the position of maximum compression—cells will be immobilized before you reach this point, and will probably be crushed if you crank the inner ring down this far. Back off the inner ring by rotating the pin counterclockwise about 90°. This position should free any cells held in the unit so that they may be recovered. This is also the position at which the unit may be loaded with fresh cells.

The newton rings between the coverslip and the platform should be centered. If they are not, you may have to rotate the coverslip assembly relative to the Inner Ring until the rings are centered. When you have the rings properly centered, make a mark on the flat top of the coverslip mount indicating the position of the pin. This will permit you to have the assembly properly positioned each time you load the unit for use.

4. Use

The profile of this design of rotocompressor is too high to permit safe rotation of a compound microscope nosepiece with high-numerical-aperture objectives. You can jam the objective into the rotocompressor if you try to shift objectives by simply rotating the nosepiece.

To load the rotocompressor, grasp the coverslip assembly and carefully slip it off the baseplate assembly. Using a dissecting microscope and a fine pipette, carefully load the cells to be observed in a small (no more than 1-mm diameter) droplet of fluid onto the middle of the platform of the baseplate. Quickly slip the coverslip assembly back onto the baseplate assembly. The droplet should touch both the platform and the coverslip, but the cells should not be trapped at this time. (If the cells are trapped or crushed by this operation, the inner ring setting is too low.) Place the assembled unit on the stage of a compound microscope and, using a low power, bring the droplet and the cells into focus. The pin on the inner ring is carefully rotated clockwise while the cells are watched. At some point, the cells will be trapped between the coverslip and the platform and will thus be flattened and immobilized. As the operator gains skill in this, he/she will be able to quickly trap a set of cells without crushing or overcompressing them. (For *Paramecium,* for example, trapping will occur when the clearance between the coverslip and the platform is about 10 μm. This will vary for different species of protozoa. Different species have differing abilities to tolerate compression; some do quite well while others lyse as soon as they are immobilized.) Once the cells are immobilized, the user can center one cell of interest in the field. Higher magnification objectives can then be used. Use caution when changing power to avoid damage to the objective or to the rotocompressor. One must first rack the nosepiece up, position the objective desired, and then carefully rack the nosepiece down. Brands of microscopes with autofocus stop safety features are especially useful for this procedure. Careful adjustments of the degree of flattening/immobilization can be accomplished by slight movement of the pin on the outer ring. Take care that the pin is pushed only laterally and that no vertical force is applied, or the specimen might be crushed. Protozoa seem to have the ability to adapt to the flattening and can start to creep out of the field of view after several minutes of observation, so readjustment of the compression might be needed occasionally.

To unload the rotocompressor, one should first rotate the pin about 90° counterclockwise. If recovery of the specimen is desired, this rotation must be done very carefully to avoid shearing the flattened, trapped specimens. If recovery of the specimens is not desired, then less care is needed for this rotation. Grasp the coverslip assembly and slide it out from the baseplate assembly. Recovery of the cells should be done quickly under a dissecting microscope: the tiny droplets will evaporate. The droplet will be divided between the platform and the coverslip, so one should search for the specimen on both faces. After recovery of the specimen, dip the corner of a Kimwipe or similar laboratory tissue (but not a facial tissue—some of these have diatoms that might scratch the glass surfaces) into hot water and carefully wipe and dry the platform and the coverslip faces.

Clean these faces with a good microscope lens cleaner. Use particular care to prevent lubricating oil, grease, or immersion oil from smearing across these surfaces. Phase contrast and DIC optical techniques are very sensitive to this kind of contamination. The unit should then be ready to be reloaded or stored.

5. Lubrication and Maintenance

All moving parts in the rotocompressor need to be properly lubricated. The threads between the outer and inner rings should be lubricated with a high-quality oil of moderate viscosity. We have used 10W-30 motor oil. If the threads start to feel sticky or gritty, they should be carefully cleaned with a solvent and relubricated. The smooth face between the inner ring and the coverslip mount should be lightly coated with a strong grease. As the unit is used, the grease needs occasional refreshing. Care should be taken not to use too much lubricant anywhere to avoid contamination of the glass surfaces.

The glass surfaces that actually touch the specimen (the platform and the lower side of the coverslip) must be properly siliconized. This helps with cleaning, but mostly siliconization assists in protecting the specimen from being sheared by the rotating glass surfaces. A good commercial siliconizing agent (we use Pierce #42800 SurfaSil or #42799 AquaSil, properly diluted) will provide a lasting coating that rarely needs refreshing.

Following are the Carolina Biological Supply Co. order numbers and prices for coverslips, taken from catalog #68, 1998, p. 1016:

Round, 12 mm, #1 thickness: D8-63-3029 $8.20/box of 100

Round, 12 mm, #2 thickness: D8-63-3049 $9.95/box of 100

Round, 25 mm, #1 thickness: D8-63-3037 $11.95/box of 100

B. The Ringlien Differential Screw Thread Compressor

This design was developed to eliminate the need for very fine pitch thread cutting, eliminate the need for lubrication, and simplify the process of changing specimens. An added bonus of the design is the nonrotational compression. This device uses the difference in pitch between two screw threads to obtain a precise control of the space between two glass coverslips. One thread used alone provides a coarse vertical and rotational adjustment and when coupled to the other thread provides a fine vertical adjustment which serves to compress the specimen sufficiently to restrict its motion for detailed observation. The sample substrate is a coverslip that is cemented to the pedestal with a soluble adhesive (Permount soluble in xylene) and remains fixed relative to the microscope stage. The vertically adjustable coverslip is mounted in an annular holder that provides a means of bending the coverslip toward the substrate to provide rigidity and a localized contact area with specimen.

The four threaded rings, pedestal, base plate, control arm, locking pin, and locating hardware are machined from stainless steel (type 303) and plastic (Delrin). The coverslips are standard 12 mm and 25 mm × 0.15 mm thick. Refer to the plan and cross-sectional views in Fig. 6.

Sectional View AA

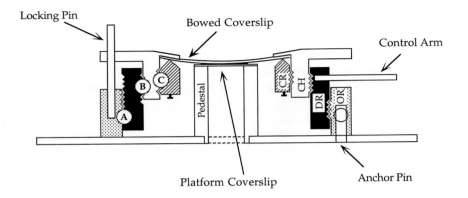

Locking Pin

Bowed Coverslip

Control Arm

Pedestal

Platform Coverslip

Anchor Pin

Plan View

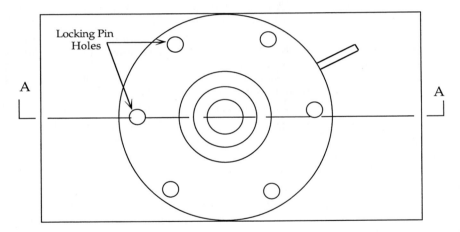

Locking Pin
Holes

A   A

Fig. 6   Schematic cross-section and plan views of the Ringlien differential screw thread compressor. (Note: Figure is not to scale. In particular, the mechanism's height has been exaggerated to give room for detail.)

Because there are no metal-on-metal contacts one avoids the need for lubricants, and the pedestal and baseplate design eliminates the need to glue the device to a glass slide, making a wider range of baseplate sizes acceptable.

1. Construction Details

Refer to the plan and sectional views (Fig. 6).

The only special tools required are the holding fixtures needed to hold the parts in a way to prevent distortion. The four threaded parts can be held for machining in machinable

5c collets. Particular attention should be given to the tolerances of mating threads A, B, and C, to prevent binding or looseness.

The stainless steel coverslip holder (CH) should be machined first since it requires the most detailed profile. Machine the stainless outer ring (OR) and the Delrin coverslip retainer (CR) followed by the Delrin drive ring (DR). Finished rings can be used to mount other rings for some operations, particularly the threading, B, of the Delrin drive ring.

Some typical dimensions of one completed model:

Coverslip holder (CH): stainless steel

Bore 3/4 inch

O.D. 2.2 inch

Thread C, 32 t.p.i., pitch diameter 1.15 inch, depth 0.25 inch

Thread B, 32 t.p.i., pitch diameter 1.30 inch, depth 0.25 inch

Threads B, C can be relieved next to the flange to provide tool clearance

Flange thickness, 0.052 reduced to 0.020 near the bore for lens clearance

Overall thickness, 0.3 inch

Coverslip retainer (CR): Delrin
Bore 3/4 inch

Thread C, 32 t.p.i., pitch diameter 1.15

Thickness, 0.25 inch. Top edge contoured to provide a pressure point which will press the glass against an angled seat to contour the glass as desired.

Drive ring (DR): Delrin

Thread B, 32 t.p.i., pitch diameter 1.30

Thread A, 20 t.p.i., pitch diameter 1.85 inch

Thickness, 0.4 inch

Control arm, 1/16 inch dia sst pin threaded or pressed into top edge of ring, 3/4 inch long

Outer ring (OR): stainless steel

Outside diameter, 2.2 inch

Thickness, 0.28 inch

Thread A, 20 t.p.i., pitch diameter 1.85 inch

Securing the outer ring to the base plate: Mating pins in the base plates and holes in the outer ring with set screws may be used, or any other clamp/screw arrangement that allows easy mounting

Locking pin: The locking pin should have a tapered seat in the outer ring to reduce unwanted play and a precise alignment with holes in the coverslip holder to prevent binding during operation of the control arm

Pedestal: The pedestal is machined from a stainless steel tube, 0.5 inch o.d. and notched to fit a 7/16 inch reamed hole in the base plate. A 1/32 inch slit is cut in the side along part of its length to allow a spring tension type of fit in the hole. The length of the pedestal is determined after the other parts have been assembled and an operating range determined (see below).

Pedestal height: The pedestal height can be determined by having as many threads engaged as possible and still allow a travel of one or two threads of coarse adjustment and 90° of control arm travel. To determine pedestal height engage threads A and B as far as they will go, then back them off about a half a turn each such that the control arm is in a convenient operating position. The height of the pedestal will be the distance from the baseplate to the underside of the coverslip minus the thickness of the pedestal coverslip. Machine the pedestal to this height and mount the coverslip. The clearance or operating space may be checked by placing an easily compressible elastic material of appropriate thickness on the pedestal and observing the compression as the control arm is operated.

2. Operating Instructions

a. Securing and Preparing the Pedestal

The pedestal is a friction fit in the base. This fit can be adjusted slightly by expanding or contracting the tension slot in the insertion end of the pedestal with an appropriate tool. Secure a 12-mm diameter glass coverslip to the top of the pedestal with an easily removable cement (Permount soluble in xylene). This is required to prevent surface tension from lifting the coverslip off the pedestal.

b. Assembling the Compressor with the Specimen in Place

Once you have mounted the pedestal, the compressor is ready to receive a specimen. With the compressor removed from the baseplate, add a drop of culture medium to the platform coverslip. Take the entire compressor assembly and mount it onto the baseplate. (*Note:* The proper working distance should already be established, this height having been adjusted when determining the pedestal height.) Under low magnification, view the specimen as you gently advance the coverslip holder (thread B), screwing it clockwise. The bowed coverslip will immediately contact the liquid medium, and a seal will form between the two coverslips. Rotate the coverslip holder while viewing the specimens until minimal contact is established (cells show retarded motion). Select one of the holes in the coverslip holder closest to the contact position and insert the locking pin through the coverslip holder and into the hole in the outer ring. The final immobilization is performed using the control arm attached to the drive ring. Advance the bowed coverslip carefully (viewing cells under magnification) using the control arm, until the target cell is secured (usually within one-quarter turn). One can now back off the low-power microscope objective, dial in the high-magnification objective (add oil if desired), and gently advance the high-mag objective into position. Relocation of the locking pin may be required to adjust the range of the control arm. The compression rate is 0.0013 mm/degree of rotation or 0.48 mm/revolution of the control handle (net travel of the differential screw system).

Note: Once a proper operating distance has been established, changing specimens becomes quite straightforward. After backing the control arm so that there is no compression, back off the coverslip holder to allow clearance. The entire compressor can be removed from the baseplate, a new specimen mounted, and the compressor simply remounted onto the baseplate. The working distance is maintained.

c. Replacing the Bowed Coverslip

Remove the bowed coverslip holder assembly by rotating it counterclockwise until it comes out of the outer ring assembly. Turn it upside-down and rotate the coverslip retaining ring counterclockwise using the two screw heads as handles. Clean the seating surface of any debris, locate a new coverslip in the recess, and replace the ring. Observe the reflection in the surface of the coverslip as the ring is tightened causing the glass to assume a bowed configuration. This requires a rather delicate touch. Coverslip glass can be made stronger by removing edge nicks with an emery stone, which serves to prevent stress sites from propagating cracks.

d. Care and Handling

This is a precision device made from stainless steel and Delrin. It may be lubricated for smoother operation but oxidized lubricants or any other debris could cause it to seize. Extreme temperature changes, due to different expansion coefficients of these materials, could also cause difficult operation.

C. Some Vital Fluorescent Dyes

Three methods for fluorescent tagging are currently being explored: labeling nuclei with fluorescent vital SYTO dyes (Molecular Probes), labeling mitochondria with MitoTracker (Molecular Probes) dyes, and GFP-tagging endogenous proteins. Nuclei can be labeled with a host of different fluorescent SYTO dyes. An effective dose can usually be made of approximately 1 μM dye in DMSO. These can be delivered directly to the growth medium and are readily taken up by living cells.

Mitochondria can be elegantly labeled with the MitoTracker dyes from Molecular Probes. In particular, MitoTracker Green (M-7514) and MitoTracker Red (M-7512) make useful probes. Stock solutions are made by dissolving 50 μg of dye in 1 ml of DMSO. Samples can be stored frozen at $-20°$C. Working strength can be achieved by diluting 10 μl of stock solution per ml (MitoTracker Green) and 2 μl of stock solution per ml of cell culture (for MitoTracker Red). Cells can be labeled for a minimum of 1 h to overnight, washed out, and viewed. These can be viewed as for fluorescein and rhodamine, respectively.

A few problems need to be addressed when working with confocal fluorescence and vital dyes. First, lethality can be an issue. When live cells encounter the high-energy excitation light, there are frequently ill effects due to excess heat (especially for the red fluorescent dyes) and the generation of free radicals. These effects can be diminished by reducing the intensity of the excitation light and reducing the exposure time. It is possible that antioxidizing agents may also help. Second, live cells (even immobilized live cells) are active. Their cytoplasmic contents are in motion. Hence, conventional confocal fluorescence microscopy, which can require up to several minutes to acquire an image, may prove problematic. Live-time confocal fluorescence microscopy may be achieved using video-rate confocal methods discussed earlier in this volume (see Matsumoto, 2002).

Acknowledgments

The principle author expresses his sincere thanks to Drs. Brian Matsumoto and Mark Sanders, whose encouragement and professional talents made this compilation possible. The work was supported by NSF RUI grant # MCB-9807555. Development of the Aufderheide rotocompressor was supported by NIH Grants AG02657 and GM34681, and by a Texas A&M Interdisciplinary Research Grant #IRI94-05.

References

Allman, G. J. (1855). On the occurrence among the infusoria of peculiar organs resembling thread-cells. *Quart. J. Microscop. Sci.* **3,** 177–179.

Aufderheide, K. J. (1986). Identification of the basal bodies and kinetodesmal fibers in living cells of *Paramecium tetraurelia* and *Paramecium sonneborni. J. Protozool.* **33**(1), 77–80.

Aufderheide, K. J. (1992). Special techniques for viewing living protozoa. *In* "Protocols in Protozoology" (J. J. Lee and A. T. Soldo, eds.), pp. E3.1–3.4. Allen Press, Lawrence, KS.

Aufderheide, K. J., Du, Q., and Fry, E. S. (1993). Directed positioning of micronuclei in *Paramecium tetraurelia* with laser tweezers: absence of detectable damage after manipulation. *J. Euk. Microbiol.* **40,** 793–796.

Aufderheide, K. J., Rotolo, T. C., and Grimes, G. W. (1999). Analyses of inverted ciliary rows in *Paramecium*. Combined light and electron microscopic observations. *Eur. J. Protistol.* **35,** 81–91.

Chilcoat, N. D., Melia, S. M., Haddad, A., and Turkewitz, A. P. (1996). Grl1p, an acidic, calcium-binding protein in *Tetrahymena thermophila* dense-core secretory granules, influences granule size, shape, content organization and release but not protein sorting or condensation. *J. Cell Biol.* **135,** 1775–1787.

Fenchel, T. (1967). The ecology of marine microbenthos. I. The quantitative importance of ciliates as compared with metazoans in various types of sediments. *Ophelia* **4,** 121–137.

Foissner, W. (1991). Basic light and scanning electron microscopic methods for taxonomic studies of ciliated protozoa. *Eur. J. Protistol.* **27,** 313–330.

Fujiu, K., and Numata, O. (1999). Localization of microtubules during macronuclear division in *Tetrahymena thermophila* and possible involvement of macronuclear microtubules in 'amitotic' chromatin distribution. *Cell Struct. Funct.* **24,** 401–404.

Fujiu, K., and Numata, O. (2000). Reorganization of microtubules in the amitotically dividing macronucleus of *Tetrahymena. Cell Motil. Cytoskeleton* **46,** 17–27.

Gaertig, J. (2000). Molecular mechanisms of microtubule organelle assembly in *Tetrahymena. J. Eukaryot. Microbiol.* **47,** 185–190.

Gaertig, J., and Fleury, A. (1992). Spatiotemporal reorganization of intracytoplasmic microtubules is associated with nuclear selection and differentiation during developmental processes in the ciliate *Tetrahymena thermophila*. *Protoplasma* **167,** 74–87.

Grimes, G. W. (1976). Laser microbeam induction of incomplete doublets of *Oxytricha fallax. Genet. Res. Camb.* **27,** 213–226.

Hanson, E. D. (1974). Methods in the cellular and molecular biology of *Paramecium. Methods. Cell Biol.* **8,** 319–365.

Hanson, E. D., and Ungerleider, R. M. (1973). The formation of the feeding organelle in *Paramecium aurelia. J. Exp. Zool.* **185,** 175–188.

Janetopoulos, C., Cole, E., Smothers, J. F., Allis, C. D., and Aufderheide, K. J. (1999). The conjusome: a novel structure in *Tetrahymena* found only during sexual reorganization. *J. Cell Sci.* **112**, 1003–1011.

Kiersnowska, M., Kaczanowski, A., and Morga, J. (2000). Macronuclear development in conjugants of *Tetrahymena thermophila*, which were artificially separated at meiotic prophase. *J. Eukaryot. Microbiol.* **47**, 139–147.

Lee, S., Wisniewski, J. C., Dentler, W. L., and Asai, D. A. (1999). Gene knockouts reveal separate functions for two cytoplasmic dyneins in *Tetrahymena thermophila*. *Mol. Biol. Cell.* **10**, 771–784.

Marsh, T. R., Cole, E. S., and Romero, D. P. (2001). The transition from conjugal development to the first vegetative cell division is dependent on RAD 51 expression in the ciliate tetrahymena thermophila. *Genetics* **157**, 1591–1598.

Matsumoto, B., and Hale, I. L. (2002). Resolution of subcellular detail in thick tissue sections: Immunohistochemical preparation and fluorescence confocal microscopy. *Methods Cell Biol.* **70**, 297–330.

Repak, A. J. (1992). Immobilization methods for protozoa. *In* "Protocols in Protozoology" (J. J. Lee and A. T. Soldo, eds.), pp. C1.1–1.14. Allen Press, Lawrence, KS.

Sambrook, J., Fritsch, E. F., and Maniatis, T. (1989). "Molecular Cloning: A Laboratory Manual," Vol. 2, pp. 10.6–10.12. Cold Spring Harbor Laboratory Press, Cold Spring Harbor, NY.

Sanders, M. A., and Salisbury, J. L. (1995). Immunofluorescence microscopy of cilia and flagella. *Methods Cell Biol.* **47**, 163–169.

Spoon, D. M. (1976). Use of thin, flexible plastic coverslips for microscopy, microcompression, and counting of aerobic microorganisms. *Trans. Am. Microsc. Soc.* **95**, 520–523.

Spoon, D. M. (1978). A new rotary microcompressor. *Trans. Am. Micros. Soc.* **97**, 412–416.

Spoon, D. M., Feise, C. O., and Youn, R. S. (1977). Poly(ethylene oxide), a new showing agent for protozoa. *J. Protozool.* **24**, 471–474.

Stuart, K. R., and Cole, E. S. (1999). Nuclear and cytoskeletal fluorescence microscopy techniques. *In* "Methods in Cell Biology" (David J. Asai and James D. Forney, eds.), Vol. 62, pp. 291–311. Academic Press, San Deigo.

Uhlig, G. (1972). Protozoa. *In* "Research Methods in Marine Biology" (C. Schlieper, ed.), pp. 129–141. University of Washington Press, Seattle.

Uhlig, G., and Heimberg, S. H. H. (1981). A new versatile compression chamber for examination of living microorganisms. *Helgoländer Meeresunter.* **34**, 251–256.

Ward, J. G., Blomberg, P., Hoffman, N., and Yao, M.-C. (1997). The intranuclear organization of normal, hemizygous and excision-deficient rRNA genes during developmental amplification in *Tetrahymena thermophila*. *Chromosoma* **106**, 233–242.

Wichterman, R. (1953). "The Biology of Paramecium." Blakiston Press, New York.

CHAPTER 9

Confocal Imaging of *Drosophila* Embryos

Stephen W. Paddock

Howard Hughes Medical Institute
Department of Molecular Biology
University of Wisconsin
Madison, Wisconsin 53706

I. Introduction

The confocal laser scanning microscope (CLSM) is used to image the three-dimensional (3-D) distribution of cells in *Drosophila* tissues. A major application of the instrument is for the visual analysis of expression patterns of genes that control the insect's development (Orenic and Carroll, 1992; Bate and Martinez Arias, 1993). Furthermore, many of the genes that are homologous to those first cloned from *Drosophila*

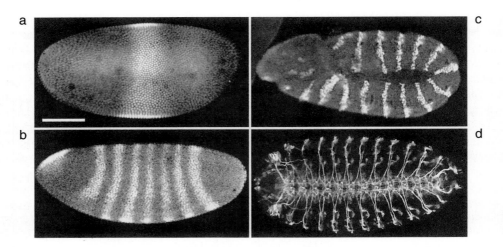

Fig. 1 Single immunofluorescence labeling of *Drosophila* embryos at different developmental stages. (a) The gap gene, *Kruppel,* is expressed in a single domain that circles the embryo at the cellular blastoderm stage; (b) the pair rule gene *hairy* is expressed in seven stripes that encircle the embryo at a slightly later time in the cellular blastoderm stage; (c) the segment polarity gene *engrailed* is expressed in 14 stripes at the germband extension stage in the posterior region of every segment; and (d) the peripheral nervous system has become differentiated at stage 12 as revealed by the 22C10 antibody. Scale bar = 50 μm.

continue to be cloned from other organisms (Nagy and Carroll, 1994; Brunetti *et al.,* 2001). Here the CLSM is used to map and to compare their patterns of expression during equivalent developmental stages (Carroll *et al.,* 1995; Grenier and Carroll, 2000) and to gain insight into their functions from the analysis of experimentally manipulated insects (Brand *et al.,* 1994; Gibson, 1999).

This chapter will focus on how the CLSM is used for imaging fixed and fluorescently labeled *Drosophila* embryos (Fig. 1; see color plate). Confocal images are now collected from most stages of the *Drosophila* life cycle including the spermatozoa (Karr, 1991; Snook and Karr, 1998), the oocytes (Theurkauf, 1994; Saunders and Cohen, 1999), the imaginal discs (Brown, 1998) and the pupal stages (Orenic *et al.,* 1993), as well as the embryos (Sullivan and Theurkauf, 1995). Some of these tissues are more amenable than others to morphological analysis, and each one presents a unique challenge for the microscopist.

The early embryos are relatively easily imaged since the cells are positioned at the surface of the embryo at the cellular blastoderm stage. Many studies have utilized conventional light microscopy; using brightfield, DIC, or even epifluorescence microscopy in order to image the embryos (Patel, 1994; Skeath, 1998). However, as the embryos develop, and the numbers of cells increase, they present more of a challenge for imaging especially those fluorescently labeled cells that are positioned deep within the volume of the embryo. The embryos can be carefully dissected and flattened out onto a slide in order to gain improved resolution (often called the "fillet" method) in the conventional wide-field light microscope.

The confocal approach is a less labor-intensive method for the elimination of signal from fluorescence from both labeled and autofluorescing cells that are positioned out of the focal plane of interest (White *et al.*, 1987). The method is also less prone to artifacts of preparation since whole mounts of the embryos can be imaged, and it is also preferred for imaging living embryos (Kiehart *et al.*, 1994; Francis-Lang *et al.*, 1998). Moreover, background can be higher in multiply labeled specimens where noise from the excitation of several fluorescent probes and from autofluorescence is increased, especially when two or more images are combined into a single multicolor image.

This chapter will cover the methods and the instrumentation used for collecting confocal images of *Drosophila* embryos, and for presenting them for publication, together with some of the current applications including multiple-label immunofluorescence, fluorescence *in situ* hybridization (FISH), and lineage tracing.

II. Preparation of the Embryos

Drosophila embryos are prepared for confocal imaging using relatively standard methods of tissue preparation for immunofluorescence (Foe and Alberts, 1983; Karr and Alberts, 1986; Langeland, 1998) and for FISH (Hughes *et al.*, 1996; Wilkie and Davis, 1998; Wilkie *et al.*, 1999). These methods can also be used as starting points for the preparation of most tissue types for CLSM analysis (Paddock, 1998). Some features of the protocols that are peculiar to the preparation of *Drosophila* embryos are included below.

A. Embryo Collection and Permeabilization

Embryos are collected on egg-lay caps, or if greater numbers are required, they are collected on a large dish from a population cage (Bate and Martinez-Arias, 1993). If embryos of a certain age are required, for example, 2 h old, the flies are usually allowed to lay on fresh egg-lay caps for 4 h before the embryos are harvested. In this way a population of embryos between 0 and 4 h of age is collected, and the required developmental stage is subsequently selected from the population of embryos in the microscope field (Campos-Ortega and Hartenstein, 1985; Foe, 1989).

Drosophila embryos, in common with the embryos of many insects, are surrounded by a tough protective outer chorion and an inner vitelline membrane, both of which are impermeable to large molecules and must be removed before antibody probes are able to penetrate evenly throughout their entire volume. The efficiency of these initial permeabilization steps is extremely important and can be a major source of poor staining. A balance must be achieved between adequate permeabilization that allows the fluorescent probes to enter the entire embryonic volume, and too much permeabilization when the embryo will fall apart. The embryos of some mutant flies may have different physical properties from the wild type and may require modifications to the fixation/extraction conditions. In those instances where the chemical extraction of the outer membranes can interfere with the probes themselves, the embryos can be devitellinized by hand (von Dassow and Schubiger, 1995). A mild sonication step is often used for preparing embryos from insects with especially resistant outer membranes (Grbic *et al.*, 1996).

B. Fluorescent Probes

Primary antibodies can be raised against recombinant proteins in *Drosophila* (Williams *et al.,* 1995). Many probes are readily available from members of the fly research community (www.ceolas.org/fly). The antibody probes can usually be used more than once, especially if they are in short supply (Brown, 1998).

Secondary antibodies and other probes, for example, fluorescent phalloidin or DNA probes, can be purchased from one of a number of companies, for example, Jackson Laboratories (www.jacksonimmuno.com), Molecular Probes (www.probes.com), or Sigma-Aldrich (www.sigma-aldrich.com). Fluorescein, lissamine rhodamine, and cyanine 5 conjugated to secondary antibodies are commonly used fluorochromes in strategies for labeling *Drosophila* embryos for the CLSM since their excitation and emission spectra are matched to the lasers and filters of most commercially available CLSMs (Brismar *et al.,* 1995). Alternative dyes to fluorescein or rhodamine include Bodipy, cyanine 3, and the Alexa series of dyes (Haugland, 1999).

The immunofluorescence protocols for staining *Drosophila* embryos generally use primary antibodies from three different animal species for triple labeling (Paddock *et al.,* 1993). Here the equivalent steps of the protocol can be performed in parallel. However, when primary antibodies from three different species are not available, these steps are performed sequentially, with many thorough washes between each primary/secondary antibody sequence in order to avoid cross-reactivity of probes (Shindler and Roth, 1996; Brown, 1998).

C. Mounting the Embryos

Embryo preparations are mounted on a standard microscope slide in a few drops of an antifade agent, for example Vectashield (Vector Laboratories Inc., Burlingame, CA; www.vectorlabs.com), and covered with a coverslip. Care should be taken to match the coverslip thickness to the objective lens. The coverslip is sealed to the glass slide using nail polish. This method of mounting causes some flattening of the embryos. To avoid this, spacers can be cut from coverslips using a diamond marker, or from narrow-gauge fishing line, and placed between the slide and the coverslip to avoid distortion of the embryos. However, care should be taken to ensure that the embryos are not mounted too deeply in the chamber at a position that is out of the range of the working distance of the objective lens. When using an inverted microscope, the embryos can be mounted so that they are against the coverslip surface.

It is extremely important for the coverslip surface to be clean and dry before viewing with the CLSM. A dirty coverslip can be the cause of poor images, and, in extreme cases, it can also cause the objective lens to become dirty or even scratched. The nail polish is allowed to dry before the slides are examined in the microscope or before the coverslips are cleaned with distilled water, if necessary. Most of the organic solvents, for example, chloroform, used for cleaning objective lenses will dissolve the nail polish and form a smear across the coverslip surface, which is extremely difficult to remove. This situation should be avoided at all costs.

If an oil immersion objective lens is to be used, the coverslip surface should be dry since any water on the coverslip surface will interfere with the light transmission through the oil. Moreover, many different brands of immersion oils do not mix; for example, Zeiss and Nikon oils are immiscible, and care should be taken when viewing the samples on different microscopes in different laboratories to be consistent with the brand of immersion oils used.

The embryo preparations should ideally be viewed as soon as is practical after their preparation since the fluorochromes can degrade with time. This usually results in an increase in the background fluorescence in the specimen. The preparations can be stored in the dark and in a refrigerator (usually in a light-tight box or wrapped in aluminum foil). The slides should be allowed to warm to room temperature before viewing in order to remove any condensation that may form on the surface of the coverslip.

III. Imaging Embryos

A. Microscope Details

All of the images presented in this chapter were collected with commercially available CLSMs—either a Bio-Rad MRC 600 or a Bio-Rad MRC 1024 (Bio-Rad Microsciences, Hercules, CA; www.microscopy.bio-rad.com). Further details can be found elsewhere (Paddock, 1999). The scan head of our system is mounted on a Nikon Optiphot upright microscope stand, which allows the embryos to be viewed directly by eye using conventional epifluorescence optics before imaging with the CLSM. Here, we have a choice of rhodamine, fluorescein, indodicarbocyanine (Cy-5), or transmitted light optics, although the Cy-5 emission is nearly invisible to the human eye but not to most digital cameras. A good rule of thumb is that if the specimen can be viewed by eye in the rhodamine or fluorescein channel, then it can be imaged with the CLSM.

The Nikon Optiphot is extremely convenient for screening embryos before imaging with the CLSM. Using this microscope stand, it is impossible to allow laser light from the CLSM to reach the eyes since the entire eyepiece is rotated out of the field in order to open the optical axis of the microscope to the laser light from the scan head. Some form of rotating stage is also useful for orienting the embryos in the microscope field although it is relatively simple to orient the digital image of the embryos after imaging. Some of the recently available CLSMs are able to change the orientation of the scanning beam to collect an image in any orientation.

The microscope and scan head are mounted on an anti-vibration table and care is taken to isolate any potential sources of vibrations from it. Vibration can cause degradation and loss of resolution of the images, and in extreme cases appears as irregular horizontal lines in the images. In most of the modern systems the laser light is transferred to the scan head via a fiber optic cable rather than mounting the laser directly on the scan head, which was the main cause of vibration artifacts in the older systems.

The krypton–argon laser is an extremely convenient light source for multiple labeling experiments since it produces three major lines around 488 nm (blue), 568 nm (yellow), and 647 nm (red). These wavelengths are matched to the excitation spectra of some of the more commonly used fluorochromes including fluorescein (exc. max. 496 nm), lissamine rhodamine (exc. max. 572 nm), and Cy-5 (exc. max. 649 nm). In addition, the peak emission spectra of these dyes (fluorescein em. 518 nm; lissamine rhodamine em. 590 nm; Cy-5 em. 672 nm) are separated sufficiently to enable efficient optical filtering for detection of the emission of the three fluorophores, and are within the range of detection of the photomultiplier tube detectors used in most CLSMs. This gives reduced chances of the emitted signals bleeding through from one channel into another.

One drawback of a krypton–argon laser as a light source is its relatively short lifetime to deliver all three wavelengths at full power. There is a tendency for the far red line to weaken before the other two, which causes problems with bleed-through and can produce a grainy and lined image quality. Many of the recently available CLSMs have multiple laser light sources that have longer lifetimes than a single krypton–argon laser: for example, a 25-mW argon (488/514 nm), a 1.5-mW green helium–neon (543 nm), and a 5-mW red diode laser (638 nm).

As a general rule an objective lens with the highest numerical aperture available for the magnification required should be used to image the region of interest in the specimen. This results in thinner optical sections and improved resolution of structures, especially in those images collected from thick and bright samples. Electronically zooming lower magnification lenses with lower numerical aperture is undesirable since the higher magnification is achieved by decreasing the area of specimen scanned. This gives inferior resolution and increased photobleaching of the fluorophore.

A Zeiss $16\times$ oil (NA 0.5) or a Nikon $10\times$ dry (NA 0.45) objective lens can be used for imaging whole embryos (Figs. 1 and 2), and a Nikon $40\times$ (NA 1.2) and a Nikon $60\times$ (NA 1.4) Plan Apochromat objective lenses are used for higher resolution analyses of individual cells within embryos (Fig. 3). The optical section thickness for the $60\times$ (NA 1.4) objective with the pinhole set at 1 mm (closed) is on the order of 0.4 μm, and for a $16\times$ (NA 0.5) objective, again with the pinhole at 1 mm, the optical section thickness is around 1.8 mm. The Nikon $4\times$ (NA 0.20) Plan Apo. lens is useful for low-power work, and especially for locating specimens in the field of the microscope. Care should be taken when changing to the higher magnification objectives since the working distance is different for this lens relative to the others, and it is possible to crush the preparation.

More recently, high-resolution, high-NA water immersion lenses have become popular for confocal imaging. Although relatively expensive, these lenses are an excellent choice since they have a longer working distance than the equivalent oil immersion lenses, and give excellent depth penetration into thick specimens.

B. Collecting the Images

The images are collected as single optical sections or as Z-series (using a stepper motor to drive the fine focus of the microscope), in the proprietary Bio-Rad "pic" format (not to be confused with the Macintosh "pict" format). Images are collected as single gray-scale images usually at a size of 512×768 pixels, which is a good aspect ratio for

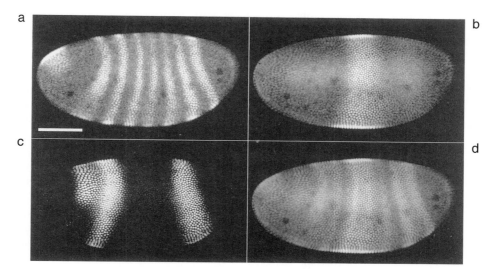

Fig. 2 Multiple wavelength imaging. Three confocal images collected sequentially from a triple-labeled embryo at the cellular blastoderm stage labeled with (a) *hairy;* (b) *Kruppel;* and (c) *giant,* together with (d) the merged image in gray scale. See Fig. 7 and the corresponding Color Plate for the three-color merged image. Scale bar = 50 μm.

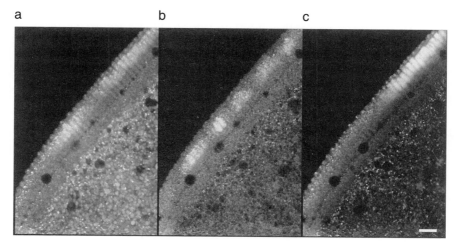

Fig. 3 Multiple wavelength imaging. Three images collected from the same embryo as Fig. 2 but using a 60× objective lens (cellular details) rather than a 16× objective lens (whole embryo), and focused along the side of the embryo. The cells are positioned around the periphery of the embryo at the cellular blastoderm stage. Note the reduced autofluorescence in the center of the embryo in the Cy-5 image (c) as compared with the background in the rhodamine (a) and the fluorescein-labeled images (b). Scale bar = 5 μm.

Drosophila embryos. Using the MRC 1024 the images can be collected up to a resolution of up to 1024 × 1280 pixels.

The MRC 1024 collects three images from a triple-labeled sample simultaneously, and places each of the 8-bit images into the red, green, and blue channels of an RGB image. The microscope usually places the rhodamine emission (red) into the red channel, the fluorescein emission (green) into the green channel, and the Cy-5 emission (far red) into the blue channel, and displays the merged RGB image in real time. Images can be saved as individual gray-scale images or as merged RGB images (Fig. 2).

The light path of the MRC 1024 system has been designed to collect images from triple-labeled specimens both automatically and simultaneously (Brelje *et al.,* 1993). This is an advantage to collecting images sequentially on a frame-by-frame basis (as was the case with the MRC 600) since the amount of laser light required for imaging, and hence photobleaching of the specimen, is reduced, as are the chances of registration artifacts caused by movement of the specimen or by filter misalignments during the imaging session. Sequential imaging has the advantage of reduced bleed-through between channels, and some confocal systems are able to collect images sequentially by rapid switching between channels on a line-by-line basis in real time.

The laser power is adjusted using neutral density filters to give maximum excitation with minimum bleaching of the fluorophore (usually between 10% and 30% laser power for fixed embryos). The gain and the black level are adjusted to full dynamic range, or to highlight features of interest in the specimen where some parts of the image will be unavoidably saturated. A lookup table can be used as an aid to this adjustment if necessary. Here, all of the black pixels in the image are displayed in one color, usually in green, and all saturated pixels (peak white) are displayed in another color, usually in red. The idea is to adjust the gain and black levels so that there are at least some red and green pixels in the image but not too many.

Some signal averaging is usually required to eliminate background noise from the photomultipliers. With fixed embryos, depending on the signal levels, between three and seven images are usually averaged with the Kalman filter using the MRC 1024.

C. Autofluorescence

Most of the autofluorescence from *Drosophila* embryos is well below the signal levels of the fluorophores used in fixed preparations. In fact, in most instances autofluorescence can be used as a low-level background signal to outline the embryo. This is achieved by adjusting the gain and black level to give contrast between the background, the tissue autofluorescence, and the labeled cells.

Background autofluorescence of most fly tissues is much reduced when imaging in red light as compared to imaging with blue or green light (Cullander, 1994). For example, if an unstained *Drosophila* embryo is imaged in the LSCM using the three different excitation wavelengths of the krypton/argon laser, then signal from background tissue autofluorescence is highest at 488 nm (fluorescein) and 568 nm (rhodamine), and most reduced at 647 nm (cyanine 5). A major source of autofluorescence in early embryos is the yolk and fat bodies in the center of the embryo. It is always good practice to check an unstained and fixed preparation for the levels of autofluorescence. If necessary, the gain, black level, pinhole setting, and laser power for imaging autofluorescence should be recorded, with the result that any signal detected above these settings in the experimental preparation will be from the fluorochromes of interest, rather from autofluorescence.

During the early cellular blastoderm stages the cells are positioned around the outside of the embryo, and most of the autofluorescence is not usually a major factor, especially when cells at the surface of the embryo and closest to the objective lens are imaged. However, if the cells on the far side of the embryo are to be imaged then the signal can be reduced by attenuation of the light from the contents of the embryo itself acting as a neutral-density filter. The laser power can be increased to counteract this effect, or a second embryo with a different orientation can be found, or multiple photon imaging can be used.

D. Bleed–Through

One of the major sources of error (and frustration) when imaging multiple-labeled specimens with the CLSM is when part of the emission of one channel appears in that of another. This phenomenon is called bleed-through. Assuming that the CLSM is working correctly, the best solution for bleed-through is to titrate the concentration of fluorescently labeled secondary antibody probes so that the preparation is tuned to the characteristics of the microscope. It is always a good idea to have a familiar test specimen on hand, for example, a preparation of mixed pollen grains that are strongly autofluorescent in all of the channels of the CLSM (available from Carolina Biological Supply; www.carolina.com), to check that the microscope is performing efficiently. If possible, an image of the test specimen should be collected at a time when the microscope has just been serviced. The image should be saved, and the settings of gain, black level, pinhole size, and laser power should be noted so that this can be compared with the performance of the microscope at any given time.

Bleed-through is most often a problem when there is a preponderance of one antigen in the specimen relative to a second one of interest. For example, if the fluorescein channel is bleeding through into the rhodamine channel, then the most concentrated antigen should be placed on the rhodamine channel. For double label experiments Cy-3 and Cy-5 secondary antibodies are good alternatives to the traditional choices of fluorescein and rhodamine since bleed-through from the Cy-3 channel to the Cy-5 channel is much reduced (Sargeant, 1994).

The problem of bleed-through and its remedies have been extensively covered (Brelje *et al.,* 1993). Some of the remedies include sequential rather than simultaneous imaging, and some commercially available CLSMs are able to sequentially scan several channels on a line-by-line basis. Alternatively, the proportion of the image that is bleeding through from one image to the other can be digitally subtracted either during imaging or after imaging. An acoustooptic tunable filter (AOTF) rather than conventional glass filters has been used in CLSMs (Hoyt, 1996; Sharma, 1999). The AOTF matches more precisely the power and wavelength of the laser to maximally excite the fluorophore. This is especially useful when the laser begins to age and the output from one of the channels is reduced in power.

IV. Image Presentation

A. Image Transfer

The images are subsequently exported from the workstation of the confocal microscope either as RAW binary or in the tagged-image file format (TIFF) to a Macintosh G4 microcomputer into Adobe Photoshop 6.0 (Adobe Systems, Mountain View, CA; www.adobe.com). In our laboratory the confocal workstation is set up as a server and the image files are accessed by any computer on the network using the file transfer program Fetch (www.dartmouth.edu/pages/softdev/fetch.html). The Bio-Rad "pic" files are batch converted to TIFF files using Graphic Converter (available from www.lemkesoft.com).

Files can also be converted in batches directly into Adobe Photoshop (version 6.0) using "Actions." Here the images are imported as RAW files, and a dialog box is filled out including the pixel dimensions, for example, 768×512, 3 channels (for a triple label image), noninterleaved, and the header size (76 bytes for a Bio-Rad image). The images are subsequently saved in Macintosh format, usually as TIFF files.

B. Image Processing

Many image manipulations can be performed using the software supplied with most confocal imaging systems, and the details of their use are covered in the manuals from the confocal companies. Images are often exported to a second software package such as Adobe PhotoShop or NIH Image (available at no charge on the World Wide Web at http://rsb.info.nih.gov/nih-image/) for manipulation and for compiling into figures, often with images from other sources, for publication.

The images can be manipulated in PhotoShop using methods previously published (Paddock *et al.*, 1997; Halder and Paddock, 1998). The colors of the individual channels in double- or triple-label images can be rearranged in PhotoShop to give the image that conveys the maximum amount of information and is the most esthetically pleasing to the eye.

The color combinations of the images can be changed in several different ways. Ideally, the colors assigned to the images collected at different emission wavelengths should be correlated with the actual colors of the specimens as they would appear in a conventional epifluorescence microscope. This is easily achieved in a red/green/blue (RGB) image of a double-label fluorescein/rhodamine image using the green and red channels, and leaving the blue channel as black. Here any overlap of expression appears as yellow in the image by simple color addition, and methods of measuring such overlap in confocal images have been developed (Manders *et al.*, 1993). Sometimes it is helpful to change these colors, for example when more than two proteins are imaged in a series of experiments, and they are color coded in a multi-panel figure (see Fig. 7a–7f and Color Plate).

Extra colors can be included in double-label RGB images by copying one of the gray-scale images into the third channel; for example, the possible combinations are green and purple (blue+red), blue and yellow (red+green), or red and light blue (green+blue). Yet more colors can be achieved by passing any of the images through a color lookup table (LUT). LUTs are especially useful for coloring single-label images, which otherwise tend to be rather flat when reproduced in a publication when produced by placing the gray-scale image into a single channel (Halder and Paddock, 1998). Using PhotoShop the brightness and contrast, the color levels, and the sharpness (usually by using Unsharp Mask) can be adjusted and matched to the images in a figure (see Fig. 7a–7f and Color Plate).

The Z-series were usually projected into a single image using software that was supplied with the CLSM or subsequently using a separate image processing workstation and third-party 3-D reconstruction software (White, 1995). Z-Series can be displayed as a montage of images, as a single image projection, or as a stereo pair, or can be further processed into a 3D reconstruction. Z-Series can also be displayed using channels in PhotoShop (Fig. 4). Here, three images from different positions within the Z-series can be placed into the red, green, and blue channels and displayed as a single RGB image since the images maintain the X, Y, and Z registration from the specimen and are the same size and pixel resolution so that color is mapped to depth. In order to display the entire Z-series in this way, it is necessary to merge three different levels of the Z-series before placing the merged images into the R, G, and B channels. For example, using a Z-series of nine images; the top three, middle three, and bottom three images would first be merged, and the resulting single images would then be placed into the R, G, and B channels. This method, like most 3-D reconstruction methods, works best with images such as neurons that have good 3-D structure but a lot of black background in the image (see Figs. 7g and 7g′ and Color Plate).

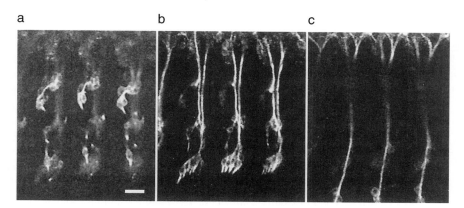

Fig. 4 Color is mapped to depth in a 22C10-labeled embryo. Three optical sections taken from a Z-series. For the merged and colored image, see Fig. 7 and Color Plate. Scale bar = 5 μm.

In a similar way, single images of three different time points can be extracted from a time-lapse movie sequence file and three-color merged, so that color is now mapped to change of position over time, and structures that have not moved will appear white.

C. Production of Hard Copies

For hard copies, images are exported from Photoshop to a Lasergraphics Personal LFR slide maker (www.lasergraphics.com), to a Tektronix Phaser 350 color laser printer that uses solid ink-jet technology, or to a Tektronix IISDX dye sublimation printer for publication quality prints (www.tek.com/Color_Printers/). The Tektronix Phaser 350 prints onto plain paper at a significant reduction in cost, and an increase in speed (up to six pages of color per minute) as compared with many dye sublimation printers.

An increasing number of journals accept electronic submission of the final publication-quality figures, which ensures that the original resolution and color balance of the images are reproduced. Care should be taken when converting RGB images to cyan/magenta/yellow/black (CMYK) images for publication purposes, especially to check that the image is not out of gamut. It may be prudent to send along the RGB image as well as the CMYK image since many printing presses will have their own values for optimizing CMYK images for the color press. Confocal images can also be compiled into a digital presentation including the use of movie presentations.

D. Image Archiving

Since the images are in a digital form they can be stored on any available medium. We currently store our final images on CDs, and in a fireproof safe. Many of the most valuable images are stored in at least two locations. Many images are quickly produced with the CLSM, and it is usually advisable to use some form of database for their organization and quick retrieval (Karten, 1998).

V. Multiple Labels

The analysis of multiple-labeled embryos using the CLSM has become essential for mapping expression patterns of developmentally important genes as an aid to the determination of their function during development (Figs. 1 and 2; also see Fig. 7 and Color Plate). Various antibodies and fluorescent probes are used in multiple label experiments to provide landmarks within *Drosophila* cells and tissues onto which proteins with unknown functions can be mapped.

Probes that label specific cellular organelles are used in conjunction with antibodies in double- or triple-labeling strategies in order to confirm the cellular localizations of proteins. For example chromatin dyes that label nuclei are used in conjunction with antibody staining to determine whether a protein is located in the nucleus, the cytoplasm, or the extracellular matrix. Examples of useful nuclear probes for staining fixed *Drosophila* tissues for analysis using the CLSM include propidium iodide (exc. 536 nm; em. 617 nm,

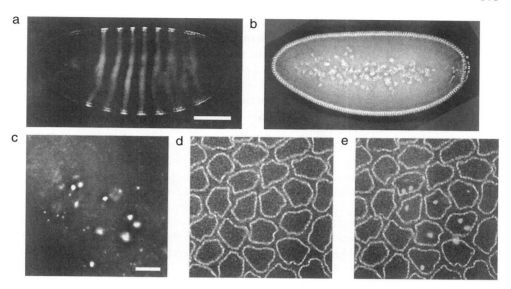

Fig. 5 Fluorescence *in situ* hybridization. Optical section produced by deconvolution through the middle of an embryo at the cellular blastoderm stage expressing *ftz* mRNA (a, c) and labeled with Alexa Fluor 488-conjugated wheat germ agglutinin (b, d), with a merged gray-scale image (e) of (c) and (d). Scale bars (a, b) = 50 μm; (c, d, e) = 5 μm.

Orsulic and Peifer, 1994), acridine orange (exc. 502 nm; em. 526 nm), and the TOTO and TOPRO group of heterodimers (e.g., TOTO3 exc. 642 nm; em. 661 nm; Haugland, 1999). Fluorochrome-conjugated wheat germ agglutinin is used to label nuclear membranes and appears as bright rings around the nuclei in favorable optical sections (Wilkie *et al.,* 1999) (Fig. 5; also see Figs. 7h and 7h' and Color Plate).

The excitation spectra of Hoechst (e.g., Hoechst 3342 exc. 346 nm; em. 460 nm) and DAPI dyes (exc. 359 nm; em. 461 nm) are too short to be excited by the krypton–argon laser and require an additional UV excitation source attached to the LSCM, which itself would need to be fitted with UV-reflecting mirrors and UV-transmitting lenses for more efficient imaging of these dyes (Bliton *et al.,* 1993). Multiple photon imaging allows the excitation of these short-wavelength dyes without specialized UV optics (Denk *et al.,* 1990; Potter, 1996) and can also be used for multiple-label imaging (Xu *et al.,* 1996; White *et al.,* 1997).

Fluorescent phalloidins are useful probes for imaging cell outlines in developing tissues (Condic *et al.,* 1991). Cells stained in this way appear as networks of bright rings in favorable optical sections of the tissues with the dividing cells appearing as slightly larger circles (Rothwell *et al.,* 1998). In this case, phalloidin is directly conjugated to a fluorophore such as fluorescein, rhodamine, Texas Red, or Alexa and, in a relatively simple one-step protocol, specifically stains the peripheral actin meshwork located directly beneath the cell membrane. It should be noted that phalloidin is soluble in methanol, and this should not be used in any fixation regime. Anti-actin antibodies can be used as substitutes for phalloidin staining should methanol be vital in the protocol.

Many fluorescently labeled or tagged antibody probes can be utilized in multiple-label protocols as landmarks for mapping experimental ones in the embryo. Some examples include the 22C10 antibody, which is used for labeling peripheral neurons (Figs. 1d, 4; also see Figs. 7g, 7g′, and Color Plate); *engrailed* antibodies, which are useful markers for the posterior of each segment (Chu-LaGraff and Doe, 1993); *distal-less,* a marker for appendage primordia (Panganiban *et al.,* 1994); and vestigial-expressing cells, which mark the wing and haltere primordia (Carroll *et al.,* 1995).

When the distributions of two of the components are known and do not overlap, they can be placed on the same channel in order to image more probes in a single sample. For example, rhodamine phalloidin (cell outlines) and propidium iodide (nuclei) may be used in tandem in this way. A single or double fluorescence image can be combined with a transmitted light image that is collected with the same scanning beam of the CLSM. DIC optics usually give the most contrast within the images of the embryos, although phase-contrast or darkfield images can also be collected. Darkfield images are often useful because they combine well with fluorescence images, because they both have relatively few gray levels. In addition, darkfield is also preferred for imaging embryonic cuticle preparations (Grenier and Carroll, 2000).

While immunofluorescence continues to be a major technique for mapping the distribution of specific macromolecules within *Drosophila* cells and tissues using the CLSM, fluorescence *in situ* hybridization (FISH) has become a powerful and complementary technique for localizing DNA and RNA sequences in cells using brighter nonradioactive fluorescent probes that can be detected with modern CLSMs (Dernburg and Sedat, 1998). Protocols have been developed for FISH on *Drosophila* embryos and imaginal discs (Hughes *et al.,* 1996; Hughes and Krause, 1998), and improved sensitivity can be achieved using the tyramide amplification system (Wilkie and Davis, 1998) (Fig. 5; also see Figs. 7h, 7h′, and Color Plate). Proteins and transcripts can be mapped in the same specimen using multiple labeling protocols that combine immunofluorescence and FISH (Goto and Hayashi, 1997) and the CLSM for imaging. Triple-label strategies have also been developed for imaging apoptosis, FISH, and nuclear staining using ToPro

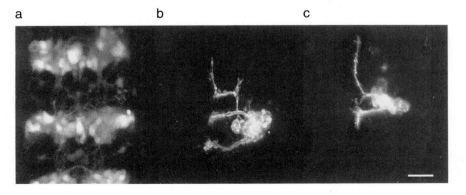

Fig. 6 Lineage tracing. Three stripes of *engrailed*–GFP (a), and two DiI-labeled neuroblasts (b) and (c).

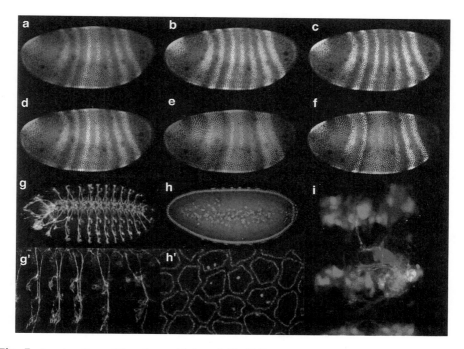

Fig. 7 Imaging *Drosophila* embryos. (a) through (f): Different color combinations of the triple-labeled embryo of Fig. 2; double-labeled images (a) through (d), and triple-labeled images (e, f). Note that (f) is the result of passing image (e) through a color LUT in Graphic Converter. (g) and (g′): Two magnifications of a 22C10-labeled embryo demonstrating color coded to depth in a Z-series from Fig. 4. (h) and (h′): *ftz* mRna in red and wheat germ agglutinin staining the nuclear envelope in green from Fig. 5. (i): *engrailed* GFP in green with DiI-labeled neuroblast clones in red and blue from Fig. 6. (See Color Plate.)

(Davis *et al.,* 1997). Cell death can also be imaged using the TUNEL method (Pazdera *et al.,* 1998).

The green fluorescent protein (GFP) is an extremely useful reporter for multiple label studies for mapping the distribution of proteins in embryos (Brand, 1995). For example, tau GFP has been used to label the peripheral nervous system in embryos in a study of *runt* and cell fate in the embryonic CNS (Dormand and Brand, 1998). An *engrailed*–GFP has been used in double-label experiments with the lineage tracer DiI to map neurogenesis in both living and fixed embryos (Schmid *et al.,* 1999). This study combined multiple-label imaging, DIC transmitted-light imaging, and 3-D reconstruction of the multiple-label images (Fig. 6; also see Figs. 7i and 7i′ and Color Plate).

VI. Conclusions

This chapter has briefly highlighted some of those methods for confocal imaging of fixed and fluorescently labeled *Drosophila* embryos. As the spatial distributions of more proteins are mapped in the embryos then more landmarks are available for improved localization of new proteins of interest. When higher resolution analysis is required, methods for integrated light and EM analysis have been developed (Deerinck *et al.,* 1994). The CLSM has become a vital tool for imaging proteins within the embryos of *Drosophila* and other insects (Carroll *et al.,* 1995; Grenier and Carroll, 2000).

Acknowledgments

I would like to thank Jim Langeland (now of Kalamazoo College) for the embryo preparations in Figs. 1 through 4, Gavin Wilkie and Ilan Davis, Institute of Cell and Molecular Biology, University of Edinburgh for Figure 5, Alice Schmid and Chris Doe, University of Oregon, for providing Fig. 6, and Sean Carroll for his support and collaboration over the years.

References

Bate, M., and Martinez-Arias, A. (1993). "The Development of *Drosophila melanogaster,*" Vols. I and II. Cold Spring Harbor Laboratory Press, Cold Spring Harbor, NY.

Bliton, C., Lechleiter, J., and Clapham, D. E. (1993). Optical modifications enabling simultaneous confocal imaging with dyes excited by ultra-violet and visible-wavelength light. *J. Microsc.* **169,** 15–26.

Brand, A. H. (1995). GFP in *Drosophila. Trends Genet.* **11,** 324–325.

Brand, A. H., Manoukian, A. S., and Perrimon, N. (1994). Ectopic expression in *Drosophila. Methods Cell Biol.* **44,** 635–654.

Brelje, T. C., Wessendorf, M. W., and Sorenson, R. L. (1993). Multicolor laser scanning confocal immunofluorescence microscopy: practical applications and limitations. *Methods Cell Biol.* **38,** 98–177.

Brismar, H., Trepte, O., and Ulfhake, B. (1995). Spectra and fluorescence lifetimes of lissamine rhodamine, tetramethylrhodamine isothiocyanate, Texas Red, and cyanine 3.18 fluorophores—influences of some environmental factors recorded with a confocal laser scanning microscope. *J. Histochem. Cytochem.* **43,** 699–707.

Brown, N. L. (1998). Imaging gene expression using antibody probes. *Methods Mol. Biol.* **122,** 75–91.

Brunetti, C. R., Selegue, J. E., Monteiro, A., French, V., Brakefield, P. M., and Carroll, S. B. (2001). The generation and diversification of butterfly eyespot color patterns. *Curr. Biol.* **11,** 1578–1585.

Campos-Ortega, J. A., and Hartenstein, V. (1985). "The embryonic Development of *Drosophila melanogaster.*" Springer, Berlin.

Carroll, S. B., Weatherbee, S. D., and Langeland, J. A. (1995). Homeotic genes and the regulation and evolution of insect wing number. *Nature* **375,** 58–61.

Chu-LaGraff, Q., and Doe, C. (1993). Neuroblast specification and formation regulated by *wingless* in the *Drosophila* CNS. *Science* **261,** 1594–1597.

Condic, M. L., Fristrom, D., and Fristrom, J. W. (1991). Apical cell shape changes during *Drosophila* imaginal leg disc elongation: a novel morphogenetic mechanism. *Development* **111,** 23–33.

Cullander, C. (1994). Imaging in the far-red with electronic light microscopy: requirements and limitations. *J. Microsc.* **176,** 281–286.

Davis, W. P., Janssen, Y. M. W., Mossman, B. T., and Taatjes, D. J. (1997). Simultaneous triple fluorescence detection of mRNA localization, nuclear DNA, and apoptosis in cultured cells using confocal scanning laser microscopy. *Histochem. Cell Biol.* **108,** 307–311.

Deerinck, T. J., Martone, M. E., Lev-Ram, V., Green, D. P. L., Tsien, R. Y., Spector, D. L., Huang, S., and Ellisman, M. H. (1994). Fluorescence photooxidation with eosin: a method for high resolution immunolocalization and *in situ* hybridization detection for light and electron microscopy. *J. Cell Biol.* **126,** 901–910.

Denk, W., Strickler, J. H., and Webb, W. W. (1990). Two-photon laser scanning fluorescence microscopy. *Science* **248,** 73–76.

Dernburg, A. F., and Sedat, J. W. (1998). Mapping three-dimensional chromosome architecture *in situ*. *Methods Cell Biol.* **53,** 187–197.

Dickinson, M. E., Bearman, G., Tille, S., Lansford, R., and Fraser, S. E. (2001). Multi-Spectral Imaging and Linear Unmixing Add a Whole New Dimension to Laser Scanning Fluoresence Microscopy. *Biotechniques* **31,** 1272–1278.

Dormand, E. L., and Brand, A. H. (1998). Runt determines cell fates in the *Drosophila* embryonic CNS. *Development* **125,** 1659–1667.

Foe, V. E. (1989). Mitotic domains reveal early commitment of cells in *Drosophila* embryos. *Development* **107,** 1–22.

Foe, V. E., and Alberts, B. M. (1983). Studies of nuclear and cytoplasmic behavior during the five mitotic cycles that precede gastrulation in *Drosophila* embryogenesis. *J. Cell Sci.* **61,** 31–70.

Francis-Lang, H., Minden, J., Sullivan, W., and Oegema, K. (1998). Live confocal analysis with fluorescently-labeled proteins. *Methods Mol. Biol.* **44,** 223–239.

Gibson, G. (1999). Developmental evolution: going beyond "just so." *Curr. Biol.* **9,** R942–R945.

Goto, S., and Hayashi, S. (1997). Cell migration within the embryonic limb primordium of *Drosophila* as revealed by a novel fluorescence method to visualize mRNA and protein. *Dev. Genes Evol.* **207,** 194–198.

Grbic, M., Nagy, L. M., Carroll, S. B., and Strand, M. (1996). Polyembryonic development: insect pattern formation in a cellularized environment. *Development* **122,** 795–804.

Grenier, J. K., and Carroll, S. B. (2000). Functional evolution of the Ultrabithorax protein. *Proc. Nat. Acad. Sci. USA* **97,** 704–709.

Halder, G., and Paddock, S. W. (1998). Presentation of confocal images. *Methods Mol. Biol.* **122,** 373–384.

Haugland, R. P. (1999). "Handbook of Fluorescent Probes and Research Chemicals," 7th ed. Molecular Probes, Inc., Eugene, OR.

Hoyt, C. (1996). Liquid crystal filters clear the way for imaging multiprobe fluorescence. *Biophotonics Int.* July/Aug., 49–51.

Hughes, S. C., Saulier Le Drean, B., Livne-Bar, I., and Krause, H. M. (1996). Fluorescence *in situ* hybridization in whole mount *Drosophila* embryos. *BioTechniques* **20,** 748–750.

Hughes, S. C., and Krause, H. M. (1998). Double labeling with FISH in *Drosophila* whole mount embryos. *BioTechniques* **24,** 530–532.

Karr, T. L. (1991). Intracellular sperm/egg interactions in *Drosophila:* a three-dimensional structural analysis of a paternal product in the developing egg. *Mech. Dev.* **34,** 101–112.

Karr, T. L., and Alberts, B. M. (1986). Organization of the cytoskeleton in early *Drosophila* embryos. *J. Cell Biol.* **102,** 1494–1509.

Karten, H. J. (1998). Information management of confocal microscopy images. *Methods Mol. Biol.* **122,** 403–420.

Kiehart, D. P., Montague, R. A., Rickoll, W. L., Foard, D., and Thomas, G. H. (1994). High-resolution microscopic methods for the analysis of cellular movements in *Drosophila* embryos. *Methods Cell Biol.* **44,** 507–532.

Kopp, A., Duncan, I. D., and Carroll, S. B. (2000). Genetic control and evolution of sexually dimorphic characters in *Drosophila*. *Nature* **408,** 553–559.

Langeland, J. A. (1998). Imaging immunolabeled *Drosophila* embryos. *Methods Mol. Biol.* **122,** 167–172.

Manders, E. M. M., Verbeek, F. J., and Aten, J. A. (1993). Measurement of co-localization of objects in dual-color confocal images. *J. Microsc.* **169,** 375–382.

Nagy, L. M., and Carroll, S. B. (1994). Conservation of *wingless* patterning functions in the short germ embryos of *Tribolium castaneum*. *Nature* **367,** 460–462.

Orenic, T. V., and Carroll, S. B. (1992). The cell biology of pattern formation during *Drosophila* development. *Int. Rev. Cytol.* **139,** 121–155.

Orenic, T. V., Held, L. I., Paddock, S. W., and Carroll, S. B. (1993). The spatial organization of epidermal structures: *hairy* establishes the geometrical patterns of *Drosophila* leg bristles by delimiting the domains of *achaete* expression. *Development* **118,** 9–20.

Orsulic, S., and Peifer, M. (1994). A method to stain nuclei of *Drosophila* for confocal microscopy. *BioTechniques* **16,** 441–447.

Paddock, S. W. ed. (1998). Confocal microscopy: Methods and protocols. *Methods Mol. Biol.* **122**, 441–446.

Paddock, S. W. (1999). Confocal laser scanning microscopy. *BioTechniques* **27**, 992–1004.

Paddock, S. W, Langeland, J. A., DeVries, P. J., and Carroll, S. B. (1993). Three-color immunofluorescence imaging of *Drosophila* embryos by laser scanning confocal microscopy. *BioTechniques* **14**, 42–48.

Paddock, S. W., Hazen, E. J., and DeVries, P. J. (1997). Methods and applications of three color confocal imaging. *BioTechniques* **22**, 120–126.

Panganiban, G., Nagy, L., and Carroll, S. B. (1994). The role of the *Distal-less* gene in the development and evolution of insect limbs. *Curr. Biol.* **4**, 671–675.

Patel, N. H. (1994). Imaging neuronal subsets and other cell types in whole-mount *Drosophila* embryos and larvae using antibody probes. *Methods Cell Biol.* **44**, 445–487.

Pazdera, T. M., Janardhan, P., and Minden, J. S. (1998). Patterned epidermal cell death in wild-type and segment polarity mutant *Drosophila* embryos. *Development* **125**, 3427–3436.

Potter, S. M. (1996). Vital imaging: Two photons are better than one. *Curr. Biol.* **6**, 1595–1598.

Rothwell, W. F., Fogarty, P., Field, C. M., and Sullivan, W. (1998). Nuclear-fallout, a *Drosophila* protein that cycles from the cytoplasm to the centrosomes, regulates cortical microfilament organization. *Development* **125**, 1295–1303.

Sargeant, P. B. (1994). Double-label immunofluorescence with the laser scanning confocal microscope using cyanine dyes. *NeuroImage* **1**, 288–295.

Saunders, C., and Cohen, R. S. (1999). Double FISH and FISH-fluorescence immunolocalization procedures for whole-mount *Drosophila* ovaries. *BioTechniques* **26**, 186–188.

Schmid, A., Chiba, A., and Doe, C. Q. (1999). Clonal analysis of *Drosophila* embryonic neuroblasts: neural cell types, axon projections and muscle targets. *Development* **126**, 4653–4689.

Sharma, D. (1999). The use of an AOTF to achieve high quality simultaneous multiple label imaging. Bio-Rad Tech. Note 04. www.microscopy.bio-rad.com

Shindler, K. S., and Roth, K. A. (1996). Double immunofluorescent staining using two unconjugated primary antisera raised in the same species. *J. Histochem. Cytochem.* **44**, 1331–1335.

Skeath, J. B. (1998). The *Drosophila* EGF receptor controls the formation and specification of neuroblasts along the dorsal–ventral axis of the *Drosophila* embryo. *Development* **125**, 3301–3312.

Snook, R. R., and Karr, T. L. (1998). Only long sperm are fertilization-competent in six sperm-heteromorphic *Drosophila* species. *Curr. Biol.* **8**, 291–294.

Sullivan, W., and Theurkauf, W. E. (1995). The cytoskeleton and morphogenesis of the early *Drosophila* embryo. *Curr. Opin. Cell Biol.* **7**, 18–22.

Theurkauf, W. E. (1994). Immunofluorescence analysis of the cytoskeleton during oogenesis and early embryogenesis. *Methods Cell Biol.* **44**, 489–505.

von Dassow, G., and Schubiger, G. (1995). How an actin network might cause fountain streaming and nuclear migration in syncytial *Drosophila* embryos. *J. Cell Biol.* **127**, 1637–1653.

White, J. G., Amos, W. B., and Fordham, M. (1987). An evaluation of confocal versus conventional imaging of biological structures by fluorescence light microscopy. *J. Cell Biol.* **105**, 41–48.

White, J., Centonze, V., Wokosin, D., and Mohler, W. (1997). Using multiphoton microscopy for the study of embryogenesis. *Microsc. Anal.* **3**, 307–308.

White, N. S. (1995). Visualization systems for multidimensional CLSM images. *In* "Handbook of Biological Confocal Microscopy," 2nd. ed. (J. B. Pawley, Ed.). Plenum Press, New York.

Wilkie, G. S., and Davis, I. (1998). High resolution and sensitive mRNA *in situ* hybridization using fluorescent tyramide amplification. Tech. Tips Online. t01458. (http://www.biomednet.com/db/tto)

Wilkie, G. S., Shermoen, A. W., O'Farrell, P. H., and Davis, I. (1999). Transcribed genes are localized according to chromosomal position within polarized *Drosophila* embryonic nuclei. *Curr. Biol.* **9**, 1263–1266.

Williams, J. A., Langeland, J. A., Thalley, B., Skeath, J. B., Carroll, S. B., Glover, David M., and Hames, B. David (1995). Production and purification of polyclonal antibodies against proteins expressed in *E. coli*. *In* "DNA Cloning: Expression Systems." pp. 15–57. IRL Press, Oxford, UK.

Xu, C., Zipfel, W., Shear, J. B., Williams, R. M., and Webb, W. W. (1996). Multiphoton fluorescence excitation: new spectral windows for biological nonlinear microscopy. *Proc. Natl. Acad. Sci. USA* **93**, 10763–10768.

CHAPTER 10

Confocal Fluorescence Microscopy of the Cytoskeleton of Amphibian Oocytes and Embryos

David L. Gard

Department of Biology
University of Utah
Salt Lake City, Utah 84112

METHODS IN CELL BIOLOGY, VOL. 70

I. Introduction

The advent of confocal microscopy and widespread availability of confocal laser-scanning microscopes (CLSM[1]) during the past decade has revolutionized fluorescence microscopy of large cells and intact tissues (Fig. 1; see color plate). Optical sectioning of large cells, such as *Xenopus* oocytes and embryos, by CLSM eliminates the out-of-focus fluorescence that degrades conventional epifluorescence images (White *et al.,* 1987), revealing cellular details that were heretofore invisible (compare Fig. 2C, 2D with 2A, 2B). Moreover, the ability to rapidly collect entire series of serial optical sections and reconstruct cellular volumes expands our view of cellular organization into three dimensions.

The following discussion is drawn from our experiences using CLSM to examine the organization of the three cytoskeletal filament systems present during oogenesis and early development in the African frog *Xenopus laevis:* microtubules (MTs), microfilaments (F-actin), and keratin filaments (KFs). As we began our investigations, we found that many of the published procedures for preparing oocytes for conventional whole-mount microscopy (Dent and Klymkowsky, 1989; Dent *et al.,* 1989; Elinson and Rowning, 1988; Klymkowsky and Hanken, 1991) or histology (Brachet *et al.,* 1970; Huchon *et al.,* 1981; Huchon and Ozon, 1985; Palecek *et al.,* 1982, 1984, 1985; Palecek and Romanovsky, 1985) often did not adequately preserve cytoskeletal structure and/or organization. Over the past 10 years, we have evaluated the preservation and staining of all three cytoskeletal systems using more than three dozen fixation formulas and schedules (Cha *et al.,* 1998; Cha and Gard, 1999; Gard, 1991, 1992, 1993a,b, 1994, 1999; Gard *et al.,* 1995a,b,c, 1997, Gard and Klymkowsky, 1998; Gard and Kropf, 1993; Roeder and Gard, 1994; Schroeder and Gard, 1992). From our experiences, we have concluded that the conditions needed to optimally preserve and visualize the three cytoskeletal elements of *Xenopus* oocytes are mutually exclusive, and that the optimal fixation and staining protocol must be determined empirically for each antibody or antigen.

---

[1] CLSM will be used interchangeably to refer to confocal laser microscopes and confocal laser scanning microscopy.

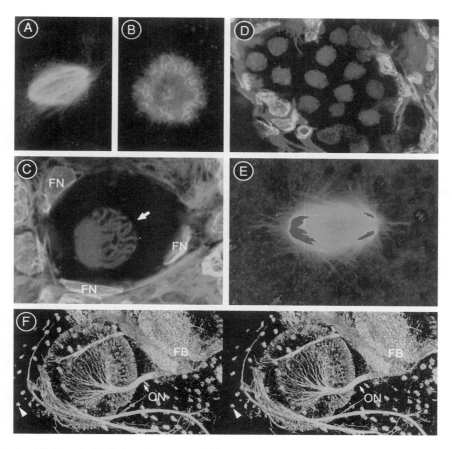

Fig. 1 (A) A longitudinal view of a transverse M1 meiotic spindle in a maturing *Xenopus* oocyte fixed in methanol and stained with monoclonal anti-α-tubulin/fluorescein-anti-mouse IgG (green) and PI (red). (B) A cross-sectional view of the metaphase plate of an M2 meiotic spindle in an unfertilized *Xenopus* egg fixed in methanol and stained with monoclonal anti-α-tubulin/fluorescein-anti-mouse IgG (green) and PI (red) (A and B reprinted from Gard, 1992). (C) A single late-stage 0 oocyte (35 μm diameter) in an ovary isolated from a juvenile frog, fixed in methanol, and stained with monoclonal anti-KF/fluorescein-anti-mouse IgG (green) and EH (red). KF staining is only apparent in the surrounding follicle cells (FN denote follicle cell nuclei). Note the characteristic "bouquet" organization of chromatin in the oocyte nucleus (arrow). A projection of three optical sections. (D) A nest of early stage 0 oocytes isolated from a juvenile frog and stained with monoclonal anti-KF/Texas Red-anti-mouse IgG (green) and YP (red). KF staining is only apparent in the follicle cells surrounding the oocytes. A projection of three optical sections. (E) The same image of an M1 spindle shown in Fig. 6D, displayed in a false color LUT in which pixel values greater than 252 appear red and less than 3 appear green. Intermediate pixel values are shown in cyan-blue. This LUT (applied in Photoshop) simulates the appearance of the "setcol" LUT of Biorad CLSMs (and is similar to the "range indicator" palette of the Zeiss LSM 510). Collecting images in such false color LUTs facilitates setting of the gain and black levels to optimize utilization of the instruments dynamic range. (F) A 3-D reconstruction of the eye of a stage 35 *Xenopus* embryo, fixed in methanol and stained with antibodies to acetylated α-tubulin, which stains "stable" MTs (Piperno *et al.,* 1987; Schulze *et al.,* 1987; Webster and Borisy, 1989). View is from behind the eye. ON denotes the optic nerve, FB is the forebrain. The numerous small foci of staining (one denoted with arrowhead) are ciliated cells in the embryonic epithelium. (See color plate.)

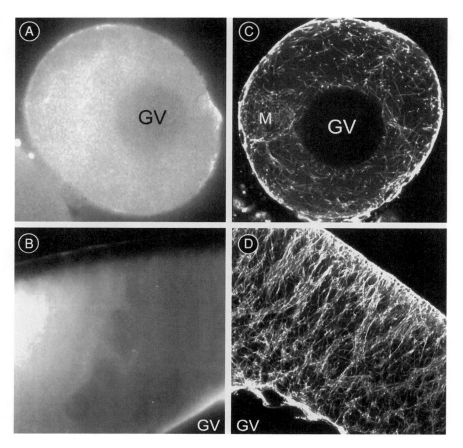

Fig. 2 A comparison of conventional and epifluorescence microscopy of microtubules in *Xenopus* oocytes. (A, B) Conventional immunofluorescence of microtubules in stage I (A) and the animal cytoplasm of a stage VI (B) *Xenopus* oocytes. Fluorescence from outside the focal plane severely degrades the images, lowering contrast and obscuring details of the microtubule organization. (C, D) Single optical sections of similar samples obtained by CLSM provide a dramatically improved view of microtubules. A and B were photographed on a Zeiss axiophot with a 63× NA 1.4 plan apochromatic objective. C and D (and all other images in this chapter) were collected using a Biorad MRC-600 fitted to a Nikon optiphot and a 60 × NA 1.4 plan apochromatic objective. GV denotes the germinal vesicle (nucleus).

The results of some of our trials and errors are discussed here, and staining schedules that we have successfully used to characterize cytoskeletal organization in *Xenopus* oocytes and early embryos are presented in Tables I–IV. Although this discussion is intended to serve as a practical guide to confocal microscopy of the cytoskeleton in amphibian oocytes, eggs, and embryos, many of the techniques should prove useful for confocal microscopy of other large cells or complex tissues. Finally, most of the procedures we have developed or used are modifications of those previously used by others. We gratefully acknowledge all of those who contributed to the cited references.

II. Fixation of *Xenopus* Oocytes and Eggs for Confocal Fluorescence Microscopy of the Cytoskeleton

A. Microtubules

Although we had successfully used conventional immunofluorescence microscopy to examine the number and organization of microtubules (MTs) in methanol-fixed blastulae (Gard *et al.,* 1990), we soon discovered that methanol and other organic solvents were inadequate for preserving MT structure and organization when oocytes were viewed at the magnifications and resolutions made possible by CLSM (Gard, 1991). When viewed by CLSM, cytoplasmic MTs in methanol- (Figs. 3A–3C) or acetone-fixed *Xenopus* oocytes and eggs appeared severely fragmented and/or entirely unrecognizable. Fixation in methanol:acetic acid or Carnoy's solution (ethanol:chloroform:acetic acid) (Clark, 1981) also proved unsatisfactory, preserving few cytoplasmic MTs and resulting in excessive cytoplasmic background. Addition of DMSO (Dent's fixative) (Dent and Klymkowsky, 1989; Dent *et al.,* 1989) (Fig. 3D) or formaldehyde (3.7%) (not shown) during methanol fixation improved MT preservation to some degree (Fig. 3E). However, oocytes fixed in Dent's fixative or methanol–formaldehyde exhibited extensive shrinkage and distortion.

Despite the poor preservation of MTs in methanol-fixed oocytes, fixation in 100% methanol has proven useful in several specialized situations. Although MTs in the interior of stage VI oocytes (Fig. 3B) or in the sperm aster of fertilized eggs (Fig. 3C) were often severely fragmented or unrecognizable, MTs in the cortex of oocytes (not shown) or fertilized eggs (Fig. 3E) were preserved to a greater degree (Elinson and Rowning, 1988; Houliston and Elinson, 1991; Schroeder and Gard, 1992). Also, spindles in maturing oocytes and cleaving blastulae appeared more resistant to fixation in methanol (Gard, 1992), perhaps because of the extreme number of MTs in these arrays. We have successfully used methanol fixation in conjunction with double staining of spindles for chromosomes and MTs (Gard, 1992) (see Figs. 1A, 1B, and color plate) and to examine the organization of cortical MTs in fertilized eggs (Schroeder and Gard, 1992; Becker, Cha, and Gard, unpublished observations).

Optimal preservation of MTs in *Xenopus* oocytes required the use aldehydes such as formaldehyde or glutaraldehyde (Gard, 1991). In our experience, the most consistent preservation and staining of MTs in both stage I (Fig. 3F) and stage VI *Xenopus* oocytes (Fig. 3G) was obtained after fixation with a modification of Karnovsky's fixative (Karnovsky, 1965) containing 3.7% formaldehyde, 0.25% glutaraldehyde, and 0.5 μM taxol in a Pipes-buffered MT assembly buffer containing 0.2% Triton X-100 (FGT fix) (Gard, 1991) (see Table I). Although we routinely include the MT stabilizing agent taxol when fixing MTs, we have also obtained excellent preservation of MTs in oocytes fixed with the combination of 3.7% formaldehyde and 0.25% glutaraldehyde without taxol (FG fix) (Gard, 1991, 1992; Gard *et al.,* 1995a; Schroeder and Gard, 1992).

We typically fix cells for 4–6 h in FGT at room temperature, followed by postfixation in 100% methanol at room temperature (we found no difference in the quality of preservation by postfixing at room temperature versus $-20°C$). Postfixation was initially

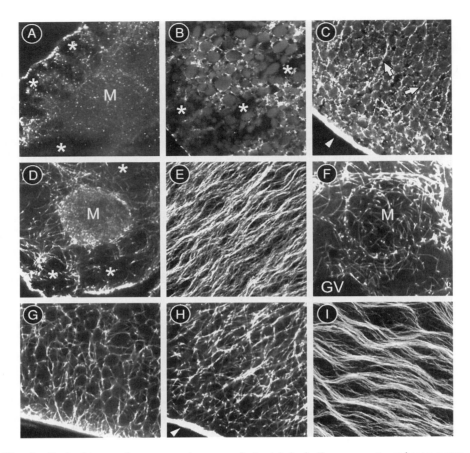

Fig. 3 Confocal immunofluorescence microscopy of microtubules in *Xenopus* oocytes and eggs: a comparison of fixatives. (A) A stage I oocyte fixed in methanol and stained with monoclonal anti-α-tubulin. Note the severe shrinkage artifacts (asterisks) and poor preservation of MTs. M denotes the mitochondrial cloud. (B) Few MTs are preserved in the vegetal cytoplasm of this stage VI oocyte fixed in methanol and stained with monoclonal anti-α-tubulin. Asterisks denote shrinkage artifacts. (C) Although MTs in the cytoplasm of fertilized eggs are poorly preserved by methanol, some hints of the radial MT array are apparent (arrows). The vegetal cortex is brightly stained (arrowhead), consistent with the dense MT array found in the egg cortex. (D) A stage I oocyte fixed in methanol plus DMSO (Dent's fixative; Dent and Klymkowsky, 1989) and stained with monoclonal anti-α-tubulin. Although DMSO improves preservation of cytoplasmic MTs, MTs in the mitochondrial cloud (M) are severely fragmented and shrinkage artifacts are apparent (asterisks). (E) A face-on view of the vegetal cortex of a fertilized egg fixed in methanol reveals that cortical MTs are reasonably well preserved after fixation in methanol. (F) A stage I oocyte fixed in FGT and stained with monoclonal anti-α-tubulin exhibits excellent preservation of MTs within and surrounding the mitochondrial cloud (M; compare with A and D). (G) A stage VI oocyte fixed in FGT and stained with anti-α-tubulin exhibits excellent MT preservation (compare to B). Electron microcopy revealed that the "filaments" observed by CLSM correspond to individual MTs (Pfeiffer and Gard, 1999). (H, I) A fertilized egg fixed in low FGT and stained with anti-α-tubulin. Both cytoplasmic (H) and cortical MTs (I) are well preserved by low FGT.

Table I

Confocal Immunofluorescence of Microtubules in *Xenopus* Oocytes[a]

1.	Oocytes and embryos should be as free of extraneous tissue or extracellular material as possible. Oocytes are isolated by collagenase digestion followed by manual removal of residual follicle material. Eggs and early embryos are de-jellied with 2% cysteine (pH 8.0) or 5 mM DTT (in 30 mM Tris-HCl pH 8.5). Remove the vitelline envelope surrounding late-stage embryos.
2.	Fix samples 4–6 h in approx 1 ml of FGT or FG fix (see text) at room temperature.
3.	Postfix in 100% methanol overnight. Samples can be stored in methanol indefinitely.
4.	Optional: Rehydrate samples in TBSN and hemisect with a sharp scalpel.
5.	Optional: Bleach with peroxide–methanol (1 : 3, see Section IV) for 24–72 h at room temperature.
6.	Rehydrate in PBS without detergent (several changes at room temperature).
7.	Incubate FGT-fixed samples[b] in 50 mM NaBH$_4$ in PBS without detergent overnight at 4°C. Do not cap tubes!
8.	Add 1 drop of TBSN/sample to break the surface tension of bubbles formed in the NaBH$_4$ solution. Carefully aspirate the NaBH$_4$, and wash samples with TBSN for 60–90 min at room temperature (3–4 washes, ∼1 ml each).
9.	Incubate samples 16–48 h with gentle rotation at 4°C in 75–150 μl of monoclonal mouse anti-tubulin, diluted in TBSN plus 2% BSA.
10.	Wash in TBSN for 30–48 h at 4C, changing buffer at 8 to 12 intervals.
11.	Incubate samples for 16–48 h with gentle rotation at 4°C in 75–150 μl of fluorescent-conjugated anti-mouse IgG diluted in TBSN plus 2% BSA.
12.	Wash with TBSN, as in step 9.
13.	Dehydrate through 3–4 changes of 100% methanol, over 60–90 min at room temperature. Samples can be stored indefinitely in methanol.
14.	Remove methanol, and add ∼1 ml of BA:BB clearing solution. Do not mix! The dehydrated oocytes will float. Gently tap tubes on bench to break the surface tension. The oocytes will clear[c] as they sink through BA:BB (takes 5–10 min). Remove the clearing solution and replace with fresh BA:BB.
15.	Mount in BA:BB (see text).

[a]All incubations and washes can be performed in 1.5-ml microcentrifuge tubes (about 5–15 oocytes per tube).

[b]Borohydride reduction can be eliminated and incubation times can be shortened for FT or MeOH-fixed samples.

[c]If samples retain a milky appearance, they may not have been adequately dehydrated. Poorly dehydrated samples can be salvaged by repeating the dehydration and clearing.

included as an intermediate step prior to bleaching with peroxide–methanol (Dent and Klymkowsky, 1989) (see later discussion). However, fixed oocytes can be stored in 100% methanol for several months without noticeable deterioration, making methanol postfixation a matter of convenience. Occasionally, the germinal vesicles (nuclei) of stage VI oocytes collapse during FGT fixation. In some cases, this appears to result from overfixation (Gard, unpublished observations). However, it also may result from rapid dehydration as the oocytes are transferred to the methanol postfix. Stepwise dehydration might reduce or eliminate this shrinkage.

Despite the excellent preservation of MTs by FGT, fixation in glutaraldehyde introduces several secondary problems. First, penetration of antibodies into glutaraldehyde-fixed oocytes is slower than penetration into samples fixed in methanol or formaldehyde alone. Fixation in glutaraldehyde also can increase nonspecific fluorescence of the cytoplasm. This "background" results from several sources, including the reaction of residual glutaraldehyde with the primary and secondary antibodies, as well as the formation of Schiff bases with cellular proteins. Glutaraldehyde-induced autofluorescence and background can be substantially reduced by treating cells with sodium borohydride (Weber *et al.,* 1978) (see later discussion). Finally, glutaraldehyde fixation has been reported to alter MT structure (Cross and Williams, 1991).

MTs can also be fixed with formaldehyde alone (not shown), at concentrations ranging from 1 to 4%. However, the preservation of MT structure and organization in formaldehyde-fixed oocytes is not as consistent as that obtained with FGT. MT preservation in formaldehyde-fixed oocytes is significantly improved by including taxol (0.1–0.5 μM; FT fix). The appearance of MTs in oocytes preserved with FT fix is also substantially improved by postfixation in methanol (see later discussion). FT followed by methanol postfixation is a viable option for immunostaining with antibodies that are incompatible with glutaraldehyde-containing fixatives. Interestingly, the nucleoplasm of stage VI oocytes is poorly fixed by formaldehyde and may "squirt" from the oocyte during hemisection.

The concentrations of formaldehyde and glutaraldehyde in FGT and FT fix can reduce or abolish staining by some antibodies. Although none of the commercial tubulin antibodies we have used have suffered from this difficulty, staining of FGT- or FT-fixed oocytes with one polyclonal antibody raised against XMAP230, a high molecular weight MT-associated protein in *Xenopus* oocytes and embryos, was inconsistent (Cha *et al.,* 1998). To circumvent this problem, we modified our standard FGT procedure by reducing the concentrations of both formaldehyde (to 0.25–0.5%) and glutaraldehyde (to 0.1%) and shortening the fixation time (to 0.5–1 h). This "low FGT" fix, when followed by postfixation in methanol, adequately preserved MTs in stage VI oocytes (not shown) and fertilized eggs (Figs. 3H and 3I) while providing for more consistent staining with anti-XMAP230 (Chat *et al.,* 1998). With some antibodies, notably those recognizing KFs (Gard *et al.,* 1997; Klymkowsky *et al.,* 1987) (see later discussion), staining was completely eliminated by even minimal concentrations of aldehydes. In these cases, we have had to resort to fixation in methanol or other organic solvents (see later discussion), despite the poor preservation of MT structure and/or organization afforded by these agents.

B. Keratin Filaments

Although combinations of formaldehyde and glutaraldehyde provided the best preservation of oocyte MTs for confocal immunofluorescence microscopy, aldehyde fixation proved to be incompatible with staining of keratin filaments (KFs) in *Xenopus* oocytes (Gard *et al.,* 1997; Klymkowsky *et al.,* 1987). Staining of oocyte KFs by several monoclonal anti-keratin antibodies (including clones C11 from Sigma and 1h5 from the Developmental Studies Hybridoma Bank) was substantially reduced or eliminated by even brief exposure (<30 s) to aldehydes, probably as a result of modification of epitopes on the antigen during aldehyde fixation. To visualize KF organization in *Xenopus* oocytes throughout oogenesis (Fig. 4), we modified the methanol-based fixation protocol

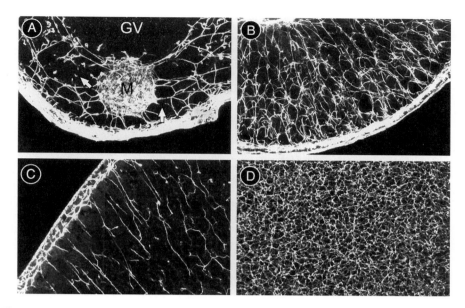

Fig. 4 Confocal immunofluorescence microscopy of keratin filaments in *Xenopus* oocytes. (A) A projection of 25 optical sections showing the association of KFs with the mitochondrial cloud of a late-stage I oocyte (~200 μm diameter) fixed in methanol and stained with monoclonal anti-KF. Arrows point to filament breaks, resulting from shrinkage during fixation in methanol. A stereo pair showing the same data set is presented in Fig. 10C (note that the 3-D rendering is reversed right-to-left). (B) A projection of 15 optical sections showing the extensive network of KFs in the vegetal cytoplasm of an early stage IV oocyte (~450–500 μm diameter) fixed in methanol and stained with monoclonal anti-KF. (C) A projection of five optical sections showing the radial organization of KFs in the animal cytoplasm of a stage VI oocyte fixed in methanol and stained with monoclonal anti-KF. Arrows denote the cortical meshwork of KFs, shown face-on in D. (D) A projection of five optical sections grazing the animal cortex of a methanol-fixed stage VI oocyte, showing the dense meshwork of cortical KFs. Electron microscopy reveals that the "filaments" observed by CLSM correspond to loosely organized bundles of 3–6 individual KFs (Chang *et al.,* 1999); Pfeiffer and Gard, unpublished observations).

described by Dent and Klymkowsky (Dent and Klymkowsky, 1989; Dent *et al.,* 1989; Gard *et al.,* 1997; Klymkowsky and Hanken, 1991; Klymkowsky *et al.,* 1987) (see Table II).

Several problems plague protocols relying on methanol- or acetone-based fixation. First, significant shrinkage and distortion of oocytes and eggs is apparent after fixation in organic solvents. Shrinkage of the GV and nucleoplasm of oocytes is commonly observed, as is the separation of the cortex from the underlying cytoplasm of both oocytes and eggs. Samples fixed in methanol, ethanol, or acetone are also quite fragile and tend to break apart during subsequent hemisection (see below) or processing. This is particularly true of methanol-fixed blastulae, which often break apart into individual cells or small clumps of cells, which makes processing and mounting problematic. Despite

Table II

Confocal Immunofluorescence Microscopy of Keratin Filaments in *Xenopus* Oocytes[a]

1. Fix oocytes 4–24 h in 100% methanol (see text).
2. Rehydrate in TBSN.
3. Optional: Carefully hemisect larger oocytes with a sharp scalpel blade.
4. Optional: Bleach with peroxide-methanol (1 : 3, see text section IV) for 24–72 h at room temperature.
5. Rehydrate in TBSN (several changes at room temperature).
6. Incubate samples 16–48 h with gentle rotation at 4°C in 75–150 μl of monoclonal mouse anti-keratin antibodies, diluted in TBSN plus 2% BSA.
7. Wash in TBSN for 30–48 h at 4°C, changing buffer at 8- to 12-h intervals.
8. Incubate samples for 16–48 h with gentle rotation at 4°C in 75–150 μl of fluorescent-conjugated anti-mouse IgG diluted in TBSN plus 2% BSA.
9. Wash with TBSN, as in step 7.
10. Dehydrate through 3–4 changes of 100% methanol, over 60–90 min at room temperature. Samples can be stored indefinitely in methanol.
11. Remove methanol, and add ~1 ml of BA:BB clearing solution. Do not mix! The dehydrated oocytes will float. Gently tap tubes on bench to break the surface tension. The oocytes will clears as they sink through BA:BB (takes 5–10 min). Remove the clearing solution and replace with fresh BA:BB.
12. Mount in BA:BB (see text).

[a] All incubations and washes can be performed in 1.5-ml microcentrifuge tubes (about 5–15 oocytes per tube).

[b] If samples retain a milky appearance, they may not have been adequately dehydrated. Poorly dehydrated samples can be salvaged by repeating the dehydration and clearing.

these problems, fixation in 100% methanol (at −20°C or RT) provided exceptional views of the keratin filament networks throughout oogenesis (Gard *et al.,* 1997; Gard and Klymkowsky, 1998) (Fig. 4).

C. Microfilaments (F-Actin)

The large pool of unpolymerized (G-) actin in the cytoplasm and nucleus of *Xenopus* oocytes and eggs (Clark and Merriam, 1978; Merriam and Clark, 1978) is brightly stained with commercial antibodies specific for cytoplasmic actin (Roeder and Gard, 1994; Gard, unpublished observations), hindering visualization of actin filament (F-actin) organization (Fig. 5A). In cultured cells and tissues, the organization of actin microfilaments is commonly examined using fluorescent conjugates of the fungal toxin phalloidin, which binds tightly and specifically to F-actin. Many of the published protocols for phalloidin staining of F-actin rely on formaldehyde fixation followed by extraction in cold methanol or acetone. However, even brief fixation or dehydration with methanol or acetone completely abolished phalloidin staining of F-actin in *Xenopus* oocytes (Roeder and Gard, 1994). We therefore developed a formaldehyde-based fixation protocol for phalloidin staining of F-actin in *Xenopus* oocytes (Roeder and Gard, 1994) (Table III).

Unfortunately, the incompatibility of phalloidin staining with fixation or dehydration in methanol or acetone prevents both peroxide–methanol bleaching of oocyte pigment and the use of benzyl alcohol:benzylbenzoate clearing solution (see later discussion).

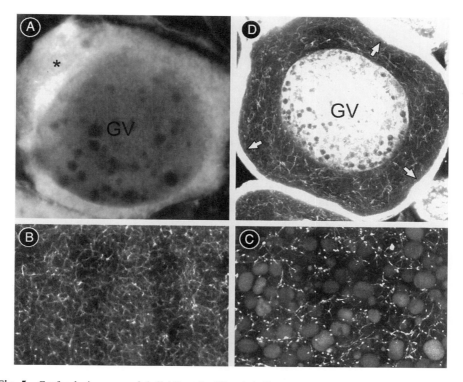

Fig. 5 Confocal microscopy of phalloidin-stained F-actin in *Xenopus* oocytes. (A) An early stage I *Xenopus* oocyte fixed in formaldehyde and stained for confocal immunofluorescence microscopy with monoclonal anti-actin antibodies. Note the bright cytoplasmic fluorescence, due to staining of cytoplasmic G-actin, which obscures any actin cables present. (B) F-actin cables in the subcortical animal cytoplasm of an albino stage VI *Xenopus* oocyte fixed in formaldehyde and stained with rhodamine–phalloidin (a projection of five sections). (C) F-actin cables in the vegetal cytoplasm of a hemisected stage VI oocyte (a projection of four sections) fixed in formaldehyde and stained with rhodamine–phalloidin. (D) CLSM reveals a complex network of actin cables in the cytoplasm of this stage I oocyte (150–175 μm diameter) fixed in formaldehyde and stained with rhodamine–phalloidin, as well as brightly stained F-actin in the GV and cortical cytoplasm (arrows) (a projection of two optical sections).

The accumulation of dense deposits of yolk during stages II–VI of oogenesis renders later stage oocytes, eggs, and embryos completely opaque, and the inability to bleach or clear the oocytes prevents visualization of actin organization more than 5–10 μm below the cell surface. Oocytes from albino females can be used for examination of microfilament distribution in the cortical and immediate subcortical cytoplasm (Fig. 5B). Although we have manually hemisected stage IV–VI oocytes to obtain cross-sectional views of F-actin organization (Roeder and Gard, 1994) (Fig. 5C), the unevenness of the cut surface and inability to section below the resulting knife damage in these uncleared oocytes makes collection of useful images difficult. We have had greater success using fluorescent phalloidin to examine F-actin organization in oogonia and previtellogenic

Table III
Phalloidin Labeling of F-Actin in *Xenopus* Oocytes[a]

1. Manually isolate[b] and defollicle stage VI *Xenopus* oocytes, or dissect ovaries from juvenile frogs.
2. Fix in 1.0–1.5% formaldehyde in fix buffer for 4–18 h at room temperature.
3. Rinse overnight in TBSN (several changes with gentle rotation).
4. Optional: Manually hemisect large oocytes with a sharp scalpel.
5. Stain with rhodamine phalloidin (Molecular Probes) diluted to 2–4 units/ml in TBSN for 6–18 h at room temperature.
6. Wash oocytes through several changes of TBSN over 24–48 h.
7. Mount ovaries from juvenile frogs or small stage I oocytes in glycerol-mounting buffer (see text). Mount larger oocytes in well slides with Tris-buffered saline (pH 8.0) containing NaN_3 or *n*-propyl gallate to retard photobleaching.

[a]All incubations and washes can be performed in 1.5-ml microcentrifuge tubes (about 5–15 oocytes per tube).
[b]Collegenase digestion appears to partially disrupt the organization of F-actin in *Xenopus* oocytes.

oocytes from both *Xenopus* (Roeder and Gard, 1994) (see Fig. 5B) and *Rana pipiens* (Gard, unpublished observations). However, the extreme difference in brightness between the brightly stained cortex and GV, compared to the much less brightly stained cytoplasmic cables, makes collecting optimal images challenging.

III. Processing *Xenopus* Oocytes and Eggs for "Whole-Mount" Confocal Microscopy

A. Handling Oocytes, Eggs, and Early Embryos

Preparation of oocytes and embryos for "whole-mount" immunofluorescence microscopy is similar in most respects to preparation of other cell or tissue samples, with two adaptations: first, the large size (up to 1.5 mm) of oocytes requires that antibody incubations and intermediate washes be lengthened considerably; and second, oocytes are not readily attached to a solid substrate (such as a coverslip or microscope slide) for processing.

To aid antibody penetration, we routinely cut, or "hemisect," larger oocytes and eggs. Cells are briefly rehydrated in Tris-buffered saline with NP-40 (TBSN) after fixation and are hand-cut with a sharp scalpel. Although hand-cutting oocytes might sound tedious, with practice we are able to successfully hemisect oocytes as small as 0.4 mm in diameter. Smaller oocytes ($<300~\mu$m) are normally processed whole.

To examine the earliest stages of oogenesis, ovaries are dissected from juvenile frogs (19–35 mm, snout-to-vent). These ovaries are fixed, and then cut so that the individual lobes can be opened and laid flat to more readily expose the oocytes to antibodies. Intact or hemisected oocytes, or pieces of ovary from juvenile frogs, are fixed and processed for microscopy in 1.5-ml microfuge tubes. As many as 10–15 stage VI oocytes (1.2-mm diameter) can be fixed and processed per tube, using 50–100 μl of antibody. Hemisected oocytes can be mounted to give views of the cortex, or cut surface, collecting images from below the knife-damaged surface (see later discussion).

B. Bleaching

The cortex of amphibian late-stage *Xenopus* oocytes (stages III–VI) and eggs contains a dense layer of pigment that obscures underlying cytoplasmic structures. To reduce or eliminate interference from cortical pigment, samples may by bleached with a combination of ~30% hydrogen peroxide (H_2O_2) and 100% methanol (1 : 3 by volume) for 24–72 h (modified from Dent and Klymkowsky, 1989). Exposure to fluorescent room lights or natural sunlight appears to speed the bleaching process. Although some pigment may remain after extensive bleaching of darkly pigmented oocytes, residual pigment often washes out during subsequent processing. The animal and vegetal hemispheres are nearly indistinguishable after bleaching. If the separated animal and vegetal hemispheres must be distinguished, they should be processed separately. Cytoskeletal organization and reactivity with anti-tubulin and anti-keratin antibodies appears unaffected by bleaching with H_2O_2. However, bleaching with methanol–H_2O_2 is incompatible with phalloidin staining (see above).

C. Borohydride Reduction of Glutaraldehyde-Fixed Oocytes

Cells fixed with glutaraldehyde or combinations of formaldehyde and glutaraldehyde should be treated with sodium borohydride to eliminate unreacted aldehydes and reduce autofluorescence (Weber *et al.,* 1978). We have used $NaBH_4$ (in PBS or TBS without detergents) at concentrations ranging from 50 mM (O/N at 4°C) to 100 mM (4–6 h at RT) without noticing any detrimental effect on MT structure, organization, or antibody reactivity. As we discovered by accident, the brisk effervescence of higher concentrations of $NaBH_4$ can physically disrupt samples. Also, to prevent foaming, $NaBH_4$ should be used in buffers lacking any detergent. Finally, containers should be left uncapped during borohydride treatment to prevent pressure buildup (which can pop the caps of microfuge tubes, sending them flying and scattering oocytes across the room). Reduction with $NaBH_4$ is not necessary for samples fixed in methanol (or other organic solvents) or formaldehyde alone.

D. Antibody Incubations

In our experience, the penetration of antibodies into FGT-fixed oocytes is limited to 75–100 μm in an overnight incubation (12–16 h at 4°C). Penetration is somewhat faster in methanol- or formaldehyde-fixed oocytes. Hemisected or intact oocytes from 0.5 to 1.2 mm in diameter are commonly incubated in antibody for 16–24 h at 4°C, with gentle rotation or agitation. We use longer incubations (36–48 h at 4°C) for small oocytes (<200 μm diameter) and ovary fragments, to allow sufficient time for uniform penetration of antibody to the center of the sample. Long incubations are especially important with samples that are to be serially sectioned by CLSM, to ensure uniform antibody penetration.

Of course, if long incubations are required for penetration of antibody into a sample, equally long incubations are required for the unbound antibody to diffuse back out of the sample. Thus, intermediate rinses with TBSN typically range from 32 to 48 h at 4°C, with several changes of buffer. Samples are rotated gently on a tissue infiltrator or other lab rotator during incubations and washes.

We have used a number of commercially available antibodies against cytoskeletal proteins with varying degrees of success (see Section VII for comments on specific antibodies). Nonspecific background staining with commercial monoclonal antibodies has been slight, and we have not found it necessary to preblock samples prior to incubation with primary antibodies. We do include bovine serum albumin and nonionic detergent (NP-40 or Triton X-100) in both primary and secondary antibody solutions, and fluorescent secondary antibodies are routinely preadsorbed with rat liver acetone powder (Sigma) (Karr and Alberts, 1986) to minimize nonspecific binding.

E. Choosing Secondary Antibodies for CLSM

Commercial CLSMs are now available with a wide array of optional lasers providing excitation lines spanning the visible spectrum from 453 nm (argon ion) to 647 nm (Kr/Ar). When combined with the introduction of a variety of new fluorochromes (Haugland, 1996; Wells and Johnson, 1994) and fluorochrome-conjugated antibodies that are compatible with laser excitation, this provides microscopists remarkable flexibility in choosing secondary antibodies for CLSM applications. We use the following guidelines to select fluorochromes for our investigations:

1. The fluorochrome must be available conjugated to the needed secondary antibody (i.e., anti-mouse IgG for the detection of mouse monoclonal antibodies to tubulin). Fluorescent-conjugated antibodies are available from a variety of commercial sources.

2. The fluorochrome must be well matched with the laser line used for excitation and emission filters available on the CLSM to be used. Spectral data for many available fluorochromes is available on Web sites maintained by Molecular Probes (http://www.probes.com) and Bio-Rad (http://fluorescence.bio-rad.com; the Bio-Rad website also includes data for commonly used lasers and filter combinations).

3. The fluorochrome should be sufficiently photostable for collecting images requiring frame averaging or the collection of multiple optical sections. In general, fluorochromes excited by longer wavelengths (i.e., rhodamines, Texas Red, etc.) are more photostable than those excited by shorter wavelengths (i.e., fluoresceins). However, some newer short-wavelength fluorochromes, such as Alexa 488 (Molecular Probes; Eugene, OR), offer remarkable photostability relative to fluorescein.

4. For dual fluorescence applications, the fluorochromes should be spectrally separated, allowing them to be collected into independent channels.

For most single label CLSM applications, we prefer longer wavelength fluorochromes such as rhodamine (543-nm line of GHe–Ne lasers), long wavelength Alexa dyes (available from Molecular Probes, Eugene, OR), or Texas Red (excited by the 568-nm line of Ar–Kr lasers). Long-wavelength dyes appear to provide somewhat better contrast

Table IV
Fluorescence Confocal Microscopy of Chromatin in *Xenopus* Oocytes

1. Fix oocytes, eggs, or pieces of ovary from juvenile frogs overnight in 100% methanol (see discussion of methanol fixation).
2. Rehydrate in TBSN (no BSA).
3. Use a sharp scalpel blade to hemisect oocytes or eggs (optional, see above), or carefully tease ovary pieces open to allow access to oocytes.
4. Incubate 16–24 h in primary antibody (we have used anti-tubulin or anti-keratin) diluted in TBSN plus 2% BSA (4°C with gentle rotation).
5. Wash 16–36 h in TBSN at 4°C (with gentle rotation), changing buffer every 8–12 h.
6. Incubate 16–24 h in secondary antibody diluted in TBSN plus 2% BSA (4°C with gentle rotation). Use fluorescein-conjugated or other suitable short-wavelength secondary antibodies for dual labeling with propidium iodide or ethidium homodimer. Use Texas-Red-conjugated or other suitable long-wavelength secondary antibodies for dual fluorescence with YO PRO.
7. Wash in TBSN as in step 5.
8. Stain oocytes with propidium iodide (1–10 μg/ml), ethidium homodimer (1–5 μg/ml), or YO PRO (1–5 μM) dissolved in 100% methanol.
9. Wash 2–3× with 100% methanol (30 min each).
10. Clear and mount in BA:BB (see above).

Alternatively, add the chromatin dye to the last TBSN wash, and dehydrate in methanol as described in Table I.

between stained cytoskeletal elements and background autofluorescence, as well as exhibiting better resistance to photobleaching.

F. Staining Chromosomes in *Xenopus* Oocytes and Eggs

Dispersal of chromatin within the large nucleus of oocytes from stage I onward makes visualization of chromatin in late-stage *Xenopus* oocytes difficult or impossible. However, we have successfully used CLSM to examine the organization of chromatin in small postmitotic (stage 0) oocytes (Gard *et al.*, 1995a, 1997) and condensed meiotic chromosomes during oocyte maturation (Gard, 1992) using propidium iodide (PI) ($\lambda_{ex} = 530$, $\lambda_{em} = 615$) (Figs. 1A, 1B) or ethidium homodimer (EH) ($\lambda_{ex} = 528$–535, $\lambda_{em} = 617$–624) (Fig. 1C, Table IV). In our experience, staining of chromosomes with PI and EH was dramatically reduced by aldehyde fixation, and was most effective in oocytes fixed with methanol alone. Although methanol does not provide optimal preservation of MTs during interphase, it often provides adequate preservation of meiotic and mitotic spindles (Gard, 1992; Gard *et al.*, 1990).

We have also used BO-PRO-3 ($\lambda_{ex} = 575$, $\lambda_{em} = 599$) and YO-PRO-1 ($\lambda_{ex} = 491$, $\lambda_{em} = 509$), two of the cyanine dyes developed and marketed as chromatin dyes by Molecular Probes (Eugene, OR), to stain chromosomes in methanol and/or aldehyde-fixed stage 0 oocytes (Gard *et al.*, 1997, Gard, unpublished observations). BO-PRO-3 exhibited excessive spillover into our fluorescein filter set, hindering its use in dual-fluorescence applications. However, YO-PRO-1, which fluoresces green when bound to DNA, proved to be useful for dual-fluorescence microscopy of chromatin and KFs or MTs when used in combination with Texas-Red secondary antibodies (Gard *et al.*,

1997; Gard, unpublished observations) (Fig. 1D). Finally, we have had some success using commercial anti-histone antibodies to stain chromosomes in stage 0 oocytes and mitotic blastomeres (Gard, unpublished observations). For a more complete discussion of nuclear stains for fluorescence and confocal microscopy, see Arndt-Jovin and Jovin (1989).

G. Clearing *Xenopus* Oocytes and Eggs for Confocal Microscopy

Accumulation of yolk during stages II–III of oogenesis in *Xenopus* (Dumont, 1972) renders late-stage oocytes, eggs, and early embryos functionally opaque. In fact, our experience has shown that it is possible to "see" (by CLSM) farther into granite, which is composed primarily of crystalline quartz grains, than into uncleared *Xenopus* oocytes! Early studies of cytoskeletal organization during late oogenesis and early development in amphibians thus relied on sectioning of oocytes and eggs embedded in paraffin or plastic (Huchon *et al.*, 1981, 1988; Huchon and Ozon, 1985; Jessus *et al.*, 1986; Kelly *et al.*, 1991; Palecek *et al.*, 1982, 1984, 1985; Palecek and Romanovsky, 1985), or restricted views to the immediate cortical regions of the cell (Elinson and Rowning, 1988; Houliston and Elinson, 1991; Huchon *et al.*, 1988).

The development of a clearing/mounting solution for *Xenopus* embryos, consisting of a mixture of benzyl alcohol (BA) and benzylbenzoate (BB), by Murray and Kirschner (Dent and Klymkowsky, 1989; Dent *et al.*, 1989), revolutionized the application of immunocytochemical techniques and fluorescence microscopy to the study of amphibian oocytes and early embryos. A 1 : 2 (vol:vol) mixture of benzyl alcohol:benzylbenzoate (BA:BB) "clears" *Xenopus* oocytes, eggs, and early embryos by matching the refractive index of the nearly crystalline yolk platelets found in these cells and embryos, rendering them nearly transparent. Although BA:BB was developed specifically for clearing *Xenopus* oocytes and eggs, it has also been found to be a useful clearing and mounting agent for a wide variety of animal, plant, and fungal tissues (Klymkowsky and Hanken, 1991; Kropf *et al.*, 1990, and references therein). Furthermore, the refractive index of BA:BB (~1.55) nearly matches that of many immersion oils (~1.52) and glass coverslips used with high-NA objectives, thus reducing the spherical aberration that distorts images and volume reconstructions obtained by CLSM (Majlof and Forsgren, 1993). Finally, BA:BB appears to afford some protection against photobleaching (see below), and samples cleared and mounted in BA:BB are stable for many months.

Despite its usefulness as a clearing and mounting agent, BA:BB has several limitations and drawbacks in its use. First, BA:BB is immiscible with aqueous solutions, limiting its use to samples that can be dehydrated in methanol, ethanol, or acetone prior to clearing/mounting. Thus, clearing with BA:BB is not compatible with phalloidin-staining of F-actin (Roeder and Gard, 1994), and phalloidin-stained oocytes must be mounted in aqueous media (glycerol-based mountant or Tris buffer). Also, some chromogenic substrates used for immunocytochemistry also are not compatible with BA:BB (Klymkowsky and Hanken, 1991). Oocytes and other samples cleared and mounted in BA:BB are often brittle and subject to mechanical damage during mounting. BA:BB dissolves many plastics, including polystyrene and cellulose acetate (and plastic computer keyboards). Polyethylene tubes are resistant to BA:BB and should be used for all processing. Care must be taken to avoid costly damage to microscopes or other equipment. Finally, although both benzyl alcohol and benzylbenzoate and their derivatives are used (in small amounts) to flavor chewing gum and other confections, in higher doses they are toxic to laboratory rodents ($LD_{50} \sim 1$–3 g/kg; Windholz, 1983, and references cited therein). Benzylbenzoate may cause skin and eye irritation in humans. Therefore, appropriate caution should be used when mounting and handling samples prepared with BA:BB.

H. Mounting *Xenopus* Oocytes and Eggs for Confocal Microscopy

Many high-power objectives have working distances of less than 150 μm, limiting the depth at which optical sections can be collected. Moreover, any movement of the sample during the collection of frame-averaged images or serial optical sections will seriously degrade image quality. Thus, it is important for oocytes to be securely mounted and immobilized in a manner that allows the region of interest to be examined.

Small stage I oocytes ($\sim$100–150 μm in diameter) or fragments of ovaries from small frogs can be dehydrated, cleared, and mounted in BA:BB between a standard microscope slide and coverslip. To minimize compression damage during mounting, we often use fingernail polish[2] to "glue" four strips cut from a #1 or #2 coverslip to a slide, forming one or two wells (Fig. 6A). A thin layer of nail polish applied to the spacers and allowed to dry thoroughly acts as a resilient "gasket." After the cleared oocytes are placed into the well with a drop of BA:BB, a #1 coverslip is carefully applied (quickly applying the coverslip at an angle allows air bubbles to escape), and sufficient pressure is applied to seat the coverslip without destroying the oocytes. After aspirating any excess mounting solution squeezed from the well, the coverslip is sealed with a *liberal* application of fingernail polish.

Larger oocytes or samples are mounted using stainless steel spacers (Fig. 6B), commercially available glass well slides (Fig. 6C), or custom-made slides (Fig. 6D), depending upon their size. For oocytes or samples 200–300 μm in diameter/thickness, we have fabricated spacers from 0.25-mm stainless steel sheet stock. These spacers, each of which contain one or two $\sim$12-mm holes, are glued to a glass slide with fingernail polish forming glass-bottomed sample wells. Once the fingernail polish has dried thoroughly, oocytes can be added to the wells in a drop of BA:BB and sealed under glass coverslips. Intact oocytes from 500 to 600 μm in diameter can be mounted in glass slides with 0.5-mm wells, which are available from several commercial suppliers. Intact stage VI oocytes or eggs (1.2 mm diameter) are mounted in slides machined from aluminum (stainless steel might provide a more durable slide). Each slide is perforated by two $\sim$12-mm holes. Nail polish is used to securely glue coverslips to one face of the slide, forming two glass-bottomed wells. After the polish is thoroughly dried, intact oocytes or eggs can be added in BA:BB and sealed under another coverslip. These slides allow us to easily examine opposite sides of intact oocytes or eggs.

---

[2] Fingernail polishes or other sealants should be tested for their compatibility with BA:BB before use. Sally Hansen's "Hard-As-Nails" ("clear," with or without nylon; Sally Hansens, Farmingdale, NY), available at many suppliers of cosmetices, works well as a sealing agent.

396

David L. Gard

A. Slides with spacers cut from coverslips.

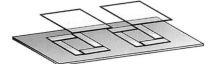

Spacers cut from coverslips, attached with fingernail polish. Allow to dry thoroughly before mounting samples.

B. Glass slides with steel spacers.

Attach steel spacers with fingernail polish and allow to dry thoroughly before mounting samples.

C. Commercially-available well slides.

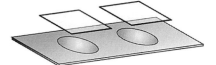

After orienting samples in BA:BB, attach coverslips with liberal application of fingernail polish. Dry thoroughly.

D. Double-sided metal slides.

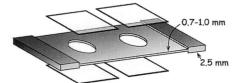

0.7–1.0 mm

2.5 mm

Attach bottom coverslips with fingernail polish and allow to dry thoroughly before mounting samples.

Fig. 6 Slides for mounting samples for confocal microscopy. (A) Mid-stage I oocytes and pieces of ovary from juvenile frogs are commonly mounted on glass slides using spacers cut from coverslips. Two sets of four spacers are "glued" to the face of a slide using clear Sally Hansen's "Hard as Nails" nail polish, forming shallow wells. A layer of nail polish is painted over the spacers. When dry, this layer forms a compressible "gasket," helping to seal the coverslip to the slide. Coverslips are cemented in place with a liberal application of nail polish around their edges. (B) Larger stage I oocytes and hemisected stage III oocytes are mounted in wells formed by cementing 0.25 mm stainless steel spacers to a glass coverslip with nail polish. As in A, a layer of nail polish is applied to the spacers which, when dry, acts as a "gasket." (C) Hemisected stage VI oocytes are often mounted in slides with 0.5-mm depressions or wells. Similar slides are available from several commercial sources. For viewing the cortex, oocytes are mounted cortical surface up. For cross-sectional views, oocytes are mounted cut face up, and optical sections are collected from below the knife-damaged region. (D) Intact oocytes (stages V–VI) and eggs can also be mounted in double-sided metal slides machined from aluminum or stainless steel. Slides are machined to provide coverslip spacing of 0.7–1.0 mm. Bottom coverslips are securely glued to one side with nail polish. After the polish is thoroughly dry (drying can be speeded by placing the slide in a fume hood or other air stream), oocytes are added and a top coverslip is affixed with nail polish. The thicker (2.5 mm) edges of the slide provide a shoulder to support the bottom coverslip above the microscope stage. When carefully used, these slides allow examination of opposite faces of the oocyte cortex.

Unfortunately, the working distance of the high-numerical-aperture objectives needed to optimally image MTs and other cytoskeletal elements limits examination of intact oocytes or eggs to the cortex or subcortical cytoplasm. To obtain cross-sectional views, we hemisect larger oocytes, eggs, and embryos with a sharp scalpel blade prior to processing and then collect optical sections from below the knife-damaged region. For cross-sectional views, hemisected oocytes and eggs are cleared and mounted in BA:BB in commercially available well slides (with 0.5-mm deep wells; Fig. 6C) with their cut face toward the coverslip. Cortical views can be obtained by mounting the cortex toward the coverslip.

Samples mounted in BA:BB and sealed with several coats of clear fingernail polish can be kept in the dark at room temperature for extended periods without appreciable fading.

As discussed above, oocytes stained with fluorescent phalloidins cannot be cleared and mounted in BA:BB, because of the incompatibility of these stains with dehydration in organic solvents. Small oocytes labeled with fluorescent phalloidin are routinely mounted either in TBSN containing NaN$_3$ or other agents to retard photobleaching (see below) or in a glycerol-based aqueous mounting medium, following the procedure outlined earlier. Despite fixation in buffers containing detergents, mounting in glycerol can cause severe osmotic distortion of larger *Xenopus* oocytes, and we commonly mount later stage oocytes in TBSN (pH 8.0) containing NaN$_3$ or propyl gallate to inhibit photobleaching.

I. Minimizing Photobleaching of Fluorescent Samples

Fading or photobleaching of fluorescent samples during observation, a problem commonly associated with fluorescence microscopy, is also encountered with laser-scanning microscopy. In practice, we have not found photobleaching to be any more severe during CLSM relative to conventional fluorescence microscopy. Several steps can be taken during sample preparation and image collection to minimize photobleaching.

First, choice of fluorochrome can play a major role in minimizing photobleaching. As a general rule, fluorochromes excited by shorter wavelengths are less stable than those excited by longer wavelengths (photons of short wavelength carry higher energies). Thus, fluorescein is bleached much more rapidly than rhodamine derivatives [some of the new short-wavelength Alexa dyes from Molecular Probes (Eugene, OR) do offer remarkable photostability].

Photobleaching of samples in aqueous solutions or other mounting agents can often be reduced by including anti-fade agents such as *p*-phenylenediamine, DABCO [1,4-diazabicyclo(2.2.2)octane; Johnson *et al.*, 1982], *n*-propyl gallate (Giloh and Sedat, 1982), or NaN$_3$ (Rost, 1991) (and references therein).

BA:BB clearing solution also appears to afford some protection against photobleaching of rhodamine. In an extreme case, we have collected more than 600 scans from a rhodamine-labeled sample without undue fading (less than 50% loss of fluorescent intensity). The rapid photobleaching of fluoresceins can be reduced by adding propyl gallate (25–50 mg/ml) to the BA:BB solution (BA:BB+PG). Propyl gallate dissolves very slowly and slightly colors the BA:BB+PG solution. This coloration does not appear to affect the fluorescence of the samples. Paradoxically, fluorescent samples mounted in BA:BB+PG are only stable for a few days, and samples mounted in BA:BB+PG should be viewed as soon as is practical.

IV. An Introduction to Confocal Microscopy and 3-D Reconstruction of the Cytoskeleton of *Xenopus* Oocytes

The following discussion is based on more than 10 years of experience using a Bio-Rad MRC-600 CLSM (Cha *et al.*, 1998, 1999; Cha and Gard, 1999; Gard, 1991, 1992, 1993a,b, 1994, 1999; Gard *et al.*, 1995a,b,c, 1997; Gard and Klymkowsky, 1998; Gard and Kropf, 1993; Roeder and Gard, 1994; Schroeder and Gard, 1992) and more limited experience with a Zeiss LSM 510. Although much of the discussion is independent of microscope brand, readers using other commercially available CLSM microscopes should consult their manufacturer's literature for descriptions of comparable hardware and software features.

A. Choosing Objectives for Confocal Microscopy of the Cytoskeleton

The resolution of cellular detail by CLSM, as with all optical methods, is limited by optical diffraction. The Rayleigh criterion defines the resolution (R) of conventional optical microscopes to be equal to the radius of the Airy disk:

$$R = 0.61 \times \lambda/\text{NA}$$

where λ is the wavelength of illumination (the emission wavelength when considering conventional fluorescence microscopy) and NA is the numerical aperture of the optical system (Inoué, 1989). Despite having diameters of only 6–25 nm, which are substantially below the resolution limit of light microscopy, individual cytoskeletal filaments (such as actin filaments, KFs, or MTs) are readily detected by immunofluorescence microscopy (Sammak and Borisy, 1988; Weber *et al.*, 1978). However, the apparent diameter of a single filament detected by conventional fluorescence microscopy equals the diameter of the Airy disk (D):

$$D = 1.22 \times \lambda/\text{NA}$$

and filaments spaced closer than R will not be resolved, but will appear as one.

From these equations, it is clear that objectives with larger NA ($\sim$1.3–1.4) provide better resolution of cytoskeletal detail. For example, an objective with an NA of 1.4 should provide images of individual filaments with apparent diameters of $\sim$0.45–0.54 μm [using λ_{em} of 520 nm (fluorescein) to 615 nm (Texas Red)], while the same filaments imaged with an objective with an NA of 0.45 would appear to be 1.4–1.7 μm in diameter. With an NA 1.4 objective, it is not possible to resolve filaments closer than 0.2–0.3 μm. However, individual filaments can often be distinguished from filament bundles by differences in their apparent brightness. Electron microscopy of *Xenopus* oocytes reveals that, although the "filaments" observed by CLSM in oocytes stained with anti-tubulin correspond to individual MTs (Pfeiffer and Gard, 1999), the keratin "filaments" actually correspond to loosely organized bundles of 3–6 individual keratin filaments (Chang *et al.*, 1999; Pfeiffer and Gard, unpublished observations). It is likely that the actin cables observed in stage I and stage VI oocytes also represent bundles of microfilaments, similar to stress fibers found in cultured fibroblasts.

Several features of the optical design of CLSMs provide a slight improvement in optical resolution relative to conventional microscopes. First, in CLSM, resolution is a function of the excitation wavelength, rather than the emission wavelength, of the fluorochrome. Additionally, confocal illumination results in a slight improvement ($\sim$30%) in lateral (XY) resolution (Majlof and Forsgren, 1993). In our experience, the apparent diameters of Texas Red-labeled MTs excited by an Ar–Kr laser and imaged with a 60$\times$ objective (NA 1.4) averaged $\sim$0.45 μm, which approaches the theoretical limit for this objective and excitation wavelength (568 nm).

We routinely use plan apochromatic objectives with the highest NA possible, typically $40\times$ to $60\times$ objectives with NA 1.3 or 1.4. Unfortunately, the shorter working distance (less than $150-200$ μm) associated with these high numerical apertures constrains examination of large cells such as oocytes. Longer working distances can be obtained by decreasing objective numerical aperture, at the expense of image resolution. As an example, we often used a $40\times$ (NA 1.0) objective when dictated by the need for greater working distance. However, images obtained with this objective were noticeably less "crisp" than those obtained with a $60\times$ (NA 1.4) objective, because of the increased diameter of the Airy disk ($0.6-0.7$ μm). Objectives providing magnifications of $10\times$ (NA 0.45) and $20\times$ (NA 0.75) provide substantially greater working distances and are useful for wide-field views of cytoskeletal organization, when resolution of individual filaments is not required.

B. Optimizing Image Scale

As with other digital imaging techniques, images collected by CLSM are composed of individual picture elements called pixels. The relationship between pixel size and objective resolution can affect the quality and/or fidelity of CLSM images. Optimum resolution and image accuracy is obtained when the image scale of the digital image provides a pixel size less than or equal to one-half of the theoretical limit of resolution (R) of the objective (Inoué, 1986). For example, the resolution of digital images collected with an objective with an NA of 1.4 and a pixel size less than ~ 0.1 μm (XY) will be limited by the objective resolution ($R = 0.2-0.25$ μm), not the pixel size. With pixel dimensions greater than $R/2$, image resolution and accuracy will be limited by the pixel size, not the objective.

The analog/digital converters in most commercial CLSMs provide images ranging from 512×512 to 2048×2048 pixels (note that this represents the pixel resolution of the digital image, not the dot pitch or resolution of the image monitor). The pixel size for an image is often provided by the controlling software during image acquisition, or can be determined by dividing the field of view by the number of pixels. At lower pixel resolutions (512×512), the pixel size may exceed the resolution of the objective being used, and image resolution will be limited by pixel size. In such cases, image quality can be improved by either (1) increasing the image scale, using the "zoom" function provided on most commercial CLSM; or (2) collecting the image at greater pixel resolutions (i.e., 1024×1024 or 2048×2048 pixels). Drawbacks of the former include decreased field of view and increased photobleaching. On the other hand, increasing the pixel resolution results in a geometric increase the size of the image file: an image collected at 2048×2048 pixels requires 16 times more memory or disk space (4.2 MB for an 8-bit image) than a 512×512 image (256 KB).

C. Optical Sectioning of Whole-Mounted Samples

The suppression of fluorescence from outside of the focal plane, often referred to as "optical sectioning," is one of the primary advantages provided by confocal microscopy applied to large biological samples (White *et al.,* 1987) (compare Figs. 2C and 2D with

Table V

Assessing Field Flatness and Optical Section Thickness Using an Inclined Reflector

1. Prepare an inclined reflective slide by cementing a coverslip (or half of a microscope slide) to a microscope slide at an angle of 5–15°. It is not necessary to know the angle.

2. Use the reflection mode to collect a confocal image of the inclined surface.

3. The image produced by a flat-field lens should appear as a straight strip (in a vertical of horizontal orientation on the display monitor, depending upon the orientation of the inclined reflector relative to the direction of scanning).

4. The image strip represents the focal "plane" of the objective and should be straight and even in width at all confocal apertures. Curvature of the strip, or broadening at its edges, is an indication that the objective is not well corrected for flatness. The width of the strip varies as the confocal aperture is changed.

5. The thickness of the optical section can be calculated by measuring the width of the image stripe and the horizontal displacement of the strip produced by a known shift in focus (stage position). The measured optical section thickness (T, in μm) is:

$$T = (W \times F)/D$$

where W equals the width of the image stripe of an inclined plane (full-width at half maximum intensity, in pixels), and D equals the horizontal displacement of the stripe (in pixels) for focus shift F (in μm).

2A and 2B). Most commercial CLSMs provide for the automated collection of multiple optical sections, commonly referred to as Z-series, through activation of a motor or piezoelectric stepper driving the microscope stage. This capability can be used to sample large volumes at defined intervals, or to collect serial optical sections representing the entire sample volume. Whether collecting individual or serial optical sections, it is useful to know the "thickness" of the optical sections, which is a function of the NA and the confocal aperture.

The thickness of optical sections collected by CLSM is a function of the NA of the objective and the confocal aperture (Majlof and Forsgren, 1993). Although the optical section thickness can be calculated from these parameters, such thicknesses might not be achieved in practice because of limitations of the instrument or sample. The apparent thickness of optical sections obtainable can easily be determined by imaging an inclined reflecting plane (see Table V), or serial sectioning submicron fluorescent beads at 0.1-μm intervals. Because the thickness of the confocal section also affects image brightness, we commonly use moderate confocal apertures that provide optical section thickness of 0.7–1 μm (with 60× NA 1.4 objective). Smaller apertures (thinner optical sections) result in dimmer images, requiring increased laser intensity or PMT gain (and the inherent problems of photobleaching and image noise; see later discussion). Larger apertures provide thicker optical sections in which details appear less crisp.

D. Optimizing Image Brightness

The intrinsic properties of the sample play a major role in determining the apparent brightness of confocal images. Unfortunately, the intrinsic brightness of the sample is often difficult to predict or control. The intrinsic brightness of the fluorescence signal can often be increased by optimizing antibody concentrations and incubation times, or by choosing a fluorochrome that is better matched to the laser excitation and emission filters of the CLSM (see earlier discussion).

In addition to the intrinsic brightness of the sample, the apparent brightness of a CLSM image is affected by several parameters of the instrument under the control of the operator: (1) the choice of objective; (2) the laser intensity; (3) the confocal aperture or pinhole; and (4) the photomultiplier sensitivity.

The brightness of an image obtained with any microscope objective is proportional to NA^2/M^2 (Inoué, 1986), where NA is the numerical aperture and M is the magnification factor of the objective.[3] Therefore, a $10\times$ (NA of 0.45) objective provides a relatively brighter image than a $40\times$ (NA of 1.3) objective, which in turn provides a brighter image than a $63\times$ (NA 1.4). However, the choice of objective is usually dictated by other characteristics, such as the need for (1) higher resolution, which is directly proportional to NA; (2) greater working distance, which is inversely proportional to NA; or (3) greater field of view, which is inversely proportional to magnification.

Most commercial CLSM allow operators to vary the laser power used to illuminate the sample, through the use of neutral density filters or acoustooptical filters. Image brightness can be increased by increasing illumination (laser intensity), at the expense of increased photobleaching. To minimize photobleaching during image collection, we typically use the minimum laser intensity possible. Inclusion of antifade agents in the mounting solution can also significantly reduce photobleaching, as discussed earlier.

The confocal aperture or pinhole affects image brightness by changing the volume of sample from which photons are collected. More photons are collected from a thicker optical section, resulting in a brighter image. Conversely, thinner optical sections, obtained by closing down the confocal aperture, provide a dimmer image (all other factors being equal).

The apparent brightness of a confocal image is also a function of the detector gain or sensitivity, with higher gains providing brighter images. However, increasing detector sensitivity also amplifies detector noise, resulting in images that appear grainy. Detector noise can be significantly reduced by averaging or filtering successive scans of the sample. However, line or scan averaging also increases photobleaching.

In practice, confocal aperture, laser intensity, and detector gain all interact to determine the apparent brightness of the collected image. These parameters must be balanced empirically against each other, and against the inherent brightness and stability of the fluorescent sample. In our experience, the best compromise between sample photostability/photobleaching and image quality is afforded by reducing the laser intensity to the minimum level that allows averaging 5–15 scans using moderate confocal apertures (providing optical sections 0.7–1 μm thick with a $60\times$ NA 1.4 objective) and moderate detector gain (50–75% of maximum) to minimize noise. When additional brightness is required, adjustments are first made to the detector gain, followed by laser intensity and/or confocal aperture.

Most commercial CLSM collect 8-, 12-, or 16-bit images, corresponding to 256, 4096, or 65,536 gray levels per pixel. In our experience, 8-bit (256 gray levels) images contain

---

[3] NA^2 is a measure of the light collected by the objective, while M^2 is a measure of the area over which the collected light is spread in the resulting image.

sufficient information for most routine monochrome images and result in smaller files (a 16-bit image requires twice the memory and disk space of an 8-bit image). For the best results, image brightness and contrast should be adjusted using the "gain" and "black level" controls (or their equivalents) to make full use of the instrument's dynamic range. To facilitate use of the entire dynamic range of our CLSM, we routinely collect images using a false-color lookup table (LUT) that maps black and white to contrasting colors such as green and red (such a display is simulated in Fig. 1E). Most commercial CLSM include a color lookup table providing such an LUT or palette (the "setcol" LUT on Bio-Rad instruments or the "Range indicator" palette on the Zeiss LSM 510).

With intrinsically faint samples, it is often not possible to collect an image spanning the entire brightness range of the instrument (especially if thin optical sections are required). Unless quantitative comparisons of fluorescence intensity are to be made, intrinsically faint images can be rescaled to make full use of the dynamic range before saving them to disk. Images that are to be analyzed quantitatively and/or compared to other images should not be rescaled, or should be saved in both "raw" and "scaled" versions.

E. Dealing with Extremes of Brightness and Contrast

Many fluorescent samples include details spanning a wide range of brightness, posing problems for both collecting and displaying images. For example, the extreme differences in apparent brightness of the meiotic spindles and surrounding astral or cytoplasmic MTs make it difficult to visualize both MT populations in a single image. Details of the astral MT's organization are often lost if image brightness and contrast (gain and black level) are optimized for the central spindle (as in Fig. 7A). Conversely, increasing the image brightness (by increasing image gain) to reveal the cytoplasmic or astral MTs often oversaturates details of the central spindle (Fig. 7B).

Several strategies can be used to deal with extremes in contrast. First collection of images with a larger dynamic range (12- or 16-bit images, corresponding to 4096 or 65,536 gray levels) will preserve details over a wider range of brightness. However, display and presentation of these images remains problematic, since few printing processes preserve details over such a large range of gray levels.

Alternatively, image contrast (or gamma) can be altered during postcollection image processing. Figure 7C is the same image as that shown in Fig. 7A, following adjustment of the color levels (Image>Adjust>Levels) in Adobe PhotoShop (Adobe Systems, Inc; Mountain View, CA) to selectively expand the middle range of pixel intensities (from 1.0 to 2.0). This processing significantly enhances the appearance of the peripheral MTs without oversaturating the central spindle.

Finally, many commercial CLSMs provide for the nonlinear mapping of detector output to the pixel brightness. Nonlinear mapping controls, referred to as "gamma" or "enhance" controls on some CLSMs, can be used to either enhance or suppress middle- or low-brightness details, enhancing the appearance of astral MTs relative to the central spindle (Fig. 7D). The settings that provide optimum contrast must be determined empirically for each image. Since nonlinear mapping alters the relative brightness of pixels in an image, it should not be used if pixel intensities will be quantified and/or compared.

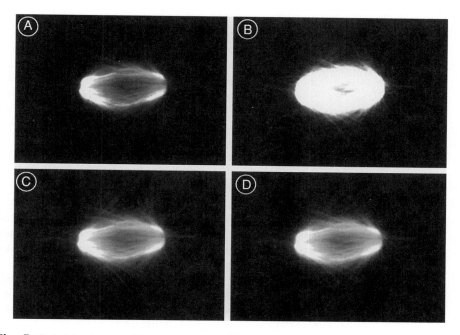

Fig. 7 Optimizing images with extreme ranges of brightness. (A) An image of a prometaphase first meiotic spindle in a maturing *Xenopus* egg, collected with gain and black levels optimized to show details of the central spindle. (B) The same spindle, collected with the gain increased to highlight astral MTs and the peripheral spindle. The central spindle is oversaturated, and most details are obscured. (C) The same data set as shown in A, after postcollection processing in Adobe PhotoShop. The "Levels" control of PhotoShop has been used to enhance the middle range of brightness (see text), revealing details of the peripheral spindle without loss of detail in the central spindle. (D) The same spindle, collected using a nonlinear photomultiplier output (enhance +2 on the Bio-Rad MRC-600). Note the enhanced detail in the peripheral microtubules without oversaturation of the central spindle.

F. Digital Processing of CLSM Images

The digital nature of CLSM images allows application of a wide variety of "image-processing" techniques for image enhancement (Inoué, 1986; Russ, 1995). In the interest of preserving an "accurate" record of our results, we try to collect the best images possible and keep subsequent postcollection manipulation or image processing to a minimum. Optimizing the image during collection includes: (1) matching the image scale and pixel size to the objective resolution; (2) optimizing optical section thickness; (3) maximizing use of the dynamic range (gray scale); and (4) optimizing contrast (using nonlinear mapping, if needed). In addition, we routinely average or filter several (5–15) successive scans to reduce detector noise, using one of the algorithms available on our CLSM.

Software for the processing and manipulation of digital images, including those collected by CLSM, is available from a wide variety of sources. Most commercial CLSM software comes with a suite of tools for image processing, including (at a minimum) adjustment of image brightness and contrast, smoothing and edge enhancement convolutions, projection of serial optical sections, and other useful operations. Much of our routine image processing, including image scaling, contrast adjustment, and minor smoothing, is now performed in Adobe PhotoShop (Adobe Systems, Inc; Mountain View, CA; http://www.adobe.com/products/photoshop/main.html). A more comprehensive suite of image processing tools, including unsharp masking, fast Fourier transforms, and tools for quantifying images, is available in the Image Processing Toolkit (Reindeer Games, Inc; Gainsville, FL; http://members.aol.com/ImagProcTK/), a library of PhotoShop plug-ins available as a companion to the *Image Processing Handbook* (Russ, 1995).

G. Presentation of Optical Sections and 3-D Reconstructions

Most commercial CLSMs allow the automated collection of serial optical sections (often referred to as Z-series). The ease with which large samples can be serial optically sectioned by CLSM often leads to collection of data sets containing large numbers of images, and the inherent problems or data storage and presentation. For example, serial optical sectioning of an entire stage I *Xenopus* oocyte requires more than 100 optical sections, resulting in a data set requiring up to 40–100 megabytes of disk space. In the recent past, the difficulties associated with saving and archiving large files limited most of our Z-series to 10–12 images. However, the recent availability of inexpensive mass storage devices (magnetooptical disks, CD-R, CD-RW, DVD, and others) has made collection and storage of large Z-series much more practical. We now routinely collect and save Z-series containing 20–100 individual sections for volume reconstruction.

The ability to collect spatial information, in the form of serial optical sections (Z-series), from large sample volumes also creates problems in data and image presentation. Several options exist for presenting volume data collected by CLSM. Perhaps the simplest option is to construct a montage consisting of one or more individual optical sections (Fig. 8A shows a single optical section). These images might show representative sections spanning the entire sample volume, or several adjacent serial optical sections containing objects of interest.

Alternatively, images with an extended depth-of-focus can be created by projecting two or more adjacent serial sections. In our experience, "maximum brightness" projection algorithms provide better contrast than those using image averaging or summation (a comparison of linear summation and maximum brightness projection algorithms is shown in Figs. 8B and 8C). The depth of focus obtainable through the projection of serial optical sections is limited only by the working distance of the objective and sample character (transparency and depth of antibody penetration). In practice, however, projection of an excessive number of optical sections often results in a confusing image (see Fig. 8D). For this reason, the optimum number of sections to project is best determined empirically for each dataset.

Finally, data collected by CLSM can be used to reconstruct the sample volume (in this case, cytoskeleton of *Xenopus* oocytes) in three dimensions (Gard, 1999), converting the two-dimensional pixels from a collection of serial optical sections into three-dimensional volume elements, or voxels. Software provided with many commercially

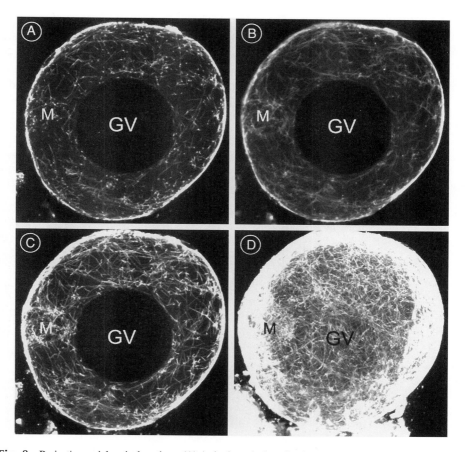

Fig. 8 Projecting serial optical sections. (A) A single optical section (approx. 1 μm thick) of a mid-stage I oocyte (85 μm diameter) stained with monoclonal antibodies to α-tubulin. MTs are apparent in the cortex, surrounding the oocyte nucleus, or germinal vesicle (GV), and throughout the cytoplasm. (B) A projection of 10 optical sections from the same oocyte, created using an arithmetic or averaging algorithm. Note the reduced contrast between the MTs and cytoplasm. (C) For comparison, the same 10 optical sections were projected using a maximum brightness algorithm, which provides much better contrast between MTs and cytoplasm. In both B and D, concentrations of MTs can be seen surrounding and penetrating the mitochondrial cloud (M). (D) A projection of 75 optical sections (out of a total of 180 sections collected at 0.5-μm intervals) from the same oocyte. The density of MTs, combined with the lack of depth information, confuses the image and nearly obscures the location of the GV and mitochondrial cloud (M).

available CLSMs provides tools for constructing stereo image pairs from *Z*-series. Newer CLSM software may include more comprehensive software for 3-D volume reconstruction and rotation, as either a standard feature or an option. Software suites including extensive volume rendering, surface extraction, and 3-D measurement tools are also available from a number of third-party sources. In our experience, data sets to be reconstructed in 3-D should be slightly oversampled in the *Z* dimension, by collecting optical sections at intervals slightly smaller than the optical section thickness. Under-sampling results in unsightly gaps or brightness contours in the reconstructed volume.

The incredible advances in processor and memory speed of the past 5 years have moved many volume rendering applications from the realm of mainframe computers and high-end graphics workstations to that of desktop personal computers. The volume reconstructions shown in Figs. 9 and 10 and in Fig. 1F were performed on typical desktop (233-MHz Pentium II with 128 MB RAM and 450-MHz Pentium III with 224 MB RAM)

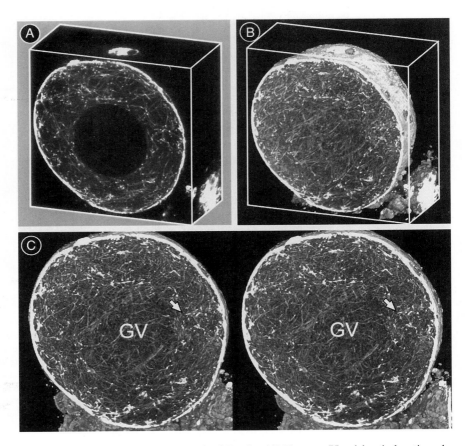

Fig. 9 3-D volume reconstruction from confocal Z-series. (A) The same 75 serial optical sections shown in Fig. 8D were reconstructed as a 3-D volume (see text) using a linear scale (0–255) for both the color lookup table (LUT) and the opacity function (α-function). The opacity of the volume obscures the MTs in the cytoplasm. (B) The same volume, reconstructed using nonlinear gray-scale LUT and α-functions, which render the background and cytoplasm transparent. The tilt of this image reveals the difference in resolution of the X, Y, and Z dimensions. (C) A stereo pair of the same volume, generated from two 3-D reconstructions differing by 6° in azimuth. Volume boundaries in A and B were added in Adobe Photoshop. Note that the volumes reconstructed in VoxBlast are reversed right-to-left (compare the position of the mitochondrial cloud M in Figs. 8A–8D and 9C).

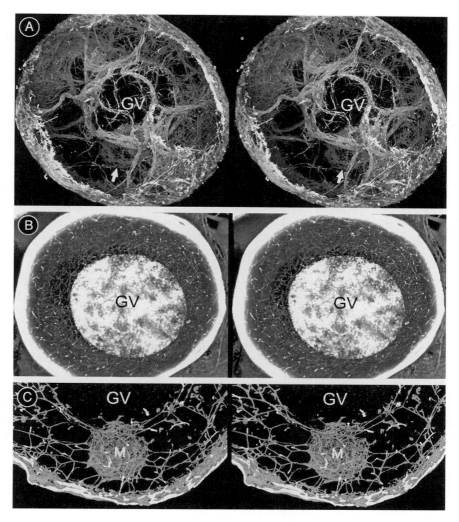

Fig. 10 Reconstructing the cytoskeletons of stage I *Xenopus* oocytes in 3-D. (A) A stereo pair (reconstructed from 75 optical sections) of a mid-stage I oocyte (110 μm in diameter) stained with monoclonal antibodies to acetylated α-tubulin (a marker for "stable" MTs (Piperno *et al.,* 1987; Schulze *et al.,* 1987; Webster and Borisy, 1989)). Note the network of bundled acetylated MTs surrounding the GV and extending to the oocyte cortex. Acetylated MTs are also associated with the mitochondrial cloud (arrow). (B) A stereo pair (reconstructed from 20 optical sections) of a mid-stage I oocyte (100 μm in diameter) stained with rhodamine–phalloidin to reveal F-actin, showing the brightly stained GV and cortex, and cytoplasmic meshwork of actin cables. (C) A stereo pair of the mitochondrial mass of a stage I oocyte (approx. 200 μm diameter) stained with monoclonal anti-keratin (reconstructed from 25 optical sections). Note the extensive network of KFs surrounding and filling the mitochondrial cloud.

or laptop (450-MHz Pentium III with 128 MB RAM) computers running VoxBlast 1.1 (Vaytek; Fairfield, IA) under Windows 9X (Windows is a registered trademark of Microsoft).

The speed of the 3-D rendering by VoxBlast is a function of size of the data set and the amount of physical RAM available. The images shown were reconstructed from datasets ranging from 10 to 100 MB (including sections interpolated to maintain the correct X–Y–Z aspect ratio). Rendering times ranged from a few seconds up to several minutes per reconstruction. Rendering speeds exceeding 5 million voxels per second were obtained when the entire data set could be loaded into physical RAM, and dropped significantly if the size of the data set required the use of virtual memory.

Three-dimensional reconstructions made using linear color lookup tables (LUTs) and opacity functions (α-functions) appear opaque (Fig. 9A) and provide little information on the cytoskeletal organization deep within the sample. Some software packages provide a "maximum brightness" opacity function, which simplifies the process of creating stereo pairs showing internal details. However, such algorithms may distort the spatial organization of the image (bright objects in the background might show through dimmer foreground objects which should obscure them). The view in Fig. 9B and the stereo images in Figs. 9 and 10 and in Fig. 1 (with its color plate) were rendered using empirically determined LUTs and α-functions which rendered the "background" cytoplasmic fluorescence transparent, thus revealing the stained cytoskeletal elements.

For publication, reconstructed volumes can be presented as stereo pairs constructed from two views differing by $\sim$5–7° in azimuth. For presentations, we often use Adobe Premiere[4] (Adobe Systems, Inc.; Mountain View, CA; http://www.adobe.com/products/premiere/main.html) to generate an animated movie in Windows AVI or Quicktime formate (Quicktime is a registered trademark of Apple Computers), depicting semitransparent oocytes rocking slowly back and forth. Many of these animations can be viewed at our Web site: http://froglab.biology.utah.edu. The reduced Z-resolution of confocal data sets (relative to XY resolution) can result in noticeable degradation in images of volumes turned more than $\sim$15–20° around their X- or Y-axes (this can be seen to some degree in Fig. 9B). We therefore limit rotations to $\pm$15–20°.

V. Conclusions

The advent and application of CLSM has revolutionized the study of cytoskeletal and cytoplasmic organization in large cells such as *Xenopus* oocytes. When combined with 3-D volume reconstruction, CLSM has extended our understanding of cytoskeletal organization in *Xenopus* oocytes into three dimensions, providing unprecedented views of the complexity of the oocyte cytoskeleton (Gard, 1999). We hope the techniques and tips presented in this chapter will prove useful to others using CLSM to examine cytoplasmic organization in *Xenopus* oocytes or other large cells or embryos.

---

[4] NIH Image and other third-party software can also be used to generate animated movies from confocal data sets.

<hr>

VI. Recipes and Reagents

Unless noted otherwise, reagents were from Sigma Chemical Corporation (St. Louis, MO).

Fix buffer:

> 80 mM K Pipes pH 6.8
> 1 mM MgCl$_2$
> 5 mM EGTA
> 0.2% Triton X-100

Formaldehyde–glutaraldehyde fix with taxol (FGT-fix):

3.7% formaldehyde	(from 37% or 18% stock)
0.25% glutaraldehyde	(from a 50% stock)
0.5 μM Taxol	(from a 100-μM stock)

Mix the above in fix buffer. Use immediately (within 1 h).

Formaldehyde can be AR grade (Mallinkrodt, Paris, KY) or EM grade (methanol-free; Ted Pella, Inc, Redding, CA). We have seen little or no difference in the quality of images collected from oocytes fixed with AR-grade or EM-grade formaldehyde or freshly prepared paraformaldehyde. Glutaraldehyde is EM grade (stored as a 50% stock at 4°C; Ted Pella, Inc., Redding, CA). Taxol is from a 100-μM stock in DMSO, stored at −20°C (CalBiochem or Sigma).

FG-fix is the identical formulation *without* Taxol. FT-fix is the identical formulation *without* glutaraldehyde.

Low-FGT fix:

0.25–0.5% formaldehyde	(from 37% or 18% stock)
0.1% glutaraldehyde	(from a 50% stock)
0.5 μM taxol	(from a 100-μM stock)

Mix the above in fix buffer. Use immediately (within 1 h). Fix for 0.5–1 h, and postfix in 100% methanol.

Formaldehyde–phalloidin fix for microfilaments:

> 1.0–1.5% formaldehyde (from 37% stock) in fix buffer.

Fix 4–18 h at room temperature, then stain with fluorescent phalloidin as described in Table III. Note that phalloidin can be added during fixation.

Peroxide–methanol bleach (Adapted from Dent and Klymkowsky, 1989):

> 1 part hydrogen peroxide (30%):2 parts 100% methanol (100%)

Peroxide is an excellent bleaching agent for oocyte pigments, clothing, and skin. Use due caution to avoid damage to clothing or painful chemical "burns."

Phosphate-buffered saline (PBS):

128 mM NaCl
2 mM KCl
8 mM Na$_2$HPO$_4$
2 mM KH$_2$PO$_4$ (pH 7.2)

Tris-buffered saline (*without detergent*) can be substituted for PBS during NaBH$_4$ reduction.

Tris-buffered saline with NP-40 (TBSN):

155 mM NaCl
10 mM Tris-Cl (pH 7.4)
0.1% NP-40

Triton X-100 can be substituted for NP-40.

Benzyl alcohol–benzyl benzoate clearing solution (BA:BB):
(From Murray and Kirschner, as cited by Dent and Klymkowsky, 1989.)

1 part benzyl alcohol:2 parts benzyl benzoate.

Inclusion of propyl gallate (25–50 mg/ml) reduces photobleaching of fluorescein-labeled samples. The slight brownish color imparted by propyl gallate does not affect clearing. However, extended storage in BA:BB with propyl gallate causes samples to fade. For this reason, long-term storage of samples in BA:BB with propyl gallate is not advised.

Most laboratory plastic wares (microcentrifuge tubes, etc.) are compatible with BA:BB, notable exceptions being cellulose acetate, polystyrene, and the plastic used in computer keyboards.

Note: BA:BB is both toxic (see text) and a skin irritant. Caution should be used to avoid contact (especially with the eyes).

Glycerol mounting solution:

10 mM Tris Cl pH 8.0
50 mM NaN$_3$ (to retard photobleaching[5]) in 90% glycerol.

VII. Cytoskeletal and Nuclear Probes for Confocal Microscopy of *Xenopus* Oocytes

In this section are listed some of the antibodies and probes we have successfully used (or not!) in our investigation of cytoskeletal organization in *Xenopus* oocytes, eggs, and

---

[5] Other antifade agents (phenylenediamine, DABCO, propyl gallate, etc.; see Giloh and Sedat, 1982; Johnson *et al.,* 1982; Rost, 1991) can be substituted.

early embryos. In the left margin, a subjective evaluation of the usefulness of the antibody for CLSM of *Xenopus* oocytes or eggs is provided, ranging from +++ (stains very well) to − (does not stain, or otherwise not useful).

A. Antibodies for Staining Microtubules

++ DM 1A: chick brain α-tubulin (mouse monoclonal IgG_1; Blose *et al.,* 1984). Stains oocyte and egg MTs. Commercially available from Sigma (St. Louis, MO. #T-9026) and ICN Biomedicals (Aurora, OH. #69125).

Most of our early work examining the distribution of MTs in oocytes and eggs was performed using DM1A from ICN. In the summer of 1997, several lot numbers of DM1A from ICN failed to stain MTs in *Xenopus* oocytes, and we switched to using the same monoclonal clone from Sigma. In the spring and summer of 2000, we began experiencing difficulties staining oocyte MTs with DM1A from Sigma, although it continues to stain cortical MTs in fertilized *Xenopus* eggs.

+/− DM 1B: chick brain β-tubulin (mouse monoclonal IgG_1; Blose *et al.,* 1984). Fair–good MT staining in initial tests, but recent results disappointing. Not used extensively. Commercially available from ICN Biomedicals (Aurora, OH).

+/− TUB 2.1: rat brain β-tubulin (mouse monoclonal IgG_1). Staining of oocyte MTs was inconsistent. Not used extensively. Commercially available from Sigma (St. Louis, MO; #T-4026).

++ YL1/2: α-tubulin (rat monoclonal IgG_{2a}; Kilmartin *et al.,* 1982). Brightly stains oocyte and egg MTs. Commercially available from Accurate Chemical and Scientific Co (Westbury, NY; #YSRT-MCAP77).

Although the epitope recognized by YL1/2 includes the C-terminal tyrosine of tyrosinated α-tubulin, we have recently begun using YL1/2 for much of our routine tubulin immunofluorescence.

+/− KMX-1: β-tubulin (mouse monoclonal IgG_{2b}). Fair–good MT staining in vegetal hemisphere of oocytes. Not used extensively. Available from Roche Biochemicals (Mannheim, Germany; #1 111 876).

B. Posttranslationally Modified Tubulins

+++ 6-11B-1: sea urchin acetylated α-tubulin (mouse monoclonal IgG_{2b}; Piperno *et al.,* 1987). Brightly stains stable MTs in *Xenopus* oocytes. Commercially available from Sigma (St. Louis, MO; #T-6793).

Posttranslational acetylation of α-tubulin on lys_{40} is associated with nondynamic, or "stable," MTs. 6-11B-1 brightly stains a large subset of MTs in all stages of postmitotic *Xenopus* oocytes. MTs of meiotic spindles are faintly stained. 6-11B-1 also stains midbodies of dividing blastomeres, ciliated cells in embryonic epithelia, and MTs of neuronal axons.

+/− TUB1A2: α-Tubulin peptide containing C-terminal Y (mouse monoclonal IgG_3; Kreis, 1987). Inconsistent staining. Commercially available from Sigma (St. Louis, MO; #T-9028).

Specifically recognizes tyrosinated α-tubulin. Tyrosination is a marker for recently assembled (dynamic) MTs. Brightly stained MTs in mitotic oogonia. Staining of postmitotic oocytes was inconsistent, with high background in animal cytoplasm of stage VI oocytes (due to large pool of unassembled tubulin?). Not used extensively.

C. Centrosomes, MTOCs, and Spindle Poles

+   GTU-88: gamma-tubulin (mouse monoclonal IgG$_1$). Stains poles of meiotic spindles in maturing oocytes and centrosomes/spindle poles in cleaving embryos. Commercially available from Sigma (St. Louis, MO; #T-6557).

D. Antibodies for Staining Keratin Filaments

+++   C11: Type I and II human cytokeratins (mouse monoclonal IgG$_1$; Bartek *et al.*, 1991). Brightly stains KFs in *Xenopus* oocytes. Commercially available from Sigma (St. Louis, MO; #C-2931).

Stains KFs in methanol-fixed oocytes from late stage I (not compatible with aldehyde fixation). In Western blots, C11 recognizes the single 54-kDa type II keratin present in *Xenopus* oocytes. When microinjected into stage VI oocytes C11 disrupts radial and cortical KF networks.

−   PCK-26: human epidermal cytokeratins (mouse monoclonal IgG$_1$; Moll *et al.*, 1981). Does not stain KFs in *Xenopus* oocytes. Commercially available from Sigma (St. Louis, MO; #C-1801). A broad-spectrum antibody recognizing many type II keratins.

++   1h5: *Xenopus* keratin (monoclonal mouse IgG$_1$, IgG$_{2a}$; Klymkowsky *et al.*, 1987). Stains KFs in *Xenopus* oocytes. Available from the Developmental Studies Hybridoma Bank, University of Iowa (Iowa City, IA; http://www.uiowa.edu/~dshbwww/).

In Western blots, 1h5 recognizes the single type II keratin of *Xenopus* oocytes. Staining similar to C11, but with some punctate cytoplasmic background.

E. Probes for Actin Distribution

+/−   Anti-actin (rabbit polyclonal antiserum). Stains the cytoplasm of *Xenopus* oocytes. Obtained from Sigma (St. Louis, MO; #A-2668).

Rhodamine–phalloidin (fluorescent-conjugated fungal toxin). Stains cortex, actin cables and GV of *Xenopus* oocytes. Commercially available from Molecular Probes (Eugene, OR; #R-415).

Requires mild fixation in formaldehyde. Not compatible with glutaraldehyde- or methanol-fixed oocytes. A stock solution of 200 units/ml was prepared in 100% methanol, according to the manufacturer's instructions. This stock was diluted 1:100 into TBSN for staining. Rhodamine–phalloidin was more photostable and seemed to provide better contrast than the fluorescein or Bodipy conjugates. Even so, the extreme contrast between the brightly stained cortex and cytoplasmic cables makes collecting images challenging.

F. Other Antibodies

We have been unable to obtain useful staining of *Xenopus* oocytes or eggs during limited testing of the following commercially available monoclonal antibodies to various cytoskeletal components:

Anti-cytoplasmic dynein IC (MAB1618; Chemicon, Temecula, CA)

Anti-cytoplasmic dynein IC (clone 70.1; Sigma #D-5167; blocks function when microinjected)

Anti-cytoplasmic dynein HC (clone 440.4; Sigma #D-1667)

Anti-kinesin (clone IBII; Sigma #K1005)

Anti-pan-cadherin (clone CH19; Sigma #C-1821)

Anti-vinculin (clone hVIN-1; Sigma #V-9131)

Anti-talin (clone 8d4; Sigma #T-3287)

Anti-β-actin (clone AC-15; Sigma #A-5441)

Anti-α-actinin (clone BM75.2; Sigma #A-5044)

Anti-plectin (clone 7A8; Sigma #P-9318)

G. Nuclear Stains

+ Propidium Iodide (intercalating DNA dye). Commercially available from Molecular Probes (Eugene, OR; #P-1304). Works best with methanol-fixed stage 0 oocytes or maturing eggs. Counterstain with fluorescein-conjugated secondary antibodies.

+ Ethidium homodimer (intercalating DNA dye). Commercially available from Molecular Probes (Eugene, OR; #E-1169). Works best with methanol-fixed stage 0 oocytes or maturing eggs. For dual fluorescence applications, counterstain with fluorescein-conjugated secondary antibodies.

Note: PI and EH also stain cytoplasmic dsRNA, which can be reduced by pretreating samples with RNAse.

+++ YO-PRO-1 (cyanine DNA dye). Commercially available from Molecular Probes (Eugene, OR; #Y-3603). Works best with methanol-fixed stage 0 oocytes, eggs, and embryos. Some reduction in staining of aldehyde-fixed oocytes. For dual fluorescence applications, counterstain with Texas-Red-conjugated secondary antibodies.

++ BO-PRO-3 (cyanine DNA dye). Commercially available from Molecular Probes (Eugene, OR; #B-3586). Works best with methanol-fixed stage 0 oocytes, eggs, and embryos. Some reduction in staining of aldehyde-fixed oocytes. Worked best for single labeling.

Acknowledgments

Over the years, a number of individuals have contributed to the observations discussed in this chapter, including M. Schroeder, A. Roeder, B. Error, A. Friend, D. Affleck, P. Jenkins, B.-J. Cha, B. Becker, and S. Romney, and the author thanks them for their contributions. Special thanks are due to Dr. Edward King for his invaluable assistance with the confocal microscope facility. The work described has been supported by grants from the National Institute of General Medical Studies, National Science Foundation, and University of Utah Research Committee.

References

Arndt-Jovin, D. J., and Jovin, T. M. (1989). *Methods Cell Biol.* **30,** 353–377.

Bartek, J., Vojtesek, B., Staskova, Z., Bartkova, J., Kerekes, Z., Rejthar, A., and Kovarik, J. (1991). A series of 14 new monoclonal antibodies to keratins: characterization and value in diagnostic histopathology. *J. Pathol.* **164,** 215–224.

Blose, S. H., Meltzer, D. I., and Feramisco, J. R. (1984). 10-nm filaments are induced to collapse in living cells microinjected with monoclonal and polyclonal antibodies against tubulin. *J. Cell Biol.* **98,** 847–858.

Brachet, J., Hanocq, F., and Van Gansen, P. (1970). A cytochemical and ultrastructural analysis of in vitro maturation in amphibian oocytes. *Dev. Biol.* **21,** 157–195.

Cha, B. J., and Gard, D. L. (1999). XMAP230 is required for the organization of cortical microtubules and patterning of the dorsoventral axis in fertilized *Xenopus* eggs. *Dev. Biol.* **205**, 275–286.

Cha, B., Cassimeris, L., and Gard, D. L. (1999). XMAP230 is required for normal spindle assembly in vivo and in vitro. *J. Cell Sci.* **112**, 4337–4346.

Cha, B. J., Error, B., and Gard, D. L. (1998). XMAP230 is required for the assembly and organization of acetylated microtubules and spindles in *Xenopus* oocytes and eggs. *J. Cell Sci.* **111**, 2315–2327.

Chang, P., Perez-Mongiovi, D., and Houliston, E. (1999). Organisation of *Xenopus* oocyte and egg cortices. *Microsc. Res. Tech.* **44**, 415–429.

Clark, G. (1981). *In* "Staining Procedures" (G. Clark, Ed.), pp. 1–26. Williams and Wilkins, Baltimore.

Clark, T. G., and Merriam, R. W. (1978). Actin in *Xenopus* oocytes. *J. Cell Biol.* **77**, 427–438.

Cross, A. R., and Williams, R. C., Jr. (1991). Kinky microtubules: bending and breaking induced by fixation in vitro with glutaraldehyde and formaldehyde. *Cell. Motil. Cytoskeleton* **20**, 272–278.

Dent, J., and Klymkowsky, M. W. (1989). Whole-mount analysis of cytoskeletal reorganization and function during oogenesis and early embryogenesis in *Xenopus*. *In* "The Cell Biology of Development" (H. Schatten and G. Schatten, Eds.), pp. 63–103. Academic Press, New York.

Dent, J. A., Polson, A. G., and Klymkowsky, M. W. (1989). A whole-mount immunocytochemical analysis of the expression of the intermediate filament protein vimentin in *Xenopus*. *Development* **105**, 61–74.

Dumont, J. N. (1972). Oogenesis in *Xenopus laevis* (Daudin). I. Stages of oocyte development in laboratory maintained animals. *J. Morphol.* **136**, 153–179.

Elinson, R. P., and Rowning, B. (1988). A transient array of parallel microtubules in frog eggs: potential tracks for a cytoplasmic rotation that specifies the dorso-ventral axis. *Dev. Biol.* **128**, 185–197.

Gard, D. L. (1991). Organization, nucleation, and acetylation of microtubules in *Xenopus laevis* oocytes: a study by confocal immunofluorescence microscopy. *Dev. Biol.* **143**, 346–362.

Gard, D. L. (1992). Microtubule organization during maturation of *Xenopus* oocytes: assembly and rotation of the meiotic spindles. *Dev. Biol.* **151**, 516–530.

Gard, D. L. (1993a). Confocal immunofluorescence microscopy of microtubules in amphibian oocytes and eggs. *Methods Cell Biol.* **38**, 241–264.

Gard, D. L. (1993b). Ectopic spindle assembly during maturation of *Xenopus* oocytes: evidence for functional polarization of the oocyte cortex. *Dev. Biol.* **159**, 298–310.

Gard, D. L. (1994). Gamma-tubulin is asymmetrically distributed in the cortex of *Xenopus* oocytes. *Dev. Biol.* **161**, 131–140.

Gard, D. L. (1999). Confocal microscopy and 3-D reconstruction of the cytoskeleton of *Xenopus* oocytes. *Microsc. Res. Tech.* **44**, 388–414.

Gard, D. L., and Klymkowsky, M. W. (1998). Intermediate filament organization during oogenesis and early development in the clawed frog, *Xenopus laevis*. *Subcell. Biochem.* **31**, 35–70.

Gard, D. L., and Kropf, D. L. (1993). Confocal immunofluorescence microscopy of microtubules in oocytes, eggs, and embryos of algae and amphibians. *Methods Cell Biol.* **37**, 147–169.

Gard, D. L., Affleck, D., and Error, B. M. (1995a). Microtubule organization, acetylation, and nucleation in *Xenopus laevis* oocytes: II. A developmental transition in microtubule organization during early diplotene. *Dev. Biol.* **168**, 189–201.

Gard, D. L., Cha, B. J., and King, E. (1997). The organization and animal–vegetal asymmetry of cytokeratin filaments in stage VI *Xenopus* oocytes is dependent upon F-actin and microtubules. *Dev. Biol.* **184**, 95–114.

Gard, D. L., Cha, B. J., and Roeder, A. D. (1995b). F-actin is required for spindle anchoring and rotation in *Xenopus* oocytes: a re-examination of the effects of cytochalasin B on oocyte maturation. *Zygote* **3**, 17–26.

Gard, D. L., Cha, B. J., and Schroeder, M. M. (1995c). Confocal immunofluorescence microscopy of microtubules, microtubule-associated proteins, and microtubule-organizing centers during amphibian oogenesis and early development. *Curr. Top. Dev. Biol.* **31**, 383–431.

Gard, D. L., Hafezi, S., Zhang, T., and Doxsey, S. J. (1990). Centrosome duplication continues in cycloheximide-treated *Xenopus* blastulae in the absence of a detectable cell cycle. *J. Cell Biol.* **110**, 2033–2042.

Giloh, H., and Sedat, J. W. (1982). Fluorescence microscopy: reduced photobleaching of rhodamine and fluorescein protein conjugates by *n*-propyl gallate. *Science* **217**, 1252–1255.

Haugland, R. P. (1996). "Handbook of Fluorescent Probes and Research Chemicals." Molecular Probes, Eugene, OR.

Houliston, E., and Elinson, R. P. (1991). Patterns of microtubule polymerization relating to cortical rotation in *Xenopus laevis* eggs. *Development* **112,** 107–117.

Huchon, D., and Ozon, R. (1985). Microtubules during germinal vesicle breakdown (GVBD) of *Xenopus* oocytes: effect of Ca^{2+} ionophore A-23187 and taxol. *Reprod. Nutr. Dev.* **25,** 465–479.

Huchon, D., Crozet, N., Cantenot, N., and Ozon, R. (1981). Germinal vesicle breakdown in the *Xenopus laevis* oocyte: description of a transient microtubular structure. *Reprod. Nutr. Dev.* **21,** 135–148.

Huchon, D., Jessus, C., Thibier, C., and Ozon, R. (1988). Presence of microtubules in isolate cortices of prophase I and metaphase II oocytes in *Xenopus laevis*. *Cell Tissue Res.* **254,** 415–420.

Inoué, S. (1986). "Video Microscopy." Plenum Press, New York.

Inoué, S. (1989). Foundations of confocal scanned imaging in light microscopy. *In* "The Handbook of Biological Confocal Microscopy" (J. B. Pawley, Ed.), pp. 1–13. IMR Press, Madison, WI.

Jessus, C., Huchon, D., and Ozon, R. (1986). Distribution of microtubules during the breakdown of the nuclear envelope of the *Xenopus* oocyte: an immunocytochemical study. *Biol. Cell* **56,** 113–120.

Johnson, G. D., Davidson, R. S., McNamee, K. C., Russell, G., Goodwin, D., and Holborow, E. J. (1982). Fading of immunofluorescence during microscopy: a study of the phenomenon and its remedy. *J. Immunol. Methods* **55,** 231–242.

Karnovsky, M. J. (1965). Formaldehyde–glutaraldehyde fixative of high osmolarity for use in electron microscopy. *J. Cell Biol.* **27,** 131a.

Karr, T. L., and Alberts, B. M. (1986). Organization of the cytoskeleton in early *Drosophila* embryos. *J. Cell Biol.* **102,** 1494–1509.

Kelly, G. M., Eib, D. W., and Moon, R. T. (1991). Histological preparation of *Xenopus laevis* oocytes and embryos. *Methods Cell Biol.* **36,** 389–417.

Kilmartin, J. V., Wright, B., and Milstein, C. (1982). Rat monoclonal antitubulin antibodies derived by using a new nonsecreting rat cell line. *J. Cell Biol.* **93,** 576–582.

Klymkowsky, M. W., and Hanken, J. (1991). Whole-mount staining of *Xenopus* and other vertebrates. *Methods Cell Biol.* **36,** 419–441.

Klymkowsky, M. W., Maynell, L. A., and Polson, A. G. (1987). Polar asymmetry in the organization of the cortical cytokeratin system of *Xenopus laevis* oocytes and embryos. *Development* **100,** 543–557.

Kreis, T. E. (1987). Microtubules containing detyrosinated tubulin are less dynamic. *EMBO J.* **6,** 2597–2606.

Kropf, D. L., Maddock, A., and Gard, D. L. (1990). Microtubule distribution and function in early *Pelvetia* development. *J. Cell Sci.* **97,** 545–552.

Majlof, L., and Forsgren, P. O. (1993). Confocal microscopy: important considerations for accurate imaging. *Methods Cell Biol.* **38,** 79–95.

Merriam, R. W., and Clark, T. G. (1978). Actin in *Xenopus* oocytes. II. Intracellular distribution and polymerizability. *J. Cell Biol.* **77,** 439–447.

Moll, R., Franke, W. W., and Schiller, D. L. (1982). The catalog of human cytokeratins: Patterns of expression in normal epithelia, tumors, and cultured cells. *Cell* **31,** 11–24.

Palecek, J., and Romanovsky, A. (1985). Detection of tubulin structures of animal tissues in paraffin sections by immunofluorescence. *Histochem. J.* **17,** 519–521.

Palecek, J., Habrova, V., and Nedvidek, J. (1984). Localization of tubulin structures in the course of amphibian germinal vesicle maturation. *Histochem. J.* **16,** 357–359.

Palecek, J., Habrova, V., Nedvidek, J., and Romanovsky, A. (1985). Dynamics of tubulin structures in *Xenopus laevis* oogenesis. *J. Embryol. Exp. Morphol.* **87,** 75–86.

Palecek, J., Ubbels, G. A., and Macha, J. (1982). An immunocytochemical method for the visualization of tubulin-containing structures in the egg of *Xenopus laevis*. *Histochemistry* **76,** 527–538.

Pfeiffer, D. C., and Gard, D. L. (1999). Microtubules in *Xenopus* oocytes are oriented with their minus-ends towards the cortex. *Cell. Motil. Cytoskeleton* **44,** 34–43.

Piperno, G., LeDizet, M., and Chang, X.-J. (1987). Microtubules containing acetylated α-tubulin in mammalian cells in culture. *J. Cell Biol.* **104,** 289–302.

Roeder, A. D., and Gard, D. L. (1994). Confocal microscopy of F-actin distribution in *Xenopus* oocytes. *Zygote* **2**, 111–124.

Rost, F. W. D. (1991). "Quantitative Fluorescence Microscopy," Cambridge University Press, Cambridge.

Russ, J. C. (1995). "The Image Processing Handbook." CRC Press, Boca Raton, FL.

Sammak, P. J., and Borisy, G. G. (1988). Detection of single fluorescent microtubules and methods for determining their dynamics in living cells. *Cell. Motil. Cytoskeleton* **10**, 237–245.

Schroeder, M. M., and Gard, D. L. (1992). Organization and regulation of cortical microtubules during the first cell cycle of *Xenopus* eggs. *Development* **114**, 699–709.

Schulze, E., Asai, D. J., Bulinski, J. C., and Kirschner, M. (1987). Posttranslational modification and microtubule stability. *J. Cell Biol.* **105**, 2167–2177.

Weber, K., Rathke, P. C., and Osborn, M. (1978). Cytoplasmic microtubular images in glutaraldehyde-fixed tissue culture cells by electron microscopy and by immunofluorescence microscopy. *Proc. Natl. Acad. Sci. USA* **75**, 1820–1824.

Webster, D. R., and Borisy, G. G. (1989). Microtubules are acetylated in domains that turn over slowly. *J. Cell Sci.* **92**, 57–65.

Wells, S., and Johnson, I. (1994). Fluorescent labels for confocal microscopy. *In* "Three-Dimensional Confocal Microscopy: Volume Investigation of Biological Systems" (J. K. Stevens, L. R. Mills, and J. E. Trogadis, Eds.), pp. 101–131. Academic Press, San Diego.

White, J. G., Amos, W. B., and Fordham, M. (1987). An evaluation of confocal versus conventional imaging of biological structures by fluorescence light microscopy. *J. Cell Biol.* **105**, 41–48.

Windholz, M. (1983). "The Merck Index." Merck and Co., Rahway, NJ.

CHAPTER 11

Confocal Fluorescence Microscopy Measurements of pH and Calcium in Living Cells

Kay-Pong Yip and Ira Kurtz

Department of Molecular Pharmacology, Physiology and Biotechnology
Brown University
Providence, Rhode Island

Division of Nephrology, Department of Medicine
UCLA School of Medicine
Los Angeles, California

I. Introduction

The biological problems which are studied with confocal fluorescence microscopy can be categorized into two groups: (1) the study of living or fixed preparations whose fluorescence properties do not change with time and (2) the study of living preparations whose fluorescence changes temporally and/or spatially in up to three dimensions. A number of investigations have focused on the former category, but there are few reports which fall into the latter category. An example of studies which involve the change of fluorescence temporally and spatially in a living cell is the measurement of ion activity such as pH and intracellular calcium using fluorescent probes. Although these probes have been used to measure ion activity in single cells using regular epifluorescence microscopy, only recently have measurements of ion activity been performed using confocal microscopy. Biological systems often are in the form of cylindrical structures, for example, gastric glands, nephrons, and blood vessels. Extracting the dynamics of the fluorescence signal from a cylindrical structure based on epifluorescence imaging is complicated by out-of-focus fluorescence, which limits interpretation of the data. One major advantage of confocal fluorescence microscopy is that the out-of-focus fluorescence can be excluded to a great extent, allowing the dynamics of specific cellular phenomena to be acquired from specific anatomical locations in a heterogeneous epithelium.

II. Measurement of pHi with BCECF

Since the introduction of carboxyfluorescein (CF) by Thomas *et al.* in 1979, fluorescent techniques for measuring intracellular pH (pHi) have virtually replaced microelectrode techniques (Thomas *et al.,* 1979). Optical measurements of pH have a rapid response time, have high sensitivity, and provide spatial information. Because of the excess leakage rate of CF, Tsien and colleagues developed BCECF in 1982 (Rink *et al.,* 1982). This probe is presently the most widely used pH probe. BCECF has a pK of approximately 6.9 and a very slow leakage rate. To load cells with the dye, the acetoxymethyl ester derivative is used which is lipid-permeable and therefore crosses cell membranes easily. Inside the cell, esterases cleave the probe, leaving the charged form of the dye trapped in the cytoplasm.

BCECF is excited at two wavelengths: 490 nm (pH-sensitive wavelength) and 440 nm (isosbestic wavelength). The 490/440 nm excitation ratio varies with pH and is linear between pH 6.5 and 7.4 in most cells. The dye is calibrated *in vivo* using the high-K^+ nigericin technique (Thomas *et al.,* 1979). Using a microfluorometer coupled to either a photomultiplier tube or a two-dimensional detector, pH can be measured in single cells (Tanasugarn *et al.,* 1984; Wang and Kurtz, 1990; Yip and Kurtz, 1995). However, if the epithelium is heterogeneous in the depth dimension, e.g., kidney tubule, gastric glands, measurements of pH in single cells becomes less accurate because of the potential acquisition of out-of-focus fluorescence information. This occurs because of the poor resolution in the z dimension of regular epifluorescence microscopes. The use of regular epifluorescence microscopy also makes it difficult to study spatial pH differences within different compartments in a single cell, i.e., cytoplasm versus nucleus, endocytotic vesicles. The cortical collecting duct, a portion of the kidney tubule studied in our laboratory, is cylindrical in shape and possesses at least two cell types: intercalated and principal cells (Fig. 1). The cylindrical shape and heterogeneous nature of this preparation necessitated the development of a new optical approach for monitoring pH in this preparation.

The confocal microscope has improved resolution in the z dimension (approximately 0.5 μm; Zimmer *et al.,* 1988; Wells *et al.,* 1989). Confocal microscopes are excited with

Fig. 1 Pseudo-Nomarski image of an isolated perfused rabbit cortical collecting duct (40× objective, 8× zoom). The small round cells are intercalated cells and the larger polygonal cells are principal cells. (Adapted from Wang and Kurtz, 1990, with permission from The American Physiological Society.)

either a laser or a white light source (xenon or mercury arc lamp). These two types of confocal microscopes differ in their sensitivity, speed of data acquisition, and wavelength selection. White-light-based systems offer a wide range of excitation wavelength choice; have a rapid rate of image acquisition (30 per second); and are less sensitive than the laser-based systems. The latter have limited excitation wavelength capability and slower image acquisition rates. Most laser-based systems utilize a low-powered air-cooled argon laser which emits at two wavelengths: 488 nm and 514 nm.

In our initial efforts to measure pH confocally, a laser-based system was chosen because of its greater sensitivity. The argon 488-nm line excites the pH-sensitive wavelength of BCECF. However, argon lasers do not emit at the isosbestic wavelength of BCECF. Other pH dyes such as carboxy-SNARF (semi-naphthorhodafluor) and carboxy-SNAFL (seminaphthofluorescein) are excited at a single laser line, 514 nm, and emit at two peak wavelengths whose intensity varies inversely with pH (*BioProbes,* 1991). Given the ability to measure two emission wavelengths using two photomultiplier tubes, the initial logical approach in measuring pH with a laser scanning confocal microscope appeared to be to use carboxySNARF-1. However, studies with the dye revealed an excessive leakage rate. In addition, the pK of the dye is approximately 7.6, making it difficult to measure pH up to 6.5. Therefore, we chose to use BCECF. In order to excite BCECF at its isosbestic wavelength a 15-mW helium–cadmium laser was also coupled to the confocal microscope. By alternating the two excitation sources, pH could be measured in real time.

III. Design of a Dual-Excitation Laser Scanning Confocal Fluorescence Microscope for Measuring pHi with BCECF

In the first design of the instrument, an MRC-500 scanning unit (Bio-Rad) was interfaced with an inverted Nikon Diaphot microscope (Wang and Kurtz, 1990). More recently an MRC-600 scanning unit has replaced the original unit (Fig. 2). The emission port was used to excite and collect emitted fluorescence from the kidney tubule perfused on the microscope stage. A polarized 25-mW argon laser (model 5425A, Ion Laser Technology) and a polarized 15-mW helium–cadmium laser (He–Cd) (model 4214B, Liconix) were coupled to the MRC-600 scanning unit. The two laser beams were combined by reflecting the 442-nm He–Cd laser with a 100% mirror (Oriel) onto a dichroic mirror (Omega Optics) inserted at an angle of 45° in front of the argon laser. The dichroic mirror reflected the 442-nm light and transmitted 488- and 514-nm light to the scanning unit. An electronic shutter (Vincent Assoc.) was placed at an angle in front of each laser to prevent laser light from reflecting backward into the laser cavity. The shutters were opened and closed alternately under computer control. The duration of shutter opening and the time between the opening of one shutter and the closing of the second shutter were software-selectable. In all experiments a 40× fluorite objective was used. This objective has a high optical throughput and minimal longitudinal chromatic aberration (Keller, 1989). The excitation filter cube in the MRC-600 scanner reflected 442- and 448-nm light to the microscope.

Photomultiplier tube no. 1 was used for fluorescence imaging and photomultiplier no. 2 was used for brightfield imaging. A fiber bundle was used to transfer transmitted light to photomultiplier tube no. 2 in the MRC-600 scanner. Software was written (1) to control the timing parameters of the shutters, (2) to digitally store fluorescence images sequentially at 442 nm and 488 nm excitation, and (3) to extract the emission ratio from selected areas of the digitized images retrospectively. A zoom factor of 3 or 4 times was used. Up to 15 excitation ratios from spatially distinct locations in the preparation (xy plane) could be monitored in real time with or without background subtraction.

The optical properties of the initial MRC-500 based system were characterized. To minimize curvature of field and radial chromatic aberration the data were acquired as close to the optical axis of the objective as possible (Wells *et al.*, 1989). A fluorite objective was used to minimize chromatic aberration (Keller, 1989). The z-axis resolution in the reflected-light mode at 488 nm with the pinhole at 0.96 nm was approximately 1 μm. In the fluorescence mode, the excitation and emitted wavelength are not the same. Chromatic aberration prevents the excitation and emission wavelengths from following the same optical path (Wells *et al.*, 1989). The emitted light will not be imaged at the detector pinhole which decreases the z-axis resolution. At 488 nm excitation with the detector pinhole closed maximally, the signal-to-noise ratio from cells loaded with BCECF was too low. Therefore, the pinhole was increased to 1.68 nm. The z-axis resolution at this pinhole size was 1.8 μm at 488 nm excitation and 1.1 μm at 442 nm excitation.

Fig. 2 (a) View of the coupling of a 25-mW argon laser (right) and a 15-mW He–Cd laser (left) to the MRC-600 scanning unit. (Adapted from Kurtz and Emmons, 1993.) (b) Close-up view of the optics in front of each laser (He–Cd laser, right; argon laser, left).

IV. Measurement of pHi in the Cortical Collecting Tubule

As depicted in Fig. 1 the cortical collecting tubule possesses more than one cell type, i.e., principal and intercalated cells. The cylindrical shape of the epithelium and its heterogeneous nature make this preparation difficult to study with conventional epifluorescence methodology. In studies of this preparation, the tubule is cannulated with micropipettes which permits the luminal as well as the basolateral solutions to be changed rapidly. Movement of the preparation is particularly problematic when using a confocal microscope. However, a laminar flow chamber and floating table have ameliorated this problem. We measure pH in single principal and intercalated cells with BCECF. An example of an experiment measuring pHi in a single principal cell is depicted in Fig. 3. Principal cells were found to have a basolateral Na^+/H^+ antiporter, a $Na^+/base$ cotransporter, and a Na^+-independent $Cl^-/base$ exchanger (Wang and Kurtz, 1990). In separate studies, these cells were discovered to have a basolateral Na^+-dependent organic transporter which transported BCECF and was stilbene-inhibitable (Emmons and Kurtz, 1994). More recently, studies from our laboratory have demonstrated that the majority of intercalated cells have both apical and basolateral $Cl^-/base$ exchangers (Emmons and Kurtz, 1994). This finding suggested that the current model of two types of intercalated cells, i.e., α-cells (basolateral $Cl^-/base$ exchange) and β-cells (apical $Cl^-/base$ exchange), needed to be modified. The cells with bilateral $Cl^-/base$ exchangers were called γ cells (Emmons and Kurtz, 1994).

The cortical collecting duct is a major site of NH_3 secretion in the nephron. However, the NH_3 permeability (pNH_3) properties of principal cells and intercalated cells and their contribution to passive NH_3 flux was unknown. Conventional approaches using fluid collection methodology could only estimate the whole tubule transcellular pNH_3 without differentiating the contribution by principal cells and intercalated cells. We determined the basolateral and apical pNH_3 of principal cells and intercalated cells by measuring the changes of pHi with fluorescence confocal microscopy. Pairs of fluorescence images

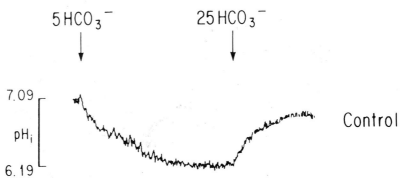

Fig. 3 Measurement of pHi in a single principal cell. Basolateral HCO_3^- was decreased from 25 to 5 mM and increased after several minutes to 25 mM. By measuring the rate of change of pHi under different experimental conditions, the mechanisms of $H^+/base$ transport across the apical and basolateral membrane were determined. (Adapted from Wang and Kurtz, 1990, with permission from The American Physiological Society.)

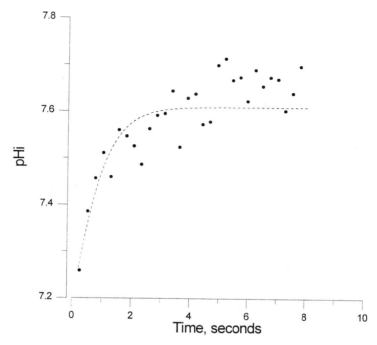

Fig. 4 Time course of changes in pHi of a principal cells when 20 m*M* NH₄Cl was added to the bath perfusate at time = 0. (Adapted from Yip and Kurtz, 1995, with permission from The American Physiological Society.)

(440 nm and 488 run excitation) were sampled and stored at 4 Hz. The time course of changes in pHi of individual cells were extracted retrospectively from the stored images (Fig. 4). The rate of cellular NH_3 influx was calculated from the time course of increase in pHi when the tubules were exposed to 20 m*M* NH₄Cl (Yip and Kurtz, 1995). After correction for membrane folding, the basolateral and apical pNH_3 of principal cells were 5.0 ± 0.7 and 34 ± 3 μm/s, respectively. The basolateral and apical pNH_3 of intercalated cells were 9.0 ± 1.0 and 47 ± 5 μm/s after membrane folding correction. The results demonstrated that the apical surface was more permeable than the basolateral surface in both cell types. In addition, intercalated cells were more permeable to NH_3 than principal cells across both membranes. The data suggested that the contribution of principal cells and intercalated cells to transtubular NH_3 secretion would be dependent, not on the difference in transcellular pNH_3 values, but on the net proton secretory rate of each cell type.

V. Measurement of Intracellular Ca²⁺ in the Perfused Afferent Arteriole

Microperfused afferent arterioles of the glomerulus have cellular heterogeneity in the z-axis dimension, i.e., a single layer of smooth muscle cells surrounding an endothelial

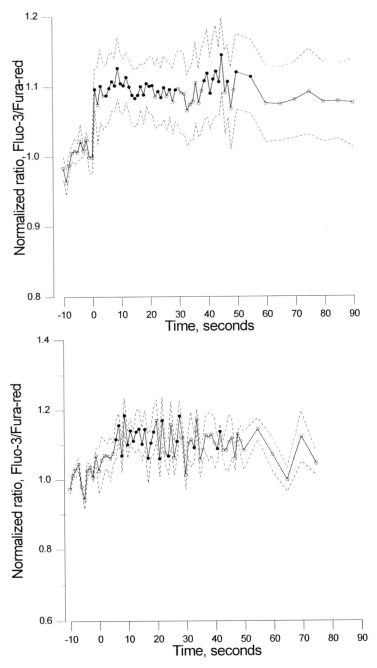

Fig. 5 Normalized time course of changes in Fluo-3/Fura-Red emission ratio of smooth muscle cells (A) and endothelial cells (B). The transmural pressure was stepped up from 80 to 120 mmHg at time 0. Dotted lines are mean ± SEM. ●, significant difference from prepressurized baseline; ○, no difference from prepressurized baseline ($n = 9$, $p < 0.05$). (Adapted from Yip and Marsh, 1996, with permission from The American Physiological Society.)

cell layer, rather than heterogeneity in the xy plane as in the cortical collecting duct. In a separate study we adapted the confocal imaging techniques developed for the perfused renal tubule experiments, to study afferent arterioles. This approach permitted the differentiation of fluorescence signals arising from vascular smooth muscle cells and endothelial cells. We determined the temporal changes of intracellular Ca^{2+} concentration ($[Ca^{2+}]i$) in smooth muscle cells and endothelial cells during myogenic constriction, triggered by a step increase in the transmural pressure. $[Ca^{2+}]i$ was measured by the emission ratio generated with a mixture of calcium indicator dyes Fluo-3 and Fura-Red, a new method introduced for use in confocal microscopy using an argon laser (Lipp and Niggli, 1993). When excited at 488 nm, Fluo-3 exhibits an increase in green fluorescence (525 nm) upon Ca^{2+} binding, while Fura-Red shows a decrease in red fluorescence (640 nm). The emission ratio of Fluo-3/Fura-Red was then can be used to monitor the changes in $[Ca^{2+}]i$. The emission ratio is independent of the changes of local dye concentration that occur during contraction and relaxation of the vessel. Both indicators were simultaneously loaded into the endothelium via a luminal perfusate, or into the vascular smooth muscle layer via the bathing solution. Confocal fluorescence images were acquired at the mid-plane of the perfused arterioles, so that the cross section of the endothelial cells and vascular smooth mucle cells could be differentiated. Because of the symmetrical geometry of the arteriole, passive dilation and subsequent constriction during the myogenic response did not change the position of the middle plane which permited scanning of the same area throughout the experiment. Local lateral dislocation of the sampling area due to the response of the vessel could be circumvented by redefining the sampling area in each image during the retrospective analysis of the acquired images. The time course of changes in the Fluo-3/Fura-Red ratio in smooth muscle and endothelium in response to a step increase in perfusion pressure is shown in Fig. 5. It had been suggested that the myogenic constriction of arterioles was due to an increase of $[Ca^{2+}]i$ in smooth muscle, plus a decrease in the production of nitric oxide from endothelium due to a decrease of $[Ca^{2+}]i$ in the endothelial cells (Yip and Marsh, 1996). The results of these studies indicated that myogenic constriction in afferent arterioles was associated with an increase of $[Ca^{2+}]i$ in vascular smooth muscle cells but did not require a simultaneous decrease of $[Ca^{2+}]i$ in endothelial cells to decreased production of nitric oxide from endothelium.

VI. Future Development

The development of confocal laser scanning microscopes designed for indicators with ultraviolet excitation wavelengths has made it possible to measure $[Ca^{2+}]i$ with indo-1 (Niggli et al., 1994). The resolution of laser scanning microscopes has been improved with the development of nonlinear laser scanning microscopes (Williams et al., 1994; Xu et al., 1996; Maiti et al., 1997). These microscopes are capable of two-photon and three-photon excitation allowing fluorescence dyes which normally require UV excitation to be excited with visible wavelengths. The elimination of UV excitation can reduce cellular damage in living tissue. Nonlinear excitation provides superior three-dimensional resolution for imaging and avoids out-of-focus information. This technique has been used to image $[Ca^{2+}]i$, NADH, and serotonin in living cells (Williams *et al.,* 1994; Xu *et al.,* 1996; Maiti *et al.,* 1997).

An important advance has been the development of near-field confocal microscopy with photon density feedback which permits diffraction-limited studies of living cells (Haydon *et al.,* 1996). By integrating wide-field fluorescence microscopy with confocal microscopy and near field microscopy, it is possible to resolve fluorescent pH and Ca^{2+} changes beneath the cell membrane. It should also be possible to detect, with a finer level of resolution than was previously possible, the behavior of fluorescently tagged molecules in the cell membrane.

Acknowledgment

This work was supported in part by NIH grants DK41212 and 851 IG-4, the Iris and B. Gerald Cantor Foundation, the Max Factor Family Foundation, the Verna Harah Foundation, the Richard and Hinda Rosenthal Foundation, and the Fredricka Taubitz Foundation. Dr. Kurtz is an Established Investigator of the American Heart Association.

References

BioProbes (1991). Molecular Probes Inc., Eugene, OR.

Emmons, C., and Kurtz, I. (1994). Functional characterization of three intercalated cell subtypes in the rabbit outer cortical collecting duct. *J. Clin. Invest.* **93,** 417–423.

Haydon, P. G., Marchese Ragona, S., Basarsky, T., Szulczewski, M., and McCloskey, M. (1996). Near-field confocal optical spectroscopy (NCOS): subdiffraction optical resolution for biological systems. *J. Microsc.* **182,** 208–216.

Keller, H. E. (1989). In "The Handbook of Biological Confocal Microscope" (J. Pawley, Ed.), pp. 69–77. IMR, Madison, WI.

Kurtz, I., and Emmons, C. (1993). Measurement of intracellular pH with a laser scanning confocal microscope. *Methods Cell. Biol.* **38,** 183–193.

Lipp, P., and Niggli, E. (1993). Ratiometric confocal Ca^{2+}-measurements with visible wavelength indicators in isolated cardiac myocytes. *Cell Calcium* **14,** 359–372.

Maiti, S., Shear, J. B., Williams, R. M., Zipfel, W. R., and Webb, W. W. (1997). Measuring serotonin distribution in live cells with three-photon excitation. *Science* **275,** 530–532.

Niggli, E, Piston, D. W., Kirby, M. S., Cheng, H., Sandison, D. R., Webb, W. W., and Lederer, W. J. (1994). A confocal laser scanning microscope designed for indicators with ultraviolet excitation wavelengths. *Am. J. Physiol.* **266 (1 Pt 1),** C303–C310.

Rink, T. J., Tsien, R. Y., and Pozzan, T. (1982). Cytoplasmic pH and free Mg^{2+} in lymphocytes. *J. Cell Biol.* **95,** 189–196.

Tanasugarn, L., McNeil, P., Reynolds, G. T., and Taylor, D. L. (1984). Microspectrofluorometry by digital image processing: measurement of cytoplasmic pH. *J. Cell Biol.* **98,** 717–724.

Thomas, J. A., Buchsbaum, R. N., Zimniak, A., and Racker, E. (1979). Intracellular pH measurements in Ehrlich ascites tumor cells utilizing spectroscopic probes generated in situ. *Biochemistry* **18,** 2210–2218.

Wang, X., and Kurtz, I. (1990). H^+/base transport in principal cells characterized by confocal fluorescence imaging. *Am. J. Physiol.* **259,** C365–C373.

Wells, K. S., Sandison, D. R., Strickler, J., and Webb, W. W. (1989). In "The Handbook of Biological Confocal Microscope" (J. Pawley, Ed.), pp. 23–35. IMR, Madison, WI.

Williams, R. M., Piston, D. W., and Webb, W. W. (1994). Two-photon molecular excitation provides intrinsic 3-dimensional resolution for laser-based microscopy and microphotochemistry. *FASEB J.* **11,** 804–813.

Xu, C., Zipfel, W., Shear, J. B., Williams, R. M., and Webb, W. W. (1996). Multiphoton fluorescence excitation: new spectral windows for biological nonlinear microscopy. *Proc. Natl. Acad. Sci. USA* **93,** 10763–10768.

Yip, K.-P., and Kurtz, I. (1995). NH, permeability of principal cells and intercalated cells measured by confocal fluorescence imaging. *Am. J. Physiol.* **269,** F545–F550.

Yip, K.-P., and Marsh, D. J. (1996). [Ca^{2+}]i in rat afferent arteriole during constriction measured with confocal fluorescence microscopy. *Am. J. Physiol.* **271,** F1004–F1011.

Zimmer, F. J., Dryer, C., and Hausen, P. (1988). The function of the nuclear envelope in nuclear protein accumulation. *J. Cell Biol.* **106,** 1435–1444.

CHAPTER 12

Confocal and Nonlinear Optical Imaging of Potentiometric Dyes

Leslie M. Loew, * **Paul Campagnola,** * **Aaron Lewis,** †
and Joseph P. Wuskell *

* Department of Physiology
Center for Biomedical Imaging Technology
University of Connecticut Health Center
Farmington, Connecticut 06030

† Division of Applied Physics
Hebrew University
Jerusalem, Israel

I. Introduction

This laboratory has been engaged in the design and synthesis of voltage-sensitive dyes for over 20 years (Loew *et al.,* 1978, 1979a,b). Several important general-purpose dyes have emerged from this effort including di-5-ASP (Loew *et al.,* 1979a), di-4-ANEPPS (Fluhler *et al.,* 1985; Loew *et al.,* 1992), di-8-ANEPPS (Bedlack *et al.,* 1992; Loew, 1994), and TMRM and TMRE (Ehrenberg *et al.,* 1988). Potentiometric dyes are designed either to measure membrane potentials in cell populations with a macroscopic system such as a spectrofluorometer or to be used in conjunction with a microscope to measure voltages associated with individual cells or organelles. The latter kinds of applications are the primary interest of this laboratory and have directed the dye chemistry development. Such experiments have been of great utility to neuroscientists interested in mapping patterns of electrical activity in complex neuronal preparations with numerous examples spanning the past 15 years (recent examples: Wu *et al.,* 1994; Prechtl *et al.,* 1997; Obaid *et al.,* 1999; Tsodyks *et al.,* 1999). In addition, the dyes have been used to map the spatial (Gross *et al.,* 1985; Bedlack *et al.,* 1992, 1994) and temporal (Shrager and Rubinstein, 1990; Zecevic, 1996) patterns of electical activity along single cell membranes and have measured potentials in fine processes and at synapses (Salzberg, 1989). We have even been able to measure the membrane potentials across the inner membranes of individual mitochondria within a single living cell (Loew *et al.,* 1993; Fink *et al.,* 1998). All of these experiments could not be accomplished with conventional electrical measurements using microelectrode or patch-clamp techniques.

The application of confocal microscopy to such measurements can significantly extend our ability to map the electrical properties of cell and organelle membranes with high spatial resolution. In addition to the obvious enhancement of resolution, especially along the optical axis, confocal microscopy offers opportunities for quantitative measurements of fluorescence intensities at each point in an image. Methods for calibrating such measurements and transforming them into spatial profiles of membrane potential will be described in this chapter. We will also review some new techniques for using nonlinear optical microscopy to probe for membrane potential in living cells. Specifically, we have developed a new nonlinear optical modality, second harmonic imaging microscopy (SHIM), that can deliver signals that are highly sensitive to membrane potential.

II. Overview of Fluorescent Methods for Measuring Membrane Potential

A. Fast Membrane Staining Potentiometric Indicators

The use of fluorescent dyes to measure membrane potential was pioneered by Cohen and his co-workers (Cohen *et al.,* 1974; Ross *et al.,* 1977; Gupta *et al.,* 1981) in an effort to develop methods for mapping activity in complex neuronal systems. Naturally, the indicators were required to respond rapidly in order to monitor the rapid voltage changes associated with action potentials. The fast potentiometric indicators are generally membrane stains that respond to changes in the electric field within the membrane via subtle conformational or electronic rearrangements. The fluorescence changes associated with an action potential correspond to 10–20% of the resting fluorescence for the best (i.e., most sensitive) of these indicators. This low sensitivity, together with the need for high enough fluorescence intensities to give good signals in the millisecond time range, have dictated that electrical activity in excitable systems could be only measured with low spatial resolution.

Until very recently only our laboratory (Loew *et al.,* 1979b; Loew and Simpson, 1981; Hassner *et al.,* 1984; Fluhler *et al.,* 1985; Tsau *et al.,* 1996) and that of Amiram Grinvald (Grinvald *et al.,* 1982, 1994; Shoham *et al.,* 1999) had been actively working on new potentiometric dye design and synthesis. The styryl class of potentiometric dyes emerged from this effort. Evidence has accumulated that styryl dyes can often operate by an electrochromic mechanism, where the energy of the separation between the ground and excited state is directly modulated by the intramembrane electric field (Loew *et al.,* 1979a; Loew and Simpson, 1981). In other words, a change in membrane potential causes a shift in the excitation or emission spectrum of the dye, permitting dual-wavelength-ratio imaging (Montana *et al.,* 1989; Bullen and Saggau, 1999). In 1995, the laboratory of Roger Tsien joined the effort to devise new potentiometric indicators (Gonzalez and Tsien, 1995, 1997). They developed a scheme based on the change in fluorescence resonance energy transfer when an anionic acceptor dye undergoes potential-dependent redistribution across a membrane and thereby changes its proximity to a donor that is fixed to the outer membrane surface. The sensitivity is high—50%/100 mV was reported—and the signals are sufficiently rapid to accurately track action potentials. However, large dye concentrations are required to achieve sufficient energy transfer efficiencies, risking dye toxic and photodynamic effects; also, the application of this dye pair technology is cumbersome, and to date only one example of the use of this idea for complex multicellular preparations has appeared (Cacciatore *et al.,* 1999). Still, this approach has great potential especially if both the donor and acceptor chromophores can be engineered into the same molecule. Another exciting new approach incorporated green fluorescent protein into the shaker potassium channel (Siegel and Isacoff, 1997). This construct can be expressed in cells and shows a 5% fluorescence change per 100-mV change in potential. This approaches the sensitivity of the existing organic dyes. Unfortunately, the time course does not reflect a response of the chromophore to the fast conformational change associated with the gating charges within the protein—it is much slower. However, this approach holds great promise because it is selective and noninvasive and because continued development could produce improved constructs with greater sensitivity and speed.

To date none of these fast membrane staining dyes have been used in conjunction with confocal imaging. The higher resolution available with confocal microscopy comes at the expense of lower fluorescence intensities. Since the fast dyes generally give relatively small changes in fluorescence in response to physiological changes of potential, long integration times will be required in confocal microscopy to achieve the requisite precision. Thus confocal microscopy of fast events such as action potentials is practical only with line scan or single-point measurements. On the other hand, static maps of membrane potential profiles could be facilitated with confocal imaging, especially now that dual-emission-ratio approaches have been described with potentiometric dyes (Beach *et al.,* 1996; Gonzalez and Tsien, 1997; Bullen and Saggau, 1999). Another area that awaits exploration is the use of potentiometric dyes in two-photon microscopy. Since the excited state that is reached by two-photon excitation is indistinguishable from that produced by absorption of a single photon, there is every reason to believe that the potential sensitivity of the emission will be the same in both cases. Evidence from our lab indicates that the styryl dyes have excellent two-photon cross sections, so two-photon microscopy should be suitable for membrane potential measurements. In Section III, we will describe a new nonlinear optical method, second harmonic imaging microscopy (SHIM), that provides the resolution of confocal or two-photon microscopy, but with a much higher sensitivity to membrane potential.

B. Slow Redistribution Potentiometric Indicators

A second class of dyes are membrane-permeant ions that equilibrate across membranes according to the Nernst equation. For a cationic dye:

$$\Delta V = -60 \log \left(\frac{[\text{dye}]_{\text{in}}}{[\text{dye}]_{\text{out}}} \right) \text{mV}. \tag{1}$$

Thus, the potential difference across the membrane, ΔV, drives an uneven distribution of dye between the cell interior and the extracellular medium. Examples of potentiometric dyes that employ a redistribution mechanism are many of the cyanines (Sims *et al.*, 1974; Reers *et al.*, 1991), oxonols (Smith *et al.*, 1976; Rink *et al.*, 1980), and rhodamines (Johnson *et al.*, 1980). It is important to note that [dye] in Eq. (1) refers to free monomeric aqueous dye. Often, the dye distribution deviates significantly from that predicted by the Nernst equation because of binding to the plasma and organelle membranes and the tendency of these compounds to form aggregates when their concentrations exceed a threshold. This is especially true of the cyanine class of dyes. Indeed, it is these features of the chemistry of the cyanines which make their spectral properties so sensitive to potential: membrane-bound dye displays enhanced fluorescence while dye aggregates have low fluorescence quantum yields. The cyanine dyes can therefore have complex spectral response to potential which depends strongly on the cell:dye ratio, the hydrophobicity of the dye, and the particular cell type. Binding to mitochondria and possible responses to mitochondrial potential changes add further complications. Still, with careful calibration protocols, these indicators have been extremely successful for studies of bulk cell populations or flow cytometry and remain the best choice for such applications. These dyes have only rarely been used for microphotometric measurements of membrane potential in single cells. An exception is the cyanine dye JC-1 (Reers *et al.*, 1991), developed for qualitative imaging of mitochondrial membrane potential.

Work in this laboratory has aimed at extending the use of permeant cationic dyes to permit quantitative imaging of membrane potential in individual cells (Ehrenberg *et al.*, 1988; Farkas *et al.*, 1989; Loew *et al.*, 1993; Loew, 1998). The idea was simply to substitute fluorescence intensities for dye concentrations in the Nernst equation to calculate the membrane potential. It became necessary, therefore, to find a dye whose fluorescence intensity both inside and outside the cell is proportional to its concentration. Thus, the cyanines which have the tendency to form nonfluorescent aggregates are generally unsuitable. Similarly, dyes that bind strongly to membranes would distribute in a more complex manner the prediction of the Nernst equation. Also, it would be desirable for the Nernstian equilibrium to be relatively rapid and fully reversible, to permit the measurement of dynamic changes at least on the time scale of seconds. Several commercially available delocalized cationic dyes were screened, including additional members of the rhodamine and cyanine class, in an effort to identify suitable "Nernstian" dyes for imaging of membrane potential (Ehrenberg *et al.*, 1987, 1988). Although none of these dyes were ideal, the data from these experiments permitted us to design and synthesize a pair of new dyes, TMRM and TMRE, which closely meet the requirements for Nernstian distributions.

Properties of Selected Potentiometric Dyes

Name	Description	ABS (nm)	EM (nm)	Structure
Di-4-ANEPPS	Versatile fast styryl dye. Most common heart dye. Electrochromic mechanism. Dual wavelength ratio is an option. t: <µs	502	723	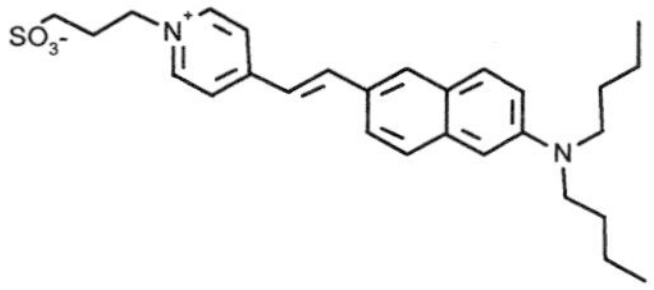
JPW-1114	Highly water soluble. Best potentiometric dye for microinjection. Also good tissue penetration when applied externally, but may wash out quickly.	529	725	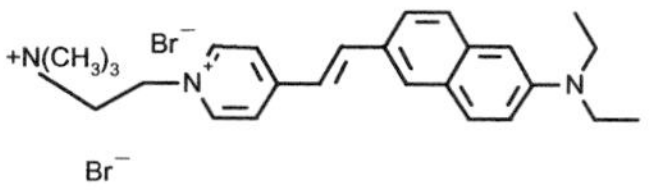
TMRE	Rhodamine redistribution dye for single cell and mitochondrial potential measurements. t: s - min	548	597	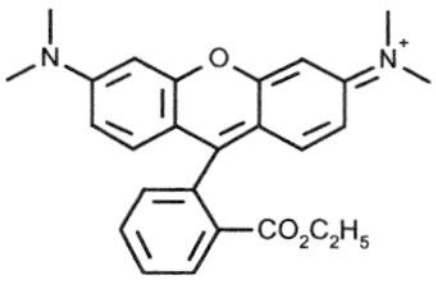
Di-8-ANEPPS	More slowly internalized than di-4-ANEPPS. Relatively insoluble in water and therefore requires Pluronic F127 to stain.	500	705	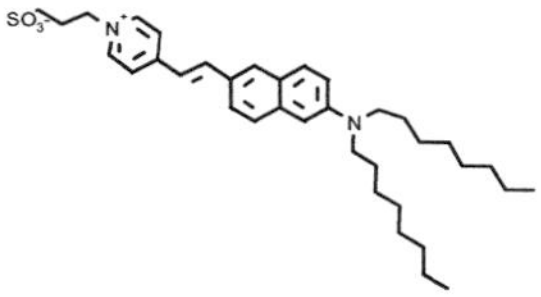
JPW 1259	Chiral ribose group enhances SHG in oriented assemblies and membranes.	502	712	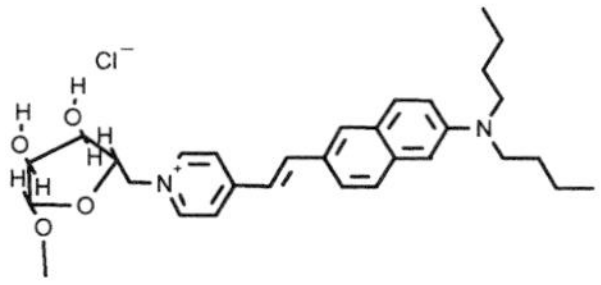
JPW-2045	Potentiometric dye for neuronal tracing and long-term studies. Highly persistent over the many hours required for retrograde transport.	513	717	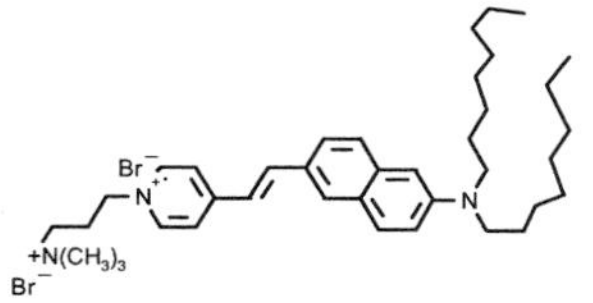
JPW-2066	Di-linoleyl-ANEPPS is oil soluble. An oil solution may be microinjected to stain the endoplasmic reticulum or other selective staining.	497	707	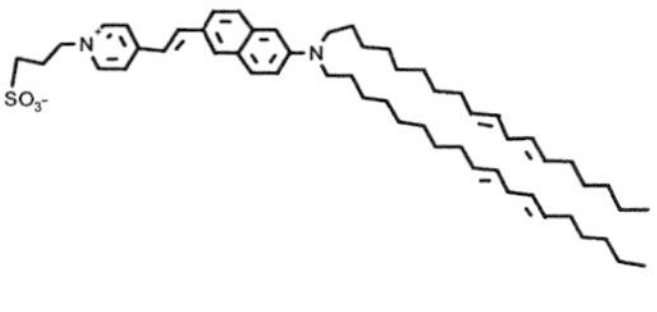

Fig. 1 Structures and properties of some popular potentiometric dyes. TMRE is an example of a good Nernstian distribution dye. JPW1259 is designed for SHIM.

Both of these dyes are members of the rhodamine class and are in fact closely related to rhodamine 123, a popular mitochondrial stain (Johnson *et al.,* 1980). They are more hydrophobic than the latter and do not contain any hydrogen bond donating groups; we believe these differences are what permit a reversible Nernstian equilibrium to be readily established with TMRE and TMRM. The time for equilibration varies somewhat with cell type over a range of ca 10 s to 3 min; equilibration is significantly faster at 37°C than at room temperature. Equilibration across mitochondrial membranes is generally much faster than across the plasma membrane of cells, possibly because of the smaller volume of this organelle; in three cell lines that were treated with high concentration of a mitochondrial uncoupler, the dyes indicate complete depolarization as fast as could be measured (<5 s). Because of their high brightness, the dyes can be used at low enough concentrations so that dye aggregation is insignificant—even in the mitochondria where dye can be concentrated up to 10,000-fold. While we have found that 50 n*M* dye is appropriate in some cells, this issue needs to be examined in the particular cell line being studied as there have been reports of significant dye aggregation and self-quenching even at this low concentration (Huser *et al.,* 1998; Scaduto and Grotyohann, 1999). The dyes are quite resistant to photobleaching and photodynamic effects (Farkas *et al.,* 1989; Scaduto and Grotyohann, 1999) but should still be used with the minimum possible light exposure. The degree of background (i.e., nonpotentiometric) binding to intracellular components is relatively low compared to other delocalized cations. Corrections for background binding must be made, however, in order to obtain accurate membrane potentials; procedures will be described later.

Structures of some of the most important potentiometric fluorescent dyes are given in Fig. 1 together with brief descriptive data. This figure serves to summarize how each class of dye can be utilized for different types of measurements.

III. Application of Laser Scanning Microscopy to Quantitative Imaging of Membrane Potential

A. Nernstian Redistribution of Permeant Cationic Dyes

1. Plasma Membrane Potential

The physical, chemical, and spectral properties of TMRE and TMRM make them appropriate dyes for microphotometric determination of membrane potential via a simple variant of the Nernst equation:

$$\Delta V = -60 \log \left(\frac{F_{in}}{F_{out}} \right) mV \tag{2}$$

The ratio of the fluorescence intensities, F_{in}/F_{out}, is taken to be equal to the ratio of the free cytosolic to extracellular dye concentrations used in the Nernst equation [Eq. (1)]. As described above, this condition is met by TMRE or TMRM after a correction for a small amount of background binding. However, in practice, limitations of the optics, rather than the dye chemistry, have restricted how readily the approach can be used. Specifically, it is difficult to determine fluorescence from a compartment as small as a cell and at the same time measure fluorescence from an equal volume in the extracellular medium. This is because a wide-field microscope will tend to dilute the in-focus intracellular fluorescence with the lower fluorescence from above or below the cell; the fluorescence from a point to the side of the cell is not so diluted because the concentration of dye is uniform in the extracellular medium at all points within the depth of the specimen. To fully appreciate this critical point it may be useful to consider an analogy with small and large cuvettes in a conventional fluorometer. If one wishes to compare the concentrations of a fluorophore in two different solutions with a fluorometer, it is necessary to measure fluorescence from solutions in cuvettes with identical dimensions. A large cuvette, with a longer path for the exciting light, will produce more fluorescence than a small cuvette for a given fluorophore concentration. In the fluorescence microscope, attempting to compare fluorescence from inside a cell to fluorescence outside is equivalent to using small and large cuvettes. Therefore, F_{in}/F_{out} may be severely underestimated. On the other hand, in trying to quantitate fluorescence from the cytosol, the presence of mitochondria above or below the plane of focus can lead to artifactually high F_{in}; this is because the high mitochondrial potential can lead to intramitochondrial dye concentrations 1000 times higher than that in the cytosol. These problems can be addressed by using small field and measuring apertures and devising complex correction procedures with a microphotometer (Ehrenberg *et al.*, 1988).

A microphotometer is capable of only single spot measurements, however. To obtain spatial information with a wide-field microscope, one must be content to monitor qualitative changes in potential whether from mitochondrial or plasma membrane (Farkas *et al.*, 1989). On the other hand, the confocal microscope offers a nearly ideal solution to this problem. The most distinctive feature of the confocal microscope—its ability to reject light originating from outside the plane of focus—is precisely what is required to measure fluorescence quantitatively from a compartment as small as the cell. This property is well demonstrated in Fig. 2A, which shows the exclusion of fluoresceinated dextran from a fibroblast. The dextran is used to mark the extracellular bathing medium surrounding the cell. With the focus centered 2 μm below the coverslip to which the cell adheres, almost no fluorescence can be measured intracellularly while extracellular fluorescence at the same focus nearly saturates the digitizer circuit. The chamber to which the coverslip is attached provides a 200-μm blanket of medium bathing the cells. A comparable image from a wide-field microscope shows almost no contrast between the intracellular and extracellular space. Effectively, the narrow depth of focus of the confocal microscope defines a "cuvette" smaller than the intracellular compartment so that equivalent volumes of dye-containing solution are sampled outside and inside the cell. Figure 2B shows the distribution of fluorescence after TMRE is allowed to equilibrate with the same cell. It is clear from the result of Fig. 2A that it is possible to measure F_{in}/F_{out} in Fig. 2B with confidence. In practice, to avoid contaminating fluorescence from out-of-focus mitochondria, F_{in} can be measured in a region that is known to be free of mitochondria, such as the nucleus.

If a region of the cell has a thickness that approaches the axial resolution of the confocal microscope, e.g., a lamellipodium, a thin neurite, or individual mitochondria, the confocal optical slice will include fluorescence from outside the pertinent region, again precluding direct application of Eq. (2). However, we have developed correction

A   B

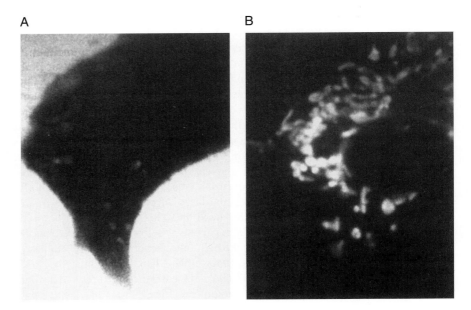

Fig. 2 Confocal images of NIH-3T3 cells. A Biorad MRC-600 laser confocal microscope was used with a Zeiss Axioskop microscope and a Nikon 60× NA 1.4 plan apochromat objective; the slow scan mode was employed with the confocal pinhole set at position 2. (A) Confocal image of a cell bathed in 4 mg/ml FITC-dextran (MW 45 kDa) which marks the extracellular medium but does not penetrate the cell. A fluorescein filter set was employed to demonstrate the ability of the confocal microscope to exclude extracellular fluorescence from an optical section centered on the middle of the cell. The intensity measured from the center of the cell is at least a factor of 10 lower than the extracellular fluorecence. This image was obtained after a series of measurements with TMRE. (B) The same cell as in (A) showing the distribution of fluorescence from TMRE. The optical setup was the same as in (A) except that a rhodamine filter set was employed. The cells were equilibrated with 0.1 μM TMRE at 37°C for 5 min and a series of 30 laser scans were accumulated into a 16-bit image. A no-light background image was subtracted and the resultant image is displayed with a linear gray scale. Punctate fluorescence, corresponding to mitochondria, is prominent; but, with this linear scale, the intensities from the cytosol, nucleus, and extracellular medium are barely perceptible. The fluorescence from these latter regions can be measured, however, by expansion of the intensity scale. To assess plasma membrane potential, the appropriate intensities are then inserted into Eq. (2).

procedures to overcome even this limitation if the geometry of the subresolution region can be approximated in a solid computational model. This approach will be illustrated for the estimation of mitochondrial membrane potential in the following section.

2. Membrane Potential of Single Mitochondria

Since mitochondria are generally below the resolution limit of a confocal microscope, the simple measurements described above will not suffice. To try to address this problem and achieve the highest possible resolution, we initially developed a procedure based on digital restoration of wide-field images (Loew *et al.,* 1993). Our strategy was to use a wide-field microscope to image a series of planes through the thickness of a cell exposed to TMRE. This 3-D image was subjected to the deconvolution algorithm developed by Carrington *et al.* (Carrington and Fogarty, 1987; Fay *et al.,* 1989; Carrington, 1990; Carrington *et al.,* 1990) in order to restore the out-of-focus light to its proper point of origin. However, the axial resolution of even the best microscope optics is never better than about 500 nm—above the diameter of typical mitochondria. So an additional procedure was developed to assess the error inherent in the measurement and thereby enable a calibration of the ratio of fluorescence between the mitochondria and the cytosol. The method used a 3-D computer model of a mitochondrion based on dimensions derived from electron micrographs. The model was first blurred with the point-spread function (PSF) of the microscope and then subjected to the restoration algorithm. The intensity in the resulting image was compared to the intensity of the original model to generate the appropriate correction factor.

Confocal microscopy of TMRE distribution allowed us to simplify this procedure by eliminating the need for 3-D restoration (Fink *et al.,* 1998). We still needed to produce 3-D computational models of mitochondria, however, in order to determine the correction factor to be applied to the mitochondrial intensities derived from the confocal images. Therefore, we analyzed electron micrographs of mitochondria in NIH 3T3 fibroblasts (Fig. 3a) and found that they were curving cylinders with quite uniform diameters averaging 350 nm, but highly variable lengths ranging from 1 to 20 μm. The cylindrical axis was generally parallel to the plane of the substrate. Since the overall lengths of the cylinders were well above the xy resolution (ca 300 nm) of the confocal microscope, a cylinder with a length of 1 μm and diameter 350 nm was chosen as an appropriate model for these mitochondria. The PSF was experimentally determined by acquiring a 3-D image of a 100-nm fluorescent microsphere (Molecular Probes, Inc., Eugene, OR) using the same confocal optics and parameters as used for TMRE imaging; in fact, the PSF is usually acquired from microspheres that are added to the same coverslip that holds the TMRE-stained cells. Convolution of the PSF with the model is achieved in 3-D either directly in the spatial domain or, in the case of very large image datasets, in the frequency domain via Fourier transfom. The PSF and the voxelized version of the cylindrical mitochondrial model are shown in Fig. 3b along with the results of the convolution. The original intensity of 10,000 units is substantially diluted with a maximal intensity in the center of the blurred image of 1300 units; the ratio of the original intensity to the maximum of the blurred intensity provides the factor of 7.6 that is used to correct the mitochondrial intensity. In Fig. 4, this correction is applied to one mitochondrion; the corrected intensity is divided by the intensity measured in an adjacent region of cytosol and plugged into the Nernst equation to determine the membrane potential. It is important to accurately determine the maximum intensity within the three-dimensional mitochondrion, so a number of planes must be sampled to be sure that the mitochondrion is correctly bracketed within the volume.

3. Methods

The primary quantity to be measured is F_{in}/F_{out} [Eq. (2)]. After equilibration with TMRE or TMRM, the intensities of light associated with the cytosol and an adjacent

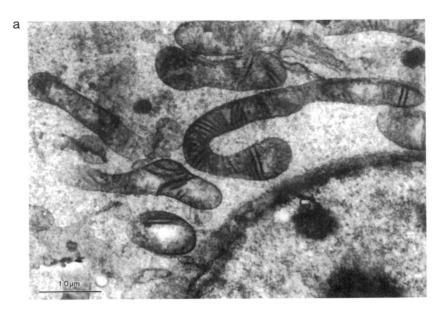

NIH 3T3 Cell
Model Mitochondrion

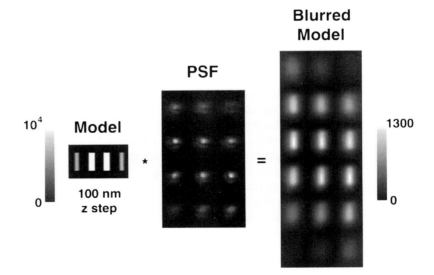

Fig. 3 Model mitochondria from analysis of electron micrographs. (a) To develop a model for a mitochondrion based on accurate dimensions, we analyzed electron micrographs like the one illustrated here. Sections cut parallel to the substrate to encompass the thickness of the mitochondria were analyzed and the mitochondria

extracellular region (plasma membrane potential) or mitochondrion and an adjacent region of cytosol (mitochondrial membrane potential) can be obtained directly from a digital confocal image. These intensities need to be corrected for dark current and stray light. Any potential-independent binding of dye to the cell must be determined and the measured intensities appropriately adjusted. If potentials in mitochondria or thin cellular processes are to be determined, corrections for convolutional blurring should be established and applied.

For the measurement to be meaningful, it is important to ensure that the cells are maintained in physiological conditions during the experiment. Temperature, pH, CO_2, and oxygen tensions should all be controlled. Practical considerations may limit how well this ideal can be met. Also, even though TMRE and TMRM are relatively benign fluorophores, the high concentrations that are attained in mitochondria can lead to phototoxic effects. These are generally thought to arise from photosensitization of singlet oxygen production and can be especially severe with the high intensity of the exciting laser light in confocal microscopes (Tsien and Waggoner, 1990).

The choice of parameters for image acquisition necessitates a decision pitting the four issues of spatial resolution, precision, minimal light exposure, and speed against each other. Each of these may be of critical importance depending on specific experimental requirements. In all cases, an objective with the highest possible numerical aperture should be used to minimize the depth of focus and achieve the greatest brightness. We have found that any one of the various $60\times$ or $63\times$ plan apochromat lenses with a numerical aperture of 1.4 provides excellent results in almost any situation. With such a lens, offered by many of the manufacturers of research microscopes, the size of the field of view can be adjusted over a range encompassing most animal cells by controlling the extent of the laser scan.

The measurement of plasma membrane potential can be achieved with rapid laser scanning over a low magnification field of view. Since the potential is rarely over $-80\,mV$, both the cytosolic fluorescence and the extracellular fluorescence can be determined within the normal 256 gray levels of an 8-bit confocal image. Care should be taken that the optical slice is centered around the middle of the cell. F_{in} and F_{out} may be determined as an average over many pixels within the respective compartments to increase the precision of the measurement. In many cells, it is impossible to find a region of cytosol that is free of contaminating fluorescence from the very bright and abundant mitochondria, even at a minimum setting of the confocal pinhole. In such cases we resort to measurement of fluorescence from the nucleus on the assumption that the large nuclear membrane pores

---

were found to be cylindrical with varying lengths but relatively uniform diameters of 350 ± 65 (SD) nm. (b) The model, based on the electron microscopy, is a cylinder with a length of 1000 nm and a diameter of 350 nm. It is sampled in a 3-D grid consistent with the pixel resolution employed in the confocal microscope—100 nm in z and 50 nm in x and y. The point-spread function (PSF) is a 3-D image of a 100-nm fluorescent bead added to the same coverslip holding the NIH 3T3 cells. The original intensity in the model is arbitrarily assigned to be 10,000 units. After convolution with the PSF, the blurred model display a maximum intensity of 1300 units. The ratio 10000/1300 provides the calibration factor of 7.6 used to correct F_{in} in Eq. (2). Taken from Fink *et al.* (1998), with permission from the Biophysical Society.

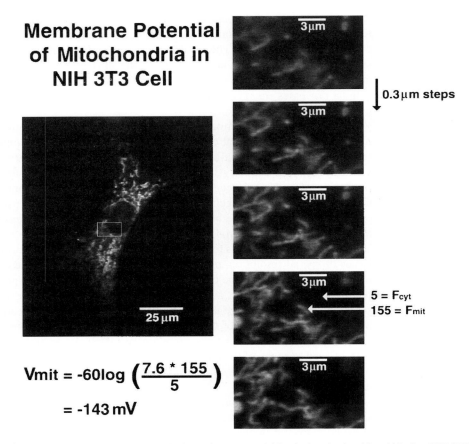

Fig. 4 Determination of mitochondrial membrane potential in single mitochondria within live NIH 3T3 fibroblasts by confocal microscopy. A Zeiss LSM410 confocal microscope equipped with a 63× NA 1.4 plan apochromat objective was used to measure TMRE fluorescence intensities from the cytosol, F_{cyt}, and the mitochondria, F_{mit}. Cells were treated with 100 nM dye and maintained on the stage at 37°. A cell is shown on the left with the indicated rectangular region zoomed by a factor of 8 on the right. A set of five focal planes were acquired at 300-nm steps to ensure that the mitochondria to be analyzed were bracketed within the imaged volume. A membrane potential calculation for a typical mitochondrion is shown using the calibration factor generated from the blurred model of Fig. 3b. Taken from Fink *et al.* (1998), with permission from the Biophysical Society.

preclude a potential difference between cytosol and nucleoplasm (this assumption may not be always valid; Mazzanti *et al.,* 1990).

The dynamic range and resolution requirements are significantly more stringent if mitochondrial membrane potential is the subject of study. With a potential as high as −180 mV (Rottenberg, 1986; Brand and Murphy, 1987), the concentration of dye, and therefore the fluorescence intensity, may be three orders of magnitude higher in a mitochondrion than in the cytosol. Both intensities must be obtained to calculate membrane potential from Eq. (2). This may be achieved by measuring the mitochondrial potential with a strongly absorbing calibrated neutral density filter in the laser light path; this filter would then be removed to permit measurement of cytosolic (or intranuclear) fluorescence. Alternatively, data may be acquired into a 16-bit image to provide sufficient dynamic range for both compartments within a single frame. Also, the small size of mitochondria, typically 0.25–2 μm in diameter, require a high pixel density and a minimum confocal pinhole for maximum 3-D resolution. It is important to be sure that the intensities are measured from mitochondria which are well centered in the plane of focus; a mitochondrion that is slightly off focus will display artificially low intensities. The acquisition of a limited 3-D dataset consisting of a series of about five focal planes obviates this problem by delineating the disposition of mitochondria along the z (i.e., optical) axis. These requirements force multiple exposure of the cell to the laser scan and necessitate long acquisition times. Therefore, phototoxicity, photobleaching, or motion artifacts may become limiting factors in the success of mitochondrial membrane potential measurements.

To determine the level of nonpotentiometric binding, the potential across plasma and mitochondrial membranes can be set to 0 by treatment of the preparation with 1 μM valinomycin in a high-potassium medium. Valinomycin is a K^+-selective ionophore which clamps the membrane potential to the potassium equilibrium potential. It will depolarize the mitochondria, which have membranes with normally low potassium permeability. It will also depolarize the plasma membrane if the extracellular medium has a K^+ concentration comparable to the cytosolic K^+. If there is residual fluorescence, it defines the level of nonpotentiometric binding and can be used as an additional correction to F_{in}/F_{out} in Eq. (2). Also, to accurately quantitate membrane potential in single mitochondria, a correction factor to F_{in}/F_{out} should be determined to account for the residual convolutional blurring of the mitochondrial intensities in the confocal image, as described in Section III.A.2. However, because of the logarithmic form of Eq. (2), *changes* in potential can be determined without such corrections. The ability of the confocal microscope to resolve single mitochondria is the key feature that allows dynamic studies of mitochondrial physiology.

B. Nonlinear Optical Imaging of Membrane Potential

1. Principles of Two-Photon Excited Fluorescence and Second Harmonic Generation

We are all familiar with linear optical phenomena such as transmittance, fluorescence, or light scattering where the output signal is directly proportional to the incident light intensity. By definition, *nonlinear* optical effects arise when an intense light beam interacts with a material to produce a signal with an intensity that is proportional to some power of the incident intensity other than order 1. A nonlinear approach that is being widely applied to microscope imaging is two-photon excitation of fluorescence (TPEF). This idea was first introduced by Winfried Denk and Watt Webb in 1990 (Denk *et al.*, 1990) and several microscope companies are now offering commercial instruments. In this method, a mode-locked laser, typically a titanium sapphire laser with 100-fs pulses at 80 MHz, is passed through the scanhead of a confocal microscope. The focused laser pulses are of sufficient intensity to allow excitation into an excited state corresponding to the summed energy of two photons. The probability of such a two-photon transition is proportional to the square of the exciting light flux and therefore has its highest probability at the focal plane where the excitation is most intense. This is illustrated schematically in the cartoon of Fig. 5a (see color plate) where the excitation pattern for a membrane stain is illustrated. Because the fluorescence only emanates from this focal plane (Fig. 5b), a pinhole is not required to achieve optical sectioning. The fluorescence excited with two photons is equivalent to the linear excitation with a single photon at half the wavelength (or twice the energy) of the incident light. Therefore, the Ti–sapphire laser should be tuned to a wavelength corresponding to approximately twice the absorbance maximum of the dye. For many dyes, the excitation wavelength dependence of TPEF does not correspond perfectly to twice the linear excitation spectrum, because the selection rules for two-photon excitation differ from those for one-photon excitation. However, once the excited state is populated its prior history is irrelevant. Therefore, the fluorescence emission spectrum is identical for TPEF and linear fluorescence. More details on the principles of TPEF and applications to microscopy can be found in the Chapter 3. This brief overview of TPEF serves here to set the stage for a description of the newer nonlinear optical technique of second harmonic generation (SHG).

SHG involves the interaction of the electromagnetic field of the light beam with a material to generate a coherent light beam at half the incident wavelength. The reader may have encountered the application of SHG in the form of inorganic crystals that are used as frequency doublers for lasers. These crystals have been employed from the very earliest stages of the development of laser optics. They are characterized as having high second-order nonlinear susceptibilities ($\chi^{(2)}$). Equation (3) is a simplified expression for the relationship, which in its full form is a complex tensor equation:

$$I(2\nu) \propto \left[\chi^{(2)} I(\nu) \right]^2. \tag{3}$$

As in TPEF, the intensity of the SHG signal is proportional to the square of the incident light intensity. Four important developments in the 1980s are essential prerequisites for the application of SHG to microscopy. The first was the realization that it should be possible to fabricate organic crystals for SHG from dyes with large excitation-induced charge transfer transitions (Nicoud and Twieg, 1987). The second was that it would be possible to measure SHG from not only in a crystal, but from an interface (Shen, 1989). The third was the realization that such an interface could be a biological membrane (Huang *et al.*, 1988; Rasing *et al.*, 1989). The fourth was the demonstration that an appropriate chromophore embedded in the membrane could be the principal and only significant contributor to the membrane second harmonic signal (Huang *et al.*, 1989; Rasing *et al.*, 1989). To understand these developments and appreciate the mechanism of contrast formation in second harmonic imaging microscopy (SHIM, Campagnola *et al.*, 1999), we will have to explore the molecular origins of $\chi^{(2)}$.

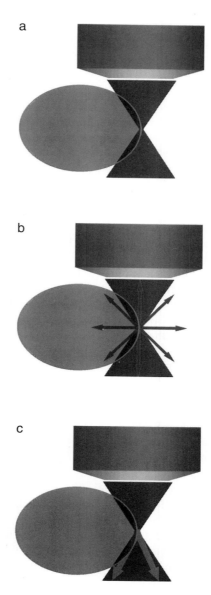

Fig. 5 Schematic illustration of the principles of microscopies derived from TPEF and SHG. A focused laser beam emanating from a microscope objective is focused onto the dye-stained cell membrane. (a) The section of membrane that interacts with the incident light beam is shown in cyan. It is confined to a region close to the plane of focus by the quadratic dependence of both TPEF and SHG. (b) Fluorescence excited by two-photon excitation propagates in all directions and is collected by the microscope objective in a typical incident light (epi) configuration. The wavelength range of the TPEF corresponds to the normal emission spectrum of the dye. (c) SHG propagates along the same path as the incident light and must be collected by a high numerical aperture condenser or a second objective (not shown) in a transmitted light configuration. The wavelength of the SHG signal is precisely half the wavelength of the incident laser light. (See Color Plate.)

It is important to appreciate that $\chi^{(2)}$ is a bulk quantity that is composed of coherent contributions from an array of molecules with strong molecular hyperpolarizabilities:

$$\chi^{(2)} = N_S \langle \beta \rangle_{OR} \tag{4}$$

where N_S is the density of the molecules and the hyperpolarizabilities, β, are expressed as an orientational average, $\langle \beta \rangle_{OR}$. If the array of molecules is centrosymmetric within dimensions comparable to the wavelength of the light, $\langle \beta \rangle_{OR}$ will vanish. So a simple homogeneous solution of dye cannot produce SHG, whereas a noncentrosymmetric crystalline array can. Likewise, a membrane-staining dye that is confined to one leaflet of the lipid bilayer is able to produce an SHG signal, whereas a dye that is equally distributed between both faces of the membrane cannot. Another consequence of Eqs. (3) and (4) is that the SHG signal is proportional to the square of the surface density of the SHG active molecules. These features make SHG clearly distinct from TPEF, as the latter is essentially independent of the distribution of dye and is linearly dependent on dye concentration or density. Thus very different image contrast patterns for these two nonlinear optical microscopies may be obtained. Several scenarios are possible. If the dye is present on only one face of the bilayer at a reasonably low surface density, images from TPEF or SHG might be identical. At very high surface densities for staining of only one leaflet of the bilayer, strong SHG would be expected, but the fluorescence signal might be self-quenched. Finally, if the dye is equally distributed across both faces of the bilayer, one might expect a strong TPEF signal, while the SHG would be extinguished. Of course, all of these relations also hold for comparison of SHG and linear fluorescence in a confocal microscope.

Another practical distinction between SHG/SHIM and TPEF can be understood by returning to Fig. 5. TPEF can be collected via the same objective used to focus the exciting light as is the norm for all forms of fluorescence microscopy (Fig. 5b). This is because the fluorescence is emitted in all directions, making it convenient to collect as much of it as possible with the same high-numerical-aperture lens that was used to excite it and which is of course automatically at the correct focus. SHG, on the other hand, copropagates coherently with the fundamental light (Fig. 5c). Transmitted light collection optics must therefore be incorporated in the microscope with numerical aperture at least as high as the objective. Good filters must be included to reject the fundamental wavelength. Furthermore, there is often collateral TPEF that can be collected in the transmitted light path; so care must be taken to choose filters that reject these wavelengths as well. Ideally, for any dye that is used to resonantly enhance the SHG, as discussed in the following section, there should be a large difference in the wavelengths of the excitation and emission spectrum (i.e., a large Stokes shift) to minimize contamination of SHG with TPEF.

2. Second Harmonic Generation Is Enhanced by Electrochromic Dyes and Is Highly Sensitive to Membrane Potential

The molecular hyperpolarizabiltity, β, can be especially large for dyes with large charge redistributions in the transition from the ground to the excited state (Dirk *et al.*,

1986). The expression governing this dependence is

$$\beta = \frac{4\mu_{eg}^2(\mu_e - \mu_g)}{3h^2(v_{eg}^2 - v^2)(v_{eg}^2 - 4v^2)} \tag{5}$$

where h is Planck's constant, μ_g and μ_e are the dipole moments of the ground and excited states, respectively, μ_{eg} is the transition moment, v is the frequency of the incident light, and hv_{eg} is the energy for the electronic transition from the ground to the excited state. Of course, this expression will become very large as $2v$ approaches v_{eg}, characteristic of a resonance enhancement effect. Other consequences of this simple expression are that β is enhanced when there is a large difference in the ground and excited state electron distribution and when the dye has a large extinction coefficient (i.e., a large μ_{eg}). These criteria can all be met by the electrochromic dyes that this laboratory has developed as was demonstrated with the detection of large SHG from a monolayer of di-4-ANEPPS in 1988 (Huang et al., 1988).

Encouraged by this success, we set out to explore the use of naphthylstyryl dyes as contrast agents for imaging second harmonic generation from cell membranes. SHG was demonstrated from both artificial (Bouevitch et al., 1993) and cellular (Ben-Oren et al., 1996) membranes stained with these dyes. The idea of second harmonic imaging microscopy (SHIM) was demonstrated with high-resolution images of single cells obtained simultaneously with TPEF images (Campagnola et al., 1999). Several naphthylstyryl (ANEP) dyes were tested, including specially designed SHG dyes with covalently linked chiral sidechains. Whereas β would be expected to be the same for all dyes with the same ANEP chromophore, one might predict that the orientationally averaged ensemble, $\langle\beta\rangle_{OR}$, might be larger for chiral dyes. Indeed, these chiral dyes provided significantly larger SHG (Bouevitch et al., 1993; Campagnola et al., 1999). Although achiral dyes produce no SHG when distributed on both faces of the membrane or when they are on the outer leaflet of two apposing membranes, the chiral dyes do provide signals even when there is significant equilibration across the membrane (Campagnola et al., 1999; Moreaux et al., 2000). Examples of SHIM and TPEF images from the same field of view of differentiated N1E-115 neuroblastoma cells are shown in Fig. 6.

The fact that the same molecular property, namely a large charge shift upon excitation, is responsible for both the sensitivity of the dye spectra to electric fields and the enhancement of SHG led us to investigate the possibility that the SHG signal might itself be sensitive to membrane potential. This idea was tested on a voltage-clamped hemispherical lipid bilayer system using a NdYag laser at 1064 nm and pulse widths in the 100-ps range (Bouevitch et al., 1993; Ben-Oren et al., 1996). The best dye gave a fractional change of ca 4%/10mV—significantly better than the 1–2% for the best fluorescence changes. Even this superior result was surpassed in our study of the neuroblastoma cells where the SHG signal was halved upon depolarization (Campagnola et al., 1999). The improvement may be connected to the use of the 800-nm light from a mode-locked Ti–sapphire laser. The mechanism of this potential sensitivity is not the well-known electric-field-induced second harmonic (EFISH) effect that operates in homogeneous solutions of organic dyes (e.g., Meyers et al., 1992); EFISH operates

Two Photon and SHIM of Neurons Stained with JPW2080

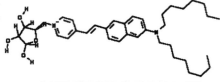

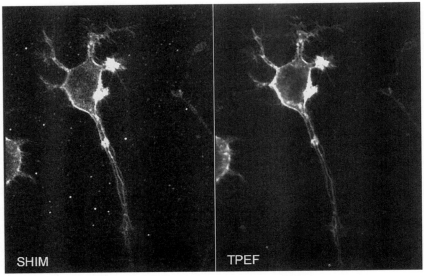

Fig. 6 TPEF and SHIM images of differentiated N1E-115 neuroblastoma cells. An 860-nm incident laser beam focused through a 40× 0.8 NA water immersion objective was used to produce these images. SHIM was collected in the transmitted light channel after excluding the fundamental and any fluorescence emission with a 450-nm low-pass dichroic filter. The TPEF signal was acquired in the epifluorescence channel using a 560-nm long-pass filter.

via an orientation of the dye molecules through interaction of the fields with the dye molecular dipoles, but in our case the dye molecules are already well oriented by the membrane and do not undergo a significant voltage-dependent reorientation (Loew and Simpson, 1981). We believe that the origin of this potential sensitivity is a consequence of additional higher order coupling terms through the imposed electric field that are still wavelength dependent. However, the details of this coupling mechanism are not thoroughly understood.

3. Methods

To obtain images like the ones in Fig. 6, a coverslip with adherent cells is exposed to 1 μM of dye in the appropriate buffered aqueous medium for 10 min at 4°C. JPW1259 (Fig. 1) is a good chiral naphthylstyryl dye for SHIM. Another important chiral dye

Ti:Sapphire In

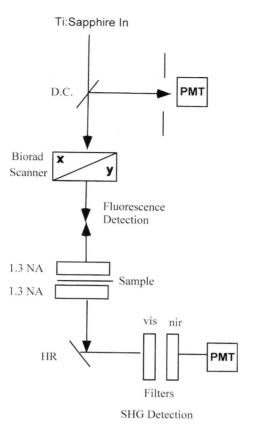

Fig. 7 Schematic of a combined TPEF and SHIM microscope. A Bio-Rad MRC600 confocal scanhead, modified with near-infrared reflecting mirrors, is used to deliver the light from a mode-locked Ti–sapphire laser to the specimen. The TPEF is descanned and collected with the pinhole fully open. The SHG is collected in a transmitted light configuration and isolated with optical filters. Use of a matching collection objective resulted in larger collection efficiency than was observed with a high-NA condenser. The signal is detected with a hybrid single photon counting/analog integration scheme and integrated by the Biorad MRC600 electronics. Taken from Campagnola *et al.* (1999), with permission from the Biophysical Society.

is JPW2080 which has octyl chains in place of the butyl chains of JPW1259. Longer alkyl chains retard equilibration of the dye across the plasma membrane. For the less soluble dyes such as JPW2080, the staining solution also contains 0.05% Pluronic F127, a nonionic polymeric surfactant available from Molecular Probes, Inc. (Eugene, OR). The staining solution is then removed and the cells washed with fresh medium before mounting on the microscope slide or chamber.

The apparatus for acquiring SHIM is based on a confocal scanhead coupled to a short-pulse-width mode-locked near-infrared laser—in our laboratory we use a Ti–sapphire laser—as is the norm for TPEF. As emphasized in the discussion of Fig. 5, the key difference is that a light collection path in the transmitted light geometry is required for SHIM. The collecting lens must have a numerical aperture at least as high as that of the lens that delivers the focussed fundamental beam. In our setup, (Fig. 7), the incident light is delivered through a Bio-Rad MRC 600 scanner that has been modified with mirrors optimized for the near-infrared. The light is focused through a high-numerical-aperture objective. The transmitted light is collected either through a second objective or through a condenser assembly that must have a numerical aperture at least as high as that of the objective that delivers the fundamental light. The second harmonic signal is separated from the fundamental via a short-pass dichroic filter before being detected by a photon-counting photomultiplier. In this regard, the naphthylstyryl dyes have the additional beneficial feature of a huge Stokes shift—the fluorescence emission is separated by 150 nm from the excitation. This obviates the task of separating the TPEF from the SHG in the transmitted light signal. In fact the TPEF signal can be detected separately and simultaneously through the epifluorescence light path of our nonlinear optical microscope. Thus, TPEF and SHIM images can be directly compared. Within the past 2 years, several microscope manufacturers have marketed combined confocal/nonlinear optical microscopes configured with Ti–sapphire lasers. These can all be readily used for SHIM by adapting the transmitted light path according to Fig. 7.

IV. Perspectives

A pair of very different approaches toward imaging membrane potential have been presented. It is worth considering some of the broader applications of these approaches that are sure to be more fully developed as the dye and instrument technologies continue to mature.

The quantitative measurement of relative dye concentration in our method for converting the distribution of Nernstian dyes into a membrane potential is just one example of the rarely exploited power of confocal microscopy for quantitative studies. The equivalence of the optical slice obtained in a confocal microscope to an intracellular cuvette was explained in Section III.A.1. This concept can readily be extended to the analysis of fluorescent probes or fluorescently labeled molecules other than Nernstian dyes, as has been described (Fink *et al.*, 1998). In that paper, in addition to membrane potential, a number of other applications of the general approach to quantitative measurements were described. This included the determination of the intracellular concentration of a calcium indicator using a calibration from a series of microscope slides containing known concentrations of indicator and imaged with the same confocal settings as the cells. Also in that paper, the method was applied to the intracellular distributions of both surface and cytoplasmic immunofluorescently labeled antigens. Subsequently (Fink *et al.*, 1999), this same strategy was used to determine the precise intracellular concentrations of a non-fluorescent caged compound that had been coinjected with a fluorescent marker. Indeed, with proper calibration of the confocal image via convolution of models with the point-spread function, such measurements can be extended to objects from which fluorescence can be detected in the confocal microscope, but that may be below its resolution limits. Of course, such an extension of the approach is precisely what permitted the measurement of Nernstian dye concentrations within individual mitochondria.

In our discussion of SHIM, we have also emphasized the particular applicability of SHG in combination with chiral voltage-sensitive dyes, to detect changes in membrane potential. Here, as well, there are emerging opportunities for extensions of SHIM. For example, there is evidence that SHG can be detected from membrane-associated green fluorescent protein and that this signal may also be sensitive to membrane potential (Khatchatouriants *et al.*, 2000). We have shown that the SHG signal can be enhanced by at least two orders of magnitude by proximal metal nanoparticles (Clark *et al.*, 2000), raising the prospect that membrane potential can be monitored at highly localized sites on a membrane (Peleg *et al.*, 1999). High-resolution SHG images of inorganic crystals have been reported (Gauderon *et al.*, 1998), raising the prospect that some highly ordered biological specimens might give intrinsic signals. Indeed, some reports of SHG signals from rat tendon have appeared (Freund *et al.*, 1986). Should it be possible to detect SHG from such ordered arrays of biological molecules without any resonance enhancement, a new imaging modality will be realized that promises high 3-D contrast without any significant photodynamic effects. The sensitivity of such intrinsic signals to the physiological state of the cells from which they originate is also an open question that deserves thorough investigation.

Finally, it should be noted that there is no theoretical reason that any voltage-sensitive dye that is used in wide-field microscopy cannot also be used to image membrane potential with high resolution through the confocal microscope. In particular, we have mapped membrane potential along the surface of cells by ratiometric imaging of electrochromic dyes (Montana *et al.*, 1989; Bedlack *et al.*, 1992; Loew, 1994). The practical reasons that made such studies difficult to transfer to the confocal microscope included the high illumination intensities, and therefore likelihood of photodynamic damage, required for the precise high-dynamic-range measurements needed to detect the small differences inherent in these measurements. Also, limited choices of wavelengths have made the optimal settings for ratiometric confocal membrane potential imaging difficult to attain. Many of these issues have been solved by the most recent generation of commercial confocals which all have much more sensitive and efficient optics, more flexible scanning mechanisms, and a wide choice of wavelength combinations.

Acknowledgments

The participation of Mei-de Wei, Charles Fink, and Frank Morgan in obtaining and analyzing the original data is gratefully acknowledged. The U.S. Public Health Service under grant GM35063, the Office of Naval Research (N0014-98-1-0703), and the National Science Foundation (DBI-9601609) supported this work.

References

Beach, J. M., McGahren, E. D., Xia, J., and Duling, B. R. (1996). Ratiometric measurement of endothelial cell depolarization in arterioles with a potential sensitive dye. *Am. J. Physiol.* **270,** H2216–H2227.

Bedlack, R. S., Wei, M.-D., and Loew, L. M. (1992). Localized membrane depolarizations and localized intracellular calcium influx during electric field-guided neurite growth. *Neuron* **9,** 393–403.

Bedlack, R. S., Wei, M.-D., Fox, S. H., Gross, E., and Loew, L. M. (1994). Distinct electric potentials in soma and neurite membranes. *Neuron* **13**, 1187–1193.

Ben-Oren, I., Peleg, G., Lewis, A., Minke, B., and Loew, L. M. (1996). Infrared nonlinear optical measurements of membrane potential in photoreceptor cells. *Biophys. J.* **71**, 1616–1620.

Bouevitch, O., Lewis, A., Pinevsky, I., Wuskell, J. P., and Loew, L. M. (1993). Probing membrane potential with non-linear optics. *Biophys. J.* **65**, 672–679.

Brand, M. D., and Murphy, M. P. (1987). Control of electron flux through the respiratory chain in mitochondria and cells. *Camb. Phil. Soc. Biol. Rev.* **62**, 141–193.

Bullen, A., and Saggau, P. (1999). High-speed, random-access fluorescence microscopy: II. Fast quantitative measurements with voltage-sensitive dyes. *Biophys. J.* **76**, 2272–2287.

Cacciatore, T. W., Brodfuehrer, P. D., Gonzalez, J. E., Jiang, T., Adams, S. R., Tsien, R. Y., Kristan, W. B. Jr., and Kleinfeld, D. (1999). Identification of neural circuits by imaging coherent electrical activity with FRET-based dyes. *Neuron* **23**, 449–459.

Campagnola, P. J., Wei, M.-D., Lewis, A., and Loew, L. M. (1999). High resolution optical imaging of live cells by second harmonic generation. *Biophys. J.* **77**, 3341–3349.

Carrington, W. A. (1990). *In* "Proceedings on Bioimaging and Two Dimensional Spectroscopy" (L. C. Smith, Ed.), pp. 72–83. SPIE, Los Angeles.

Carrington, W., and Fogarty, K. E. (1987). *In* "Proceedings of the 13th annual Northeast Bioengineering Conference, Philadelphia" (K. Foster, Ed.), pp. 108–111. IEEE, New York.

Carrington, W., Fogarty, K. E., and Fay, F. S. (1990). *In* "Non-invasive Techniques in Cell Biology" (K. Foster, Ed.), pp. 53–72. Wiley-Liss, New York.

Clark, H. A., Campagnola, P. J., Wuskell, J. P., Lewis, A., and Loew, L. M. (2000). Second harmonic generation properties of fluorescent polymer encapsulated gold nanoparticles. *J. Am. Chem. Soc.* **122**, 10234–10235.

Cohen, L. B., Salzberg, B. M., Davila, H. V., Ross, W. N., Landowne, D., Waggoner, A. S., and Wang, C. H. (1974). Changes in axon fluorescence during activity: Molecular probes of membrane potential. *J. Membr. Biol.* **19**, 1–36.

Denk, W., Strickler, J. H., and Webb, W. W. (1990). Two-photon laser scanning fluorescence microscopy. *Science* **248**, 73–76.

Dirk, C. W., Twieg, R. J., and Wagniere, J. (1986). The contribution of pi electrons to second harmonic generation in organic molecule. *J. Am. Chem. Soc.* **108**, 5387–5395.

Ehrenberg, B., Wei, M. D., and Loew, L. M. (1987). Nernstian dye distribution reports membrane potential in individual cells. *In* "Membrane Proteins" (S. C. Goheen, Ed.), pp. 279–294. Bio-Rad Laboratories, Richmond, CA.

Ehrenberg, B., Montana, V., Wei, M.-D., Wuskell, J. P., and Loew, L. M. (1988). Membrane potential can be determined in individual cells from the Nernstian distribution of cationic dyes. *Biophys. J.* **53**, 785–794.

Farkas, D. L., Wei, M., Febbroriello, P., Carson, J. H., and Loew, L. M. (1989). Simultaneous imaging of cell and mitochondrial membrane potential. *Biophys. J.* **56**, 1053–1069.

Fay, F. S., Carrington, W., and Fogarty, K. E. (1989). *Journal of Microscopy* **153**, 133–149.

Fink, C., Morgan, F., and Loew, L. M. (1998). Intracellular fluorescent probe concentrations by confocal microscopy. *Biophys. J.* **75**, 1648–1658.

Fink, C. C., Slepchenko, B., and Loew, L. M. (1999). Determination of time-dependent inositol 1,4,5-trisphosphate concentrations during calcium release in a smooth muscle cell. *Biophys. J.* **77**, 617–628.

Fluhler, E., Burnham, V. G., and Loew, L. M. (1985). Spectra, membrane binding and potentiometric responses of new charge shift probes. *Biochemistry* **24**, 5749–5755.

Freund, I., Deutsch, M., and Sprecher, A. (1986). Connective tissue polarity. Optical second-harmonic microscopy, crossed-beam summation, and small-angle scattering in rat-tail tendon. *Biophys. J.* **50**, 693–712.

Gauderon, R., Lukins, P. B., and Sheppard, C. J. R. (1998). Three-dimensional second-harmonic generation imaging with femtosecond laser pulses. *Opt. Lett.* **23**, 1209–1211.

Gonzalez, J. E., and Tsien, R. Y. (1995). Voltage sensing by fluorescence resonance energy transfer in single cells. *Biophys. J.* **69**, 1272–1280.

Gonzalez, J. E., and Tsien, R. Y. (1997). Improved indicators of membrane potential that use fluorescence resonance energy transfer. *Chem. Biol.* **4,** 269–277.

Grinvald, A. S., Hildesheim, R., Farber, I. C., and Anglister, J. (1982). Improved fluorescent probes for the measurement of rapid changes in membrane potential. *Biophys. J.* **39,** 301–308.

Grinvald, A., Lieke, E. E., Frostig, R. D., and Hildesheim, R. (1994). Cortical point-spread function and long-range lateral interactions revealed by real-time optical imaging of macaque monkey primary visual cortex. *J. Neurosci.* **14,** 2545–2568.

Gross, D., Loew, L. M., and Webb, W. W. (1985). Spatially resolved optical measurement of membrane potential. *Biophys. J.* **47,** 270.

Gupta, R. K., Salzberg, B. M., Grinvald, A., Cohen, L. B., Kamino, K., Lesher, S., Boyle, M. B., Waggoner, A. S., and Wang, C. (1981). Improvements in optical methods for measuring rapid changes in membrane potential. *J. Membr. Biol.* **58,** 123–137.

Hassner, A., Birnbaum, D., and Loew, L. M. (1984). Charge shift probes of membrane potential. Synthesis. *J. Org. Chem.* **49,** 2546–2551.

Huang, J. Y., Lewis, A., and Loew, L. M. (1988). Non-linear optical properties of potential sensitive styryl dyes. *Biophys. J.* **53,** 665–670.

Huser, J., Rechenmacher, C. E., and Blatter, L. A. (1998). Imaging the permeability pore transition in single mitochondria. *Biophys. J.* **74,** 2129–2137.

Johnson, L. V., Walsh, M. L., and Chen, L. B. (1980). Localization of mitochondria in living cells with rhodamine 123. *Proc. Natl. Acad. Sci. USA* **77,** 990–994.

Khatchatouriants, A., Lewis, A., Rothman, Z., Loew, L., and Treinin, M. (2000). GFP is a selective non-linear optical sensor of electrophysiological processes in *Caenorhabditis elegans. Biophys. J.* **79,** 2345–2352.

Loew, L. M. (1994). Voltage sensitive dyes and imaging neuronal activity. *Neuroprotocols* **5,** 72–79.

Loew, L. M. (1998). Measuring membrane potential in single cells with confocal microscopy. *In* "Cell Biology: A Laboratory Handbook" (J. E. Celis, Ed.), Vol. 3, pp. 375–379. Academic Press, San Diego.

Loew, L. M., and Simpson, L. (1981). Charge shift probes of membrane potential. A probable electrochromic mechanism for ASP probes on a hemispherical lipid bilayer. *Biophys. J.* **34,** 353–365.

Loew, L. M., Bonneville, G. W., and Surow, J. (1978). Charge shift optical probes of membrane potential. Theory. *Biochemistry* **17,** 4065–4071.

Loew, L. M., Scully, S., Simpson, L., and Waggoner, A. S. (1979a). Evidence for a charge-shift electrochromic mechanism in a probe of membrane potential. *Nature* **281,** 497–499.

Loew, L. M., Simpson, L., Hassner, A., and Alexanian, V. (1979b). An unexpected blue shift caused by differential solvation of a chromophore oriented in a lipid bilayer. *J. Am. Chem. Soc.* **101,** 5439–5440.

Loew, L. M., Cohen, L. B., Dix, J., Fluhler, E. N., Montana, V., Salama, G., and Wu, J.-Y. (1992). A naphthyl analog of the aminostyryl pyridinium class of potentiometric membrane dyes shows consistent sensitivity in a variety of tissue, cell, and model membrane preparations. *J. Membr. Biol.* **130,** 1–10.

Loew, L. M., Tuft, R. A., Carrington, W., and Fay, F. S. (1993). Imaging in 5 dimensions: Time dependent membrane potentials in individual mitochondria. *Biophys. J.* **65,** 2396–2407.

Mazzanti, M., DeFelice, L. J., Cohen, J., and Malter, H. (1990). Ion channels in the nuclear envelope. *Nature* **343,** 764–767.

Meyers, F., Brédas, J. L., and Zyss, J. (1992). Electronic structure and nonlinear optical properties of push–pull polyenes: theoretical investigation of benzodithia polyenals and dithiolene polyenals. *J. Am. Chem. Soc.* **114,** 2914–2921.

Montana, V., Farkas, D. L., and Loew, L. M. (1989). Dual wavelength ratiometric fluorescence measurements of membrane potential. *Biochemistry* **28,** 4536–4539.

Moreaux, L., Sandre, O., Blanchard-desce, M., and Mertz, J. (2000). Membrane imaging by simultaneous second-harmonic generation and two-photon microscopy. *Opt. Lett.* **25,** 320–322.

Nicoud, J. F., and Twieg, R. J. (1987). Design and synthesis of organic molecular compounds for efficient second-harmonic generation. *In* "Nonlinear Optical Properties of Organic Molecules and Crystals" (D. S. Chemla, and J. Zyss, Eds.), Vol. 1, pp. 227–296. Academic Press, New York.

Obaid, A. L., Koyano, T., Lindstrom, J., Sakai, T., and Salzberg, B. M. (1999). Spatiotemporal patterns of activity in an intact mammalian network with single-cell resolution: optical studies of nicotinic activity in an enteric plexus. *J. Neurosci.* **19**, 3073–3093.

Peleg, G., Lewis, A., Linial, M., and Loew, L. M. (1999). Nonlinear optical measurement of membrane potential around single molecules at selected cellular sites. *Proc. Natl. Acad. Sci. USA* **96**, 6700–6704.

Prechtl, J. C., Cohen, L. B., Pesaran, B., Mitra, P. P., and Kleinfeld, D. (1997). Visual stimuli induce waves of electrical activity in turtle cortex. *Proc. Natl. Acad. Sci. USA* **94**, 7621–7626.

Rasing, T., Huang, J., Lewis, A., Stehlin, T., and Shen, Y. R. (1989). *In situ* determination of induced dipole moments of pure and membrane-bound retinal chromophores. *Phys. Rev. A* **40**, 1684–1687.

Reers, M., Smith, T. W., and Chen, L. B. (1991). J-Aggregate formation of a carbocyanine as a quantitative fluorescent indicator of membrane potential. *Biochemistry* **30**, 4480–4486.

Rink, T. J., Montecucco, C., Hesketh, T. R., and Tsien, R. Y. (1980). Lymphocyte membrane potential assessed woth fluorescent probes. *Biochim. Biophys. Acta* **595**, 15–30.

Ross, W. N., Salzberg, B. M., Cohen, L. B., Grinvald, A., Davila, H. V., Waggoner, A. S., and Wang, C. H. (1977). Changes in absorption, fluorescence, dichroism, and birefringence in stained giant axons: Optical measurement of membrane potential. *J. Membr. Biol.* **33**, 141–183.

Rottenberg, H. (1986). Energetics of proton transport and secondary transport. *Methods. Enzymol.* **125**, 1–15.

Salzberg, B. M. (1989). Optical recording of voltage changes in nerve terminals and in fine neuronal processes. *Ann. Rev. Physiol.* **51**, 507–526.

Scaduto, R. C., Jr., and Grotyohann, L. W. (1999). Measurement of mitochondrial membrane potential using fluorescent rhodamine derivatives [In Process Citation]. *Biophys. J.* **76**, 469–477.

Shen, Y. R. (1989). Surface properties probed by second-harmonic and sum-frequency generation. *Nature* **337**, 519–525.

Shoham, D., Glaser, D. E., Arieli, A., Kenet, T., Wijnbergen, C., Toledo, Y., Hildesheim, R., and Grinvald, A. (1999). Imaging cortical dynamics at high spatial and temporal resolution with novel blue voltage-sensitive dyes. *Neuron* **24**, 791–802.

Shrager, P., and Rubinstein, C. T. (1990). Optical measurement of conduction in single demyelinated axons. *J. Gen. Physiol.* **95**, 867–890.

Siegel, M. S., and Isacoff, E. Y. I. N. (1997). A genetically encoded optical probe of membrane voltage. *Neuron* **19**, 735–741.

Sims, P. J., Waggoner, A. S., Wang, C.-H., and Hoffman, J. F. (1974). Studies on the mechanism by which cyanine dyes measure membrane potential in red blood cells and phosphotidylcholine. *Biochemistry* **13**, 3315–3330.

Smith, J. C., Russ, P., Cooperman, B. S., and Chance, B. (1976). Synthesis, structure determination, spectral properties, and energy-linked spectral responses of the extrinsic probe oxonol V. *Biochemistry* **15**, 5094–5105.

Tsau, Y., Wenner, P., O'Donovan, M. J., Cohen, L. B., Loew, L. M., and Wuskell, J. P. (1996). Dye screening and signal-to-noise ratio for retrogradely transported voltage-sensitive dyes. *J. Neurosci. Methods* **170**, 121–129.

Tsien, R. Y., and Waggoner, A. (1990). Fluorophores for confocal microscopy: photophysics and photochemistry. *In* "Handbook of Biological Confocal Microscopy" (J. B. Pawley, Ed.), pp. 169–178. Plenum Press, New York.

Tsodyks, M., Grinvald, T. K., and Arieli, A. (1999). Linking spontaneous activity of single cortical neurons and the underlying functional architecture. *Science* **286**, 1943–1946.

Wu, J.-Y., Cohen, L. B., and Falk, C. X. (1994). Neuronal activity during different behaviors in *Aplysia:* a distributed organization? *Science* **263**, 820–822.

Zecevic, D. (1996). Multiple spike-initiation zones in single neurons revealed by voltage-sensitive dyes. *Nature* **381**, 322–325.

CHAPTER 13

Measurement of Intracellular Ca^{2+} Concentration

Nicolas Demaurex, * **Serge Arnaudeau,** * **and Michal Opas** [†]

*Department of Physiology
University of Geneva Medical Center
CH-1211, Geneva 4, Switzerland

[†] Department of Laboratory Medicine and Pathobiology
University of Toronto
Toronto, Ontario, M5S 1A8 Canada

I. Introduction

Intracellular organelles play a major role in calcium signaling, as the calcium stored within intracellular compartments provides a pool of rapidly releasable calcium (Berridge, 1993), allows restriction of the calcium signal to specific subcellular regions (Berridge *et al.,* 1999), and controls the calcium permeability of the plasma membrane (Putney and McKay, 1999). The main source of the calcium released during activation is the endoplasmic reticulum (ER) or specialized subcompartments within the ER network

(Pozzan *et al.*, 1994; Meldolesi and Pozzan, 1998), but increasing evidence suggests that other organelles also participate in calcium signaling (Pozzan *et al.*, 1994). Mitochondria take up calcium very efficiently and modulate calcium oscillations and waves by their intimate connection with the calcium sources (Babcock and Hille, 1998; Rutter *et al.*, 1998), and the Golgi complex also appear to be a *bona fide* Ins(1,4,5)P$_3$-sensitive calcium store (Pinton *et al.*, 1998). In addition to its role in calcium signaling, the calcium stored within intracellular compartments has numerous effects on the function of the organelles themselves. The free calcium concentration within the lumen of the ER affects protein synthesis, sorting, and degradation (Sitia and Meldolesi, 1992), whereas the calcium content of the nuclear envelope, a specialized domain of the ER, regulates the opening of nuclear pores (Stehno-Bittel *et al.*, 1995; Perez-Terzic *et al.*, 1996). In the Golgi apparatus, calcium affects glycosylation as well as the sorting of proteins between the constitutive and regulated pathway (Orci *et al.*, 1987; Carnell and Moore, 1994). In the mitochondria, a rise in lumenal calcium activates several dehydrogenases, increasing the level of NAD(P)H and the production of ATP (Pralong *et al.*, 1994; Rizzuto *et al.*, 1994; Hajnoczky *et al.*, 1995) to meet the cell energy demand.

Our understanding of the calcium homeostasis of intracellular compartments is still incomplete, because the highly dynamic Ca^{2+} signals occurring within organelles are difficult to measure. Trapped fluorescent dyes such as Mag-fura-2 and mag-indo-1 have been used to measure calcium within the ER and mitochondria (Hofer and Machen, 1993; Hirose and Iino, 1994; Tse *et al.*, 1994; Golovina and Blaustein, 1997; Hofer *et al.*, 1998). However, these dyes are not specifically targeted and their selectivity for calcium over magnesium is poor, making quantification of the calcium changes uncertain. The cationic probe rhod2 has been used to measure calcium within mitochondria (Hajnoczky *et al.*, 1995; Simpson and Russell, 1996; Hoth *et al.*, 1997; Babcock *et al.*, 1997; Kaftan *et al.*, 2000), but its targeting depends on the negative membrane potential of this organelle. The calcium-sensitive photoprotein aequorin, on the other hand, can be molecularly addressed to specific compartments (Rizzuto *et al.*, 1993, 1994). This approach has been used to measure calcium signals within the mitochondria (Rizzuto *et al.*, 1992, 1994; Rutter *et al.*, 1996; Montero *et al.*, 2000), the ER (Montero *et al.*, 1995; Button and Eidsath, 1996; Barrero *et al.*, 1997; Alonso *et al.*, 1999; Pinton *et al.*, 2000), the nucleus (Brini *et al.*, 1994), the Golgi complex (Pinton *et al.*, 1998), and the subplasmalemmal space (Marsault *et al.*, 1997). However, the weak luminescence of the photoprotein and its irreversible consumption upon binding to the Ca^{2+} ion precludes single-cell imaging, limiting the use of this approach to study the redistribution and functional interactions of these organelles during calcium signals.

Recent improvements in confocal technology permit the use of a confocal microscope as an effective tool to both measure concentration and visualize distribution of several ions. Here, we discuss ways to measure intracellular [Ca^{2+}] (pCa$_i$) with a typical, present-day confocal microscope. Such a microscope may consist of a scanner capable of simultaneous dual excitation and simultaneous collection of dual emissions. The scan time for a full frame is about 1 s in a "standard" mode or 2 s in a "slow" mode, where the pixel dwell time is greater and consequently, the signal/noise ratio is higher. The simultaneous dual excitation/dual emission mode of operation requires an [FITC/Texas

Red]-type filter module in which the two emission wavelengths are separated at 560 nm with a dichroic mirror. The light source is likely to be a mixed gas Krypton-Argon laser. The technicalities behind confocal microscopy can be found in an excellent overview compiled by J. B. Paweley (1995).

The confocal microscope effectively discriminates against out-of-focus information, which permits for mapping of ion distribution at the subcellular level (Opas, 1997). The need for efficient optical sectioning applies also to ion ratio imaging (Monck *et al.*, 1992; Lipp and Niggli, 1993; Seksek and Bolard, 1996), and this is where the confocal method is at its best. The very ability to clearly image planes buried deep in the sample may cause substantial chromatic aberrations, such as shift of focus for different emissions in multi-wavelength imaging. This is being partly remedied by the renaissance of water immersion objectives, which, however, may carry a substantial price tag. Confocal microscopy, while alleviating several problems inherent to wide-field microscopy, introduces new problems. The most vexing problem of today's confocal microscopy is a severe limitation in the number of wavelengths available for fluorochrome excitation. Probe selection is limited if Ca^{2+}-sensitive fluorochromes are excited at wavelengths close to those emitted by standard, affordable lasers. Unfortunately, there has neither been a dramatic reduction in the price of UV lasers or two-photon excitation microscopes (Pawley, 1995), nor a major breakthrough in the synthesis of ratio Ca^{2+}-sensitive fluorochromes excitable with visible light. Finally, fluorescent [Ca^{2+}] indicators are rather poor fluorochromes and are not photostable. To collect their emissions and their changes in time, fast and efficient detectors are required, such as the charge-couple devices of the wide-field microscopes. For all its inherent difficulties, a microscopic approach to spectrometry has been steadily increasing in popularity. A microscope allows us to select particular cells, to map specific molecules at the subcellular level and follow temporal kinetics of the map with variable sampling frequency (Farkas *et al.*, 1993). An emerging, powerful approach of manufacturing organelle-specific fluorescent probes (De Giorgi *et al.*, 1996) allows selecting specific sites of investigation (Rizzuto *et al.*, 1996). Finally, multi-parameter microscopy permits correlative investigations of ions, membranes, cytoskeletal elements, etc. (Opas and Kalnins, 1985; DeBiasio *et al.*, 1987; Farkas *et al.*, 1993; Zhou and Opas, 1994; Ferri *et al.*, 1997). In this chapter we attempt to overview two manners in which pCa$_i$ can be measured. The first is a traditional use of visible-light [Ca^{2+}] indicators, while the other is an emerging technology, which employs proteins that have been genetically modified to become [Ca^{2+}] indicators.

II. Ratio Measurement of pCa$_i$ with Visible–Light [Ca^{2+}] Indicators

Conditions prevailing within a cell are far from uniform. Many variables have to be compensated for, if meaningful information is to be obtained. Fortunately, most of these variables can be dealt with summarily by using ratiometric measurements (Bright *et al.*, 1989; Dunn *et al.*, 1994). Fluorescence ratiometry takes advantage of the dual spectral sensitivity of a ratio probe to the measured parameter, intracellular [Ca^{2+}] in the present case. Briefly, while fluorescent properties of the probe at one wavelength are parameter-sensitive, the emission (or excitation) of the probe at a distinct, well-separated wavelength must be either parameter-insensitive, inversely sensitive or exhibit a different sensitivity profile. Emission intensities for the two wavelengths are then divided by each other and thus the resulting ratio becomes normalized for inhomogeneities in probe distribution and concentration, and in the system geometry. To obtain meaningful data, the ratio values must be calibrated against a standard, which is usually a graded change in the parameter.

A. Probes

Fluorescent Ca^{2+} indicators cover a spectrum from UV to far red. Thus, superficially, pCa_i measurements on a UV-equipped wide-field microscope appear relatively straightforward. However, all commercially available indicators that exhibit a spectral shift upon Ca^{2+} binding (necessary for ratio measurements) require UV excitation. None of the commercially available Ca^{2+} indicators that are excited with visible light exhibit a spectral shift. They all change emission intensity upon Ca^{2+} binding. This makes pCa_i measurement with a standard confocal microscope difficult. Useful information regarding fluorescent Ca^{2+} indicators that are excited with visible light can be found at http://www. probes.com/handbook/sections/2003.html[1] and http://fluorescence.bio-rad.com, as well in several reviews (Kao, 1994; Opas, 1997; Williams *et al.*, 1999; Takahashi *et al.*, 1999). A simple, but more expensive, way around this problem is to use either UV laser or multiphoton confocal microscopy. To do the measurements on a tight budget, it is necessary to mix dyes. So far, the most popular combination has been a mixture of Fluo-3 and Fura Red, first described by Lipp and Niggli (1993). This mixture can be used on a microscope equipped with either an argon or krypton–argon laser. Fluo-3 has FITC-like spectral properties and its fluorescence intensity increases with an increase in Ca^{2+} concentration (Minta *et al.*, 1989). Fura Red is a ratiometric dye that is excited at ~440 and at ~490 nm and emits with a large Stokes shift at ~660 nm. It is unusual in that its emission intensity decreases with an increase in Ca^{2+} concentration. Calcium Green-1 has also been used in the dye mixture as it has spectral properties similar to those of Fluo-3 (Eberhard and Erne, 1991), except that it is more fluorescent at low Ca^{2+} concentrations. There are two forms of Calcium Green-1 available that differ in Ca^{2+} binding properties: the affinity of Calcium Green-1 for Ca^{2+} ($K_d = 190$ nM) is lower than that of Fluo-3 ($K_d = 390$ nM), while the affinity Calcium Green-2 ($K_d = 550$ nM) is higher. We would argue against using pH indicator dyes (Whitaker *et al.*, 1991) as a Ca^{2+}-insensitive volumetric reference, as it is not certain whether or not pH changes accompany those of pCa. Reported dye proportions for loading the mixture of Fluo-3:Fura Red acetoxylmethyl esters vary from 3 : 4 to 1 : 10, while concentrations range from 1 to 100 μM for Fluo-3 and 4 to 300 μM for Fura Red (Schild *et al.*, 1994; Diliberto *et al.*, 1994; Floto *et al.*, 1995; Lipp *et al.*, 1996; Ingalls *et al.*, 1998; Blatter and Niggli, 1998); we load Calcium Green-1:Fura Red at an 8.3:4.6 μM ratio.

[1]We have no financial interest in any of the companies we refer to.

Table I

Composition of the Buffers Used for Dye Loading (A), pCa$_i$ Measurement (B), and Calibration (C)

	A	B	C
NaCl, mM	140	140	—
KCl, mM	4	4	100
CaCl$_2$, mM	1	1	—
MgCl$_2$, mM	1	1	—
HEPES, mM	10	10	—
Glucose, mM	10	10	—
Ionomycin, μM	—	—	2
pH	7.2	7.2	7.2
MOPS, mM	—	—	10
EGTA/CaEGTA, mM	—	—	0 → 10

B. Loading, Measurement, and Calibration

Meaningful data can be collected only when the probe is where we think it is. Overloaded probes will compartmentalize immediately. Also, most probes loaded into the cytosol will compartmentalize eventually, meaning that they end up in various cell compartments such as mitochondria, endoplasmic reticulum, and, most often, in all sorts of vesicles. This may lead to a variety of measurement errors. Therefore extreme caution has to be exercised while loading cells with fluorescent probes, and especially so when using acetoxylmethyl esters. Probe concentration, cell density, time of loading, temperature, and the presence or absence of serum and particular amino acids in the medium will affect cell loading. DMSO can be used as a vehicle for loading cells[2] with a mixture of acetoxylmethyl esters of Calcium Green-1and Fura Red in the loading buffer.[3] Acetoxylmethyl esters have been designed to be lipophilic (Tsien, 1981) and thus are hardly soluble in aqueous media. The presence of a dispersing agent, such as Pluronic F-127 in the range of 1–4 μl per ml of stain solution is required to facilitate loading. Higher dye concentrations and/or longer incubation times will cause compartmentalization of the dye, which may cause measurement errors. The effect of presence or absence of serum on efficacy of loading and compartmentalization of loaded dye should be checked first. Next, temperature of loading should be experimented with. While some

[2]These conditions have been optimized for sparse retinal pigment epithelial cells of the chick embryo. We load cells with a mixture of acetoxylmethyl esters of Calcium Green-1 (8 μM) and Fura Red (5 μM) in the loading buffer in presence of 0.02% Pluronic F-127 (1 μl of 20% stock solution per ml of stain solution) for 30–40 min at room temperature. The dye mixture comes from a stock solution in anhydrous DMSO, which was either prepared immediately before use or used from aliquots stored at −20°C. Aliquots are only good for 1 month in DMSO. The final concentration of DMSO in the culture medium should not exceed 0.1%. After loading cells should be incubated for 15–20 min in α-MEM with 10 mM Hepes containing 2% serum to deesterify the probe and recover.

[3]Table I shows the composition of the buffers used for dye loading (A), pCa$_i$ measurement (B), and calibration (C).

cell types will load quickly and well in 37°C, others may need more time at a lower temperature to load uniformly. To ascertain a uniform diffusion of dyes into thicker specimens, refrigerated loading might be necessary (Williams *et al.,* 1999). However, the loading temperature affects the activity of extracellular esterases, which is also a factor dictating how much time and what concentration of dye can be used to load the specimen. If incubation in a temperature lower than room temperature is required, it might be necessary to warm up the specimen for up to 30 min at 37°C to allow for dye deesterification. To asses the degree of dye compartmentalization, a simple procedure of double extraction can be followed: an image of loaded cells is compared to that of the same cells after mild permeabilization with digitonin or saponin (Takahashi *et al.,* 1999), after which all the cell membranes are open with Triton X-100 to reveal the background image (Kao, 1994). Fortunately, it has been reported that Fluo-3, Calcium Green, and Fura Red have little propensity for vesicular and mitochondrial localization and the majority of loaded dyes end up in the cytoplasm and nucleus (O'Malley, 1994; Diliberto *et al.,* 1994; O'Malley *et al.,* 1999). Obviously, the proper loading of cells is one of the more serious problems in using a combination of dyes (Floto *et al.,* 1995). A final difficulty is that the dyes leak from the cytoplasm, apparently by an active extrusion mechanism. This loss of the probe can be effectively attenuated by low temperature as well as by anion transport inhibitors, such as probenecid and sulfinpyrazone (Merritt *et al.,* 1990; Kao, 1994; Takahashi *et al.,* 1999). A very effective remedy against both dye compartmentalization and extrusion is to use dyes that are conjugated to dextrans (Stricker and Whitaker, 1999). However, dextrans have to be microinjected or otherwise forced into the cell. Several ways of doing this have been described by McNeil (1989). The conjugates are commercially available in several varieties and sizes (Haugland, 1998, and http://www.probes.com/handbook/sections/2004.html). Calcium Green-1-conjugated dextran has been co-microinjected with Texas Red-conjugated dextran (Chang and Meng, 1995; McDougall and Sardet, 1995) or with tetramethylrhodamine-conjugated dextran (Gresham *et al.,* 1989; Fontanilla and Nuticelli, 1998). In an elegant approach Xu and Heath microinjected single 10-kDa dextran species conjugated to both Calcium Green-1 and Texas Red Calcium (Xu and Heath, 1998). Finally, O'Malley (1999) provides an extensive description of the use of microinjected Calcium Green-1-conjugated dextran to perform calibrated pCa_i measurements (O'Malley *et al.,* 1999).

After the cells are loaded, both Calcium Green-1 and Fura Red are excited by the 488-nm laser band and their emissions are simultaneously collected in the green channel (Calcium Green-1 at ∼525 nm) and in the red channel (Fura Red at ≥615 nm). One emission is then divided by the other to obtain the ratio value.[4] This is usually followed with

---

[4]We collect images (∼20 cells per image) with a Zeiss Plan-Neofluar objective (40 × , multiimmersion) at either room temperature or 37°C. This is done using *only* the 488-nm laser band, the [FITC/Texas Red]-type filter set (e.g., K1 and K2 in a Bio-Rad MRC 600), and simultaneous dual-emission recording feature of a confocal microscope to collect intracellular emission of Calcium Green-1 from the green channel and of Fura Red from the red channel. The gain ranges of both channels should cover the expected dynamic range of both emissions, i.e., while the weakest emission is still recorded, the strongest emission should not saturate the photomultiplier. In regard to this it is important to keep in mind that, with higher intracellular $[Ca^{2+}]$, Calcium Green-1 images will get *brighter* and Fura Red images will get *darker*.

multiplication by a constant factor. Any software that performs arithmetic operations on image files can be used for this purpose. The measurements must be calibrated as the calibration data are used to generate a standard pCa_i curve for each experiment. Unfortunately, the widely used calibration procedure based on Grynkiewicz's equation (Grynkiewicz et al., 1985) is not applicable to cells loaded with a mixture of acetoxyl-methyl esters of two dyes. Even after modification of Grynkiewicz's equation to accommodate the two different $Ca^{2+} K_d$ values of the dyes, its applicability remains problematic (Floto et al., 1995). Therefore, a calibration procedure similar to that used for pH_i measurements should be implemented (Opas and Dziak, 1999). This can be achieved by using a series of solutions of different Ca^{2+} concentrations containing a Ca^{2+} ionophore, ionomycin (Marks and Maxfield, 1991). A series of measurements[5] is taken of the dye-loaded Ca^{2+}-permeable cells, which have been flushed with the graded $[Ca^{2+}]$ solutions to generate a standard pCa curve for each experiment (Kao, 1994; Williams et al., 1999; Takahashi et al., 1999).

C. Pitfalls

One of the major expected difficulties in using a mixture of two different fluorochromes is the differential photobleaching of green versus red dye giving rise to the ratio bias. Surprisingly, differential photobleaching of Fluo-3 and Fura Red has not been a detectable problem (Lipp and Niggli, 1993; Schild et al., 1994; Floto et al., 1995). Another problem that can be fairly acute when performing simultaneous collection of dual emissions is that of a "bleed-through" ("cross-talk") of fluorescence signals between channels. This is due to the fact that a fraction of photons emitted by the green fluorochrome will still pass through red channel filters and, less often, vice versa. Standard filter sets for simultaneous dual collection do not discriminate well against the bleed-through. Several manufacturers sell filter sets that are customized for this purpose. It is also possible to eliminate bleed-through from the green channel by setting a fairly high threshold for the red channel (Camacho et al., 2000). This, however, may require increasing concentration of Fura Red, as its quantum efficiency is not high and, moreover, its emission decreases as Ca^{2+} concentration increases. Consequently, an additional problem of increased Ca^{2+} buffering by Fura Red occurs, which further limits the accuracy of the measurement

---

[5]Calibration of pCa_i should be done immediately after the end of each experiment under the same microscopic working conditions as those for pCa_i measurements but using a fresh batch of cells loaded with the dye mixture. We calibrate the pCa_i measurement by using solutions of different Ca^{2+} concentrations (Ca^{2+} buffer kits are available from, e.g., Molecular Probes, Inc.) containing 2–10 μM ionomycin to flush the dye-loaded cells. A series of measurements is then performed in the graded $[Ca^{2+}]$ solutions to generate a standard pCa curve. Between each measurement, cells have to be washed twice in buffer for the next-point calibration and allowed 3–5 min equilibration. We normally take at least five different image pairs of images from different areas of the cell monolayer for each calibration point. Some cell lines are very sensitive to any change in conditions of their medium (such as a difference in pCa) and do not survive the complete calibration procedure very well. It might be advisable to load a few smaller coverslips under exactly the same conditions (we were doing it in one Petri dish) and use one or two of them for each calibration point instead of using one coverslip for the calibration of all the points. In our experience, the entire experiment, including calibration, should be completed within not more than 3 h; otherwise a fresh batch of cells must be loaded.

(Floto *et al.,* 1995). Typically, however, bleed-through can be corrected by linear image operations. Briefly, under the same loading conditions, Fura Red is omitted from the procedure and, under the same microscopical conditions, Calcium Green-1 emission is collected through both its own channel and the channel normally used to collect Fura Red images. Division of the "green" channel image by the "red" channel image gives the bleed-through factor that can be subtracted from the "red" channel image acquired in the actual measurement using both dyes. Unfortunately, this bleed-through compensation technique is not perfect. Because of small local variations, it is impossible to avoid pixels that are either over- or undercompensated (Brelje *et al.,* 1994). Finally, imaging of very thick specimens (~hundreds of micrometers), such as *Xenopus* eggs, can produce artifactual measurements as shown by Fontanilla and Nuticelli (1998). Using Calcium Green-1-conjugated dextran co-microinjected with tetramethylrhodamine-conjugated dextran, they showed that because of differential absorption of either excitation or emission or both, ratio values varied by ~25% between the egg surface and 600 μm into the egg cytoplasm.

In summary, it is clear that UV-excitable fluorescent Ca^{2+} indicators are the most convenient and reliable tools for pCa$_i$ measurements. A mixture of dyes, at least one of which is Ca^{2+}-sensitive, is the next best choice if a confocal microscope with visible light excitation is used. Necessarily, dye loading and calibration are the serious problem areas for visible light confocal microscopy of intracellular Ca^{2+} indicators. One of the major improvements in confocal imaging that rapidly gathers recognition is two-photon excitation (Denk *et al.,* 1990). Briefly, a fluorochrome instead of a high-energy photon is excited simultaneously by two photons of low energy. Because the probability of excitation is high only in a region of high photon density, object layers that are out of focus are not bleached. Photodamage, photobleaching, and background fluorescence are reduced, compared to one-photon excitation. This is of prime importance in intracellular Ca^{2+} imaging (Kuba and Nakayama, 1998; Koester *et al.,* 1999; Piston and Knobel, 1999). These advantages of two-photon microscopy are particularly useful when imaging of intracellular Ca^{2+} is done with novel and exciting genetically engineered, protein-based fluorescent Ca^{2+} indicators (Miyawaki *et al.,* 1997, 1999).

III. Measurement of Organellar pCa$_i$ with Genetically Engineered [Ca^{2+}] Indicators

The "cameleons" indicators developed in the group of R. Y. Tsien are based on green fluorescent proteins (GFP) and calmodulin (Miyawaki *et al.,* 1997, 1999) and appear better suited for calcium measurements within organelles. The bright fluorescence of the GFP, combined with the selective targeting conferred by the addition of retention sequences, allows visualizing the calcium dynamics in organelles of living cells by fluorescence ratio imaging. Yellow cameleons (YC) are currently the preferred constructs (Fan *et al.,* 1999; Emmanouilidou *et al.,* 1999; Jaconi *et al.,* 2000; Foyouzi-Youssefi *et al.,* 2000; Yu and Hinkle, 2000) as they are brighter and more photostable than the original blue-green cameleons. These indicators use fluorescence resonance energy transfer (FRET) between a cyan GFP donor and a yellow GFP acceptor, which are linked together with calmodulin (CaM) and the calmodulin-binding peptide M13. Binding of Ca^{2+} to CaM increases the affinity of CaM for M13, resulting in a conformational change that brings the two GFP mutants in close proximity, increasing FRET and causing a shift in fluorescence emission from 480 nm to 535 nm. Because the ratio of the 535-nm to 480-nm emissions is directly proportional to the amount of Ca^{2+} linked to YC, the probes behave as ratiometric Ca^{2+} indicators and can be imaged using wide-field or multiphoton microscopy. Provided that a calibration curve is performed, YC then allows quantitative measurements of the free $[Ca^{2+}]$ within the organelles where they are expressed. Furthermore, the calcium affinity of calmodulin can be adjusted by molecular engineering, making it possible to match the calcium concentration within the organelle of interest. Whereas the original YC2 can report $[Ca^{2+}]$ variations between 100 nM and 10 μM, site-directed mutation of CaM has generated two other YCs with lower affinities for Ca^{2+}: YC3, suitable for monitoring $[Ca^{2+}]$ changes from 1 to 100 μM, and YC4, for measuring $[Ca^{2+}]$ changes in the range 10 μM to 1 mM (Miyawaki *et al.*, 1997).

A. Targeting

The beauty of the genetically encoded fluorescent protein indicators is that cells will produce the indicator and address them to specific compartments. The cell sorting machinery can thus be exploited to address YC in different location by fusion with specific targeting signals, retention sequences, or parts of adaptor proteins. At present, YCs have been successfully targeted to:

• *Cytosol:* YC indicators devoid of targeting signal display a uniform distribution within the cytosol, with only a slight labeling of the nucleus, consistent with the nuclear exclusion of the 74-kDa protein. Granules and other organelles appear in "negative" in the fluorescence images (Fig. 1a; also see Color Plate). Compared to chemical indicators, the cytosolic cameleon offers three main advantages: (1) The staining is purely cytosolic and provides a more precise picture of the Ca^{2+} changes occurring within this compartment; (2) Ca^{2+} changes can be resolved accurately well above the micromolar level, whereas fluorescent chelators typically saturate above 1 μM; and (3) long-term measurements can be performed as the fluorescent protein is constantly resynthesized by the cells and does not leak across the plasma membrane. The only limiting factor in this case is the amount of illumination that cells can withstand without incurring irreversible damage.

• *Nucleus:* Nuclear localization was achieved by adding a nuclear targeting sequence from SV40 (MPKKKRKV) to YC. In this case the fluorescence is exclusively restricted to nuclei and excluded from nucleoli (Miyawaki *et al.*, 1997).

• *ER:* The addition of a calreticulin signal sequence (MLLSVPLLLGLLGLAAAD) and a KDEL retention signal restricts the expression of YC to the ER (Miyawaki *et al.*,

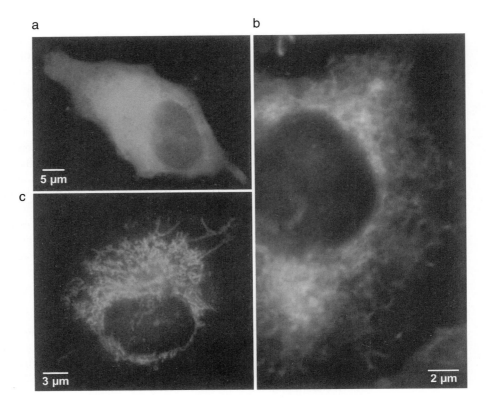

Fig. 1 Cellular distribution of three different YCs. Images of HeLa cells transiently transfected with YC2 (a), $YC4_{ER}$ (b) and $YC2_{mit}$ (c). In (b) the cell was also stained with MitoTracker Red (Molecular Probes) to show the distribution of mitochondria relative to the ER. Fluorescence images were obtained with a cooled CCD camera using 535DF25 (for YCs) and LP 590 (for mitotracker) emission filters. (See Color Plate.)

1997). However, care must be taken not to overexpress the protein as swelling of part of the ER can be observed in highly fluorescent cells (see below). This suggests that accumulation of large amounts of YC within the lumen of the ER might have detrimental effects on cellular function. We[6] used this probe to determine the free calcium concentration within the lumen of the ER, which in HeLa cells averaged 360 μM at rest and decreased

[6]**Gene transfer into the cells:** Plate cells on 25-mm glass coverslips in 35-mm petri dishes. Since transfection efficiency is sensitive to culture confluence, we have standardized the seeding protocol at $1-2 \times 10^5$ cells per petri dish in 2 ml of the appropriate complete growth medium. Cells are grown at 37°C in a CO_2 incubator until they reach 60–80% of confluence (this will take 24–48 h depending on cell types). Cell transfection can be performed using two standard methods.

1. Ca^{2+} phosphate coprecipitation protocol

In a polystyrene 12×75 mm sterile tube prepare the transfection mixture by adding per 35 mm petri dish (all the reagents must be at room temperature):

to 100 μM upon stimulation with Ca^{2+} mobilizing agonists. Similar values were obtained in other cell types measured with cameleons at 37°C, while lower values were observed at room temperature that could be accounted for by a reduced Ca^{2+} pumping activity (Miyawaki *et al.,* 1997, 1999). The $YC4_{ER}$ probe has been used to study the role of the anti-apoptotic protein Bcl-2 in ER Ca^{2+} homeostasis, (Foyouzi-Youssefi *et al.,* 2000) and the turnover of Ca^{2+} in the ER during signaling (Yu and Hinkle, 2000). We have also used $YC4_{ER}$ in conjunction with red shifted dyes to study the interactions between organelles (Fig. 1b).

---

2 μg DNA

Bidistilled H_2O so that the final volume H_2O/DNA $= 219 \mu l$

31 μl 2 M $CaCl_2$

and mix with the micropipettor or with finger tapping.

Then add quickly 250 μl of 2× HBS and bubble with automatic pipettor for 5–15 s or simply close the tube, inverse it and shake it vigorously for 10–15 s.

Add this HBS/DNA mixture dropwise onto the cells without removing the culture medium. You can observe the cells under a microscope to verify that very small black particles are precipitating.

Gently rock the petri dish in a crosslike manner to evenly distribute the DNA/$CaPO_4$ particles (circular movements concentrate the particles in the middle).

Return the cells to the incubator for 6–8 h before changing the medium with a fresh one.

Solutions (all the reagents must be of culture grade when possible):

2× HBS:50 mM Hepes, pH 7.05; 10 mM KCl; 12 mM D(+)-glucose; 280 mM NaCl; 1.5 mM Na_2HPO_4. The final pH must be exactly at 7.05. Filtrate at 0.2 μm, aliquot and stock at −20°C. Avoid freeze/thaw cycles. Let the aliquot thaw at room temperature and shake it to allow a uniform mixing. Even at −20°C this solution deteriorates after 6 months/1 year. If it becomes more difficult to produce small $CaPO_4$ precipitates change the solution for a fresh one.

$CaCl_2$: Prepare a 2 M solution and filtrate at 0.2 μm, aliquot and stock at −20°C.

2. Cationic lipid-based reagent

All the quantities and times that we give in this section were established with Transfast reagent (Promega). Prepare the following medium in a polystyrene 12 × 75 mm sterile tube by adding per 35 mm petri dish:

1 ml of culture medium (serum and antibacterial agents free) prewarmed at 37°C.

2 μg of DNA

6 μl of Transfast previously prepared as indicated by the manufacturer

Mix gently and incubate at room temperature for 10–15 min to allow DNA/liposomes complexes to form.

Mix again and add this transfection medium onto the cells after complete removal of the culture medium. Return the cells with this medium into the incubator for 1 h.

Following this incubation add 2 ml of complete growth medium without removing the transfection mixture and return the cells to the incubator.

Replace the medium with fresh complete growth medium at 8 h following the start of transfection.

The same quantities of DNA/transfection reagent were successfully tested with other cationic lipid-based reagents (DAC30 from Eurogentec and DOSPER from Boehringer) with incubation period with liposomes complexes for 6–8 h in presence of serum. Experiments with YC can be done between 2 and 5 days after gene transfer into the cells.

• *Mitochondria:* Targeting of YC to mitochondria has been obtained by incorporation of the mitochondrial targeting sequence from the complex VIII of the cytochrome oxidase (Fig. 1c). Import of this protein in the mitochondria appears to be a saturable process, as a slight labeling of the cytosol is frequently observed. This implies that cells must be visually screened and selected for proper staining of mitochondria before performing the Ca^{2+} measurements. Although this procedure restricts its use to microscopy-based assays, the $YC2_{mit}$ construct is at present the only indicator that is selectively targeted to mitochondria and provides sufficient signal to allow imaging of the Ca^{2+} changes occurring within the mitochondrial matrix with adequate spatial and temporal resolution. This construct has been used to study the flow of Ca^{2+} between mitochondria and the endoplasmic reticulum (Jaconi *et al.,* 2000; Arnaudeau *et al.,* 2001).

• *Secretory granule surface:* Targeting to the surface of large dense-core secretory vesicles (LDCVs) was obtained by fusing YC to the carboxyl terminus of the transmembrane LDCV protein phogrin (Emmanouilidou *et al.,* 1999). This allowed imaging of Ca^{2+} changes occurring around secretory granules, revealing that vesicles close to the plasma membrane—and hence from the Ca^{2+} source—experienced significantly larger Ca^{2+} increases than deeper vesicles.

In addition to these already existing cameleons, further molecular engineering (Fig. 2) should allow targeting of the probes to:

• *Golgi:* YC expression into the Golgi complex can in theory be obtained by fusing in frame the cDNA encoding the sialyltransferase, a Golgi resident glycosylation enzyme. This strategy was successful used with aequorin (Pinton *et al.,* 1998).

• *Secretory granule lumen:* The approach used to address a pH-sensitive GFP to the lumen of secretory granules by fusion to the lumenal domain of synaptobrevin-2 (Miesenbock *et al.,* 1998) can be adapted to YC to measure $[Ca^{2+}]$ changes within granules.

• *Plasma membrane:* YC probes can be targeted to the inner face of the plasma membrane by fusion to SNAP25, a polypeptide that participates in the docking of vesicles to the plasma membrane. Using this strategy, GFP and aequorin have been successfully targeted to the subplasmalemmal space (Marsault *et al.,* 1997).

B. Expression

Measurements with YC need gene transfer into the cells. For these cDNA transfections we currently use either Ca^{2+} phosphate (for HEK-293 cells) or reagent based on cationic lipids (for HeLa cells). These transfections are made directly on cells plated on 25-mm glass coverslips at $\sim 60\%$ confluence. Concentration of cDNA was established for YC2, $YC4_{ER}$, and $YC2_{mit}$ at 2 μg per 35-mm petri dish. Ca^{2+} measurements can be made with YC between 2 and 5 days after gene transfer into the cells. While attempting to optimize the efficacy of transfection of YC2, $YC4_{ER}$, and $YC2_{mit}$, we realized that YC is correctly addressed only when expressed at low levels. When high-efficiency transfection was achieved, correct targeting of $YC4_{ER}$ or $YC2_{mit}$ was only observed in weakly fluorescent

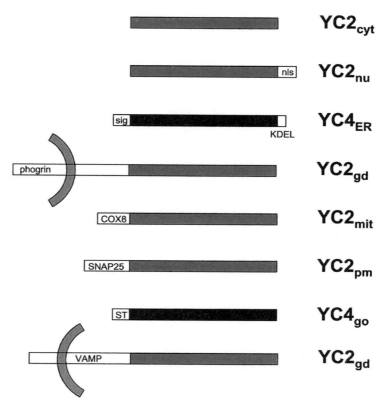

Fig. 2 Existing and potential cameleon constructs. The Ca^{2+} sensor in the various chimeras is the high-affinity YC2 or the low-affinity YC4 cameleon. Probes targeted to the cytosol (YC2$_{cyt}$), the nucleus (YC2$_{nu}$), the ER (YC4$_{ER}$), the secretory granule surface (YC2$_{gd}$), and the mitochondria (YC2$_{mit}$) are already available (Miyawaki *et al.*, 1997, 1999; Fan *et al.*, 1999; Emmanouilidou *et al.*, 1999; Jaconi *et al.*, 2000; Foyouzi-Youssefi *et al.*, 2000; Yu and Hinkle, 2000; Allen *et al.*, 1999). Cameleon directed to the Golgi complex (YC2$_{go}$), the secretory granule lumen (YC2$_{gl}$), or plasma membrane (YC2$_{pm}$) could be generated. The arcs show the probe orientation in the granule membrane.

cells, and the probe was visibly mistargeted in most, if not all, of the bright cells (Fig. 3a). In addition, the average fluorescence level of cells with correct YC4$_{ER}$ targeting closely matched that of our stable YC4$_{ER}$ cell line, which does not contain mistargeted cells (Fig. 3b). As increasing the efficacy of transfection increases the number of mistargeted cells, stable or low-efficiency transfections are preferable to ensure good addressing of YC. In addition, because each YC can bind four Ca^{2+}, high concentrations of YC can significantly affect Ca^{2+} dynamics by their buffering power. Miyawaki *et al.* reported that whereas [Ca^{2+}]$_c$ oscillations can be observed in cells expressing YC3.1 at concentrations between 40 μM and 150 μM, these were never observed in cells with [YC3.1] $>$300 μM (Miyawaki *et al.*, 1999).

a

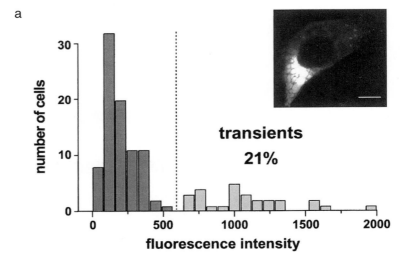

b

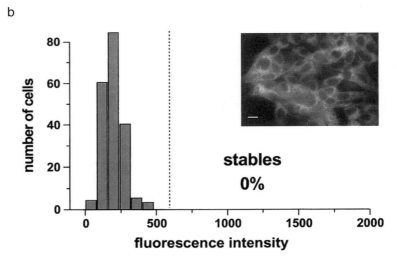

Fig. 3 Effect of YC4_{ER} expression level on targeting of the probe. Fluorescence intensity distribution of HEK-293 cells expressing YC4_{ER}. (a) Cells transfected transiently at high efficiency (5 μg/ml cDNA) distributed in two populations: weakly fluorescent cells that displayed proper retention of the probe in the ER, and bright cells in which swelling of the ER or leakage of the probe to the cytosol was clearly visible (inset). (b) In contrast, no mistargeting was observed in cells stably expressing YC4_{ER}, which had low fluorescence levels. Scale bar $= 10\ \mu$m.

C. Measurements[7]

For confocal measurements YC can be excited by a He–Cd laser or the 458-nm line of the 25-mW argon laser. YC can also be used with two-photon excitation in the wavelength range 770–810 nm which selectively excites the ECFP (Fan *et al.*, 1999). In this case the emitted light is split by a 510 dichroic mirror and the 480DF30 and 535DF25 filters are placed in front of two PMTs for simultaneous measurements of the emission wavelengths. For ratio fluorescence imaging with a wide-field microscope

[7]**Preparation of cells and measurement of fluorescence ratio:**

Place the cell into the Leiden perfusion chamber (MSC-TD coverslip dish from Harvard Apparatus), orienting the coverslip so that cells face upward. Ensure that good seal is achieved between the coverslip and the O-ring and close the chamber.

Add 1 ml of incubation medium.

Wipe away excess medium from the underside of the coverslip and check for leaks. It is important that this side be well dried; otherwise the remaining medium will make an emulsion with the oil immersion.

When using a mercury burner turn on the lamp power supply, then start the arc lamp. It is important to always ignite the lamp before turning on anything else, to prevent damage to the computer, the detectors (camera or PMTs), or other sensitive subsystems due to power surges.

Turn on the computer and other external hardware devices.

Verify that the temperature controller of the microincubator is set to the appropriate value (37°C, for example).

Turn off room lights to eliminate stray light from entering the optical system. Red light can be used for dim illumination of the room if a red blocking filter (550CFSP-Omega Optical) is mounted in front of the camera (or detectors).

Ensure that the correct driver is chosen for the CCD camera and that parameters are configured to the manufacturer specifications.

Load the software designed for time-lapse ratio fluorescence measurements and initiate a new experiment. We use Metafluor (Universal Imaging). With the illumination control panel install the correct hardware drivers for the external devices and subsequently configure them to the appropriate settings. Through the use of "Metadevices," Metafluor allows the user to create a variety of hardware configuration with different names. For example, you can configure a Metadevice that simultaneously controls the monochromator for excitation wavelengths (defined as wavelength control) and the filter wheel for emission wavelengths (defined as intensity control).

In the "configure acquisition" window set experiment specific parameters such as exposure time, pixel binning, Metadevice selection, and filter selection for each wavelength. Keep in mind that wavelength 1 is always the numerator of ratio 1. After configuring these settings save them as a protocol file that can be retrieved for subsequent use.

In the "configure experiment" menu choose the wavelengths that you want to acquire and ensure that these were selected so that their images will be displayed and updated continuously.

Observe the sample using transmission light to bring the cells in focus, then switch to fluorescence illumination to select cells that correctly express the YC.

Open the camera port and acquire a set of images. If necessary reset the "configure acquisition" parameters to optimize the acquisition region.

In the "reference image" menu acquire a set of images with the illumination shutter closed to subtract the background noise of the camera. Then reopen the shutter and acquire updated images with background subtraction.

By clicking on the "region" button you can select the regions of interest from which data will be extracted to construct time-lapse graphs of the fluorescence emissions and of the ratio during the experiment. Within the regions selected, data will only be extracted from pixels with absolute values higher than the preset lower

we use a Axiovert S100TV microscope and a 100×, 1.3 NA oil-immersion objective (Carl Zeiss AG, Feldbach, Switzerland). Excitation at 430 ± 10 nm is delivered by a Deltaram monochromator (Photon Technology International Inc., Monmouth Junction, NJ) through a 455DRLP dichroic mirror (Omega Optical, Brattleboro, VT). Dual emission is performed with a filter wheel (Ludl Electronic Products, Hawthorn, NY) that alternately changes two emission filters (480DF30 and 535DF25, Omega) placed in front of a back-illuminated CCD frame transfer camera (Princeton Instruments, Trenton, NJ) attached to the bottom port of the microscope. Because cells with proper YC targeting are typically not very bright (see earlier discussion), image acquisition of a single wavelength can take between 0.5 and 2 s to give a good signal-to-noise ratio using a medium intensity of illumination. Improvement in acquisition speed cannot be achieved by increasing illumination intensity due to EYFP photochromism (Miyawaki *et al.*, 1997). This photochromism causes an exponential decrease of the 535-nm fluorescence emission, whereas the 480-nm emission remains stable, resulting in a decrease in the 535/480 nm ratio that mimics a decrease in Ca^{2+} concentration. This artifact can be corrected by subtracting an exponential fit performed using to the first 1–2 min of the ratio trace, but it is preferable to circumvent the problem by reducing the illumination below the threshold level causing photochromism.

D. Calibration

In situ calibration curves of fluorescent indicators can vary significantly from those obtained *in vitro*. We encourage experimenters to establish their own *in vivo* calibration curve on each setup. The complete calibration curve can be obtained by incubating cells sequentially with solutions of known Ca^{2+} concentrations buffered with 5 mM EGTA and 5 mM HEEDTA using the calculation described in Bers *et al.* (1994). For cytosolic Ca^{2+} calibration of YC, equilibrium with these external Ca^{2+}-containing solutions is achieved by using a high concentration of ionomycin (10 μM). For calibration of YC targeted to organelles, the calibration is more problematic as in this case, the Ca^{2+} clamp must be imposed first in the cytosol and then in the organelles. To favor Ca^{2+} equilibration between the external medium and the cytosol, the permeability of the plasma membrane

---

threshold. Once the regions are defined, return to the image windows and set the threshold levels to fine tune the areas of interest.

Return to the "configure experiment" and configure the save interval and data logging. These determine which images will be recorded in the hard disk and what type of data, extracted from the image will be saved.

From the "experiment control panel" set the time-lapse interval for data collection and the number of images you want to acquire. Select "save image" and "log data" and initiate the collection of experimental results.

Using the "event mark" window, markers can be placed throughout the experiment at key events. These markers facilitate tracking the course of the experiment when the captured images are replayed.

Usually we prefer to save all the images of an experiment with coarse regions and no data logging just to follow ratio changes during the experiment. Fine-tuning of the regions and threshold for data logging are done during playback of the experiment after subtraction of the background fluorescence (defined outside the cells).

can be increased by adding digitonin (5 μg/ml). However, higher concentrations should not be used to preserve the organelle integrity and to avoid the loss of YC in the external medium. After obtaining the calibration curves, the apparent K_d^* and Hill coefficient (n) can be extracted and used to transform the emission ratio (R) in [Ca^{2+}]:

$$[Ca^{2+}] = K_d^*[(R - R_{min})/(R_{max} - R)]^{(1/n)}.$$

However, because of cell-to-cell heterogeneity (particularly evident with ER targeted YC), R_{max} and R_{min} have to be systematically determined at the end of each experiment. This is performed by sequentially equilibrating the cells with solutions containing 10 to 20 μM ionomycin and 20 mM Ca^{2+} or 20 mM EGTA.

E. Pitfalls

One of the major problem encountered with the original YC is their pH sensitivity. The fluorescence of the yellow GFP mutant (S65G/S72A/T203Y) increases with pH within the physiological pH range, with a pKa of 7.1 (Llopis *et al.*, 1998). Because the cyan fluorescence of the cameleon is much less pH sensitive, pH changes within the range 6–8 cause changes in ratio that mimic [Ca^{2+}] changes. Thus any pH decrease sensed by the YC can be interpreted as a [Ca^{2+}] decrease and vice versa. pH changes can be recognized from Ca^{2+} changes, however, as pH causes parallel changes in both 535-nm and 480-nm emission, whereas Ca^{2+} causes opposite changes in the cyan and yellow emission. However, this pH sensitivity makes YC unreliable in compartments undergoing concomitant changes in Ca^{2+} and pH, unless the pH is determined independently. To circumvent this pH dependency, the group of R. Tsien generated new cameleon probes by mutating the EYFP acceptor protein (Miyawaki *et al.*, 1999). These improved cameleons (YC2.1 and YC3.1) are much less pH sensitive and can be readily targeted to the ER and mitochondria by adding the appropriate retention sequence. However, folding of the modified probes is much less efficient at 37°C and a prolonged incubation period at 30°C is required to ensure protein expression in the ER. In addition, the dynamic range of the YC4.1$_{ER}$ probe was reduced ($R_{max}/R_{min} = 1.3$), impairing its ability to measure Ca^{2+} changes accurately in the ER. Thus, despite their pH dependency, we prefer to use the original YC for Ca^{2+} measurements in the ER and the mitochondria. The alkaline pH of mitochondria minimizes the pH interference in this organelle, whereas the pH of the ER has been shown to remain stable during stimulation with Ca^{2+}-mobilizing agonists (Kim *et al.*, 1998). The stability of the organelle pH can be verified by harnessing the built-in pH dependency of the cameleon probes. By directly exciting the pH-sensitive EYFP of YC4$_{ER}$ (ex/em: 480/535 nm), Ca^{2+}-dependent FRET is bypassed and the probe reports only pH changes. Using this approach, the pH changes occurring during the Ca^{2+} measurements can be assessed independently provided that the imaging system allows multiple excitations/emissions.

The other disadvantages of the cameleon are listed in Table II. They include the photochromism of the EYFP acceptor, which requires that illumination be kept at low

Table II
Comparative Advantages of Cameleon over Chemical Dyes

Advantages	Disadvantages
Molecular targeting	Low-level transfection required
Bright fluorescence	Photochromism
Tunable Ca^{2+} affinity	pH dependency
Quantitative (ratio)	Low dynamic range
Multiple combinations possible	Resolution limited by organelle motility

levels. This, together with the need for low-level transfections (discussed earlier), limits the intensity of the signal that has to be imaged and hence the final spatiotemporal resolution that can be achieved. When using a CCD camera and wide-field microscopy, the sequential acquisition of two images of sufficient S/N typically takes between 0.5 and 2 s. Consequently, the emission ratio is calculated with wavelengths acquired at best at 500-ms intervals, a time resolution that does not allow capture of fast Ca^{2+} transients. The spatial resolution also decreases with long exposures, especially when imaging very motile organelles such as mitochondria. The use of a beam splitter to simultaneous capture the two wavelengths on the CCD can significantly improve both the spatial and temporal resolution, but optimal signal detection might require the use of video-rate scanning two-photon excitation (Fan *et al.*, 1999). Finally, the relatively low dynamic range of the probes ($R_{max}/R_{min} = 1.5$–2.2) requires that a careful *in situ* calibration be performed to preserve the quantitative nature of the ratiometric measurements.

Acknowledgments

M.O.'s research has been supported by grants from the Ontario Heart and Stroke Foundation and the Medical Research Council of Canada. N.D. is a fellow from the Prof. Dr. Max Cloëtta Foundation and his research is supported by operating grant No. 31-46859.96 from the Swiss National Science Foundation (NSF).

References

Allen, G. J., Kwak, J. M., Chu, S. P., Llopis, J., Tsien, R. Y., Harper, J. F., and Schroeder, J. I. (1999). Cameleon calcium indicator reports cytoplasmic calcium dynamics in *Arabidopsis* guard cells. *Plant J.* **19**, 735–747.

Alonso, M. T., Barrero, M. J., Michelena, P., Carnicero, E., Cuchillo, I., Garcia, A. G., Garcia-Sancho, J., Montero, M., and Alvarez, J. (1999). Ca^{2+}-induced Ca^{2+} release in chromaffin cells seen from inside the ER with targeted aequorin. *J. Cell Biol.* **144**, 241–154.

Arnaudeau, S., Kelley, W. L., Walsh, J. V. Jr., and Demaurex, N. (2001). Mitochondria recycle Ca^{2+} to the endoplasmic reticulum and prevent the depletion of neighboring endoplasmic reticulum regions. *J. Biol. Chem.* **276**, 29430–29439.

Babcock, D. F., and Hille, B. (1998). Mitochondrial oversight of cellular Ca^{2+} signaling. *Curr. Opin. Neurobiol.* **8**, 398–404.

Babcock, D. F., Herrington, J., Goodwin, P. C., Park, Y. B., and Hille, B. (1997). Mitochondrial participation in the intracellular Ca^{2+} network. *J. Cell Biol.* **136**, 833–844.

Barrero, M. J., Montero, M., and Alvarez, J. (1997). Dynamics of $[Ca^{2+}]$ in the endoplasmic reticulum and cytoplasm of intact HeLa cells. A comparative study. *J. Biol. Chem.* **272**, 27694–27699.

Berridge, M. J. (1993). Inositol trisphosphate and calcium signalling. *Nature* **361**, 315–325.

Berridge, M. J., Lipp, P., and Bootman, M. (1999). Calcium signalling. *Curr. Biol.* **9**, R157–159.

Bers, D. M., Patton, C. W., and Nuccitelli, R. (1994). A practical guide to the preparation of Ca²⁺ buffers. *Methods Cell Biol.* **40**, 3–29.

Blatter, L. A., and Niggli, E. (1998). Confocal near-membrane detection of calcium in cardiac myocytes. *Cell Calcium* **23**, 269–279.

Brelje, T. C., Wessendorf, M. W., and Sorenson, R. L. (1994). Multicolor laser scanning confocal immunofluorescence microscopy: practical application and limitations. *Methods Cell Biol.* **38**, 97–181.

Bright, G. R., Fisher, G. W., Rogowska, J., and Taylor, D. L. (1989). Fluorescence ratio imaging microscopy. *Methods Cell Biol.* **30**, 157–192.

Brini, M., Marsault, R., Bastianutto, C., Pozzan, T., and Rizzuto, R. (1994). Nuclear targeting of aequorin. A new approach for measuring nuclear Ca²⁺ concentration in intact cells. *Cell Calcium* **16**, 259–268.

Button, D., and Eidsath, A. (1996). Aequorin targeted to the endoplasmic reticulum reveals heterogeneity in luminal Ca²⁺ concentration and reports agonist- or IP3–induced release of Ca²⁺. *Mol. Biol. Cell* **7**, 419–434.

Camacho, L., Parton, R., Trewavas, A. J., and Malhó, R. (2000). Imaging [Ca²⁺]c distribution and oscillations in pollen tubes with confocal microscopy. A comparison of different dyes and loading methods. *Protoplasma* **212**, 162–173.

Carnell, L., and Moore, H. P. (1994). Transport via the regulated secretory pathway in semi-intact PC12 cells: role of intra-cisternal calcium and pH in the transport and sorting of secretogranin II. *J. Cell Biol.* **127**, 693–705.

Chang, D. C., and Meng, C. (1995). A localized elevation of cytosolic free calcium is associated with cytokinesis in the zebrafish embryo. *J. Cell Biol.* **131**, 1539–1545.

DeBiasio, R., Bright, G. R., Ernst, L. A., Waggoner, A. S., and Taylor, D. L. (1987). Five-parameter fluorescence imaging: Wound healing of living Swiss 3T3 cells. *J. Cell Biol.* **105**, 1613–1622.

De Giorgi, F., Brini, M., Bastianutto, C., Marsault, R., Montero, M., Pizzo, P., Rossi, R., and Rizzuto, R. (1996). Targeting aequorin and green fluorescent protein to intracellular organelles. *Gene* **173**, 113–117.

Denk, W., Strickler, J. H., and Webb, W. W. (1990). Two-photon laser scanning fluorescence microscopy. *Science* **248**, 73–76.

Diliberto, P. A., Wang, X. F., and Herman, B. (1994). Confocal imaging of Ca²⁺ in cells. *Methods Cell Biol.* **40**, 243–262.

Dunn, K. W., Mayor, S., Myers, J. N., and Maxfield, F. R. (1994). Applications of ratio fluorescence microscopy in the study of cell physiology. *FASEB J.* **8**, 573–582.

Eberhard, M., and Erne, P. (1991). Calcium binding to fluorescent calcium indicators: Calcium green, calcium orange and calcium crimson. *Biochem. Biophys. Res. Commun.* **180**, 209–215.

Emmanouilidou, E., Teschemacher, A. G., Pouli, A. E., Nicholls, L. I., Seward, E. P., and Rutter, G. A. (1999). Imaging Ca²⁺ concentration changes at the secretory vesicle surface with a recombinant targeted cameleon. *Curr. Biol.* **9**, 915–918.

Fan, G. Y., Fujisaki, H., Miyawaki, A., Tsay, R. K., Tsien, R. Y., and Ellisman, M. H. (1999). Video-rate scanning two-photon excitation fluorescence microscopy and ratio imaging with cameleons. *Biophys. J.* **76**, 2412–2420.

Farkas, D. L., Baxter, G., DeBiasio, R. L., Gough, A., Nederlof, M. A., Pane, D., Pane, J., Patek, D. R., Ryan, K. W., and Taylor, D. L. (1993). Multimode light microscopy and the dynamics of molecules, cells, and tissues. *Annu. Rev. Physiol.* **55**, 785–817.

Ferri, G. L., Gaudio, R. M., Castello, I. F., Berger, P., and Giro, G. (1997). Quadruple immunofluorescence: A direct visualization method. *J. Histochem. Cytochem.* **45**, 155–158.

Floto, R. A., Mahaut-Smith, M. P., Somasundaram, B., and Allen, J. M. (1995). IgG-induced Ca²⁺ oscillations in differentiated U937 cells; a study using laser scanning confocal microscopy and co-loaded Fluo-3 and Fura-Red fluorescent probes. *Cell Calcium* **18**, 377–389.

Fontanilla, R. A., and Nuticelli, R. (1998). Characterization of the sperm-induced calcium wave in *Xenopus* eggs using confocal microscopy. *Biophys. J.* **75**, 2079–2087.

Foyouzi-Youssefi, R., Arnaudeau, S., Borner, C., Kelley, W. L., Tschopp, J., Lew, D. P., Demaurex, N., and Krause, K. H. (2000). Bcl-2 decreases the free Ca^{2+} concentration within the endoplasmic reticulum. *Proc. Natl. Acad. Sci. USA* **97**, 5723–5728.

Golovina, V. A., and Blaustein, M. P. (1997). Spatially and functionally distinct Ca^{2+} stores in sarcoplasmic and endoplasmic reticulum. *Science* **275**, 1643–1648.

Gresham, H. D., Goodwin, J. L., Allen, P. M., Anderson, D. C., and Brown, E. J. (1989). *J. Cell Biol.* **108**, 1935–1943.

Grynkiewicz, G., Poenie, M., and Tsien, R. Y. (1985). A new generation of Ca^{2+} indicators with greatly improved fluorescence properties. *J. Biol. Chem.* **260**, 3440–3450.

Hajnoczky, G., Robb-Gaspers, L. D., Seitz, M. B., and Thomas, A. P. (1995). Decoding of cytosolic calcium oscillations in the mitochondria. *Cell* **82**, 415–424.

Haugland, R. P. (1998). "Handbook of Fluorescent Probes and Research Chemicals." Molecular Probes, Inc., Eugene, OR.

Hirose, K., and Iino, M. (1994). Heterogeneity of channel density in inositol-1,4,5-trisphosphate-sensitive Ca^{2+} stores. *Nature* **372**, 791–794.

Hofer, A. M., and Machen, T. E. (1993). Technique for in situ measurement of calcium in intracellular inositol 1,4,5-trisphosphate-sensitive stores using the fluorescent indicator mag-fura-2. *Proc. Natl. Acad. Sci. USA* **90**, 2598–2602.

Hofer, A. M., Fasolato, C., and Pozzan, T. (1998). Capacitative Ca^{2+} entry is closely linked to the filling state of internal Ca^{2+} stores: a study using simultaneous measurements of ICRAC and intraluminal $[Ca^{2+}]$. *J. Cell Biol.* **140**, 325–334.

Hoth, M., Fanger, C. M., and Lewis, R. S. (1997). Mitochondrial regulation of store-operated calcium signaling in T lymphocytes. *J. Cell Biol.* **137**, 633–648.

Ingalls, C. P., Warren, G. L., Williams, J. H., Ward, C. W., and Armstrong, R. B. (1998). E-C coupling failure in mouse EDL muscle after in vivo eccentric contractions. *J. Appl. Physiol.* **85**, 658–676.

Jaconi, M., Bony, C., Richards, S. M., Terzic, A., Arnaudeau, S., Vassort, G., and Puceat, M. (2000). Inositol 1,4,5-trisphosphate directs Ca^{2+} flow between mitochondria and the endoplasmic/sarcoplasmic reticulum: A role in regulating cardiac autonomic Ca^{2+} spiking. *Mol. Biol. Cell* **11**, 1845–1858.

Kaftan, E. J., Xu, T., Abercrombie, R. F., and Hille, B. (2000). Mitochondria shape hormonally-induced cytoplasmic calcium oscillations and modulate exocytosis. *J. Biol. Chem.* **275**, 25465–25470.

Kao, J. P. Y., (1994). Practical aspects of measuring $[Ca^{2+}]$ with fluorescent indicators. *Methods Cell Biol.* **40**, 155–181.

Kim, J. H., Johannes, L., Goud, B., Antony, C., Lingwood, C. A., Daneman, R., and Grinstein, S. (1998). Noninvasive measurement of the pH of the endoplasmic reticulum at rest and during calcium release. *Proc. Natl. Acad. Sci. USA* **95**, 2997–3002.

Koester, H. J., Baur, D., Uhl, R., and Hell, S. W. (1999). Ca^{2+} fluorescence imaging with pico- and femtosecond two-photon excitation: Signal and photodamage. *Biophys. J.* **77**, 2226–2236.

Kuba, K., and Nakayama, S. (1998). Two-photon laser-scanning microscopy: tests of objective lenses and Ca^{2+} probes. *Neurosci. Res.* **32**, 281–294.

Lipp, P., and Niggli, E. (1993). Ratiometric confocal Ca^{2+}-measurements with visible wavelength indicators in isolated cardiac myocytes. *Cell Calcium* **14**, 359–372.

Lipp, P., Luscher, C., and Niggli, E. (1996). Photolysis of caged compounds characterized by ratiometric confocal microscopy: a new approach to homogeneously control and measure the calcium concentration in cardiac myocytes. *Cell Calcium* **19**, 255–266.

Llopis, J., McCaffery, J. M., Miyawaki, A., Farquhar, M. G., and Tsien, R. Y. (1998). Measurement of cytosolic, mitochondrial, and Golgi pH in single living cells with green fluorescent proteins. *Proc. Natl. Acad. Sci. USA* **95**, 6803–6808.

Marks, P. W., and Maxfield, F. R. (1991). Preparation of solutions with free calcium concentration in the nanomolar range using 1,2-bis(*o*-aminophenoxy)ethane-*N,N,N',N'*-tetraacetic acid. *Anal. Biochem.* **193**, 61–71.

Marsault, R., Murgia, M., Pozzan, T., and Rizzuto, R. (1997). Domains of high Ca^{2+} beneath the plasma membrane of living A7r5 cells. *EMBO J.* **16**, 1575–1581.

McDougall, A., and Sardet, C. (1995). Function and characteristics of repetitive calcium waves associated with meiosis. *Curr. Biol.* **5,** 318–328.

McNeil, P. L. (1989). Incorporation of macromolecules into living cells. *Methods Cell Biol.* **29,** 153–173.

Meldolesi, J., and Pozzan, T. (1998). The heterogeneity of ER Ca^{2+} stores has a key role in nonmuscle cell signaling and function. *J. Cell Biol.* **142,** 1395–1398.

Merritt, J. E., McCarthy, S. A., Davies, M. P. A., and Moores, K. E. (1990). Use of fluo-3 to measure cytosolic Ca^{2+} in platelets and neutrophils. *Biochem. J.* **269,** 513–519.

Miesenbock, G., De Angelis, D. A., and Rothman, J. E. (1998). Visualizing secretion and synaptic transmission with pH-sensitive green fluorescent proteins. *Nature* **394,** 192–195.

Minta, A., Kao, J. P. Y., and Tsien, R. Y. (1989). Fluorescent indicators for cytosolic calcium based on rhodamine and fluorescein chromophores. *J. Biol. Chem.* **264,** 8171–8178.

Miyawaki, A., Griesbeck, O., Heim, R., and Tsien, R. Y. (1999). Dynamic and quantitative Ca^{2+} measurements using improved cameleons. *Proc. Natl. Acad. Sci. USA* **96,** 2135–2140.

Miyawaki, A., Llopis, J., Heim, R., McCaffery, J. M., Adams, J. A., Ikura, M., and Tsien, R. Y. (1997). Fluorescent indicators for Ca^{2+} based on green fluorescent proteins and calmodulin. *Nature* **388,** 882–887.

Monck, J. R., Oberhauser, A. F., Keating, T. J., and Fernandez, J. M. (1992). Thin-section ratiometric Ca^{2+} images obtained by optical sectioning of fura-2 loaded mast cells. *J. Cell Biol.* **116,** 745–759.

Montero, M., Alonso, M. T., Carnicero, E., Cuchillo-Ibanez, I., Albillos, A., Garcia, A. G., Garcia-Sancho, J., and Alvarez, J. (2000). Chromaffin-cell stimulation triggers fast millimolar mitochondrial Ca^{2+} transients that modulate secretion. *Nat. Cell Biol.* **2,** 57–61.

Montero, M., Brini, M., Marsault, R., Alvarez, J., Sitia, R., Pozzan, T., and Rizzuto, R. (1995). Monitoring dynamic changes in free Ca^{2+} concentration in the endoplasmic reticulum of intact cells. *EMBO J.* **14,** 5467–5475.

O'Malley, D. M. (1994). Calcium permeability of the neuronal nuclear envelope: evaluation using confocal volumes and intracellular perfusion. *J. Neurosci.* **14,** 5741–5758.

O'Malley, D. M., Burbach, B. J., and Adams, P. R. (1999). Fluorescent calcium indicators: subcellular behavior and use in confocal imaging. *Methods Mol. Biol.* **122,** 261–303.

Opas, M. (1997). Measurement of intracellular pH and pCa with a confocal microscope. *Trends Cell Biol.* **7,** 75–80.

Opas, M., and Dziak, E. (1999). Intracellular pH and pCa measurements. *Methods Mol. Biol.* **122,** 305–313.

Opas, M., and Kalnins, V. I. (1985). Multiple labeling of cellular constituents by combining surface reflection interference and fluorescence microscopy. *Exp. Cell Biol.* **53,** 241–251.

Orci, L., Ravazzola, M., Amherdt, M., Perrelet, A., Powell, S. K., Quinn, D. L., and Moore, H. P. (1987). The trans-most cisternae of the Golgi complex: a compartment for sorting of secretory and plasma membrane proteins. *Cell* **51,** 1039–1051.

Pawley, J. B., Ed. (1995). "Handbook of Biological Confocal Microscopy." Plenum Press, New York.

Perez-Terzic, C., Pyle, J., Jaconi, M., Stehno-Bittel, L., and Clapham, D. E. (1996). Conformational states of the nuclear pore complex induced by depletion of nuclear Ca^{2+} stores. *Science* **273,** 1875–1877.

Pinton, P., Ferrari, D., Magalhaes, P., Schulze-Osthoff, K., Di Virgilio, F., Pozzan, T., and Rizzuto, R. (2000). Reduced loading of intracellular Ca^{2+} stores and downregulation of capacitative Ca^{2+} influx in bcl-2-overexpressing cells. *J. Cell Biol.* **148,** 857–862.

Pinton, P., Pozzan, T., and Rizzuto, R. (1998). The Golgi apparatus is an inositol 1,4,5-trisphosphate-sensitive Ca^{2+} store, with functional properties distinct from those of the endoplasmic reticulum. *EMBO J.* **17,** 5298–5308.

Piston, D. W., and Knobel, S. M. (1999). Quantitative imaging of metabolism by two-photon excitation microscopy. *Methods Enzymol.* **307,** 351–368.

Pozzan, T., Rizzuto, R., Volpe, P., and Meldolesi, J. (1994). Molecular and cellular physiology of intracellular calcium stores. *Physiol. Rev.* **74,** 595–636.

Pralong, W. F., Spat, A., and Wollheim, C. B. (1994). Dynamic pacing of cell metabolism by intracellular Ca^{2+} transients. *J. Biol. Chem.* **269,** 27310–27314.

Putney, J. W. Jr., and McKay, R. R. (1999). Capacitative calcium entry channels. *Bioessays* **21,** 38–46.

Rizzuto, R., Bastianutto, C., Brini, M., Murgia, M., and Pozzan, T. (1994). Mitochondrial Ca^{2+} homeostasis in intact cells. *J. Cell Biol.* **126**, 1183–1194.

Rizzuto, R., Brini, M., De Giorgi, F., Rossi, R., Heim, R., Tsien, R. Y., and Pozzan, T. (1996). Double labelling of subcellular structures with organelle- targeted GFP mutants in vivo. *Curr. Biol.* **6**, 183–188.

Rizzuto, R., Brini, M., and Pozzan, T. (1993). Intracellular targeting of the photoprotein aequorin: a new approach for measuring, in living cells, Ca^{2+} concentrations in defined cellular compartments. *Cytotechnology* **11**, S44–46.

Rizzuto, R., Brini, M., and Pozzan, T. (1994). Targeting recombinant aequorin to specific intracellular organelles. *Methods Cell Biol.* **40**, 339–358.

Rizzuto, R., Simpson, A. W., Brini, M., and Pozzan, T. (1992). Rapid changes of mitochondrial Ca^{2+} revealed by specifically targeted recombinant aequorin. *Nature* **358**, 325–327.

Rutter, G. A., Burnett, P., Rizzuto, R., Brini, M., Murgia, M., Pozzan, T., Tavare, J. M., and Denton, R. M. (1996). Subcellular imaging of intramitochondrial Ca^{2+} with recombinant targeted aequorin: significance for the regulation of pyruvate dehydrogenase activity. *Proc. Natl. Acad. Sci. USA* **93**, 5489–5494.

Rutter, G. A., Fasolato, C., and Rizzuto, R. (1998). Calcium and organelles: a two-sided story. *Biochem. Biophys. Res. Commun.* **253**, 549–557.

Schild, D., Jung, A., and Schultens, H. A. (1994). Localization of calcium entry through calcium channels in olfactory receptor neurones using a laser scanning microscope and the calcium indicator dyes Fluo-3 and Fura-Red. *Cell Calcium* **15**, 341–348.

Seksek, O., and Bolard, J. (1996). Nuclear pH gradient in mammalian cells revealed by laser microspectrofluorimetry. *J. Cell Sci.* **109**, 257–262.

Simpson, P. B., and Russell, J. T. (1996). Mitochondria support inositol 1,4,5-trisphosphate-mediated Ca^{2+} waves in cultured oligodendrocytes. *J. Biol. Chem.* **271**, 33493–33501.

Sitia, R., and Meldolesi, J. (1992). Endoplasmic reticulum: a dynamic patchwork of specialized subregions. *Mol. Biol. Cell.* **3**, 1067–1072.

Stehno-Bittel, L., Perez-Terzic, C., and Clapham, D. E. (1995). Diffusion across the nuclear envelope inhibited by depletion of the nuclear Ca^{2+} store. *Science* **270**, 1835–1838.

Stricker, S. A., and Whitaker, M. (1999). Confocal laser scanning microscopy of calcium dynamics in living cells. *Microsc. Res. Tech.* **46**, 356–369.

Takahashi, A., Camacho, P., Lechleiter, J. D., and Herman, B. (1999). Measurement of intracellular calcium. *Physiol. Rev.* **79**, 1089–1125.

Tse, A., Tse, F. W., and Hille, B. (1994). Calcium homeostasis in identified rat gonadotrophs. *J. Physiol. (Lond.)* **477**, 511–525.

Tsien, R. Y. (1981). A non-destructive technique for loading calcium buffers and indicators into cells. *Nature* **290**, 527–528.

Whitaker, J. E., Haugland, R. P., and Prendergast, F. G. (1991). Spectral and photophysical studies of benzo[*c*]xanthene dyes: dual emission pH sensors. *Anal. Biochem.* **194**, 330–344.

Williams, D. A., Bowser, D. N., and Petrou, S. (1999). Confocal Ca^{2+} imaging of organelles, cells, tissues, and organs. *Methods Enzymol.* **307**, 441–469.

Xu, H., and Heath, M. C. (1998). Role of calcium in signal transduction during the hypertensive response caused by basidiospore-derived infection of the cowpea rust fungus. *Plant Cell* **10**, 585–597.

Yu, R., and Hinkle, P. M. (2000). Rapid turnover of calcium in the endoplasmic reticulum during signaling: Studies with cameleon calcium indicators. *J. Biol. Chem.* **275**, 23648–23653.

Zhou, Y., and Opas, M. (1994). Cell shape, intracellular pH and FGF responsiveness during transdifferentiation of retinal pigment epithelium into neuroepithelium *in vitro*. *Biochem. Cell Biol.* **72**, 257–265.

CHAPTER 14

Running and Setting Up a Confocal Microscope Core Facility

Susan DeMaggio

President, Flocyte Associates
Irvine, California 92612

I. Introduction

The use of core facilities for the supply and support of large shared instruments is a growing trend that will benefit all researchers, as well as those funding the projects. In these days of quickly developing technologies, no one researcher could possibly keep up to date with all the software and hardware for the applications and techniques used in his or her lab. Individual researchers don't always have time to learn all the finer points of using the new technologies, or even to research what new technologies out there are the ones they need, or that could give them the desired results. It is our service as core managers to develop and maintain facilities equipped with these fast changing technologies and to allow the researcher to concentrate on the science.

There are many facets of developing a core facility with shared equipment. In this chapter we will cover all aspects of core management, including planning the layout of your facility, procurement of instruments, scheduling use, training of users and operators, instrument care and maintenance, data handling, ancillary services such as protocols and sample preparation, as well as governmental issues that affect recharging. We will also cover some resources available to core managers to assist us in the job we are to do.

II. Planning Your Facility

The first step, and most likely, the impetus for starting this whole process, is the establishment of the need for confocal imaging and a user base anxious to have the technology available to them. Only when the needs are defined, is it possible to choose the microscope that will fill them. In putting together a shared facility, it is important to choose either an instrument with a lot of flexibility or one which can be augmented in the future to add all the possible capabilities you *might* need in a core. Polling the potential users, or grant applicants, for their projected needs is the most vital next step. If possible, when soliciting researchers to write the grant proposal, speak to users who will have varied applications, and cover all the possible configurations. This will help you justify the most important features for your users on the initial purchase and justify the widest variety of options.

A general idea of the requirements of your users and the basic instrument you will need is important before seeking funding, since the prices of instruments vary considerably. Funding for large equipment is available as shared instrument grants such as the National Institute of Health Shared Instrument Grant and NSF program or from individual or corporate donors. Many times matching funds from institutions are necessary in order for these grants to be funded. Strong institutional support often is the difference between a winning grant proposal and a losing one. Make sure that you understand the instrument you are requesting, and how it will support your research needs. If you sound knowledgeable about the process and the instrument, you are more apt to be funded. Before writing the grant proposal, check with your local microscope sales representatives and find a microscope you can see in action. Ask the operators, the people who actually operate the hardware and software on a daily basis, questions such as: How stable is this system? How easy is it to learn and operate this software? How long before you feel competent operating this scope? Have you needed any repairs? How was the service? Is technical assistance easy to get? Answers to this kind of question can help you rule out one or another manufacturer. Service in different parts of the country can also vary—not all technicians are created equal, so to speak—but it can be a guide.

Once you have chosen the instrument to purchase, it is vital to ensure that the environment is adequate, that you have allowed enough room to accommodate the instrument. The microscope may require a large vibration isolation table, and an access room behind the instrument for service and maintenance. You might need additional desktop workspace for the computer. Room for sample preparation nearby is helpful, but this should be far enough away from the microscope that work there will not interfere with the imaging. This allows both parts of the process to take place simultaneously.

Lasers add a complexity to the room design. The room may need extra ventilation or may need access to ceiling crawlspace or an outside vent in which to put the air-cooled laser cooling fans. This will allow you to vent the hot air out of the room, away from the lasers, the computer components, any labile or delicate samples, and you, all of which should be kept cool. If the laser is water cooled, you will need to address which water circulation or chilling system will work best. One of the three varieties circulates the water and cools it with ambient air. The second circulates water through the laser and dumps it into a drain. Another circulates the water, continually recycling it to a large chilling unit outside the building or on the roof, chilling the laser more quickly and more efficiently.

Lighting is also a concern when dealing with fluorescent techniques. It is helpful to have an incandescent light source as well as the fluorescent overhead lights so you can control them with a rheostat and keep the room dimly lit during acquisition. The lighting supplies to the confocal are also a drain on the electrical circuitry. It is necessary to consider if you have enough separate circuits that the lasers and arc lamps will not interfere with the computer and electronic components when turning them off and on. For an extra measure of safety, turn the light sources on before the computer parts and off after—do not light them while delicate electronics are on. They have been known to spike and damage your computer.

III. Scheduling

If your facility offers services to several laboratories or even different schools or colleges within your university, it is advisable to have a system of scheduling that will allow users to access it from their own computers or via the Internet. This eliminates the need for users to call *you* for an appointment or for them to walk to your laboratory door and physically write on a calendar. Programs such as Corporate Time or Windows Outlook have the capability for scheduling resources and allow users to connect to them from their computers. As general rule of thumb, the user who is properly signed up on the computer, *not the one who arrived first,* is the one with access, should there be a discrepancy. Computer calendar documentation is also a tool for confirmation of use, in case your logging system fails, or if there is question of who was using the microscope last or from which laboratory they came. Many of the newer systems include a login feature, which allows you to set up users with passwords and thus protect data and limit use by untrained users. These are important features, and it is advisable to take advantage of them when setting up your operations. An added benefit is that you have an accurate record of use and discrepancies are rare, since the computer has documented the use. Another side benefit is that the billing returns often increase when the computer takes care of the logging.

In some institutions, it is necessary to give priority to certain groups in scheduling. If this is the case, those people may have direct access to the calendar and others may have to submit an e-mail request for use. Depending on the demand for your instrument, this may cause unnecessary delays, or may be a vital step in keeping peace.

IV. Training Operators and Users

After your microscope is installed, it is necessary to become trained to operate the scope and in turn to train frequent users. All major manufacturers give you a training class or sessions with the purchase of your instrument. Parts of this training can be adapted for your frequent users, or used as an outline for the important issues to cover in your training. As a service facility manager, you may be asked to image all types of specimens for your researchers, especially if they just need one or two pictures to prove a point for a reviewer, or to document a cellular event for their own records. You will often be presented with research problems that you have never encountered before. For this reason it is beneficial to have access to a technical and applications support unit for your microscope. With the technical assistance of the manufacturer, you can face

---

Optical Biology Core Facility
User Request Form

User Name:_____ P.I._____

Email address:_____

LOGIN:_____

Cancer Center Member?:   yes   no   (Circle 1)

Phone:_____**Lab Address:**_____

Peer - Reviewed Grant Agency:_____**Grant No.**_____

University Account and Fund _____-_____-_____-_____

Training Date:_____**Part II**_____

Instrument: ☐ Confocal Microscope   ☐ MRC 600   ☐ MRC 1024 UV
                        ☐ FACScan   ☐ Sorting
                        ☐ CytoFluor   ☐ Video Imaging
Dye or Probe:_____

    Excitation wavelength need for dyes:

    351   363   488   514   529   568   647

Laser: Argon/ air cooled   Krypton/Argon   Argon / Water (UV)

Application:_____

Brief abstract or description of project from Grant - (Include the need for the instrument):

Fig. 1   User request form for use of the optical biology core facilities.

problems you really have never seen before. Do not let inexperience scare you away. All of us at one time knew nothing about confocal imaging. Special training courses are also given at research institutes such as Cold Spring Harbor, Hopkins Marine Station, Woods Hole Marine Station, and the University of British Columbia, which holds a class in 3-D microscopy of living cells during the summer. These courses will enhance your understanding of imaging and help you offer better service.

Your frequent users will want to have access to the facilities and to take images at times when you are not available, so developing a thorough training for your users is imperative. They need to be able to accomplish their research objectives but not necessarily know everything there is to know about the microscope, so each training is tailored to the needs of that researcher. An interview with or request form filled out by the prospective user will help you establish the topics of specific interest, the research objectives of the project, and the background or experience level of the user. This type of form identifies the needs of each researcher and will help you tailor his or her training. It also helps you compile your database of users, so that communication, billing, and reporting the various uses of your facility are easier to accomplish. For an example of such a request, see Fig. 1. Before training on your system, users should be comfortable on computers or at least be able to function easily on your platform, should know the science they are working with, and should understand basic principles of fluorescence. Then you can cover these basic topics in confocal imaging:

1. The theory of confocal microscopy will give users a better idea of its capabilities and limitations. Explain to them how your system and other confocal technologies work.

2. Explain the features with which your system is equipped—such as the excitation wavelengths available on your lasers, and what dyes, stains, and probes are appropriate for those wavelengths. A table similar to that in Fig. 2, designed for an MRC 1024 UV system from BioRad, may be useful for deciding which laser and filter setup is appropriate for inexperienced microscopists. There are also Web sites available, such as one from Bio-Rad (http://fluorescence.bio-rad.com/) and one from Molecular Probes (http://www.probes.com/handbook/toc.html), that will help users understand fluorescence microscopy and determine good combinations of dyes for specific excitations as well.

3. Basic care and use of the microscope itself should be addressed, since no matter how fancy the devices attached, the images will not be good if the optics and microscope are not treated with respect and cleanliness. A more thorough discussion of instrument care follows.

4. Data collection options are a critical issue. It is important to decide where to store the images. Do you allow images to be left on your drives? How about specific methods and settings? What are your guidelines for transporting images to other computers? The most fundamental lesson is: *back it up* and have two copies of everything, whenever possible.

5. Software basic operation training should cover the options needed by that particular researcher, but also offer a taste of options the researcher might possibly want to use at

Laser	Laser Line	Default Method	Emission Filter Choices*	Dyes
Argon Ion Water Cooled	351	Blue / Green	455 / 30 405 / 35 460 LP	AMCA / FITC Alexa 350 (Em 422)
Argon Ion Water Cooled	351	Emission Ratioing	455 / 30 405 / 35 460 LP	Indo - 1
Argon Ion Water Cooled	363	Vital DNA BFP/GFP FRET	455 / 30 460 LP	Hoechst / DAPI* BFP, EBFP
Krypton Argon Mixed Gas	488	Fluorescent In Situ Triple Labeling Traditional DNA	522 / 35 540 / 30 585 / LP OG 515	FITC / PI FITC/DTAF/Bodipy/Cy3 PI / TO / AO GFP, EGFP Alexa 488 (Em519) Oregon Green (Em 524)
Krypton Argon Mixed Gas	488	Cellular Calcium Membrane Labels Neural Tracer	522 / 35 540 / 30 585 / LP OG 515	Ca^{++} Green / Fluo-3 Di A / Di O Lucifer Yellow
Krypton Argon Mixed Gas	488	Reflection	522 / 35 540 / 30 585 / LP OG 515	Reflectance
Krypton Argon Mixed Gas	488	3 Color Transmission	522 / 35 540 / 30 585 / LP OG 515	Transmission
Krypton Argon Mixed Gas	568	Cellular Calcium Membrane / Neural Tracer	605 / 32	Ca^{++} Crimson Di I
Krypton Argon Mixed Gas	568	Traditional DNA Triple Labeling	605 / 32	Ethidium Bromide / 7AAD / PI TxRed/LissRhod/TRITC Alexa 568 (Em 603) DsRed (Em 583)
Krypton Argon Mixed Gas	647	Triple Labeling	680 / 32	Cy 5, Cy 7

- Band filters - first number is the center of the band / the second the width – ie 455 / 30 means a band from 440 – 470, 15 nm above and below the center.
- LP indicates Long Pass filters
- OG – blue absorbing Long pass green filter

Fig. 2 Excitation and emission wavelengths, filters, and associated days.

some time in the future. Instrument use will increase not only when users know how to get the best out of the software, but when they discover new ways to use it.

6. Turn-on and turn-off procedures can be printed on quick reference charts near the instrument or as the home page on the Web browsers for each workstation. These guidelines are important in order to best maintain long life of your light sources and limit surge damage to computer and electronic elements.

7. One of the most difficult areas will be to instill a knowledge in your users of how to recognize emergencies, problems with the system that need immediate attention by an expert, and whom to call in such an emergency.

Of course, the bulk of your training will be on the specific software and operation of your particular instrument. Use your operators' training with the instrument manufacturer as a guide to design a training outline for your users, to cover all aspects of the operation of the confocal. However, limit user training primarily to those things that are necessary for daily operation. For example, perhaps not everyone needs to know the details of alignment on your system, especially if it is stable and needs alignment only rarely. Tailoring individual training to the needs of the researchers allows them to concentrate on getting the appropriate images they desire, and getting them quickly. However, it is also useful to present other features available on your system. This allows researchers to consider all capabilities of the system and may perhaps inspire them to expand their use to take advantage of those options in a future experiment.

V. Instrument Care

Keeping the microscope in good condition is a critical issue. No matter how fancy the electronics or computer enhancement of a system, if the optics are not cared for and kept clean, the images will not be good. Information on the importance of keeping the instrument clean, directions to where the cleaning supplies are kept, and techniques for cleaning the lenses should be included as a part of each training. Supplies are always kept within sight of the microscope. It is also a good idea to check the microscope weekly for cleanliness, and give it a thorough cleaning yourself when needed. Oil left on inverted objectives can run down into the internal parts and cloud lenses or, worse yet, corrode the objectives and ruin them. Immersion oil is very corrosive and can cause permanent damage if allowed to drain down into the microscope parts. Keep your eyes open for damage caused by users and misuse of the microscope. Don't be afraid to ask the last users for their input on the performance of the microscope when they used the instrument last. It might be possible to catch misunderstandings of microscope care, or methods of handling the equipment before permanent damage occurs. Waiting until the damage is done and the microscope is in need of costly repair limits the use of the scope for everyone and decreases the income derived from it.

The quality of images produced on your microscope can be monitored by the preparation of a slide or two containing fluorescent tapes or filter paper, and beads. Such slides are available from your microscope manufacturer or commercial suppliers, or they can be reproduced in your own laboratory. Fluorescent controls can be made from safranin dye-impregnated filter paper, mounted with an antiquench mounting medium on a glass slide. Fluorescent beads of different colors and intensities from commercial sources can also be mounted on slides in antiquench media and used over and over for controls. Images of these controls at specified settings can be stored in a default folder, allowing you to recheck the laser intensity and the optics weekly, or when there is a question about the brightness of images. Simply reset the settings to those used in the controls and compare the images. There are also readings of voltage and power of the laser that you can read and record, to determine if the laser is losing power and in need of repair. Check with your microscope or laser manufacturer for appropriate levels and how to check them with a voltmeter.

Accurate measurements can be made with the confocal microscope by checking the calibration of each objective with a slide micrometer. Image the micrometer at different magnifications and store the information and image in a file. The confocal software package stores the calibrations, and if each image is saved with the appropriate lens chosen, you can measure structures within the images accurately. As a recheck, measure another micrometer as if it were an unknown.

VI. Data Storage, Analysis, and Handling

Storage options for images are also a huge topic of interest for core managers. Volumes of images can often be as large as 100 MB and hard-disk space on your computer can be quickly consumed by images left there. Magnetooptical drives and the advent of Jaz and Zip disks have simplified this issue, and the addition of one or more of these drives allows your users to transport images to other offline workstations for analysis. Images can even be burned onto CDs when taken to an offline system uninterrupted by other users or the network. These three options have revolutionized image storage, since they appear to be stable for many years, allow users to take their images to other workstations, and hold very large volumes of information at much less expense than previous methods. When using Zip or Jaz disks, however, it is best not to fill them to capacity, for they have been known to crash and lose all data when filled too full. Also, keep a second backup, two copies of everything important.

When multiple users have access to the system, the need to save images to individual folders or directories and to individual disks is even more critically important. Each file is important to that researcher, representing hours of work and sometimes immeasurable cost to procure. Some samples are precious and cannot be duplicated, and some require hundreds of dollars worth of sample, antibodies, and/or reagents to prepare. For these reasons, individual media are recommended for storage. Each user can keep his or her own Zip or Jaz disk in his or her possession, ensuring that all images are safe. Settings and methods can and should also be backed up onto a Zip or Jaz disk in case of system crashes. As core manager, you can decide if you want to be responsible for that level of backup, or if you want users to be responsible to back up their own images. In that case you should recommend very strongly to your users that they take that responsibility seriously.

In most instances, images can be taken from the confocal in formats that allow the users to view and change them in offline software. Most software packages available read the common confocal formats, but if not, it is also possible to open files as raw images. Confocal Assistant, NIH Image and Scion, and Adobe PhotoShop, to mention a few, make good offline additions to your analysis capabilities. Offline analysis will allow your machine a higher level of use for imaging and increase your throughput of users. Where heavy use is a problem, you might encourage users to use these auxiliary programs on their own computers.

VII. Core Issues

Keeping track of the use of your facility is imperative with some grant funding, and for billing, but it has other benefits for everyone. It helps you know what needs your researchers have by showing what resources they are using the most. This will help you in deciding what to upgrade or update first, and what services you need to offer in the future.

It is simple enough to prepare your own forms, such as the example in Fig. 1 for record keeping, including information useful for your particular purposes, usage requests, billing, or demographics. Logs for daily use can also be used for keeping track of the microscope use by "frequent users" and of problems encountered in imaging, by including a column for comments. Quality control and maintenance on your instruments can also be monitored using personalized log sheets prepared by computer.

Confocal software systems often offer personalized features for each account, such as security logins to monitor usage, individual subdirectories to keep images safely within the user's control, and the ability for researchers to save their own methods and settings. This allows them to repeat an experiment with the same conditions, or will give them a starting place when beginning a new experiment. Information is often stored in the headers of files, indicating conditions as they were documented in the computer at the time of collection, so it is important to make sure the settings are correct when collecting the image. Six years from now, you might not remember which objective you used! Measurements taken then would be erroneous, if the correct setting was not saved originally.

Databases constructed to maintain records of all users, their usage, billing charges for each instrument for each type of uses, on and off campus, and records of all facilities help in the compilation of figures for reports and billing. With just a click you can pull up files within a time period, for just one instrument, or by one user. The time it takes to prepare the database will be well worth your while, when you need to get a statistic about usage at a moment's notice some Friday afternoon!

Charging and recharging for the use of your facility is also a primary concern. In many laboratories there is subsidy available to pay for salaries or equipment and the only goal of the recharge is to maintain the equipment and pay for supplies. Whichever scenario is the case at your institution, preparing the recharge account is not an easy task. You must justify the amount of your charge, considering all the expenses it is to cover, as well as the amount of recharge the market will bear, the standard in your part of the world, and what your clients will be able to afford. Corporations are often immune from this problem and offer the services to anyone in the company who needs them, at no charge to individual accounts. In the university setting, the situation is a little different, where the most frequent users need to bear the burden of keeping the instrument going—with recharge money, generated by their frequent use. Along with these considerations, many granting agencies have guidelines about the amount of recharges you can assess, and whom you can charge. The National Institutes of Health (NIH) dictate that anyone using NIH granted equipment must offer it to all researchers on the campus at the same rate. The National Cancer Institute (NCI) says that anyone belonging to the cancer center supported by one of their cancer center grants is to have priority on CCC resources over other researchers, on and off campus. Researchers who contribute to the writing and support of grants for instruments often expect free use of facilities for a period of time. It is possible to meet all of these expectations using creative management, and still keep all granting agencies and your users happy. Surveys of current recharge schedules are often available through resources such as those listed in Fig. 3.

Resource	URL	Email address
International Society for Analytical Cytology (ISAC)	www.isac-net.org	ISAC@isac-net.org
Microscopy Society of America (MSA)	www.microscopy.org/	Zaluzec@Microscopy.com
Local Affiliates of Microscopy Society of America (MSA)	www.msa.microscopy.com/	Zaluzec@Microscopy.com
Purdue Cytometry Group	www.cyto.purdue.edu/	jpr@flowcyt.cyto.purdue.edu
Technical help in Microscopy and BioImaging	www.olympus-biosystems.com/technical.html	info@olympus-biosystems.com
The Bio-Rad Fluorescence Database	http://fluorescence.bio-rad.com/	microscopy@bio-rad.com
MICROSCOPY & IMAGING RESOURCES ON THE WWW	http://www.pharmacy.Arizona.EDU/centers/tox center/swehsc/exp path/m-i onw3.html	Doug-cromey@ns.arizona.edu
Recruitment Specialist for Microscopists	www.flocyte.com	flocyte@cox.net

Fig. 3   Internet resources for confocal microscopy.

VIII. Ancillary Services

Depending on your institutional needs and personal experience, you might also offer services which complement your imaging capabilities. Consultation on project design, grant proposal preparation, the writing of methods and materials, assistance with staining and specimen preparation, appropriate protocols, and reagent sources are all services needed by researchers.

IX. Resources

As your level of expertise increases, you will feel more and more comfortable in assisting your users with their project setups. Remember, as new users come to you, that you most likely have a better knowledge of your system than they do, and will be able to help them get the results they need for their research. You are the first valuable resource!

Resources such as the international organizations, listservs, and user groups are organized and available for the purposes of assisting laboratories utilizing highly technical instrumentation such as the confocal microscope. The International Society for Analytical Cytology (ISAC), The Microscopy Society of America (MSA), and its local affiliates are all available for advanced training and up-to-the-minute techniques. The microscopy listserv, the cytometry listserv, and the confocal microscopy listserv all serve as arenas for exploring new ideas and presenting questions regarding microscopy itself. The core managers' listserv deals with issues particular to the running of shared resources. All of these resources are available at no charge over the Internet and will be of immeasurable assistance when you are operating your facility. ISAC has a core managers' workgroup, organized to handle issues for core, shared resources, that is available to all ISAC members. URLs and e-mail addresses for the listservs and societies are available in Fig. 3. This is by no means an exhaustive list, but will be a starting place for your reference.

X. Summary

Confocal shared resource facilities are not only useful, but the way to go! With the equipment and its maintenance costing more and more each year, and the technology growing and expanding all the time, it is difficult, if not impossible, for individual researchers to keep pace. However, core facilities can offer the latest equipment and services for the newest techniques only when they are maintained properly and supported by researchers. Policies for the encouragement of core facilities are important considerations in the budgets of university departments. Researchers should be offered incentives to use core facilities rather than purchase their own instruments, equipment which often is neglected within just a couple of years. Core facility managers are trained and skilled operators of these highly technical pieces of equipment and are prepared to keep your facilities in the best possible condition—for the best possible results.

INDEX

VOLUMES IN SERIES

Founding Series Editor
DAVID M. PRESCOTT

Volume 1 (1964)
Methods in Cell Physiology
Edited by David M. Prescott

Volume 2 (1966)
Methods in Cell Physiology
Edited by David M. Prescott

Volume 3 (1968)
Methods in Cell Physiology
Edited by David M. Prescott

Volume 4 (1970)
Methods in Cell Physiology
Edited by David M. Prescott

Volume 5 (1972)
Methods in Cell Physiology
Edited by David M. Prescott

Volume 6 (1973)
Methods in Cell Physiology
Edited by David M. Prescott

Volume 7 (1973)
Methods in Cell Biology
Edited by David M. Prescott

Volume 8 (1974)
Methods in Cell Biology
Edited by David M. Prescott

Volume 9 (1975)
Methods in Cell Biology
Edited by David M. Prescott

Volume 10 (1975)
Methods in Cell Biology
Edited by David M. Prescott

Volume 11 (1975)
Yeast Cells
Edited by David M. Prescott

Volume 12 (1975)
Yeast Cells
Edited by David M. Prescott

Volume 13 (1976)
Methods in Cell Biology
Edited by David M. Prescott

Volume 14 (1976)
Methods in Cell Biology
Edited by David M. Prescott

Volume 15 (1977)
Methods in Cell Biology
Edited by David M. Prescott

Volume 16 (1977)
Chromatin and Chromosomal Protein Research I
Edited by Gary Stein, Janet Stein, and
Lewis J. Kleinsmith

Volume 17 (1978)
Chromatin and Chromosomal Protein Research II
Edited by Gary Stein, Janet Stein, and
Lewis J. Kleinsmith

Volume 18 (1978)
Chromatin and Chromosomal Protein Research III
Edited by Gary Stein, Janet Stein, and
Lewis J. Kleinsmith

Volume 19 (1978)
Chromatin and Chromosomal Protein Research IV
Edited by Gary Stein, Janet Stein, and
Lewis J. Kleinsmith

Volume 20 (1978)
Methods in Cell Biology
Edited by David M. Prescott

Advisory Board Chairman
KEITH R. PORTER

Series Editor
LESLIE WILSON

Volume 38 (1993)
Cell Biological Applications of Confocal Microscopy
Edited by Brian Matsumoto

Volume 39 (1993)
Motility Assays for Motor Proteins
Edited by Jonathan M. Scholey

Volume 40 (1994)
A Practical Guide to the Study of Calcium in Living Cells
Edited by Richard Nuccitelli

Volume 41 (1994)
Flow Cytometry, Second Edition, Part A
Edited by Zbigniew Darzynkiewicz, J. Paul Robinson, and Harry A. Crissman

Volume 42 (1994)
Flow Cytometry, Second Edition, Part B
Edited by Zbigniew Darzynkiewicz, J. Paul Robinson, and Harry A. Crissman

Volume 43 (1994)
Protein Expression in Animal Cells
Edited by Michael G. Roth

Volume 44 (1994)
Drosophila melanogaster: **Practical Uses in Cell and Molecular Biology**
Edited by Lawrence S. B. Goldstein and Eric A. Fyrberg

Volume 45 (1994)
Microbes as Tools for Cell Biology
Edited by David G. Russell

Volume 46 (1995)
Cell Death
Edited by Lawrence M. Schwartz and Barbara A. Osborne

Volume 47 (1995)
Cilia and Flagella
Edited by William Dentler and George Witman

Volume 48 (1995)
Caenorhabditis elegans: **Modern Biological Analysis of an Organism**
Edited by Henry F. Epstein and Diane C. Shakes

Volume 49 (1995)
Methods in Plant Cell Biology, Part A
Edited by David W. Galbraith, Hans J. Bohnert, and Don P. Bourque

Volume 50 (1995)
Methods in Plant Cell Biology, Part B
Edited by David W. Galbraith, Don P. Bourque, and Hans J. Bohnert

Volume 51 (1996)
Methods in Avian Embryology
Edited by Marianne Bronner-Fraser

Volume 52 (1997)
Methods in Muscle Biology
Edited by Charles P. Emerson, Jr. and H. Lee Sweeney

Volume 53 (1997)
Nuclear Structure and Function
Edited by Miguel Berrios

Volume 54 (1997)
Cumulative Index

Volume 55 (1997)
Laser Tweezers in Cell Biology
Edited by Michael P. Sheez

Volume 56 (1998)
Video Microscopy
Edited by Greenfield Sluder and David E. Wolf

Volume 57 (1998)
Animal Cell Culture Methods
Edited by Jennie P. Mather and David Barnes

Volume 58 (1998)
Green Fluorescent Protein
Edited by Kevin F. Sullivan and Steve A. Kay

Volume 59 (1998)
The Zebrafish: Biology
Edited by H. William Detrich III, Monte Westerfield, and Leonard I. Zon

Volume 60 (1998)
The Zebrafish: Genetics and Genomics
Edited by H. William Detrich III, Monte Westerfield, and Leonard I. Zon

Volume 61 (1998)
Mitosis and Meiosis
Edited by Conly L. Rieder

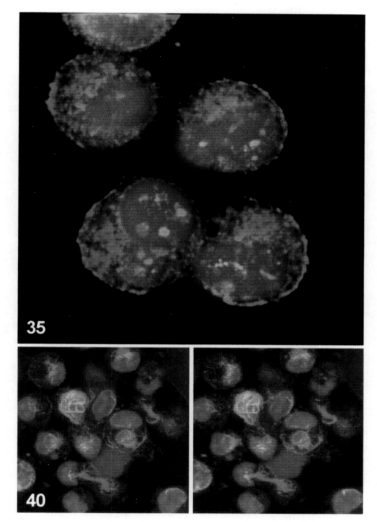

Ch. 1, Fig. 35. Confocal fluorescence image of a single *xy* optical section of tissue culture cells double-labeled to show topoisomerase II (green) and nuclei (red). Topoisomerase II was stained with a rabbit polyclonal anti-topoisomerase II antibody followed by a FITC-conjugated goat anti-rabbit IgG. Positions of the nuclei were revealed using propidium iodide. The image was generated on a Bio-Rad MRC-1024 confocal laser scanning microscope using a 60X 1.4 NA objective on a Nikon Diaphot. A separate channel was used to image each fluorochrome, and the images were later merged using Confocal Assistant 4.02. The image shows that topoisomerase II localizes to both the cytoplasm and nuclei.   **Ch. 1, Fig. 40.**   Stereo pair confocal fluorescence image of the same tissue culture cells in Fig. 1.39 double-labeled for microtubules (green) and nuclei (red). The stereo pair was generated by merging the two projections using Confocal Assistant 4.02. The three-dimensional relationship of microtubules and nuclei is easier to observe in the merged color stereo image.

Ch. 2, Fig. 6. Autofluorescence of pollen grain of Mallow. This multicolor autofluorescence was captured in ca. 0.6 sec with a chilled, 3-CCD color video camera (Dage-MTI Model 330, Michigan City, IN) mounted on the CSU-10 and illuminated with 488-nm laser light. The video signal, after conversion from RGB to standard (Y/C) format through a scan converter (Truevision Vid I/O, Pinnacle Systems, Mt. View, CA), was captured onto a Sony ED-Beta VCR. For this illustration, a single frame of the VCR output was captured and printed with a Sony Mavigraph video printer. The same full-color image is seen through the eyepiece of the CSU-10 in real time.

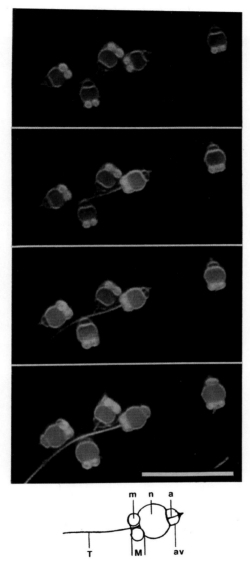

Ch. 2, Fig. 11. Selected frames from CSU-10 confocal video sequences showing striking photodynamic changes in Eosin-B-stained *Spisula* (clam) sperm exposed to the 488-nm (fluorescence exciting) laser beam. Top to bottom panels: 13, 14, 15, 31 s after exposure to the laser beam (0.01 mW per μm^2 at the specimen). Except for the fourth sperm from the left, much of the tail is invisible owing to the shallow depth of field of the confocal optics. Scale bar = 10 μm. Bottom: Schematic diagram of *Spisula* sperm. A, Acrosome: av, acrosomal vesicle; M, midpiece; m, mitochondrion; n, nucleus; T, sperm tail (part only shown). *Technical details.* Sperm were suspended in seawater containing 0.1% Eosin B and observed through a Nikon E-800 upright microscope equipped with a 100×/1.4 NA Plan Apo oil-immersion lens and a CSU-10. The weak fluorescence of the sperm parts excited by the 488-nm laser beam was integrated on chip for 20 frames (0.66 s) on a chilled, 3-chip color camera (Dage-MTI Model 330, Michigan City, IN) attached to the CSU-10 and observed continuously on an RGB color monitor. The RGB output through the monitor was converted by a scan converter (Truevision Vid I/O) to Y/C format, color balance and background levels adjusted with a video processor (Elite Video BVP-4 Plus, www.elitevideo.com), and recorded to an ED-Beta VCR. (From Inoué *et al.*, 1997.)

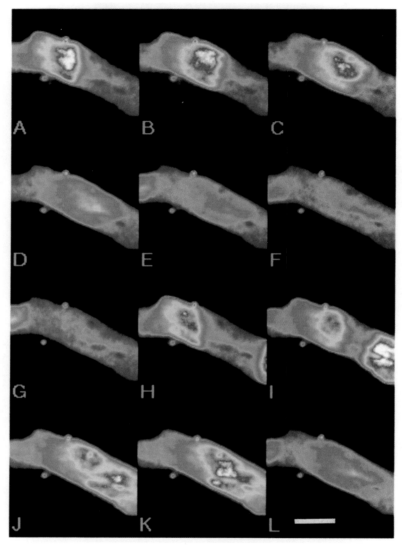

Ch. 2, Fig. 12. Mouse cardiac myocyte. Waves and sparks of elevated cytosolic Ca^{2+} are presented in pseudocolor in this heart muscle and cell injected with Fluo 3. (Ca^{2+} concentrations rise in order of purple, light-blue, dark-blue, green, yellow, orange, red, and white.) Frame intervals each 200 ms between panels A-F. Then after 1.4 s, panel G-L show the approach and annihilation of two calcium waves. Frame intervals: G–H = 360 ms, H–I = 160 ms, I–J = 120 ms, J–K = 40 ms, K–L = 40 ms. Scale bar = 5 μm.[7] *Technical details.* The image from the CSU-10 (modified by addition of motorized filter wheel as a part of the "UltraView" system distributed by PerkinElmer Wallac), viewed through an Olympus 60/1.4 oil-immersion objective lens, was intensified with a Videoscope Intensifier. The images, each exposed for 25 ms, were captured at 25 frames per second onto a PerkinElmer FKI-300 cooled-CCD camera equipped with a Kodak Olympics interline chip.

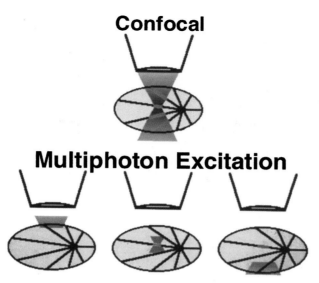

Ch. 3, Fig. 1. Diagrammatic representation of the volume of excitation for confocal and multiphoton imaging. Upper panel depicts illumination used in confocal imaging: all fluorescent structures in the beam path are excited; light emanating from above and below the plane of focus must be eliminated by the confocal blocking aperture. Lower three panels depict passage of short pulse of infrared light through a sample. Lower three panels depict passage of short pulse of infrared light through the sample. Only at the point of focus is the photon density sufficient to elicit multiphoton excitation. Flourescence continues to be emitted for the duration of the excited state lifetime of the fluorophore after the passage of the short-pulse excitation. (Reproduced from Fig. 97.1, "Cells: A Laboratory Manual," Vol. 2, Spector, Goldman, and Leinwand, Eds. Cold Spring Harbor Laboratory Press, New York.)

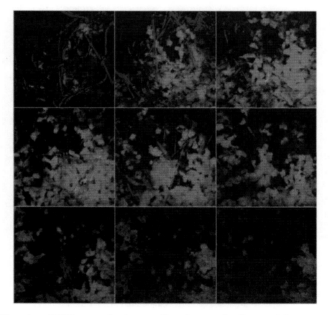

Ch. 3, Fig. 4. Intravital imaging of GFP-expressing tumor cells and connective tissue autofluorescence. GFP-expressing MTLn3 cells were injected in the mannary fat pads of Fischer 344 rats. Tumors which formed after 2 weeks postinjection were viewed intra-vitally with multiphoton microscopy using 900-nm femtosecond pulses. The GFP images were collected in one channel that selected for 500–550 nm emission. The images of connective tissue were collected simultaneously in the second channel that selected for 350–480 nm emission. The images were collected at 10-μm intervals through the volume of the sample. Image degradation was significant at depths beyond 60 μm. (Specimen courtesy of J. B. Wyckoff and J. S. Condeelis, Albert Einstein College of Medicine.)

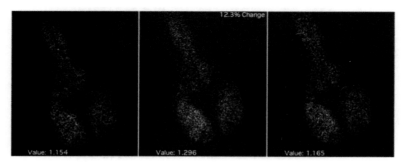

Ch. 3, Fig. 5. Ratioed images of cameleon FRET detection using multiphoton excitation imaging. ycam2 cameleon (provided by R. Tsien) was expressed in pharyngeal muscles cells of *C. elegans* worms using the myo-2 promotor. Illumination with 800-nm light primarily excited the CFP moiety of the cameleon. Emission was collected between 380 and 485 nm for CFP and between 535 and 550 nm for YFP. Intensity measurements in regions of interest of the images were ratioed to determine the binding of free calcium by cameleon. At least a 10% change in YFP/CFP emission ratio from baseline to peak was correlated with opening of the terminal bulb lumen. (Specimen courtesy of E. Maryon, B. Saari, and P. Anderson, University of Wisconsin.)

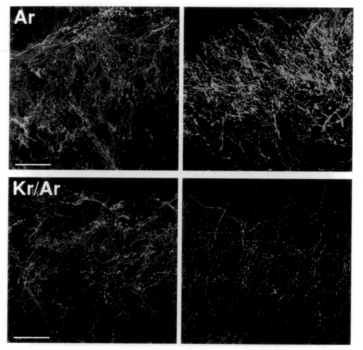

Ch. 5, Fig. 7. Comparison of two-color imaging, using simultaneous excitation of fluorescein and Lissamine rhodamine with the 514-nm line of an argon ion laser (Fig. 10) and sequential excitation with the 488- and 568-nm lines of a krypton–argon laser (Fig. 17). Rat spinal cord sections were stained with goat antiserotonin (5-HT) and rabbit antisubstance P (SP) antisera, and visualized with the combination of fluorescein (FITC)-conjugated donkey anti-goat IgG and Lissamine rhodamine (LRSC)-conjugated donkey anti-rabbit IgG (left), or Lissamine rhodamine-conjugated anti-goat IgG and fluorescein-conjugated donkey anti-rabbit IgG (right). All secondary antibodies were used as 1:100 dilutions. With the argon ion laser (Ar), a separate population of fibers appears to be stained with the first combination of secondary antibodies (top, left), but they appear to coexist in the same fibers when the fluorophores were switched (top, right). In the latter case, the bright SP staining with fluorescein contributes more to the red fluorescence signal than the weaker 5-HT staining with Lissamine rhodamine. The considerable differences observed between these two images demonstrate how the choice of fluorophores can dramatically affect conclusions, using the argon ion laser. In contrast, when the same specimens were examined with the krypton–argon ion laser, separate populations of fibers are observed for both combinations of fluorophores (bottom, left and right). Each image is a projection of either 16 (top) or 20 (bottom) optical sections acquired at 1.0 μm. Nikon ×60, NA 1.4.

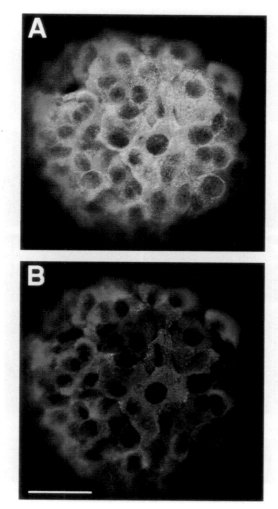

Ch. 5, Fig. 8. Merged display of the observed and "corrected" images shown in Fig. 26 of a triple-labeled rat islet of Langerhans, using an argon ion laser. The green (green), red (red), and far red (blue) fluorescence images are shown as collected (A) and after correction for cross-talk between the images (B). Although the cross-talk compensation dramatically improves the separation of the fluorescence signals, several cells still appear to be possibly double-labeled. These images were manually shifted to correct for the misalignment of the individual images resulting from the use of separate filter sets. Scale bar = 25 μm.

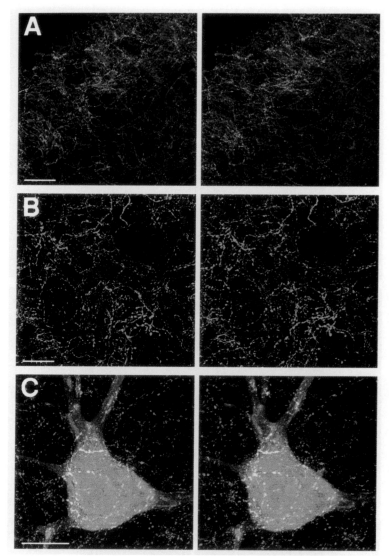

Ch. 5, Fig. 9. Stereo images of double-labeled rat spinal cord specimens, using a krypton–argon ion laser. Rat spinal cord sections were stained with goat antiserotonin and rabbit antisubstance P antisera, followed by fluorescein-conjugated donkey anti-goat (green) and Lissamine rhodamine-conjugated donkey anti-rabbit (red) IgGs (multiple species adsorbed). In the dorsal horn (A), the neurotransmitters are observed in separate populations of fibers. In contrast, in the ventral horn (B) the coexistence of serotonin and substance P is observed. Note that the nerve fibers appear to be positioned around large cells resembling motor neurons. To examine whether these cells might be motor neurons, the motor neurons were stained by retrograde labeling from the sciatic nerve with hydroxystilbamidine (Fluoro-Gold). In spinal cord sections, individual fluorogold-labeled motor neurons were intracellularly injected with Lucifer yellow, and subsequently stained with a rat antisubstance P monoclonal antibody and cyanine 5.18-conjugated donkey anti-rat IgG. This allowed the visualization of both the motor neuron (yellow) and the substance P fibers (blue) wrapping around its surface with the confocal microscope (C). The stereo projections were constructed from 25, 29, and 51 optical sections, respectively, acquired at 0.8-μm intervals. Scale bars = 25 μm. Nikon ×60, NA 1.4.

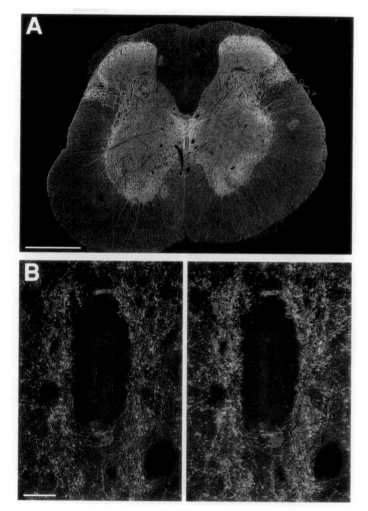

Ch. 5, Fig. 10. Merged images of a triple-labeled rat spinal cord, using a krypton–argon laser. Spinal cord sections were stained with goat antiserotonin antiserum and fluorescein-conjugated donkey anti-goat IgG (green), a rat antisubstance P monoclonal antibody and cyanine 3.18-conjugated donkey anti-rat IgG (red), and rabbit anti-Met-enkephalin antiserum and cyanine 5.18-conjugated anti-rabbit IgG (blue). The fluorophore-conjugated secondary antibodies were adsorbed against serum proteins from multiple species to reduce cross-reactivity. (A) A low-magnification image obtained by combining images collected from two fields of view. Laser illumination intensity was 4 mW for the imaging of all three fluorophores. Scale bar = 100 μm. Olympus ×4, NA 0.13. (B) A stereo pair of the region around the central canal of the spinal cord at higher magnification. The stereo projections were constructed from 17 images acquired at 0.6-μm intervals. Laser illumination intensity was 0.4 mW for the imaging of all three fluorophores. Scale bar = 25 μm. Nikon ×60, NA 1.4. (Specimen courtesy of J. Wang and R. Elde.)

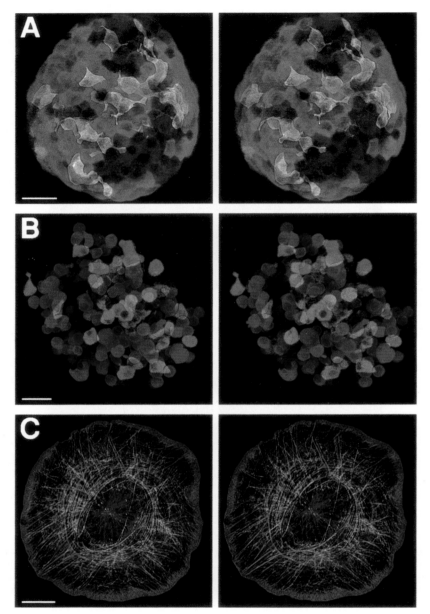

Ch. 5, Fig. 11. Stereo images of double- and triple-labeled specimens, using a krypton–argon ion laser. (A) Triple-labeled rat islet of Langerhans. This stereo pair was formed from the merging of the images of the insulin (green), somatostatin (red), and glucogon (blue) shown in Fig. 32. In this merged image, the somatostatin-containing δ cells can be seen to be positioned between the glucagon-containing α cells on the surface of the islet and the insulin-containing β cells that comprise the majority of cells in the islet. Scale bar = 25 μm. (B) Double-labeled human islet of Langerhans. Islets were stained with rabbit anti-somatostatin antiserum and cyanine 3.18-conjugated donkey anti-rabbit IgG (red), and a mouse anti-glucagon monoclonal antibody and cyanine 5.18-conjugated donkey anti-mouse IgG (blue). The stereo projections were constructed from 62 optical sections acquired at 1.0-μm intervals. Scale bar = 25 μm. Olympus ×40, NA 0.85. (C) A non-neuronal cell from a chicken dorsal root ganglion culture. Actin filaments were stained with fluorescein-conjugated phalloidin (red) and microtubules with a mouse antitubulin monoclonal antibody and cyanine 3.18-conjugated donkey anti-mouse IgG (green). The stereo projections were constructed from 12 optical sections acquired at 0.6-μm intervals. Scale bar = 10 μm. Nikon ×60, NA 1.4. (Specimen courtesy of P. Letourneau.)

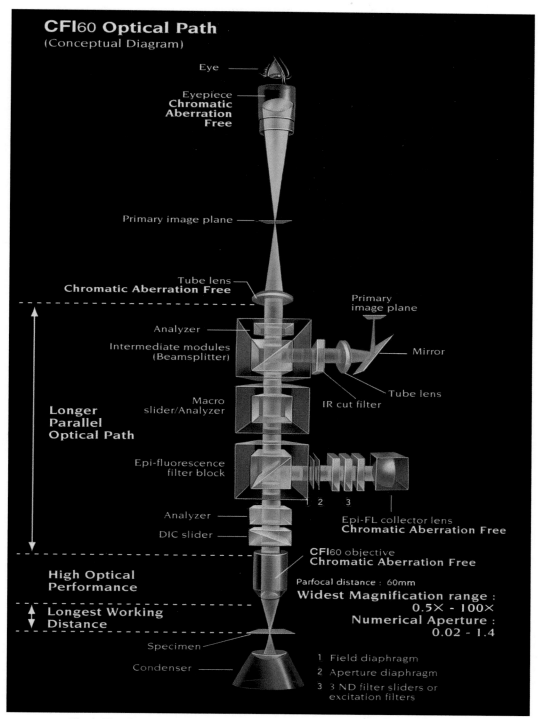

Ch. 6, Fig. 1. Nikon CFI optical path (conceptual diagram). Courtesy of Nihoshi (2000).

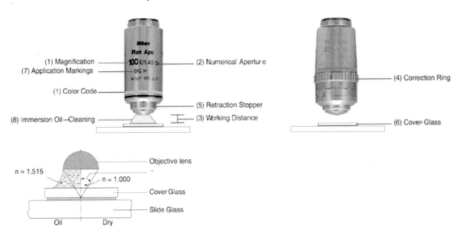

Ch. 6, Fig. 2. (*Top*) Picture of two objective lenses showing all markings. (*Bottom*) Comparison of dry (air) versus immersion objective in relation to specimen slide and cover glass. Courtesy of Nihoshi (2000).

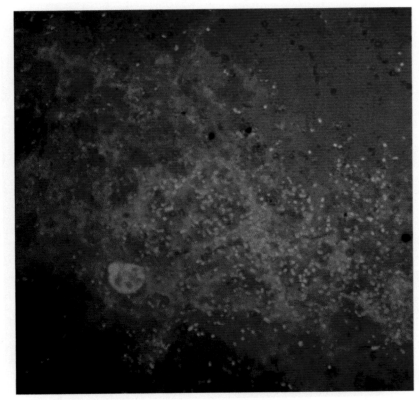

Ch. 6, Fig. 3A. (A) Sulfate-reducing bacterium, *Pseudomonas aeruginosa*, stained with acridine orange grown in 0.1× TSB on a zirconium coupon. Imaged with a Plan Apochromat 100×/1.40 oil immersion objective.

Ch. 6, Fig. 3B.   (B) Sulfate-reducing bacterium, *Pseudomonas aeruginosa* stained with acridine orange grown in water with mineral salts on aluminum coupon. Imaged with a Plan Apochromat 100×/1.40 oil immersion objective. Courtesy of Bruhn (2000).

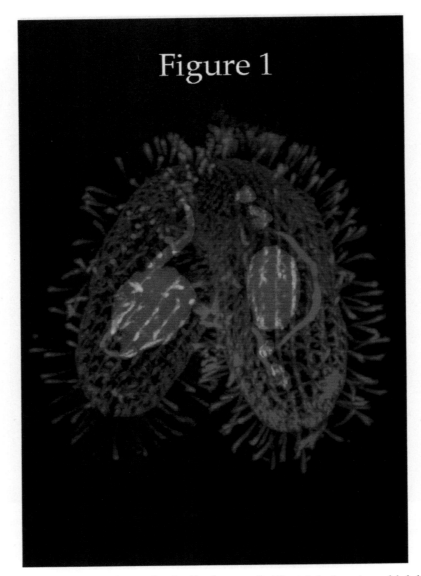

Ch. 8, Fig. 1. Confocal fluorescence image of mating *Tetrahymena* cells. Microtubules (green) were labeled with a beta-tubulin monoclonal antiserum (12G10 kindly provided by Dr. Joseph Frankel, University of Iowa) and visualized with Cy-5 conjugated goat anti-mouse secondary antiserum (Amersham). Nuclei (red) were visualized with Sytox nuclear stain (Molecular Probes, Inc.). Cells were scanned under a Bio-Rad MRC1000/1024 laser confocal microscope operated by Dr. Mark Sanders, Program Director of the Imaging Center, Department of Genetics and Cell Biology, University of Minnesota.

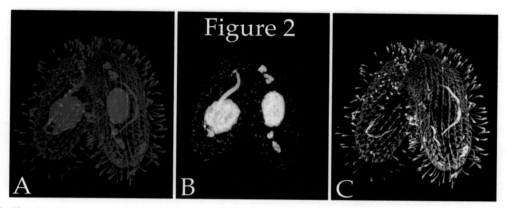

Ch. 8, Fig. 2. The same mating pair shown in Fig. 1, with the various fluorescent channels separated. (A) The entire image, (B) the nuclei, and (C) the microtubules.

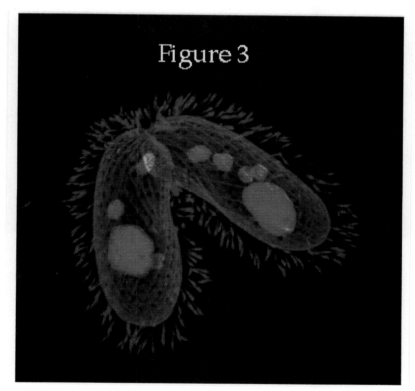

Ch. 8, Fig. 3. Mating *Tetrahymena* viewed under laser confocal microscopy. Microtubules (blue) were labeled with 12G10 beta tubulin antiserum (courtesy of Dr. Joseph Frankel) and a Cy-5 conjugated secondary antiserum. The subcortical layer (pink) was labeled with a polyclonal antiserum directed against TCBP-25 (courtesy of Dr. Osamu Numata) and a Texas Red conjugated secondary antiserum. Nuclei (green) have been labeled with Sytox.

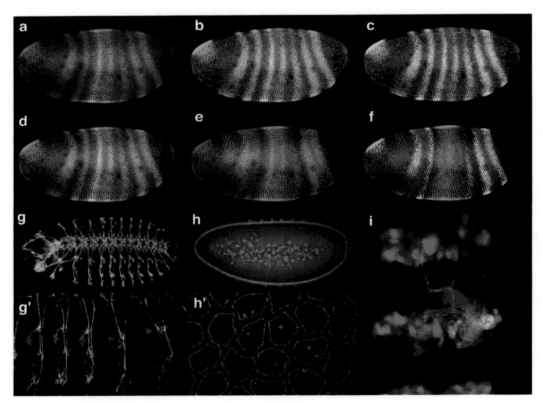

Ch. 9, Fig. 7. Imaging *Drosophila* embryos. (a) through (f): Different color combinations of the triple-labeled embryo of Fig. 9.2; double-labeled images (a) through (d), and triple-labeled images (e, f). Note that (f) is the result of passing image (e) through a color LUT in Graphic Converter. (g) and (g'): Two magnifications of a 22C10-labeled embryo demonstrating color coded to depth in a Z-series from Fig. 9.4. (h) and (h'): *ftz* mRna in red and wheat germ agglutinin staining the nuclear envelope in green from Fig. 9.5. (i): *engrailed* GFP in green with DiI-labeled neuroblast clones in red and blue from Fig. 9.6.

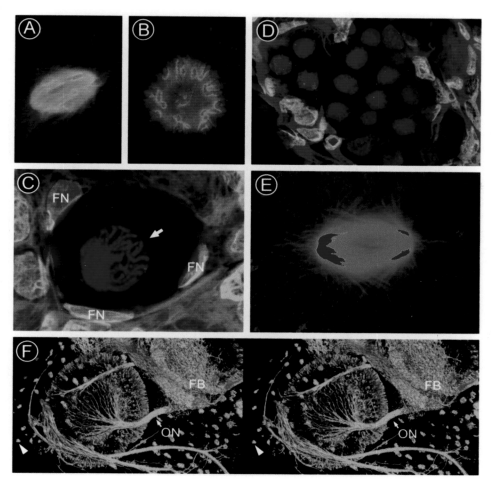

Ch. 10, Fig. 1. (A) A longitudinal view of a transverse M1 meiotic spindle in a maturing *Xenopus* oocyte fixed in methanol and stained with monoclonal anti-α-tubulin/fluorescein-anti-mouse IgG (green) and PI (red). (B) A cross-sectional view of the metaphase plate of an M2 meiotic spindle in an unfertilized *Xenopus* egg fixed in methanol and stained with monoclonal anti-α-tubulin/fluorescein-anti-mouse IgG (green) and PI (red) (A and B reprinted from Gard, 1992). (C) A single late-stage 0 oocyte (35 μm diameter) in an ovary isolated from a juvenile frog, fixed in methanol, and stained with monoclonal anti-KF/fluorescein-anti-mouse IgG (green) and EH (red). KF staining is only apparent in the surrounding follicle cells (FN denote follicle cell nuclei). Note the characteristic "bouquet" organization of chromatin in the oocyte nucleus (arrow). A projection of three optical sections. (D) A nest of early stage 0 oocytes isolated from a juvenile frog and stained with monoclonal anti-KF/Texas Red-anti-mouse IgG (green) and YP (red). KF staining is only apparent in the follicle cells surrounding the oocytes. A projection of three optical sections. (E) The same image of an M1 spindle shown in Fig. 10.6D, displayed in a false color LUT in which pixel values greater than 252 appear red and less than 3 appear green. Intermediate pixel values are shown in cyan-blue. This LUT (applied in Photoshop) simulates the appearance of the "setcol" LUT of Biorad CLSMs (and is similar to the "range indicator" palette of the Zeiss LSM 510). Collecting images in such false color LUTs facilitates setting of the gain and black levels to optimize utilization of the instruments dynamic range. (F) A 3-D reconstruction of the eye of a stage 35 *Xenopus* embryo, fixed in methanol and stained with antibodies to acetylated α-tubulin, which stains "stable" MTs (Piperno *et al.*, 1987; Schulze *et al.*, 1987; Webster and Corisy, 1989). View is from behind the eye. ON denotes the optic nerve, FB is the forebrain. The numerous small foci of staining (one denoted with arrowhead) are ciliated cells in the embryonic epithlium.

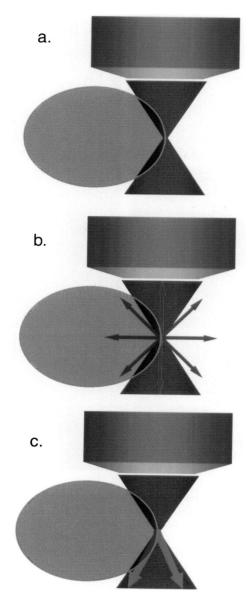

Ch. 12, Fig. 5. Schmatic illustration of the principles of microscopies derived from TPEF and SHG. A focused laser beam emanating from a microscope objective is focused onto the dye-stained cell membrane. (a) The section of membrane that interacts with the incident light beam is shown in cyan. It is confined to a region close to the plane of focus by the quadratic dependence of both TPEF and SHG. (b) Fluorescence excited by two-photon excitation propagates in all directions and is collected by the microscope objective in a typical incident light (epi) configuration. The wavelength range of the TPEF corresponds to the normal emission spectrum of the dye. (c) SHG propagates along the same path as the incident light and must be collected by a high numerical aperature condenser or a second objective (not shown) in a transmitted light configuration. The wavelength of the SHG signal is precisely half the wavelength of the incident laser light.

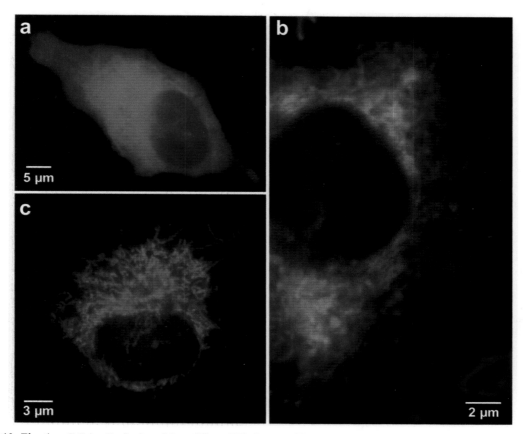

Ch. 13, Fig. 1. Cellular distribution of three different YCs. Images of HeLa cells transiently transfected with YC2 (a), $YC4_{ER}$ (b), and $YC2_{mit}$ (c). In (b) the cell was also stained with MitoTracker Red (Molecular Probes) to show the distribution of mitochondria relative to the ER. Fluorescence images were obtained with a cooled CCD camera using 535DF25 (for YCs) and LP 590 (for MitoTracker) emission filters.